ROOM 19A
HIGHVIEW CARE
HOME

✱ PROPERTY OF FRANCES SMITH

KEY TO MAP PAGES OCEANIA, ASIA, EUROPE
(see back endpapers for Africa, North America and South America)

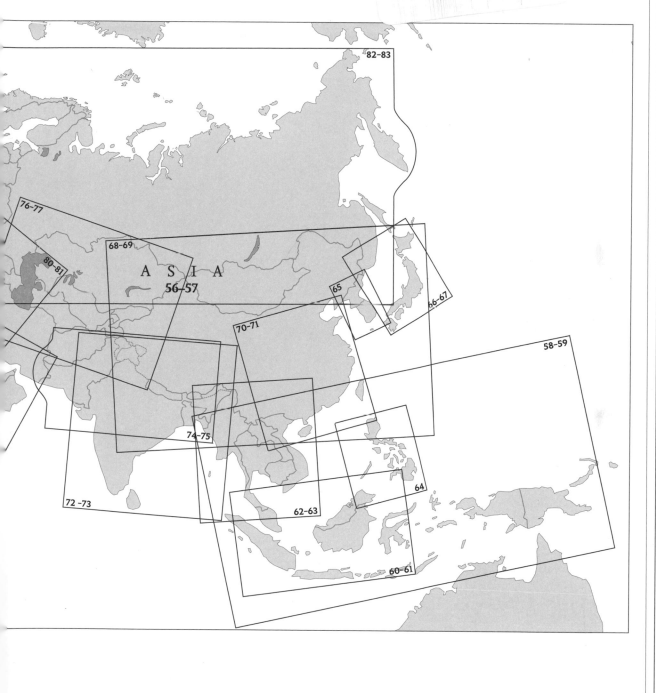

D1180663

82-83

76-77

68-69

80-81

A S I A
56-57

65

66-67

70-71

58-59

74-75

72-73

62-63

64

60-61

DESKTOP
ATLAS
OF THE WORLD

Published by Times Books
An imprint of HarperCollins Publishers
Westerhill Road
Bishopbriggs
Glasgow G64 2QT
www.harpercollins.co.uk

First published 2006
Second Edition 2009
Third Edition 2012

Fourth Edition 2015
Reprinted with changes 2016

Copyright © HarperCollins Publishers 2015

Maps © Collins Bartholomew Ltd 2015

The Times® is a registered trademark of Times Newspapers Ltd

All rights reserved. No part of this publication may be reproduced, stored in a
retrieval system, or transmitted, in any form or by any means, electronic, mechanical,
photocopying, recording or otherwise without the prior written permission of the
publisher and copyright owners.

The contents of this publication are believed correct at the time of printing.
Nevertheless the publisher can accept no responsibility for errors or omissions,
changes in the detail given or for any expense or loss thereby caused.

HarperCollins does not warrant that any website mentioned in this title will be provided
uninterrupted, that any website will be error free, that defects will be corrected, or that the
website or the server that makes it available are free of viruses or bugs. For full terms and
conditions please refer to the site terms provided on the website.

A catalogue record for this book is available from the British Library

ISBN 978-0-00-810498-6

10 9 8 7 6 5 4 3
Printed in Hong Kong

All mapping in this atlas is generated from Collins Bartholomew digital databases.
Collins Bartholomew, the UK's leading independent geographical information supplier,
can provide a digital, custom, and premium mapping service to a variety of markets.
For further information:
tel: +44 (0) 208 307 4515
e-mail: collinsbartholomew@harpercollins.co.uk
or visit our website at: www.collinsbartholomew.com

If you would like to comment on any aspect of this atlas,
please contact us at the above address or online.
www.timesatlas.com
email: timesatlases@harpercollins.co.uk
 facebook.com/thetimesatlas
@TimesAtlas

THE TIMES

DESKTOP
ATLAS
OF THE WORLD

TIMES BOOKS
LONDON

CONTENTS

All independent countries and populated dependent and disputed territories are included in this list of the states and territories of the world; the list is arranged in alphabetical order by the conventional name form. For independent states, the full name is given below the conventional name, if this is different; for territories, the status is given. The capital city name is the same form as shown on the reference maps.

The statistics used for the area and population are the latest available and include estimates. The information on languages and religions is based on the latest information on 'de facto' speakers of the language or 'de facto' adherents to the religion. The information available on languages and religions varies greatly from country to country. Some countries include questions in censuses, others do not, in which case best estimates are used. The order of the languages and religions reflect their relative importance within the country; generally, languages or religions are included when more than one per cent of the population are estimated to be speakers or adherents.

Membership of selected international organizations is shown for each independent country. Territories are not shown as having separate memberships of these organizations.

ABBREVIATIONS

CURRENCIES

| CFA | Communauté Financière Africaine |
| CFP | Comptoirs Français du Pacifique |

ORGANIZATIONS

APEC	Asia-Pacific Economic Cooperation
ASEAN	Association of Southeast Asian Nations
CARICOM	Caribbean Community
CIS	Commonwealth of Independent States
COMM.	The Commonwealth
EU	European Union
OECD	Organisation for Economic Co-operation and Development
OPEC	Organization of the Petroleum Exporting Countries
SADC	Southern African Development Community
UN	United Nations

Abkhazia
Disputed Territory (Georgia)

Area Sq Km	8 700	Map page	81
Area Sq Miles	3 360		
Population	180 000		
Capital	Sokhumi (Aq"a)		

AFGHANISTAN
Islamic Republic of Afghanistan

Area Sq Km	652 225	Religions	Sunni Muslim, Shi'a Muslim
Area Sq Miles	251 825		
Population	30 552 000	Currency	Afghani
Capital	Kābul	Organizations	UN
Languages	Dari, Pashto (Pashtu), Uzbek, Turkmen	Map page	76–77

ALBANIA
Republic of Albania

Area Sq Km	28 748	Religions	Sunni Muslim, Albanian Orthodox, Roman Catholic
Area Sq Miles	11 100		
Population	3 173 000		
Capital	Tirana (Tiranë)	Currency	Lek
Languages	Albanian, Greek	Organizations	UN
		Map page	109

ALGERIA
People's Democratic Republic of Algeria

Area Sq Km	2 381 741	Religions	Sunni Muslim
Area Sq Miles	919 595	Currency	Algerian dinar
Population	39 208 000	Organizations	OPEC, UN
Capital	Algiers (Alger)	Map page	114–115
Languages	Arabic, French, Berber		

American Samoa
United States Unincorporated Territory

Area Sq Km	197	Religions	Protestant, Roman Catholic
Area Sq Miles	76		
Population	55 000	Currency	United States dollar
Capital	Fagatogo	Map page	49
Languages	Samoan, English		

ANDORRA
Principality of Andorra

Area Sq Km	465	Religions	Roman Catholic
Area Sq Miles	180	Currency	Euro
Population	79 000	Organizations	UN
Capital	Andorra la Vella	Map page	104
Languages	Catalan, Spanish, French		

ANGOLA
Republic of Angola

Area Sq Km	1 246 700	Religions	Roman Catholic, Protestant, traditional beliefs
Area Sq Miles	481 354		
Population	21 472 000		
Capital	Luanda	Currency	Kwanza
Languages	Portuguese, Bantu, other local languages	Organizations	OPEC, SADC, UN
		Map page	120

Anguilla

United Kingdom Overseas Territory

Area Sq Km	155	**Religions**	Protestant, Roman Catholic
Area Sq Miles	60		
Population	14 000	**Currency**	East Caribbean dollar
Capital	The Valley	**Map page**	147
Languages	English		

ANTIGUA AND BARBUDA

Area Sq Km	442	**Religions**	Protestant, Roman Catholic
Area Sq Miles	171		
Population	90 000	**Currency**	East Caribbean dollar
Capital	St John's	**Organizations**	CARICOM, Comm., UN
Languages	English, creole		
		Map page	147

ARGENTINA

Argentine Republic

Area Sq Km	2 766 889	**Religions**	Roman Catholic, Protestant
Area Sq Miles	1 068 302		
Population	41 446 000	**Currency**	Argentinian peso
Capital	Buenos Aires	**Organizations**	UN
Languages	Spanish, Italian, Amerindian languages	**Map page**	152–153

ARMENIA
Republic of Armenia

Area Sq Km	29 800	**Religions**	Armenian Orthodox
Area Sq Miles	11 506	**Currency**	Dram
Population	2 977 000	**Organizations**	CIS, UN
Capital	Yerevan (Erevan)	**Map page**	81
Languages	Armenian, Kurdish		

Aruba

Self-governing Netherlands Territory

Area Sq Km	193	**Religions**	Roman Catholic, Protestant
Area Sq Miles	75		
Population	103 000	**Currency**	Aruban florin
Capital	Oranjestad	**Map page**	147
Languages	Papiamento, Dutch, English		

Ascension
Part of St Helena, Ascension and Tristan da Cunha

Area Sq Km	88	**Religions**	Protestant, Roman Catholic
Area Sq Miles	34		
Population	826	**Currency**	Pound sterling
Capital	Georgetown	**Map page**	113
Languages	English		

AUSTRALIA
Commonwealth of Australia

Area Sq Km	7 692 024	**Religions**	Protestant, Roman Catholic, Orthodox
Area Sq Miles	2 969 907		
Population	23 343 000	**Currency**	Australian dollar
Capital	Canberra	**Organizations**	APEC, Comm., OECD, UN
Languages	English, Italian, Greek	**Map page**	50–51

Australian Capital Territory (Federal Territory)

Area Sq Km	2 358	**Population**	379 600
Area Sq Miles	910	**Capital**	Canberra

Jervis Bay Territory (Territory)

Area Sq Km	73	**Population**	378
Area Sq Miles	28		

New South Wales (State)

Area Sq Km	800 642	**Population**	7 348 900
Area Sq Miles	309 130	**Capital**	Sydney

Northern Territory (Territory)

Area Sq Km	1 349 129	**Population**	236 900
Area Sq Miles	520 902	**Capital**	Darwin

Queensland (State)

Area Sq Km	1 730 648	**Population**	4 610 900
Area Sq Miles	668 207	**Capital**	Brisbane

South Australia (State)

Area Sq Km	983 482	**Population**	1 662 200
Area Sq Miles	379 725	**Capital**	Adelaide

Tasmania (State)

Area Sq Km	68 401	**Population**	512 400
Area Sq Miles	26 410	**Capital**	Hobart

Victoria (State)

Area Sq Km	227 416	**Population**	5 679 600
Area Sq Miles	87 806	**Capital**	Melbourne

Western Australia (State)

Area Sq Km	2 529 875	**Population**	2 472 700
Area Sq Miles	976 790	**Capital**	Perth

AUSTRIA
Republic of Austria

Area Sq Km	83 855	**Religions**	Roman Catholic, Protestant
Area Sq Miles	32 377		
Population	8 495 000	**Currency**	Euro
Capital	Vienna (Wien)	**Organizations**	EU, OECD, UN
Languages	German, Croatian, Turkish	**Map page**	102–103

AZERBAIJAN
Republic of Azerbaijan

Area Sq Km	86 600	**Religions**	Shi'a Muslim, Sunni Muslim, Russian and Armenian Orthodox
Area Sq Miles	33 436		
Population	9 413 000		
Capital	Baku (Bakı)	**Currency**	Azerbaijani manat
Languages	Azeri, Armenian, Russian, Lezgian	**Organizations**	CIS, UN
		Map page	81

© Collins Bartholomew Ltd

Azores (Arquipélago dos Açores)
Autonomous Region of Portugal

Area Sq Km	2 300	Religions	Roman Catholic, Protestant
Area Sq Miles	888		
Population	247 066	Currency	Euro
Capital	Ponta Delgada	Map page	112
Languages	Portuguese		

THE BAHAMAS
Commonwealth of The Bahamas

Area Sq Km	13 939	Religions	Protestant, Roman Catholic
Area Sq Miles	5 382		
Population	377 000	Currency	Bahamian dollar
Capital	Nassau	Organizations	CARICOM, Comm., UN
Languages	English, creole		
		Map page	146–147

BAHRAIN
Kingdom of Bahrain

Area Sq Km	691	Religions	Shi'a Muslim, Sunni Muslim, Christian
Area Sq Miles	267		
Population	1 332 000	Currency	Bahraini dinar
Capital	Manama (Al Manāmah)	Organizations	UN
		Map page	79
Languages	Arabic, English		

BANGLADESH
People's Republic of Bangladesh

Area Sq Km	143 998	Religions	Sunni Muslim, Hindu
Area Sq Miles	55 598	Currency	Taka
Population	156 595 000	Organizations	Comm., UN
Capital	Dhaka (Dacca)	Map page	75
Languages	Bengali, English		

BARBADOS

Area Sq Km	430	Religions	Protestant, Roman Catholic
Area Sq Miles	166		
Population	285 000	Currency	Barbados dollar
Capital	Bridgetown	Organizations	CARICOM, Comm., UN
Languages	English, creole		
		Map page	147

BELARUS
Republic of Belarus

Area Sq Km	207 600	Religions	Belarusian Orthodox, Roman Catholic
Area Sq Miles	80 155		
Population	9 357 000	Currency	Belarusian rouble
Capital	Minsk	Organizations	CIS, UN
Languages	Belarusian, Russian	Map page	88–89

BELGIUM
Kingdom of Belgium

Area Sq Km	30 520	Religions	Roman Catholic, Protestant
Area Sq Miles	11 784		
Population	11 104 000	Currency	Euro
Capital	Brussels (Brussel/ Bruxelles)	Organizations	EU, OECD, UN
		Map page	100
Languages	Dutch (Flemish), French (Walloon), German		

BELIZE

Area Sq Km	22 965	Religions	Roman Catholic, Protestant
Area Sq Miles	8 867		
Population	332 000	Currency	Belize dollar
Capital	Belmopan	Organizations	CARICOM, Comm., UN
Languages	English, Spanish, Mayan, creole		
		Map page	147

BENIN
Republic of Benin

Area Sq Km	112 620	Religions	Traditional beliefs, Roman Catholic, Sunni Muslim
Area Sq Miles	43 483		
Population	10 323 000		
Capital	Porto-Novo	Currency	CFA franc
Language	French,Fon, Yoruba, Adja, other local languages	Organization	UN
		Map page	114

Bermuda
United Kingdom Overseas Territory

Area Sq Km	54	Religions	Protestant, Roman Catholic
Area Sq Miles	21		
Population	65 000	Currency	Bermuda dollar
Capital	Hamilton	Map page	125
Languages	English		

BHUTAN
Kingdom of Bhutan

Area Sq Km	46 620	Religions	Buddhist, Hindu
Area Sq Miles	18 000	Currency	Ngultrum, Indian rupee
Population	754 000		
Capital	Thimphu	Organizations	UN
Languages	Dzongkha, Nepali, Assamese	Map page	75

BOLIVIA
Plurinational State of Bolivia

Area Sq Km	1 098 581	Religions	Roman Catholic, Protestant, Baha'i
Area Sq Miles	424 164		
Population	10 671 000		
Capital	La Paz/Sucre	Currency	Boliviano
Languages	Spanish, Quechua, Aymara	Organizations	UN
		Map page	152

Bonaire
Netherlands Special Municipality

Area Sq Km	288	Religions	Roman Catholic, Protestant
Area Sq Miles	111		
Population	17 408	Currency	United States dollar
Capital	Kralendijk	Map page	147
Languages	Dutch, English, Papiamento, Spanish		

Bonin Islands (Ogasawara-shotō)
part of Japan

Area Sq Km	104	Religions	Shintoist, Buddhist, Christian
Area Sq Miles	40		
Population	2 783	Currency	Yen
Capital	Ōmura	Map page	69
Languages	Japanese		

BOSNIA AND HERZEGOVINA

Area Sq Km	51 130	Religions	Sunni Muslim, Serbian
Area Sq Miles	19 741		Orthodox, Roman
Population	3 829 000		Catholic, Protestant
Capital	Sarajevo	Currency	Marka
Languages	Bosnian, Serbian,	Organizations	UN
	Croatian	Map page	109

BOTSWANA
Republic of Botswana

Area Sq Km	581 370	Religions	Traditional beliefs,
Area Sq Miles	224 468		Protestant, Roman
Population	2 021 000		Catholic
Capital	Gaborone	Currency	Pula
Languages	English, Setswana,	Organizations	Comm., SADC, UN
	Shona, other local	Map page	120
	languages		

BRAZIL
Federative Republic of Brazil

Area Sq Km	8 514 879	Religions	Roman Catholic,
Area Sq Miles	3 287 613		Protestant
Population	200 362 000	Currency	Real
Capital	Brasília	Organizations	UN
Languages	Portuguese	Map page	150–151

BRUNEI
Brunei Darussalam

Area Sq Km	5 765	Religions	Sunni Muslim,
Area Sq Miles	2 226		Buddhist, Christian
Population	418 000	Currency	Brunei dollar
Capital	Bandar Seri Begawan	Organizations	APEC, ASEAN,
Languages	Malay, English,		Comm., UN
	Chinese	Map page	61

BULGARIA
Republic of Bulgaria

Area Sq Km	110 994	Religions	Bulgarian Orthodox,
Area Sq Miles	42 855		Sunni Muslim
Population	7 223 000	Currency	Lev
Capital	Sofia	Organizations	EU, UN
Languages	Bulgarian, Turkish,	Map page	110
	Romany,		
	Macedonian		

BURKINA FASO

Area Sq Km	274 200	Religions	Sunni Muslim,
Area Sq Miles	105 869		traditional beliefs,
Population	16 935 000		Roman Catholic
Capital	Ouagadougou	Currency	CFA franc
Languages	French, Moore	Organizations	UN
	(Mossi), Fulani, other	Map page	114
	local languages		

BURUNDI
Republic of Burundi

Area Sq Km	27 835	Religions	Roman Catholic,
Area Sq Miles	10 747		traditional beliefs,
Population	10 163 000		Protestant
Capital	Bujumbura	Currency	Burundian franc
Languages	Kirundi (Hutu,	Organizations	UN
	Tutsi), French	Map page	119

CAMBODIA
Kingdom of Cambodia

Area Sq Km	181 000	Religions	Buddhist, Roman
Area Sq Miles	69 884		Catholic, Sunni
Population	15 135 000		Muslim
Capital	Phnom Penh	Currency	Riel
Languages	Khmer, Vietnamese	Organizations	ASEAN, UN
		Map page	63

CAMEROON
Republic of Cameroon

Area Sq Km	475 442	Religions	Roman Catholic,
Area Sq Miles	183 569		traditional beliefs,
Population	22 254 000		Sunni Muslim,
Capital	Yaoundé		Protestant
Languages	French, English,	Currency	CFA franc
	Fang, Bamileke,	Organizations	Comm., UN
	other local languages	Map page	118

CANADA

Area Sq Km	9 984 670	Religions	Roman Catholic,
Area Sq Miles	3 855 103		Protestant, Eastern
Population	35 182 000		Orthodox, Jewish
Capital	Ottawa	Currency	Canadian dollar
Languages	English, French,	Organizations	APEC, Comm.,
	other local languages		OECD, UN
		Map page	126–127

Alberta (Province)

Area Sq Km	661 848	Population	3 965 339
Area Sq Miles	255 541	Capital	Edmonton

British Columbia (Province)

Area Sq Km	944 735	Population	4 650 004
Area Sq Miles	364 764	Capital	Victoria

Manitoba (Province)

Area Sq Km	647 797	Population	1 277 339
Area Sq Miles	250 116	Capital	Winnipeg

New Brunswick (Province)

Area Sq Km	72 908	Population	754 039
Area Sq Miles	28 150	Capital	Fredericton

Newfoundland and Labrador (Province)

Area Sq Km	405 212	Population	513 568
Area Sq Miles	156 453	Capital	St John's

Northwest Territories (Territory)

Area Sq Km	1 346 106	Population	43 349
Area Sq Miles	519 734	Capital	Yellowknife

Nova Scotia (Province)

Area Sq Km	55 284	Population	945 015
Area Sq Miles	21 345	Capital	Halifax

Nunavut (Territory)

Area Sq Km	2 093 190	Population	34 023
Area Sq Miles	808 185	Capital	Iqaluit (Frobisher Bay)

© Collins Bartholomew Ltd

CANADA

Ontario (Province)

Area Sq Km	1 076 395	Population	13 583 710
Area Sq Miles	415 598	Capital	Toronto

Prince Edward Island (Province)

Area Sq Km	5 660	Population	145 763
Area Sq Miles	2 185	Capital	Charlottetown

Québec (Province)

Area Sq Km	1 542 056	Population	8 099 095
Area Sq Miles	595 391	Capital	Québec

Saskatchewan (Province)

Area Sq Km	651 036	Population	1 093 880
Area Sq Miles	251 366	Capital	Regina

Yukon Territory (Territory)

Area Sq Km	482 443	Population	36 418
Area Sq Miles	186 272	Capital	Whitehorse

Canary Islands (Islas Canarias)
Autonomous Community of Spain

Area Sq Km	7 447	Religions	Roman Catholic
Area Sq Miles	2 875	Currency	Euro
Population	2 254 944	Map page	114
Capital	Santa Cruz de Tenerife/Las Palmas		
Languages	Spanish		

CAPE VERDE (CABO VERDE)
Republic of Cabo Verde

Area Sq Km	4 033	Religions	Roman Catholic, Protestant
Area Sq Miles	1 557		
Population	499 000	Currency	Cape Verde escudo
Capital	Praia	Organizations	UN
Languages	Portuguese, creole	Map page	46

Cayman Islands
United Kingdom Overseas Territory

Area Sq Km	259	Religions	Protestant, Roman Catholic
Area Sq Miles	100		
Population	58 000	Currency	Cayman Islands dollar
Capital	George Town	Map page	146
Languages	English		

CENTRAL AFRICAN REPUBLIC

Area Sq Km	622 436	Religions	Protestant, Roman Catholic, traditional beliefs, Sunni Muslim
Area Sq Miles	240 324		
Population	4 616 000		
Capital	Bangui		
Languages	French, Sango, Banda, Baya, other local languages	Currency	CFA franc
		Organizations	UN
		Map page	118

Ceuta
Autonomous City of Ceuta

Area Sq Km	19	Religions	Roman Catholic, Muslim
Area Sq Miles	7		
Population	84 504	Currency	Euro
Capital	Ceuta	Map page	106
Languages	Spanish, Arabic		

CHAD
Republic of Chad

Area Sq Km	1 284 000	Religions	Sunni Muslim, Roman Catholic, Protestant, traditional beliefs
Area Sq Miles	495 755		
Population	12 825 000		
Capital	Ndjamena	Currency	CFA franc
Languages	Arabic, French, Sara, other local languages	Organizations	UN
		Map page	115

Chatham Islands
part of New Zealand

Area Sq Km	963	Religions	Protestant
Area Sq Miles	372	Currency	New Zealand dollar
Population	610	Map page	49
Capital	Waitangi		
Languages	English		

CHILE
Republic of Chile

Area Sq Km	756 945	Religions	Roman Catholic, Protestant
Area Sq Miles	292 258		
Population	17 620 000	Currency	Chilean peso
Capital	Santiago	Organizations	APEC, OECD, UN
Languages	Spanish, Amerindian languages	Map page	152–153

CHINA
People's Republic of China

Area Sq Km	9 606 802	Religions	Confucian, Taoist, Buddhist, Christian, Sunni Muslim
Area Sq Miles	3 709 186		
Population	1 369 993 000		
Capital	Beijing (Peking)	Currency	Yuan, Hong Kong dollar, Macao pataca
Languages	Mandarin (Putonghua), Wu, Cantonese, Hsiang, regional languages	Organizations	APEC, UN
		Map page	68–69

Anhui (Province)

Area Sq Km	139 000	Population	59 680 000
Area Sq Miles	53 668	Capital	Hefei

Bejing (Municipality)

Area Sq Km	16 800	Population	20 186 000
Area Sq Miles	6 487	Capital	Beijing (Peking)

Chongqing (Municipality)

Area Sq Km	23 000	Population	29 190 000
Area Sq Miles	8 880	Capital	Chongqing

Fujian (Province)

Area Sq Km	121 400	Population	37 200 000
Area Sq Miles	46 873	Capital	Fuzhou

Gansu (Province)

Area Sq Km	453 700	**Population**	25 642 000
Area Sq Miles	175 175	**Capital**	Lanzhou

Guangdong (Province)

Area Sq Km	178 000	**Population**	105 048 000
Area Sq Miles	68 726	**Capital**	Guangzhou (Canton)

Guangxi Zhuangzu Zizhiqu (Autonomous Region)

Area Sq Km	236 000	**Population**	46 450 000
Area Sq Miles	91 120	**Capital**	Nanning

Guizhou (Province)

Area Sq Km	176 000	**Population**	34 687 000
Area Sq Miles	67 954	**Capital**	Guiyang

Hainan (Province)

Area Sq Km	34 000	**Population**	8 773 000
Area Sq Miles	13 127	**Capital**	Haikou

Hebei (Province)

Area Sq Km	187 700	**Population**	72 405 000
Area Sq Miles	72 471	**Capital**	Shijiazhuang

Heilongjiang (Province)

Area Sq Km	454 600	**Population**	38 340 000
Area Sq Miles	175 522	**Capital**	Harbin

Henan (Province)

Area Sq Km	167 000	**Population**	93 880 000
Area Sq Miles	64 479	**Capital**	Zhengzhou

Hong Kong (Special Administrative Region)

Area Sq Km	1 075	**Population**	7 112 000
Area Sq Miles	415	**Capital**	Hong Kong

Hubei (Province)

Area Sq Km	185 900	**Population**	57 575 000
Area Sq Miles	71 776	**Capital**	Wuhan

Hunan (Province)

Area Sq Km	210 000	**Population**	65 956 000
Area Sq Miles	81 081	**Capital**	Changsha

Jiangsu (Province)

Area Sq Km	102 600	**Population**	78 988 000
Area Sq Miles	39 614	**Capital**	Nanjing

Jiangxi (Province)

Area Sq Km	166 900	**Population**	44 884 000
Area Sq Miles	64 440	**Capital**	Nanchang

Jilin (Province)

Area Sq Km	187 000	**Population**	27 494 000
Area Sq Miles	72 201	**Capital**	Changchun

Liaoning (Province)

Area Sq Km	147 400	**Population**	43 830 000
Area Sq Miles	56 911	**Capital**	Shenyang

Macao (Special Administrative Region)

Area Sq Km	17	**Population**	557 000
Area Sq Mile	7	**Capital**	Macao

Nei Mongol Zizhiqu (Inner Mongolia) (Autonomous Region)

Area Sq Km	1 183 000	**Population**	24 817 000
Area Sq Miles	456 759	**Capital**	Hohhot

Ningxia Huizu Zizhiqu (Autonomous Region)

Area Sq Km	66 400	**Population**	6 395 000
Area Sq Miles	25 637	**Capital**	Yinchuan

Qinghai (Province)

Area Sq Km	721 000	**Population**	5 682 000
Area Sq Miles	278 380	**Capital**	Xining

Shaanxi (Province)

Area Sq Km	205 600	**Population**	37 426 000
Area Sq Miles	79 383	**Capital**	Xi'an

Shandong (Province)

Area Sq Km	153 300	**Population**	96 370 000
Arca Sq Miles	59 189	**Capital**	Jinan

Shanghai (Municipality)

Area Sq Km	6 300	**Population**	23 475 000
Area Sq Miles	2 432	**Capital**	Shanghai

Shanxi (Province)

Area Sq Km	156 300	**Population**	35 930 000
Area Sq Miles	60 348	**Capital**	Taiyuan

Sichuan (Province)

Area Sq Km	569 000	**Population**	80 500 000
Area Sq Miles	219 692	**Capital**	Chengdu

Tianjin (Municipality)

Area Sq Km	11 300	**Population**	13 550 000
Area Sq Miles	4 363	**Capital**	Tianjin

Xinjiang Uygur Zizhiqu (Sinkiang) (Autonomous Region)

Area Sq Km	1 600 000	**Population**	22 087 000
Area Sq Miles	617 763	**Capital**	Ürümqi

Xizang Zizhiqu (Tibet) (Autonomous Region)

Area Sq Km	1 228 400	**Population**	3 033 000
Area Sq Miles	474 288	**Capital**	Lhasa

Yunnan (Province)

Area Sq Km	394 000	**Population**	46 308 000
Area Sq Miles	152 124	**Capital**	Kunming

Zhejiang (Province)

Area Sq Km	101 800	**Population**	54 630 000
Area Sq Miles	39 305	**Capital**	Hangzhou

© Collins Bartholomew Ltd

Christmas Island
Australian External Territory

Area Sq Km	135	Religions	Buddhist, Sunni
Area Sq Miles	52		Muslim, Protestant,
Population	2 072		Roman Catholic
Capital	The Settlement	Currency	Australian dollar
Languages	English	Map page	58

Cocos (Keeling) Islands
Australian External Territory

Area Sq Km	14	Religions	Sunni Muslim,
Area Sq Miles	5		Christian
Population	550	Currency	Australian dollar
Capital	West Island	Map page	58
Languages	English		

COLOMBIA
Republic of Colombia

Area Sq Km	1 141 748	Religions	Roman Catholic,
Area Sq Miles	440 831		Protestant
Population	48 321 000	Currency	Colombian peso
Capital	Bogotá	Organizations	UN
Languages	Spanish, Amerindian	Map page	150
	languages		

COMOROS
Union of the Comoros

Area Sq Km	1 862	Religions	Sunni Muslim, Roman
Area Sq Miles	719		Catholic
Population	735 000	Currency	Comoros franc
Capital	Moroni	Organizations	UN
Languages	Shikomor (Comorian),	Map page	121
	French, Arabic		

CONGO
Republic of the Congo

Area Sq Km	342 000	Religions	Roman Catholic,
Area Sq Miles	132 047		Protestant, traditional
Population	4 448 000		beliefs, Sunni Muslim
Capital	Brazzaville	Currency	CFA franc
Languages	French, Kongo,	Organizations	UN
	Monokutuba, other	Map page	118
	local languages		

CONGO, DEMOCRATIC REPUBLIC OF THE

Area Sq Km	2 345 410	Religions	Christian, Sunni
Area Sq Miles	905 568		Muslim
Population	67 514 000	Currency	Congolese franc
Capital	Kinshasa	Organizations	SADC, UN
Languages	French, Lingala,	Map page	118–119
	Swahili, Kongo,		
	other local languages		

Cook Islands
Self-governing New Zealand Overseas Territory

Area Sq Km	293	Religions	Protestant, Roman
Area Sq Miles	113		Catholic
Population	21 000	Currency	New Zealand dollar
Capital	Avarua	Map page	49
Languages	English, Maori		

COSTA RICA
Republic of Costa Rica

Area Sq Km	51 100	Religions	Roman Catholic,
Area Sq Miles	19 730		Protestant
Population	4 872 000	Currency	Costa Rican colón
Capital	San José	Organizations	UN
Languages	Spanish	Map page	146

CÔTE D'IVOIRE (IVORY COAST)
Republic of Côte d'Ivoire

Area Sq Km	322 463	Religions	Sunni Muslim, Roman
Area Sq Miles	124 504		Catholic, traditonal
Population	20 316 000		beliefs, Protestant
Capital	Yamoussoukro	Currency	CFA franc
Languages	French, creole, Akan,	Organizations	UN
	other local languages	Map page	114

Crimea
Disputed Territory (Ukraine)

Area Sq Km	27 000	Map page	91
Area Sq Miles	10 400		
Population	2 348 600		
Capital	Simferopol'		

CROATIA
Republic of Croatia

Area Sq Km	56 538	Religions	Roman Catholic,
Area Sq Miles	21 829		Serbian Orthodox,
Population	4 290 000		Sunni Muslim
Capital	Zagreb	Currency	Kuna
Languages	Croatian, Serbian	Organizations	EU, UN
		Map page	109

CUBA
Republic of Cuba

Area Sq Km	110 860	Religions	Roman Catholic,
Area Sq Miles	42 803		Protestant
Population	11 266 000	Currency	Cuban peso
Capital	Havana (La Habana)	Organizations	UN
Languages	Spanish	Map page	146

Curaçao
Self-governing Netherlands Territory

Area Sq Km	444	Religions	Roman Catholic,
Area Sq Miles	171		Protestant
Population	159 000	Currency	Caribbean guilder
Capital	Willemstad	Map page	147
Languages	Dutch, Papiamento		

CYPRUS
Republic of Cyprus

Area Sq Km	9 251	**Religions**	Greek Orthodox,
Area Sq Miles	3 572		Sunni Muslim
Population	1 141 000	**Currency**	Euro
Capital	Nicosia (Lefkosia)	**Organizations**	Comm., EU, UN
Languages	Greek, Turkish,	**Map page**	80
	English		

CZECH REPUBLIC

Area Sq Km	78 864	**Religions**	Roman Catholic,
Area Sq Miles	30 450		Protestant
Population	10 702 000	**Currency**	Czech koruna
Capital	Prague (Praha)	**Organizations**	EU, OECD, UN
Languages	Czech, Moravian,	**Map page**	102–103
	Slovakian		

DENMARK
Kingdom of Denmark

Area Sq Km	43 075	**Religions**	Protestant
Area Sq Miles	16 631	**Currency**	Danish krone
Population	5 619 000	**Organizations**	EU, OECD, UN
Capital	Copenhagen	**Map page**	93
	(København)		
Languages	Danish		

DJIBOUTI
Republic of Djibouti

Area Sq Km	23 200	**Religions**	Sunni Muslim,
Area Sq Miles	8 958		Christian
Population	873 000	**Currency**	Djibouti franc
Capital	Djibouti	**Organizations**	UN
Languages	Somali, Afar, French,	**Map page**	117
	Arabic		

DOMINICA
Commonwealth of Dominica

Area Sq Km	750	**Religions**	Roman Catholic,
Area Sq Miles	290		Protestant
Population	72 000	**Currency**	East Caribbean dollar
Capital	Roseau	**Organizations**	CARICOM, Comm.,
Languages	English, creole		UN
		Map page	147

DOMINICAN REPUBLIC

Area Sq Km	48 442	**Religions**	Roman Catholic,
Area Sq Miles	18 704		Protestant
Population	10 404 000	**Currency**	Dominican peso
Capital	Santo Domingo	**Organizations**	UN
Languages	Spanish, creole	**Map page**	147

Easter Island (Isla de Pascua)
part of Chile

Area Sq Km	171	**Religions**	Roman Catholic
Area Sq Miles	66	**Currency**	Chilean peso
Population	5 306	**Map page**	157
Capital	Hanga Roa		
Languages	Spanish		

EAST TIMOR (TIMOR-LESTE)
Democratic Republic of Timor-Leste

Area Sq Km	14 874	**Religions**	Roman Catholic
Area Sq Miles	5 743	**Currency**	United States dollar
Population	1 133 000	**Organisations**	UN
Capital	Dili	**Map page**	59
Languages	Portuguese, Tetun,		
	English		

ECUADOR
Republic of Ecuador

Area Sq Km	272 045	**Religions**	Roman Catholic
Area Sq Miles	105 037	**Currency**	United States dollar
Population	15 738 000	**Organizations**	OPEC, UN
Capital	Quito	**Map page**	150
Languages	Spanish, Quechua,		
	Amerindian		
	languages		

EGYPT
Arab Republic of Egypt

Area Sq Km	1 000 250	**Religions**	Sunni Muslim, Coptic
Area Sq Miles	386 199		Christian
Population	82 056 000	**Currency**	Egyptian pound
Capital	Cairo (Al Qāhirah)	**Organizations**	UN
Languages	Arabic	**Map page**	116

EL SALVADOR
Republic of El Salvador

Area Sq Km	21 041	**Religions**	Roman Catholic,
Area Sq Miles	8 124		Protestant
Population	6 340 000	**Currency**	El Salvador colón,
Capital	San Salvador		United States dollar
Languages	Spanish	**Organizations**	UN
		Map page	146

EQUATORIAL GUINEA
Republic of Equatorial Guinea

Area Sq Km	28 051	**Religions**	Roman Catholic,
Area Sq Miles	10 831		traditional beliefs
Population	757 000	**Currency**	CFA franc
Capital	Malabo	**Organizations**	UN
Languages	Spanish, French,	**Map page**	118
	Fang		

ERITREA
State of Eritrea

Area Sq Km	117 400	**Religions**	Sunni Muslim,
Area Sq Miles	45 328		Coptic Christian
Population	6 333 000	**Currency**	Nakfa
Capital	Asmara	**Organizations**	UN
Languages	Tigrinya, Tigre	**Map page**	116

ESTONIA
Republic of Estonia

Area Sq Km	45 200	**Religions**	Protestant, Estonian
Area Sq Miles	17 452		and Russian Orthodox
Population	1 287 000	**Currency**	Euro
Capital	Tallinn	**Organizations**	EU, OECD, UN
Languages	Estonian, Russian	**Map page**	88

© Collins Bartholomew Ltd

ETHIOPIA
Federal Democratic Republic of Ethiopia

Area Sq Km	1 133 880	**Religions**	Ethiopian Orthodox,
Area Sq Miles	437 794		Sunni Muslim,
Population	94 101 000		traditional beliefs
Capital	Addis Ababa	**Currency**	Birr
	(Ādīs Ābeba)	**Organizations**	UN
Languages	Oromo, Amharic,	**Map page**	117
	Tigrinya, other local		
	languages		

Falkland Islands (Islas Malvinas)
United Kingdom Overseas Territory

Area Sq Km	12 170	**Religions**	Protestant, Roman
Area Sq Miles	4 699		Catholic
Population	2 931	**Currency**	Falkland Islands
Capital	Stanley		pound
Languages	English	**Map page**	153

Faroe Islands
Self-governing Danish Territory

Area Sq Km	1 399	**Religions**	Protestant
Area Sq Miles	540	**Currency**	Danish krone
Population	49 000	**Map page**	94
Capital	Tórshavn		
Languages	Faroese, Danish		

FIJI
Republic of Fiji

Area Sq Km	18 330	**Religions**	Christian, Hindu,
Area Sq Miles	7 077		Sunni Muslim
Population	881 000	**Currency**	Fiji dollar
Capital	Suva	**Organizations**	Comm., UN
Languages	English, Fijian,	**Map page**	49
	Hindi		

FINLAND
Republic of Finland

Area Sq Km	338 145	**Religions**	Protestant, Greek
Area Sq Miles	130 559		Orthodox
Population	5 426 000	**Currency**	Euro
Capital	Helsinki	**Organizations**	EU, OECD, UN
	(Helsingfors)	**Map page**	92–93
Languages	Finnish, Swedish,		
	Sami		

FRANCE
French Republic

Area Sq Km	543 965	**Religions**	Roman Catholic,
Area Sq Miles	210 026		Protestant, Sunni
Population	64 291 000		Muslim
Capital	Paris	**Currency**	Euro
Languages	French, German	**Organizations**	EU, OECD, UN
	dialects, Italian,	**Map page**	104–105
	Arabic, Breton		

French Guiana
French Overseas Department

Area Sq Km	90 000	**Religions**	Roman Catholic
Area Sq Miles	34 749	**Currency**	Euro
Population	249 000	**Map page**	151
Capital	Cayenne		
Languages	French, creole		

French Polynesia
French Overseas Territory

Area Sq Km	3 265	**Religions**	Protestant, Roman
Area Sq Miles	1 261		Catholic
Population	277 000	**Currency**	CFP franc
Capital	Papeete	**Map page**	49
Languages	French, Tahitian,		
	other Polynesian		
	languages		

GABON
Gabonese Republic

Area Sq Km	267 667	**Religions**	Roman Catholic,
Area Sq Miles	103 347		Protestant, traditonal
Population	1 672 000		beliefs
Capital	Libreville	**Currency**	CFA franc
Languages	French, Fang, other	**Organizations**	UN
	local languages	**Map page**	118

Galapagos Islands (Islas Galápagos)
part of Ecuador

Area Sq Km	8 010	**Religions**	Roman Catholic
Area Sq Miles	3 093	**Currency**	United States dollar
Population	25 124	**Map page**	125
Capital	Puerto Baquerizo		
	Moreno		
Languages	Spanish		

THE GAMBIA
Republic of The Gambia

Area Sq Km	11 295	**Religions**	Sunni Muslim,
Area Sq Miles	4 361		Protestant
Population	1 849 000	**Currency**	Dalasi
Capital	Banjul	**Organizations**	UN
Languages	English, Malinke,	**Map page**	114
	Fulani, Wolof		

Gaza
Disputed Territory

Area Sq Km	363	**Religions**	Sunni Muslim, Shi'a
Area Sq Miles	140		Muslim
Population	1 701 437	**Currency**	Israeli shekel
Capital	Gaza	**Map page**	80
Languages	Arabic		

GEORGIA

Area Sq Km	69 700	**Religions**	Georgian Orthodox,
Area Sq Miles	26 911		Russian Orthodox,
Population	4 341 000		Sunni Muslim
Capital	Tbilisi	**Currency**	Lari
Languages	Georgian, Russian,	**Organizations**	UN
	Armenian, Azeri,	**Map page**	81
	Ossetian, Abkhaz		

GERMANY
Federal Republic of Germany

Area Sq Km	357 022	**Religions**	Protestant, Roman
Area Sq Miles	137 847		Catholic
Population	82 727 000	**Currency**	Euro
Capital	Berlin	**Organizations**	EU, OECD, UN
Languages	German, Turkish	**Map page**	102

GHANA
Republic of Ghana

Area Sq Km	238 537	**Religions**	Christian, Sunni
Area Sq Miles	92 100		Muslim, traditional
Population	25 905 000		beliefs
Capital	Accra	**Currency**	Cedi
Languages	English, Hausa,	**Organizations**	Comm., UN
	Akan, other local	**Map page**	114
	languages		

Gibraltar
United Kingdom Overseas Territory

Area Sq Km	7	**Religions**	Roman Catholic,
Area Sq Miles	3		Protestant, Sunni
Population	29 000		Muslim
Capital	Gibraltar	**Currency**	Gibraltar pound
Languages	English, Spanish	**Map page**	106

GREECE
Hellenic Republic

Area Sq Km	131 957	**Religions**	Greek Orthodox,
Area Sq Miles	50 949		Sunni Muslim
Population	11 128 000	**Currency**	Euro
Capital	Athens (Athina)	**Organizations**	EU, OECD, UN
Languages	Greek	**Map page**	111

Greenland
Self-governing Danish Territory

Area Sq Km	2 175 600	**Religions**	Protestant
Area Sq Miles	840 004	**Currency**	Danish krone
Population	57 000	**Map page**	127
Capital	Nuuk (Godthåb)		
Languages	Greenlandic, Danish		

GRENADA

Area Sq Km	378	**Religions**	Roman Catholic,
Area Sq Miles	146		Protestant
Population	106 000	**Currency**	East Caribbean dollar
Capital	St George's	**Organizations**	CARICOM, Comm.,
Languages	English, creole		UN
		Map page	147

Guadeloupe
French Overseas Department

Area Sq Km	1 780	**Religions**	Roman Catholic
Area Sq Miles	687	**Currency**	Euro
Population	466 000	**Map page**	147
Capital	Basse-Terre		
Languages	French, creole		

Guam
United States Unincorporated Territory

Area Sq Km	541	**Religions**	Roman Catholic
Area Sq Miles	209	**Currency**	United States dollar
Population	165 000	**Map page**	59
Capital	Hagåtña		
Languages	Chamorro, English,		
	Tagalog		

GUATEMALA
Republic of Guatemala

Area Sq Km	108 890	**Religion**	Roman Catholic,
Area Sq Miles	42 043		Protestant
Population	15 468 000	**Currency**	Quetzal, United
Capital	Guatemala City		States dollar
Languages	Spanish, Mayan	**Organizations**	UN
	languages	**Map page**	146

Guernsey
United Kingdom Crown Dependency

Area Sq Km	78	**Religions**	Protestant, Roman
Area Sq Miles	30		Catholic
Population	65 578	**Currency**	Pound sterling
Capital	St Peter Port	**Map page**	95
Languages	English, French		

GUINEA
Republic of Guinea

Area Sq Km	245 857	**Religions**	Sunni Muslim,
Area Sq Miles	94 926		traditional beliefs,
Population	11 745 000		Christian
Capital	Conakry	**Currency**	Guinea franc
Languages	French, Fulani,	**Organizations**	UN
	Malinke, other local	**Map page**	114
	languages		

GUINEA-BISSAU
Republic of Guinea-Bissau

Area Sq Km	36 125	**Religions**	Traditional beliefs,
Area Sq Miles	13 948		Sunni Muslim,
Population	1 704 000		Christian
Capital	Bissau	**Currency**	CFA franc
Languages	Portuguese, crioulo,	**Organizations**	UN
	other local languages	**Map page**	114

GUYANA
Co-operative Republic of Guyana

Area Sq Km	214 969	**Religions**	Protestant, Hindu,
Area Sq Miles	83 000		Roman Catholic,
Population	800 000		Sunni Muslim
Capital	Georgetown	**Currency**	Guyana dollar
Languages	English, creole,	**Organizations**	CARICOM, Comm.,
	Amerindian		UN
	languages	**Map page**	150

HAITI
Republic of Haiti

Area Sq Km	27 750	**Religions**	Roman Catholic,
Area Sq Miles	10 714		Protestant, Voodoo
Population	10 317 000		
Capital	Port-au-Prince	**Currency**	Gourde
Languages	French, creole	**Organizations**	CARICOM, UN
		Map page	147

HONDURAS
Republic of Honduras

Area Sq Km	112 088	**Religions**	Roman Catholic,
Area Sq Miles	43 277		Protestant
Population	8 098 000		
Capital	Tegucigalpa	**Currency**	Lempira
Languages	Spanish, Amerindian	**Organizations**	UN
	languages	**Map page**	147

© Collins Bartholomew Ltd

HUNGARY

Area Sq Km	93 030	Religions	Roman Catholic, Protestant
Area Sq Miles	35 919		
Population	9 955 000	Currency	Forint
Capital	Budapest	Organizations	EU, OECD, UN
Languages	Hungarian	Map page	103

 ICELAND
Republic of Iceland

Area Sq Km	102 820	Religions	Protestant
Area Sq Miles	39 699	Currency	Icelandic króna
Population	330 000	Organizations	OECD, UN
Capital	Reykjavík	Map page	92
Languages	Icelandic		

 INDIA
Republic of India

Area Sq Km	3 166 620	Religions	Hindu, Sunni Muslim, Shi'a Muslim, Sikh, Christian
Area Sq Miles	1 222 632		
Population	1 252 140 000		
Capital	New Delhi	Currency	Indian rupee
Languages	Hindi, English, many regional languages	Organizations	Comm., UN
		Map page	72–73

 INDONESIA
Republic of Indonesia

Area Sq Km	1 919 445	Religions	Sunni Muslim, Protestant, Roman Catholic, Hindu, Buddhist
Area Sq Miles	741 102		
Population	249 866 000		
Capital	Jakarta		
Languages	Indonesian, other local languages	Currency	Rupiah
		Organizations	APEC, ASEAN, UN
		Map page	58–59

 IRAN
Islamic Republic of Iran

Area Sq Km	1 648 000	Religions	Shi'a Muslim, Sunni Muslim
Area Sq Miles	636 296		
Population	77 447 000	Currency	Iranian rial
Capital	Tehrān	Organizations	OPEC, UN
Languages	Farsi, Azeri, Kurdish, regional languages	Map page	81

 IRAQ
Republic of Iraq

Area Sq Km	438 317	Religions	Shi'a Muslim, Sunni Muslim, Christian
Area Sq Miles	169 235		
Population	33 765 000	Currency	Iraqi dinar
Capital	Baghdād	Organizations	OPEC, UN
Languages	Arabic, Kurdish, Turkmen	Map page	81

 IRELAND

Area Sq Km	70 282	Religions	Roman Catholic, Protestant
Area Sq Miles	27 136		
Population	4 627 000	Currency	Euro
Capital	Dublin (Baile Átha Cliath)	Organizations	EU, OECD, UN
		Map page	97
Languages	English, Irish		

 Isle of Man
United Kingdom Crown Dependency

Area Sq Km	572	Religions	Protestant, Roman Catholic
Area Sq Miles	221		
Population	86 000	Currency	Pound sterling
Capital	Douglas	Map page	98
Languages	English		

 ISRAEL
State of Israel

Area Sq Km	22 072	Religions	Jewish, Sunni Muslim, Christian, Druze
Area Sq Miles	8 522		
Population	7 733 000	Currency	Shekel
Capital	Jerusalem* (Yerushalayim) (El Quds)	Organizations	OECD, UN
		Map page	80
Languages	Hebrew, Arabic		

*De facto capital. Disputed.

 ITALY
Italian Republic

Area Sq Km	301 245	Religions	Roman Catholic
Area Sq Miles	116 311	Currency	Euro
Population	60 990 000	Organizations	EU, OECD, UN
Capital	Rome (Roma)	Map page	108–109
Languages	Italian		

 JAMAICA

Area Sq Km	10 991	Religions	Protestant, Roman Catholic
Area Sq Miles	4 244		
Population	2 784 000	Currency	Jamaican dollar
Capital	Kingston	Organizations	CARICOM, Comm., UN
Languages	English, creole		
		Map page	146

 JAPAN

Area Sq Km	377 727	Religions	Shintoist, Buddhist, Christian
Area Sq Miles	145 841		
Population	127 144 000	Currency	Yen
Capital	Tōkyō	Organizations	APEC, OECD, UN
Languages	Japanese	Map page	66–67

 Jersey
United Kingdom Crown Dependency

Area Sq Km	116	Religions	Protestant, Roman Catholic
Area Sq Miles	45		
Population	99 000	Currency	Pound sterling
Capital	St Helier	Map page	95
Languages	English, French		

JORDAN
Hashemite Kingdom of Jordan

Area Sq Km	89 206	Religions	Sunni Muslim, Christian
Area Sq Miles	34 443		
Population	7 274 000	Currency	Jordanian dinar
Capital	'Ammān	Organizations	UN
Languages	Arabic	Map page	80

Juan Fernández Islands
part of Chile

Area Sq Km	179	Religions	Roman Catholic, Protestant
Area Sq Miles	69		
Population	909	Currency	Chilean peso
Capital	San Juan Bautista	Map page	157
Languages	Spanish, Amerindian languages		

KAZAKHSTAN
Republic of Kazakhstan

Area Sq Km	2 717 300	Religions	Sunni Muslim, Russian Orthodox, Protestant
Area Sq Miles	1 049 155		
Population	16 441 000	Currency	Tenge
Capital	Astana (Akmola)	Organizations	CIS, UN
Languages	Kazakh, Russian, Ukrainian, German, Uzbek, Tatar	Map page	76–77

KENYA
Republic of Kenya

Area Sq Km	582 646	Religions	Christian, traditional beliefs
Area Sq Miles	224 961		
Population	44 354 000	Currency	Kenyan shilling
Capital	Nairobi	Organizations	Comm., UN
Languages	Swahili, English, other local languages	Map page	119

KIRIBATI
Republic of Kiribati

Area Sq Km	717	Religions	Roman Catholic, Protestant
Area Sq Miles	277		
Population	102 000	Currency	Australian dollar
Capital	Bairiki	Organizations	Comm., UN
Languages	Gilbertese, English	Map page	49

KOSOVO
Republic of Kosovo

Area Sq Km	10 908	Religions	Sunni Muslim, Serbian Orthodox
Area Sq Miles	4 212		
Population	1 815 606	Currency	Euro
Capital	Prishtinë (Priština)	Map page	109
Languages	Albanian, Serbian		

KUWAIT
State of Kuwait

Area Sq Km	17 818	Religions	Sunni Muslim, Shi'a Muslim, Christian, Hindu
Area Sq Miles	6 880		
Population	3 369 000	Currency	Kuwaiti dinar
Capital	Kuwait (Al Kuwayt)	Organizations	OPEC, UN
Languages	Arabic	Map page	78

KYRGYZSTAN
Kyrgyz Republic

Area Sq Km	198 500	Religions	Sunni Muslim, Russian Orthodox
Area Sq Miles	76 641		
Population	5 548 000	Currency	Kyrgyz som
Capital	Bishkek (Frunze)	Organizations	CIS, UN
Languages	Kyrgyz, Russian, Uzbek	Map page	77

LAOS
Lao People's Democratic Republic

Area Sq Km	236 800	Religions	Buddhist, traditional beliefs
Area Sq Miles	91 429		
Population	6 770 000	Currency	Kip
Capital	Vientiane (Viangchan)	Organizations	ASEAN, UN
Languages	Lao, other local languages	Map page	62–63

LATVIA
Republic of Latvia

Area Sq Km	64 589	Religions	Protestant, Roman Catholic, Russian Orthodox
Area Sq Miles	24 938		
Population	2 050 000	Currency	Euro
Capital	Rīga	Organizations	EU, UN
Languages	Latvian, Russian	Map page	88

LEBANON
Lebanese Republic

Area Sq Km	10 452	Religions	Shi'a Muslim, Sunni Muslim, Christian
Area Sq Miles	4 036		
Population	4 822 000	Currency	Lebanese pound
Capital	Beirut (Beyrouth)	Organizations	UN
Languages	Arabic, Armenian, French	Map page	80

LESOTHO
Kingdom of Lesotho

Area Sq Km	30 355	Religions	Christian, traditional beliefs
Area Sq Miles	11 720		
Population	2 074 000	Currency	Loti, South African rand
Capital	Maseru		
Languages	Sesotho, English, Zulu	Organizations	Comm., SADC, UN
		Map page	123

LIBERIA
Republic of Liberia

Area Sq Km	111 369	Religions	Traditional beliefs, Christian, Sunni Muslim
Area Sq Miles	43 000		
Population	4 294 000		
Capital	Monrovia	Currency	Liberian dollar
Languages	English, creole, other local languages	Organizations	UN
		Map page	114

LIBYA
State of Libya

Area Sq Km	1 759 540	Religions	Sunni Muslim
Area Sq Miles	679 362	Currency	Libyan dinar
Population	6 202 000	Organizations	OPEC, UN
Capital	Tripoli (Ṭarābulus)	Map page	115
Languages	Arabic, Berber		

LIECHTENSTEIN
Principality of Liechtenstein

Area Sq Km	160	Religions	Roman Catholic, Protestant
Area Sq Miles	62		
Population	37 000	Currency	Swiss franc
Capital	Vaduz	Organizations	UN
Languages	German	Map page	105

© Collins Bartholomew Ltd

LITHUANIA
Republic of Lithuania

Area Sq Km	65 200	**Religions**	Roman Catholic,
Area Sq Miles	25 174		Protestant, Russian
Population	3 017 000		Orthodox
Capital	Vilnius	**Currency**	Euro
Languages	Lithuanian, Russian,	**Organizations**	EU, UN
	Polish	**Map page**	88

Lord Howe Island
part of Australia

Area Sq Km	17	**Religions**	Protestant,
Area Sq Miles	6		Roman Catholic
Population	373	**Currency**	Australian dollar
Languages	English	**Map page**	51

LUXEMBOURG
Grand Duchy of Luxembourg

Area Sq Km	2 586	**Religions**	Roman Catholic
Area Sq Miles	998	**Currency**	Euro
Population	530 000	**Organizations**	EU, OECD, UN
Capital	Luxembourg	**Map page**	100
Languages	Letzeburgish,		
	German, French		

MACEDONIA (F.Y.R.O.M.)
Republic of Macedonia

Area Sq Km	25 713	**Religions**	Macedonian Orthodox,
Area Sq Miles	9 928		Sunni Muslim
Population	2 107 000	**Currency**	Macedonian denar
Capital	Skopje	**Organizations**	UN
Languages	Macedonian,	**Map page**	111
	Albanian, Turkish		

MADAGASCAR
Republic of Madagascar

Area Sq Km	587 041	**Religions**	Traditional beliefs,
Area Sq Miles	226 658		Christian, Sunni
Population	22 925 000		Muslim
Capital	Antananarivo	**Currency**	Malagasy ariary,
Languages	Malagasy, French		Malagasy franc
		Organizations	SADC, UN
		Map page	121

Madeira
Autonomous Region of Portugal

Area Sq Km	779	**Religions**	Roman Catholic,
Area Sq Miles	301		Protestant
Population	263 091	**Currency**	Euro
Capital	Funchal	**Map page**	114
Languages	Portuguese		

MALAWI
Republic of Malawi

Area Sq Km	118 484	**Religions**	Christian, traditional
Area Sq Miles	45 747		beliefs, Sunni Muslim
Population	16 363 000	**Currency**	Malawian kwacha
Capital	Lilongwe	**Organizations**	Comm., SADC, UN
Languages	Chichewa, English,	**Map page**	121
	other local languages		

MALAYSIA

Area Sq Km	332 965	**Religions**	Sunni Muslim,
Area Sq Miles	128 559		Buddhist, Hindu,
Population	29 717 000		Christian,
Capital	Kuala Lumpur/		traditional beliefs
	Putrajaya	**Currency**	Ringgit
Languages	Malay, English,	**Organizations**	APEC, ASEAN,
	Chinese, Tamil,		Comm., UN
	other local languages	**Map page**	60–61

MALDIVES
Republic of the Maldives

Area Sq Km	298	**Religions**	Sunni Muslim
Area Sq Miles	115	**Currency**	Rufiyaa
Population	345 000	**Organizations**	Comm., UN
Capital	Male	**Map page**	56
Languages	Divehi (Maldivian)		

MALI
Republic of Mali

Area Sq Km	1 240 140	**Religions**	Sunni Muslim,
Area Sq Miles	478 821		traditional beliefs,
Population	15 302 000		Christian
Capital	Bamako	**Currency**	CFA franc
Languages	French, Bambara,	**Organizations**	UN
	other local languages	**Map page**	114

MALTA
Republic of Malta

Area Sq Km	316	**Religions**	Roman Catholic
Area Sq Miles	122	**Currency**	Euro
Population	429 000	**Organizations**	Comm., EU, UN
Capital	Valletta	**Map page**	84
Languages	Maltese, English		

MARSHALL ISLANDS
Republic of the Marshall Islands

Area Sq Km	181	**Religions**	Protestant, Roman
Area Sq Miles	70		Catholic
Population	53 000	**Currency**	United States dollar
Capital	Delap-Uliga-Djarrit	**Organizations**	UN
Languages	English, Marshallese	**Map page**	48

Martinique
French Overseas Department

Area Sq Km	1 079	**Religions**	Roman Catholic,
Area Sq Miles	417		traditional beliefs
Population	404 000	**Currency**	Euro
Capital	Fort-de-France	**Map page**	147
Languages	French, creole		

MAURITANIA
Islamic Republic of Mauritania

Area Sq Km	1 030 700	**Religions**	Sunni Muslim
Area Sq Miles	397 955	**Currency**	Ouguiya
Population	3 890 000	**Organizations**	UN
Capital	Nouakchott	**Map page**	114
Languages	Arabic, French,		
	other local languages		

MAURITIUS
Republic of Mauritius

Area Sq Km	2 040	Religions	Hindu, Roman
Area Sq Miles	788		Catholic, Sunni
Population	1 244 000		Muslim
Capital	Port Louis	Currency	Mauritius rupee
Languages	English, creole,	Organizations	Comm., SADC, UN
	Hindi, Bhojpuri,	Map page	113
	French		

Mayotte
French Overseas Department

Area Sq Km	373	Religions	Sunni Muslim,
Area Sq Miles	144		Christian
Population	222 000	Currency	Euro
Capital	Dzaoudzi	Map page	121
Languages	French, Mahorian		
	(Shimaore), Kibushi		

Melilla
Autonomous City of Melilla

Area Sq Km	13	Religions	Roman Catholic,
Area Sq Miles	5		Muslim
Population	83 762	Currency	Euro
Capital	Melilla	Map page	114
Languages	Spanish, Arabic		

MEXICO
United Mexican States

Area Sq Km	1 972 545	Religions	Roman Catholic,
Area Sq Miles	761 604		Protestant
Population	122 332 000	Currency	Mexican peso
Capital	Mexico City	Organizations	APEC, OECD, UN
Languages	Spanish, Amerindian	Map page	144–145
	languages		

MICRONESIA, FEDERATED STATES OF

Area Sq Km	701	Religions	Roman Catholic,
Area Sq Miles	271		Protestant
Population	104 000	Currency	United States dollar
Capital	Palikir	Organizations	UN
Languages	English, Chuukese,	Map page	48
	Pohnpeian, other		
	local languages		

MOLDOVA
Republic of Moldova

Area Sq Km	33 700	Religions	Romanian Orthodox,
Area Sq Miles	13 012		Russian Orthodox
Population	3 487 000	Currency	Moldovan leu
Capital	Chişinău (Kishinev)	Organizations	CIS, UN
Languages	Romanian,	Map page	90
	Ukrainian, Gagauz,		
	Russian		

MONACO
Principality of Monaco

Area Sq Km	2	Religions	Roman Catholic
Area Sq Miles	1	Currency	Euro
Population	38 000	Organizations	UN
Capital	Monaco-Ville	Map page	105
Languages	French, Monégasque,		
	Italian		

MONGOLIA

Area Sq Km	1 565 000	Religions	Buddhist,
Area Sq Miles	604 250		Sunni Muslim
Population	2 839 000	Currency	Tugrik (tögrög)
Capital	Ulan Bator	Organizations	UN
	(Ulaanbaatar)	Map page	68–69
Languages	Khalka (Mongolian),		
	Kazakh, other local		
	languages		

MONTENEGRO

Area Sq Km	13 812	Religions	Montenegrin,
Area Sq Miles	5 333		Orthodox,
Population	621 000		Sunni Muslim
Capital	Podgorica	Currency	Euro
Languages	Serbian,	Organizations	UN
	(Montenegrin),	Map page	109
	Albanian		

Montserrat
United Kingdom Overseas Territory

Area Sq Km	100	Religions	Protestant, Roman
Area Sq Miles	39		Catholic
Population	4 922	Currency	East Caribbean dollar
Capital	Brades*	Organizations	CARICOM
Languages	English	Map page	147

* Temporary capital. Official capital Plymouth abandoned in 1997 due to volcanic activity.

MOROCCO
Kingdom of Morocco

Area Sq Km	446 550	Religions	Sunni Muslim
Area Sq Miles	172 414	Currency	Moroccan dirham
Population	33 008 000	Organizations	UN
Capital	Rabat	Map page	114
Languages	Arabic, Berber,		
	French		

MOZAMBIQUE
Republic of Mozambique

Area Sq Km	799 380	Religions	Traditional beliefs,
Area Sq Miles	308 642		Roman Catholic,
Population	25 834 000		Sunni Muslim
Capital	Maputo	Currency	Metical
Languages	Portuguese, Makua,	Organizations	Comm., SADC, UN
	Tsonga, other local	Map page	121
	languages		

MYANMAR (Burma)
Republic of the Union of Myanmar

Area Sq Km	676 577	Religions	Buddhist, Christian,
Area Sq Miles	261 228		Sunni Muslim
Population	53 259 000	Currency	Kyat
Capital	Nay Pyi Taw	Organizations	ASEAN, UN
Languages	Burmese, Shan,	Map page	62–63
	Karen, other local		
	languages		

Nagorno-Karabakh
Disputed Territory (Azerbaijan)

Area Sq Km	6 000	Map page	81
Area Sq Miles	2 317		
Population	146 600		
Capital	Xankändi (Stepanakert)		

© Collins Bartholomew Ltd

NAMIBIA

Republic of Namibia

Area Sq Km	824 292	**Religions**	Protestant, Roman Catholic
Area Sq Miles	318 261		
Population	2 303 000	**Currency**	Namibian dollar
Capital	Windhoek	**Organizations**	Comm., SADC, UN
Languages	English, Afrikaans, German, Ovambo, other local languages	**Map page**	121

NAURU

Republic of Nauru

Area Sq Km	21	**Religions**	Protestant, Roman Catholic
Area Sq Miles	8		
Population	10 000	**Currency**	Australian dollar
Capital	Yaren	**Organizations**	Comm., UN
Languages	Nauruan, English	**Map page**	48

NEPAL

Federal Democratic Republic of Nepal

Area Sq Km	147 181	**Religions**	Hindu, Buddhist, Sunni Muslim
Area Sq Miles	56 827		
Population	27 797 000	**Currency**	Nepalese rupee
Capital	Kathmandu	**Organizations**	UN
Languages	Nepali, Maithili, Bhojpuri, English, other local languages	**Map page**	75

NETHERLANDS

Kingdom of the Netherlands

Area Sq Km	41 526	**Religions**	Roman Catholic, Protestant, Sunni Muslim
Area Sq Miles	16 033		
Population	16 759 000	**Currency**	Euro
Capital	Amsterdam/ The Hague ('s-Gravenhage)	**Organizations**	EU, OECD, UN
Languages	Dutch, Frisian	**Map page**	100

New Caledonia

French Overseas Territory

Area Sq Km	19 058	**Religions**	Roman Catholic, Protestant, Sunni Muslim
Area Sq Miles	7 358		
Population	256 000	**Currency**	CFP franc
Capital	Nouméa	**Map page**	48
Languages	French, other local languages		

NEW ZEALAND

Area Sq Km	270 534	**Religions**	Protestant, Roman Catholic
Area Sq Miles	104 454		
Population	4 506 000	**Currency**	New Zealand dollar
Capital	Wellington	**Organizations**	APEC, Comm., OECD, UN
Languages	English, Maori	**Map page**	54

NICARAGUA
Republic of Nicaragua

Area Sq Km	130 000	**Religions**	Roman Catholic, Protestant
Area Sq Miles	50 193		
Population	6 080 000	**Currency**	Córdoba
Capital	Managua	**Organizations**	UN
Languages	Spanish, Amerindian languages	**Map page**	146

NIGER

Republic of Niger

Area Sq Km	1 267 000	**Religions**	Sunni Muslim, traditional beliefs
Area Sq Miles	489 191		
Population	17 831 000	**Currency**	CFA franc
Capital	Niamey	**Organizations**	UN
Languages	French, Hausa, Fulani, other local languages	**Map page**	115

NIGERIA
Federal Republic of Nigeria

Area Sq Km	923 768	**Religions**	Sunni Muslim, Christian, traditional beliefs
Area Sq Miles	356 669		
Population	173 615 000	**Currency**	Naira
Capital	Abuja	**Organizations**	Comm., OPEC, UN
Languages	English, Hausa, Yoruba, Ibo, Fulani, other local languages	**Map page**	115

Niue

Self-governing New Zealand Overseas Territory

Area Sq Km	258	**Religions**	Christian
Area Sq Miles	100		
Population	1 460	**Currency**	New Zealand dollar
Capital	Alofi	**Map page**	48
Languages	English, Nivean		

Norfolk Island

Australian External Territory

Area Sq Km	35	**Religions**	Protestant, Roman Catholic
Area Sq Miles	14		
Population	2 302	**Currency**	Australian dollar
Capital	Kingston	**Map page**	48
Languages	English		

Northern Mariana Islands

United States Commonwealth

Area Sq Km	477	**Religions**	Roman Catholic
Area Sq Miles	184		
Population	54 000	**Currency**	United States dollar
Capital	Capitol Hill	**Map page**	59
Languages	English, Chamorro, other local languages		

NORTH KOREA

Democratic People's Republic of Korea

Area Sq Km	120 538	**Religions**	Traditional beliefs, Chondoist, Buddhist
Area Sq Miles	46 540		
Population	24 895 000	**Currency**	North Korean won
Capital	P'yŏngyang	**Organizations**	UN
Languages	Korean	**Map page**	65

NORWAY
Kingdom of Norway

Area Sq Km	323 878	**Religions**	Protestant, Roman Catholic
Area Sq Miles	125 050		
Population	5 043 000	**Currency**	Norwegian krone
Capital	Oslo	**Organizations**	OECD, UN
Languages	Norwegian, Sami	**Map page**	92–93

OMAN
Sultanate of Oman

Area Sq Km	309 500	Religions	Ibadhi Muslim, Sunni
Area Sq Miles	119 499		Muslim
Population	3 632 000	Currency	Omani riyal
Capital	Muscat (Masqaṭ)	Organizations	UN
Languages	Arabic, Baluchi,	Map page	79
	Indian languages		

 PAKISTAN
Islamic Republic of Pakistan

Area Sq Km	881 888	Religions	Sunni Muslim, Shi'a
Area Sq Miles	340 497		Muslim, Christian,
Population	182 143 000		Hindu
Capital	Islamabad	Currency	Pakistani rupee
Languages	Urdu, Punjabi,	Organizations	Comm., UN
	Sindhi, Pashto	Map page	74
	(Pashtu), English,		
	Balochi		

 PALAU
Republic of Palau

Area Sq Km	497	Religions	Roman Catholic,
Area Sq Miles	192		Protestant, traditional
Population	21 000		beliefs
Capital	Melekeok	Currency	United States dollar
	(Ngerulmud)	Organizations	UN
Languages	Palauan, English	Map page	59

PANAMA
Republic of Panama

Area Sq Km	77 082	Religions	Roman Catholic,
Area Sq Miles	29 762		Protestant, Sunni
Population	3 864 000		Muslim
Capital	Panama City	Currency	Balboa
Languages	Spanish, English,	Organizations	UN
	Amerindian	Map page	146
	languages		

 PAPUA NEW GUINEA
Independent State of Papua New Guinea

Area Sq Km	462 840	Religions	Protestant, Roman
Area Sq Miles	178 704		Catholic, traditional
Population	7 321 000		beliefs
Capital	Port Moresby	Currency	Kina
Languages	English, Tok Pisin	Organizations	APEC, Comm., UN
	(creole), other local	Map page	59
	languages		

PARAGUAY
Republic of Paraguay

Area Sq Km	406 752	Religions	Roman Catholic,
Area Sq Miles	157 048		Protestant
Population	6 802 000	Currency	Guaraní
Capital	Asunción	Organizations	UN
Languages	Spanish, Guaraní	Map page	152

 PERU
Republic of Peru

Area Sq Km	1 285 216	Religions	Roman Catholic,
Area Sq Miles	496 225		Protestant
Population	30 376 000	Currency	Nuevo sol
Capital	Lima	Organizations	APEC, UN
Languages	Spanish, Quechua,	Map page	150
	Aymara		

 PHILIPPINES
Republic of the Philippines

Area Sq Km	300 000	Religions	Roman Catholic,
Area Sq Miles	115 831		Protestant, Sunni
Population	98 394 000		Muslim, Aglipayan
Capital	Manila	Currency	Philippine peso
Languages	English, Filipino,	Organizations	APEC, ASEAN, UN
	Tagalog, Cebuano,	Map page	64
	other local languages		

 Pitcairn Islands
United Kingdom Overseas Territory

Area Sq Km	45	Religions	Protestant
Area Sq Miles	17	Currency	New Zealand dollar
Population	50	Map page	49
Capital	Adamstown		
Languages	English		

POLAND
Republic of Poland

Area Sq Km	312 683	Religions	Roman Catholic,
Area Sq Miles	120 728		Polish Orthodox
Population	38 217 000	Currency	Złoty
Capital	Warsaw (Warszawa)	Organizations	EU, OECD, UN
Languages	Polish, German	Map page	103

 PORTUGAL
Portuguese Republic

Area Sq Km	88 940	Religions	Roman Catholic,
Area Sq Miles	34 340		Protestant
Population	10 608 000	Currency	Euro
Capital	Lisbon (Lisboa)	Organizations	EU, OECD, UN
Languages	Portuguese	Map page	106

Puerto Rico
United States Commonwealth

Area Sq Km	9 104	Religions	Roman Catholic,
Area Sq Miles	3 515		Protestant
Population	3 688 000	Currency	United States dollar
Capital	San Juan	Map page	147
Languages	Spanish, English		

QATAR
State of Qatar

Area Sq Km	11 437	Religions	Sunni Muslim
Area Sq Miles	4 416	Currency	Qatari riyal
Population	2 169 000	Organizations	OPEC, UN
Capital	Doha (Ad Dawḩah)	Map page	79
Languages	Arabic		

 Réunion
French Overseas Department

Area Sq Km	2 551	Religions	Roman Catholic
Area Sq Miles	985	Currency	Euro
Population	875 000	Map page	113
Capital	St-Denis		
Languages	French, creole		

© Collins Bartholomew Ltd

Rodrigues Island
part of Mauritius

Area Sq Km	104	Religions	Christian
Area Sq Miles	40	Currency	Rupee
Population	38 240	Map page	159
Capital	Port Mathurin		
Languages	English, creole		

ROMANIA

Area Sq Km	237 500	Religions	Romanian Orthodox, Protestant, Roman Catholic
Area Sq Miles	91 699		
Population	21 699 000		
Capital	Bucharest (Bucureşti)	Currency	Romanian leu
Languages	Romanian, Hungarian	Organizations	EU, UN
		Map page	110

RUSSIA

Area Sq Km	17 075 400	Religions	Russian Orthodox, Sunni Muslim, Protestant
Area Sq Miles	6 592 849		
Population	142 834 000		
Capital	Moscow (Moskva)	Currency	Russian rouble
Languages	Russian, Tatar, Ukrainian, other local languages	Organizations	APEC, CIS, UN
		Map page	82–83

RWANDA
Republic of Rwanda

Area Sq Km	26 338	Religions	Roman Catholic, traditional beliefs, Protestant
Area Sq Miles	10 169		
Population	11 777 000		
Capital	Kigali	Currency	Rwandan franc
Languages	Kinyarwanda, French, English	Organizations	Comm., UN
		Map page	119

Saba
Netherlands Special Municipality

Area Sq Km	13	Religions	Roman Catholic, Protestant
Area Sq Miles	5		
Population	1 991	Currency	United States dollar
Capital	Bottom	Map page	147
Languages	Dutch, English		

St-Barthélémy
French Overseas Collectivity

Area Sq Km	21	Religions	Roman Catholic
Area Sq Miles	8	Currency	Euro
Population	9 072	Map page	147
Capital	Gustavia		
Languages	French		

St Helena, Ascension and Tristan da Cunha
United Kingdom Overseas Territory

Area Sq Km	121	Religions	Protestant, Roman Catholic,
Area Sq Miles	47		
Population	5 366	Currency	St Helena pound
Capital	Jamestown	Map page	113
Languages	English		

ST KITTS AND NEVIS
Federation of St Kitts and Nevis

Area Sq Km	261	Religions	Protestant, Roman Catholic
Area Sq Miles	101		
Population	54 000	Currency	East Caribbean dollar
Capital	Basseterre	Organizations	CARICOM, Comm., UN
Languages	English, creole		
		Map page	147

ST LUCIA

Area Sq Km	616	Religions	Roman Catholic, Protestant
Area Sq Miles	238		
Population	182 000	Currency	East Caribbean dollar
Capital	Castries	Organizations	CARICOM, Comm., UN
Languages	English, creole		
		Map page	147

St-Martin
French Overseas Collectivity

Area Sq Km	54	Religions	Roman Catholic
Area Sq Miles	21	Currency	Euro
Population	37 630	Map page	147
Capital	Marigot		
Languages	French		

St Pierre and Miquelon
French Territorial Collectivity

Area Sq Km	242	Religions	Roman Catholic
Area Sq Miles	93	Currency	Euro
Population	6 312	Map page	131
Capital	St-Pierre		
Languages	French		

ST VINCENT AND THE GRENADINES

Area Sq Km	389	Religions	Protestant, Roman Catholic
Area Sq Miles	150		
Population	109 000	Currency	East Caribbean dollar
Capital	Kingstown	Organizations	CARICOM, Comm., UN
Languages	English, creole		
		Map page	147

SAMOA
Independent State of Samoa

Area Sq Km	2 831	Religions	Protestant, Roman Catholic
Area Sq Miles	1 093		
Population	190 000	Currency	Tala
Capital	Apia	Organizations	Comm., UN
Languages	Samoan, English	Map page	49

SAN MARINO
Republic of San Marino

Area Sq Km	61	Religions	Roman Catholic
Area Sq Miles	24	Currency	Euro
Population	31 000	Organizations	UN
Capital	San Marino	Map page	108
Languages	Italian		

SÃO TOMÉ AND PRÍNCIPE
Democratic Republic of São Tomé and Príncipe

Area Sq Km	964	Religions	Roman Catholic,
Area Sq Miles	372		Protestant
Population	193 000	Currency	Dobra
Capital	São Tomé	Organizations	UN
Languages	Portuguese, creole	Map page	113

 ## SAUDI ARABIA
Kingdom of Saudi Arabia

Area Sq Km	2 200 000	Religions	Sunni Muslim, Shi'a
Area Sq Miles	849 425		Muslim
Population	28 829 000	Currency	Saudi Arabian riyal
Capital	Riyadh (Ar Riyāḍ)	Organizations	OPEC, UN
Languages	Arabic	Map page	78–79

 ## SENEGAL
Republic of Senegal

Area Sq Km	196 720	Religions	Sunni Muslim,
Area Sq Miles	75 954		Roman Catholic,
Population	14 133 000		traditional beliefs
Capital	Dakar	Currency	CFA franc
Languages	French, Wolof,	Organizations	UN
	Fulani, other local	Map page	114
	languages		

 ## SERBIA
Republic of Serbia

Area Sq Km	88 361	Religions	Roman Catholic,
Area Sq Miles	34 116		Serbian Orthodox,
Population	7 181 505		Sunni Muslim
Capital	Belgrade (Beograd)	Currency	Serbian dinar
Languages	Serbian, Hungarian	Organizations	UN
		Map page	109

 ## SEYCHELLES
Republic of the Seychelles

Area Sq Km	455	Religions	Roman Catholic,
Area Sq Miles	176		Protestant
Population	93 000	Currency	Seychelles rupee
Capital	Victoria	Organizations	Comm., SADC, UN
Languages	English, French,	Map page	113
	creole		

 ## SIERRA LEONE
Republic of Sierra Leone

Area Sq Km	71 740	Religions	Sunni Muslim,
Area Sq Miles	27 699		traditional beliefs
Population	6 092 000	Currency	Leone
Capital	Freetown	Organizations	Comm., UN
Languages	English, creole,	Map page	114
	Mende, Temne,		
	other local languages		

SINGAPORE
Republic of Singapore

Area Sq Km	639	Religions	Buddhist, Taoist, Sunni
Area Sq Miles	247		Muslim, Christian,
Population	5 412 000		Hindu
Capital	Singapore	Currency	Singapore dollar
Languages	Chinese, English,	Organizations	APEC, ASEAN,
	Malay, Tamil		Comm., UN
		Map page	60

 ## Sint Eustatius
Netherlands Special Municipality

Area Sq Km	21	Religions	Protestant, Roman
Area Sq Miles	8		Catholic
Population	3 897	Currency	United States dollar
Capital	Oranjestad	Map page	147
Languages	Dutch, English,		
	Spanish		

 ## Sint Maarten
Self-governing Netherlands Territory

Area Sq Km	34	Religions	Protestant, Roman
Area Sq Miles	13		Catholic
Population	45 000	Currency	Caribbean guilder
Capital	Philipsburg	Map page	147
Languages	Dutch, English		

 ## SLOVAKIA
Slovak Republic

Area Sq Km	49 035	Religions	Roman Catholic,
Area Sq Miles	18 933		Protestant, Orthodox
Population	5 450 000	Currency	Euro
Capital	Bratislava	Organizations	EU, OECD, UN
Languages	Slovak,	Map page	103
	Hungarian, Czech		

 ## SLOVENIA
Republic of Slovenia

Area Sq Km	20 251	Religions	Roman Catholic,
Area Sq Miles	7 819		Protestant
Population	2 072 000	Currency	Euro
Capital	Ljubljana	Organizations	EU, OECD, UN
Languages	Slovenian, Croatian,	Map page	108–109
	Serbian		

 ## SOLOMON ISLANDS

Area Sq Km	28 370	Religions	Protestant, Roman
Area Sq Miles	10 954		Catholic
Population	561 000	Currency	Solomon Islands dollar
Capital	Honiara	Organizations	Comm., UN
Languages	English, creole,	Map page	48
	other local languages		

SOMALIA
Federal Republic of Somalia

Area Sq Km	637 657	Religions	Sunni Muslim
Area Sq Miles	246 201	Currency	Somali shilling
Population	10 496 000	Organizations	UN
Capital	Mogadishu	Map page	117
	(Muqdisho)		
Languages	Somali, Arabic		

Somaliland
Disputed Territory (Somalia)

Area Sq Km	140 000	Map page	117
Area Sq Miles	54 054		
Population	3 500 000		
Capital	Hargeysa		

© Collins Bartholomew Ltd

SOUTH AFRICA

Area Sq Km	1 219 080	Religions	Protestant, Roman
Area Sq Miles	470 689		Catholic, Sunni
Population	52 776 000		Muslim, Hindu
Capital	Pretoria (Tshwane)/	Currency	Rand
	Cape Town/	Organizations	Comm., SADC, UN
	Bloemfontein	Map page	122–123
Languages	Afrikaans, English,		
	nine official local		
	languages		

 ## SOUTH KOREA
Republic of Korea

Area Sq Km	99 274	Religions	Buddhist, Protestant,
Area Sq Miles	38 330		Roman Catholic
Population	49 263 000	Currency	South Korean won
Capital	Seoul (Sŏul)	Organizations	APEC, OECD, UN
Languages	Korean	Map page	65

South Ossetia
Disputed Territory (Georgia)

Area Sq Km	4 000	Map page	81
Area Sq Miles	1 544		
Population	70 000		
Capital	Tskhinvali		

 ## SOUTH SUDAN
Republic of South Sudan

Area Sq Km	644 329	Religions	Traditional beliefs,
Area Sq Miles	248 775		Christian
Population	11 296 000	Currency	South Sudan pound
Capital	Juba	Organizations	UN
Languages	English, Arabic,	Map page	117
	Dinka, Nuer, other		
	local languages		

 ## SPAIN
Kingdom of Spain

Area Sq Km	504 782	Religions	Roman Catholic
Area Sq Miles	194 897	Currency	Euro
Population	46 927 000	Organizations	EU, OECD, UN
Capital	Madrid	Map page	106–107
Languages	Spanish (Castilian),		
	Catalan, Galician,		
	Basque		

 ## SRI LANKA
Democratic Socialist Republic of Sri Lanka

Area Sq Km	65 610	Religions	Buddhist, Hindu,
Area Sq Miles	25 332		Sunni Muslim,
Population	21 273 000		Roman Catholic
Capital	Sri Jayewardenepura	Currency	Sri Lankan rupee
	Kotte	Organizations	Comm., UN
Languages	Sinhalese, Tamil,	Map page	73
	English		

 ## SUDAN
Republic of the Sudan

Area Sq Km	1 861 484	Religions	Sunni Muslim,
Area Sq Miles	718 725		traditional beliefs,
Population	37 964 000		Christian
Capital	Khartoum	Currency	Sudanese pound
Languages	Arabic, English,		(Sudani)
	Nubian, Beja, Fur,	Organizations	UN
	other local languages	Map page	116–117

 ## SURINAME
Republic of Suriname

Area Sq Km	163 820	Religions	Hindu, Roman
Area Sq Miles	63 251		Catholic, Protestant,
Population	539 000		Sunni Muslim
Capital	Paramaribo	Currency	Suriname guilder
Languages	Dutch,	Organizations	CARICOM, UN
	Surinamese,	Map page	151
	English, Hindi		

 ## Svalbard
part of Norway

Area Sq Km	61 229	Religions	Protestant
Area Sq Miles	23 641	Currency	Norwegian krone
Population	2 637	Map page	82
Capital	Longyearbyen		
Languages	Norwegian		

 ## SWAZILAND
Kingdom of Swaziland

Area Sq Km	17 364	Religions	Christian,
Area Sq Miles	6 704		traditional beliefs
Population	1 250 000	Currency	Emalangeni,
Capital	Mbabane		South African rand
Languages	Swazi, English	Organizations	Comm., SADC, UN
		Map page	123

 ## SWEDEN
Kingdom of Sweden

Area Sq Km	449 964	Religions	Protestant,
Area Sq Miles	173 732		Roman Catholic
Population	9 571 000	Currency	Swedish krona
Capital	Stockholm	Organizations	EU, OECD, UN
Languages	Swedish, Sami	Map page	92–93

 ## SWITZERLAND
Swiss Confederation

Area Sq Km	41 293	Religions	Roman Catholic,
Area Sq Miles	15 943		Protestant,
Population	8 078 000	Currency	Swiss franc
Capital	Bern	Organizations	OECD, UN
Languages	German, French,	Map page	105
	Italian, Romansch		

 ## SYRIA
Syrian Arab Republic

Area Sq Km	184 026	Religions	Sunni Muslim, Shi'a
Area Sq Miles	71 052		Muslim, Christian
Population	21 898 000	Currency	Syrian pound
Capital	Damascus (Dimashq)	Organizations	UN
Languages	Arabic, Kurdish,	Map page	80
	Armenian		

 ## TAIWAN
Republic of China

Area Sq Km	36 179	Religions	Buddhist, Taoist,
Area Sq Miles	13 969		Confucian, Christian
Population	23 344 000	Currency	Taiwan dollar
Capital	Taibei	Organizations	APEC
Languages	Mandarin	Map page	71
	(Putonghua), Min,		
	Hakka, other local		
	languages		

The People's Republic of China claims Taiwan as its 23rd province

TAJIKISTAN
Republic of Tajikistan

Area Sq Km	143 100	Religions	Sunni Muslim
Area Sq Miles	55 251	Currency	Somoni
Population	8 208 000	Organizations	CIS, UN
Capital	Dushanbe	Map page	77
Languages	Tajik, Uzbek, Russian		

TANZANIA
United Republic of Tanzania

Area Sq Km	945 087	Religions	Shi'a Muslim, Sunni
Area Sq Miles	364 900		Muslim, traditional
Population	49 253 000		beliefs, Christian
Capital	Dodoma	Currency	Tanzanian shilling
Languages	Swahili, English,	Organizations	Comm., SADC, UN
	Nyamwezi, other	Map page	119
	local languages		

THAILAND
Kingdom of Thailand

Area Sq Km	513 115	Religions	Buddhist, Sunni
Area Sq Miles	198 115		Muslim
Population	67 011 000	Currency	Baht
Capital	Bangkok	Organizations	APEC, ASEAN, UN
	(Krung Thep)	Map page	62–63
Languages	Thai, Lao, Chinese,		
	Malay, Mon-Khmer		
	languages		

TOGO
Togolese Republic

Area Sq Km	56 785	Religions	Traditional beliefs,
Area Sq Miles	21 925		Christian, Sunni
Population	6 817 000		Muslim
Capital	Lomé	Currency	CFA franc
Languages	French, Ewe, Kabre,	Organizations	UN
	other local languages	Map page	114

Tokelau
New Zealand Overseas Territory

Area Sq Km	10	Religions	Christian
Area Sq Miles	4	Currency	New Zealand dollar
Population	1 411	Map page	49
Capital	none		
Languages	English, Tokelauan		

TONGA
Kingdom of Tonga

Area Sq Km	748	Religions	Protestant, Roman
Area Sq Miles	289		Catholic
Population	105 000	Currency	Pa'anga
Capital	Nuku'alofa	Organizations	Comm., UN
Languages	Tongan, English	Map page	49

Transnistria
Disputed Territory (Moldova)

Area Sq Km	4 200	Map page	90
Area Sq Miles	1 546		
Population	520 000		
Capital	Tiraspol		

TRINIDAD AND TOBAGO
Republic of Trinidad and Tobago

Area Sq Km	5 130	Religions	Roman Catholic,
Area Sq Miles	1 981		Hindu, Protestant,
Population	1 341 000		Sunni Muslim
Capital	Port of Spain	Currency	Trinidad and Tobago
Languages	English, creole,		dollar
	Hindi	Organizations	CARICOM, Comm.,
			UN
		Map page	147

Tristan da Cunha
Part of St Helena, Ascension and Tristan da Cunha

Area Sq Km	98	Religions	Protestant, Roman
Area Sq Miles	38		Catholic
Population	259	Currency	Pound sterling
Capital	Settlement of	Map page	113
	Edinburgh		
Languages	English		

TUNISIA
Tunisian Republic

Area Sq Km	164 150	Religions	Sunni Muslim
Area Sq Miles	63 379	Currency	Tunisian dinar
Population	10 997 000	Organizations	UN
Capital	Tunis	Map page	115
Languages	Arabic, French		

TURKEY
Republic of Turkey

Area Sq Km	779 452	Religions	Sunni Muslim, Shi'a
Area Sq Miles	300 948		Muslim
Population	74 933 000	Currency	Lira
Capital	Ankara	Organizations	OECD, UN
Languages	Turkish, Kurdish	Map page	80

TURKMENISTAN

Area Sq Km	488 100	Religions	Sunni Muslim,
Area Sq Miles	188 456		Russian Orthodox
Population	5 240 000	Currency	Turkmen manat
Capital	Aşgabat (Ashkhabad)	Organizations	UN
Languages	Turkmen, Uzbek,	Map page	76
	Russian		

Turks and Caicos Islands
United Kingdom Overseas Territory

Area Sq Km	430	Religions	Protestant
Area Sq Miles	166	Currency	United States dollar
Population	33 000	Map page	147
Capital	Grand Turk		
	(Cockburn Town)		
Languages	English		

TUVALU

Area Sq Km	25	Religions	Protestant
Area Sq Miles	10	Currency	Australian dollar
Population	10 000	Organizations	Comm., UN
Capital	Vaiaku	Map page	49
Languages	Tuvaluan, English		

© Collins Bartholomew Ltd

UGANDA
Republic of Uganda

Area Sq Km	241 038	**Religions**	Roman Catholic, Protestant, Sunni Muslim, traditional beliefs
Area Sq Miles	93 065		
Population	37 579 000		
Capital	Kampala		
Languages	English, Swahili, Luganda, other local languages	**Currency**	Ugandan shilling
		Organizations	Comm., UN
		Map page	119

UKRAINE
Republic of Ukraine

Area Sq Km	603 700	**Religions**	Ukrainian Orthodox, Ukrainian Catholic, Roman Catholic
Area Sq Miles	233 090		
Population	45 239 000		
Capital	Kiev (Kyiv)	**Currency**	Hryvnia
Languages	Ukrainian, Russian	**Organizations**	UN
		Map page	90–91

UNITED ARAB EMIRATES
Federation of Emirates

Area Sq Km	77 700	**Religions**	Sunni Muslim, Shi'a Muslim
Area Sq Miles	30 000		
Population	9 346 000	**Currency**	United Arab Emirates dirham
Capital	Abu Dhabi (Abū Ẓaby)		
		Organizations	OPEC, UN
Languages	Arabic, English	**Map page**	79

Abu Dhabi (Abū Ẓabī) (Emirate)

Area Sq Km	67 340	**Population**	1 628 000
Area Sq Miles	26 000	**Capital**	Abu Dhabi (Abū Ẓabī)

Ajman (Emirate)

Area Sq Km	259	**Population**	250 000
Area Sq Miles	100	**Capital**	Ajman

Dubai (Emirate)

Area Sq Km	3 885	**Population**	1 722 000
Area Sq Miles	1 500	**Capital**	Dubai

Fujairah (Emirate)

Area Sq Km	1 165	**Population**	152 000
Area Sq Miles	450	**Capital**	Fujairah

Ra's al Khaymah (Emirate)

Area Sq Km	1 684	**Population**	241 000
Area Sq Miles	650	**Capital**	Ra's al Khaymah

Sharjah (Emirate)

Area Sq Km	2 590	**Population**	1 017 000
Area Sq Miles	1 000	**Capital**	Sharjah

Umm al Qaywayn (Emirate)

Area Sq Km	777	**Population**	56 000
Area Sq Miles	300	**Capital**	Umm al Qaywayn

UNITED KINGDOM
of Great Britain and Northern Ireland

Area Sq Km	243 609	**Religions**	Protestant, Roman Catholic, Muslim
Area Sq Miles	94 058		
Population	63 136 000	**Currency**	Pound sterling
Capital	London	**Organizations**	Comm., EU, OECD, UN
Languages	English, Welsh, Gaelic		
		Map page	94–95

England (Constituent country)

Area Sq Km	130 433	**Population**	53 493 700
Area Sq Miles	50 360	**Capital**	London

Northern Ireland (Province)

Area Sq Km	13 576	**Population**	1 823 600
Area Sq Miles	5 242	**Capital**	Belfast

Scotland (Constituent country)

Area Sq Km	78 822	**Population**	5 313 600
Area Sq Miles	30 433	**Capital**	Edinburgh

Wales (Principality)

Area Sq Km	20 778	**Population**	3 074 100
Area Sq Miles	8 022	**Capital**	Cardiff

UNITED STATES OF AMERICA
Federal Republic

Area Sq Km	9 826 635	**Religions**	Protestant, Roman Catholic, Sunni Muslim, Jewish
Area Sq Miles	3 794 085		
Population	320 051 000		
Capital	Washington D.C.	**Currency**	United States dollar
Languages	English, Spanish	**Organizations**	APEC, OECD, UN
		Map page	132–133

Alabama (State)

Area Sq Km	135 765	**Population**	4 822 023
Area Sq Miles	52 419	**Capital**	Montgomery

Alaska (State)

Area Sq Km	1 717 854	**Population**	731 449
Area Sq Miles	663 267	**Capital**	Juneau

Arizona (State)

Area Sq Km	295 253	**Population**	6 553 255
Area Sq Miles	113 998	**Capital**	Phoenix

Arkansas (State)

Area Sq Km	137 733	**Population**	2 949 131
Area Sq Miles	53 179	**Capital**	Little Rock

California (State)

Area Sq Km	423 971	**Population**	38 041 430
Area Sq Miles	163 696	**Capital**	Sacramento

Colorado (State)
Area Sq Km	269 602	Population	5 187 582
Area Sq Miles	104 094	Capital	Denver

Connecticut (State)
Area Sq Km	14 356	Population	3 590 347
Area Sq Miles	5 543	Capital	Hartford

Delaware (State)
Area Sq Km	6 446	Population	917 092
Area Sq Miles	2 489	Capital	Dover

District of Columbia (District)
Area Sq Km	176	Population	632 323
Area Sq Miles	68	Capital	Washington

Florida (State)
Area Sq Km	170 305	Population	19 317 568
Area Sq Miles	65 755	Capital	Tallahassee

Georgia (State)
Area Sq Km	153 910	Population	9 919 945
Area Sq Miles	59 425	Capital	Atlanta

Hawaii (State)
Area Sq Km	28 311	Population	1 392 313
Area Sq Miles	10 931	Capital	Honolulu

Idaho (State)
Area Sq Km	216 445	Population	1 595 728
Area Sq Miles	83 570	Capital	Boise

Illinois (State)
Area Sq Km	149 997	Population	12 875 255
Area Sq Miles	57 914	Capital	Springfield

Indiana (State)
Area Sq Km	94 322	Population	6 537 334
Area Sq Miles	36 418	Capital	Indianapolis

Iowa (State)
Area Sq Km	145 744	Population	3 074 186
Area Sq Miles	56 272	Capital	Des Moines

Kansas (State)
Area Sq Km	213 096	Population	2 885 905
Area Sq Miles	82 277	Capital	Topeka

Kentucky (State)
Area Sq Km	104 659	Population	4 380 415
Area Sq Miles	40 409	Capital	Frankfort

Louisiana (State)
Area Sq Km	134 265	Population	4 601 893
Area Sq Miles	51 840	Capital	Baton Rouge

Maine (State)
Area Sq Km	91 647	Population	1 329 192
Area Sq Miles	35 385	Capital	Augusta

Maryland (State)
Area Sq Km	32 134	Population	5 884 563
Area Sq Miles	12 407	Capital	Annapolis

Massachusetts (State)
Area Sq Km	27 337	Population	6 646 144
Area Sq Miles	10 555	Capital	Boston

Michigan (State)
Area Sq Km	250 493	Population	9 883 360
Area Sq Miles	96 716	Capital	Lansing

Minnesota (State)
Area Sq Km	225 171	Population	5 379 139
Area Sq Miles	86 939	Capital	St Paul

Mississippi (State)
Area Sq Km	125 433	Population	2 984 926
Area Sq Miles	48 430	Capital	Jackson

Missouri (State)
Area Sq Km	180 533	Population	6 021 988
Area Sq Miles	69 704	Capital	Jefferson City

Montana (State)
Area Sq Km	380 837	Population	1 005 141
Area Sq Miles	147 042	Capital	Helena

Nebraska (State)
Area Sq Km	200 346	Population	1 855 525
Area Sq Miles	77 354	Capital	Lincoln

Nevada (State)
Area Sq Km	286 352	Population	2 758 931
Area Sq Miles	110 561	Capital	Carson City

New Hampshire (State)
Area Sq Km	24 216	Population	1 320 718
Area Sq Miles	9 350	Capital	Concord

New Jersey (State)
Area Sq Km	22 587	Population	8 864 590
Area Sq Miles	8 721	Capital	Trenton

© Collins Bartholomew Ltd

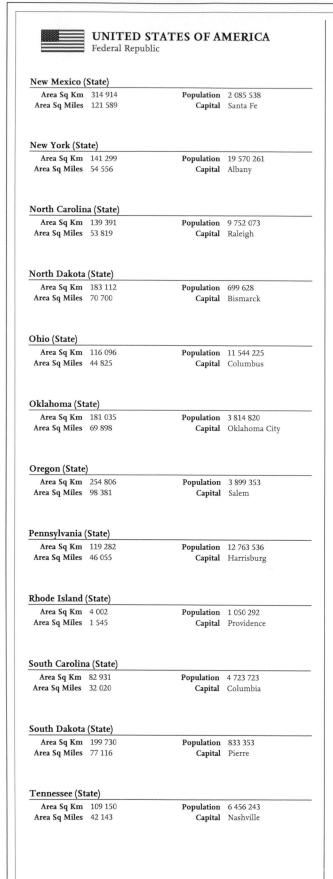

UNITED STATES OF AMERICA
Federal Republic

New Mexico (State)

Area Sq Km	314 914	**Population**	2 085 538
Area Sq Miles	121 589	**Capital**	Santa Fe

New York (State)

Area Sq Km	141 299	**Population**	19 570 261
Area Sq Miles	54 556	**Capital**	Albany

North Carolina (State)

Area Sq Km	139 391	**Population**	9 752 073
Area Sq Miles	53 819	**Capital**	Raleigh

North Dakota (State)

Area Sq Km	183 112	**Population**	699 628
Area Sq Miles	70 700	**Capital**	Bismarck

Ohio (State)

Area Sq Km	116 096	**Population**	11 544 225
Area Sq Miles	44 825	**Capital**	Columbus

Oklahoma (State)

Area Sq Km	181 035	**Population**	3 814 820
Area Sq Miles	69 898	**Capital**	Oklahoma City

Oregon (State)

Area Sq Km	254 806	**Population**	3 899 353
Area Sq Miles	98 381	**Capital**	Salem

Pennsylvania (State)

Area Sq Km	119 282	**Population**	12 763 536
Area Sq Miles	46 055	**Capital**	Harrisburg

Rhode Island (State)

Area Sq Km	4 002	**Population**	1 050 292
Area Sq Miles	1 545	**Capital**	Providence

South Carolina (State)

Area Sq Km	82 931	**Population**	4 723 723
Area Sq Miles	32 020	**Capital**	Columbia

South Dakota (State)

Area Sq Km	199 730	**Population**	833 353
Area Sq Miles	77 116	**Capital**	Pierre

Tennessee (State)

Area Sq Km	109 150	**Population**	6 456 243
Area Sq Miles	42 143	**Capital**	Nashville

Texas (State)

Area Sq Km	695 622	**Population**	26 059 203
Area Sq Miles	268 581	**Capital**	Austin

Utah (State)

Area Sq Km	219 887	**Population**	2 855 287
Area Sq Miles	84 899	**Capital**	Salt Lake City

Vermont (State)

Area Sq Km	24 900	**Population**	626 011
Area Sq Miles	9 614	**Capital**	Montpelier

Virginia (State)

Area Sq Km	110 784	**Population**	8 185 867
Area Sq Miles	42 774	**Capital**	Richmond

Washington (State)

Area Sq Km	184 666	**Population**	6 897 012
Area Sq Miles	71 300	**Capital**	Olympia

West Virginia (State)

Area Sq Km	62 755	**Population**	1 855 413
Area Sq Miles	24 230	**Capital**	Charleston

Wisconsin (State)

Area Sq Km	169 639	**Population**	5 726 398
Area Sq Miles	65 498	**Capital**	Madison

Wyoming (State)

Area Sq Km	253 337	**Population**	576 412
Area Sq Miles	97 814	**Capital**	Cheyenne

URUGUAY
Oriental Republic of Uruguay

Area Sq Km	176 215	**Religions**	Roman Catholic,
Area Sq Miles	68 037		Protestant, Jewish
Population	3 407 000	**Currency**	Uruguayan peso
Capital	Montevideo	**Organizations**	UN
Languages	Spanish	**Map page**	153

UZBEKISTAN
Republic of Uzbekistan

Area Sq Km	447 400	**Religions**	Sunni Muslim, Russian
Area Sq Miles	172 742		Orthodox
Population	28 934 000	**Currency**	Uzbek som
Capital	Tashkent	**Organizations**	CIS, UN
Languages	Uzbek, Russian,	**Map page**	76–77
	Tajik, Kazakh		

VANUATU
Republic of Vanuatu

Area Sq Km	12 190	Religions	Protestant, Roman
Area Sq Miles	4 707		Catholic, traditional
Population	253 000		beliefs
Capital	Port Vila	Currency	Vatu
Languages	English, Bislama	Organizations	Comm., UN
	(creole), French	Map page	48

 ## VATICAN CITY
Vatican City State or Holy See

Area Sq Km	0.5	Religions	Roman Catholic
Area Sq Miles	0.2	Currency	Euro
Population	800	Map page	108
Capital	Vatican City		
Languages	Italian		

 ## VENEZUELA
Bolivarian Republic of Venezuela

Area Sq Km	912 050	Religions	Roman Catholic,
Area Sq Miles	352 144		Protestant
Population	30 405 000	Currency	Bolívar fuerte
Capital	Caracas	Organizations	OPEC, UN
Languages	Spanish, Amerindian	Map page	150
	languages		

 ## VIETNAM
Socialist Republic of Vietnam

Area Sq Km	329 565	Religions	Buddhist, Taoist,
Area Sq Miles	127 246		Roman Catholic,
Population	91 680 000		Cao Dai, Hoa Hoa
Capital	Ha Nôi (Hanoi)	Currency	Dong
Languages	Vietnamese, Thai,	Organizations	APEC, ASEAN, UN
	Khmer, Chinese,	Map page	62–63
	other local languages		

 ## Virgin Islands (U.K.)
United Kingdom Overseas Territory

Area Sq Km	153	Religions	Protestant, Roman
Area Sq Miles	59		Catholic
Population	28 000	Currency	United States dollar
Capital	Road Town	Map page	147
Languages	English		

 ## Virgin Islands (U.S.)
United States Unincorporated Territory

Area Sq Km	352	Religions	Protestant,
Area Sq Miles	136		Roman Catholic
Population	107 000	Currency	United States dollar
Capital	Charlotte Amalie	Map page	147
Languages	English, Spanish		

Wallis and Futuna Islands
French Overseas Territory

Area Sq Km	274	Religions	Roman Catholic
Area Sq Miles	106	Currency	CFP franc
Population	13 000	Map page	49
Capital	Matā'utu		
Languages	French, Wallisian,		
	Futunian		

West Bank
Disputed Territory

Area Sq Km	5 860	Religions	Sunni Muslim, Jewish,
Area Sq Miles	2 263		Shi'a Muslim, Christian
Population	2 719 112	Currency	Jordanian dinar,
Capital	none		Israeli shekel
Languages	Arabic, Hebrew	Map page	80

 ## Western Sahara
Disputed Territory (Morocco)

Area Sq Km	266 000	Religions	Sunni Muslim
Area Sq Miles	102 703	Currency	Moroccan dirham
Population	567 000	Map page	114
Capital	Laâyoune		
Languages	Arabic		

 ## YEMEN
Republic of Yemen

Area Sq Km	527 968	Religions	Sunni Muslim, Shi'a
Area Sq Miles	203 850		Muslim
Population	24 407 000	Currency	Yemeni riyal
Capital	Şan'ā'	Organizations	UN
Languages	Arabic	Map page	78–79

 ## ZAMBIA
Republic of Zambia

Area Sq Km	752 614	Religions	Christian, traditional
Area Sq Miles	290 586		beliefs
Population	14 539 000	Currency	Zambian kwacha
Capital	Lusaka	Organizations	Comm., SADC, UN
Languages	English, Bemba,	Map page	120–121
	Nyanja, Tonga,		
	other local languages		

 ## ZIMBABWE
Republic of Zimbabwe

Area Sq Km	390 759	Religions	Christian, traditional
Area Sq Miles	150 873		beliefs
Population	14 150 000	Currency	US dollar and other
Capital	Harare		currencies
Languages	English, Shona,	Organizations	SADC, UN
	Ndebele	Map page	121

© Collins Bartholomew Ltd

ANTARCTICA
Total Land Area
12 093 000 sq km
4 669 107 sq miles
(excluding ice shelves)

OCEANIA
Total land area
8 844 516 sq km
3 414 868 sq miles
(includes New Guinea and
Pacific Island nations)

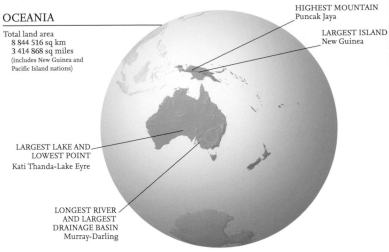

HIGHEST MOUNTAIN
Puncak Jaya

LARGEST ISLAND
New Guinea

LARGEST LAKE AND
LOWEST POINT
Kati Thanda-Lake Eyre

LONGEST RIVER
AND LARGEST
DRAINAGE BASIN
Murray-Darling

HIGHEST MOUNTAIN
Mount Vinson
4 897 m /16 066 ft

HIGHEST MOUNTAINS	metres	feet
Mount Vinson	4 897	16 066
Mt Tyree	4 852	15 918
Mt Kirkpatrick	4 528	14 855
Mt Markham	4 351	14 275
Mt Sidley	4 285	14 058
Mt Minto	4 165	13 665

HIGHEST MOUNTAINS	metres	feet	LARGEST ISLANDS	sq km	sq miles	LARGEST LAKES	sq km	sq miles	LONGEST RIVERS	km	miles
Puncak Jaya	4 884	16 023	New Guinea	808 510	312 166	Kati Thanda-Lake Eyre	0–8 900	0–3 436	Murray-Darling	3 672	2 282
Puncak Trikora	4 730	15 518	South Island (Te Waipounamu)	151 215	58 384	Lake Torrens	0–5 780	0–2 232	Darling	2 844	1 767
Puncak Mandala	4 700	15 420	North Island (Te Ika-a-Māui)	115 777	44 701				Murray	2 375	1 476
Puncak Yamin	4 595	15 075							Murrumbidgee	1 485	923
Mt Wilhelm	4 509	14 793	Tasmania	67 800	26 178				Lachlan	1 339	832
Mt Kubor	4 359	14 301							Cooper Creek	1 113	692

ASIA
Total Land Area
45 036 492 sq km
17 388 590 sq miles

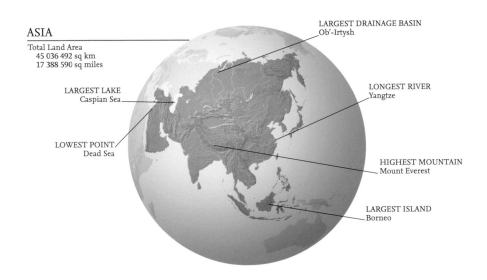

LARGEST DRAINAGE BASIN
Ob'-Irtysh

LONGEST RIVER
Yangtze

LARGEST LAKE
Caspian Sea

LOWEST POINT
Dead Sea

HIGHEST MOUNTAIN
Mount Everest

LARGEST ISLAND
Borneo

HIGHEST MOUNTAINS	metres	feet	LARGEST ISLANDS	sq km	sq miles	LARGEST LAKES	sq km	sq miles	LONGEST RIVERS	km	miles
Mt Everest	8 848	29 028	Borneo	745 561	287 861	Caspian Sea	371 000	143 243	Yangtze	6 380	3 965
K2	8 611	28 251	Sumatra	473 606	182 859	Lake Baikal	30 500	11 776	Ob'-Irtysh	5 568	3 460
Kangchenjunga	8 586	28 169	Honshū	227 414	87 805	Lake Balkhash	17 400	6 718	Yenisey-Angara-Selenga	5 550	3 449
Lhotse	8 516	27 939	Celebes	189 216	73 056	Aral Sea	17 158	6 625	Yellow	5 464	3 395
Makalu	8 463	27 765	Java	132 188	51 038	Ysyk-Köl	6 200	2 394	Irtysh	4 440	2 759
Cho Oyu	8 201	26 906	Luzon	104 690	40 421						

EUROPE

Total Land Area
9 908 599 sq km
3 825 710 sq miles

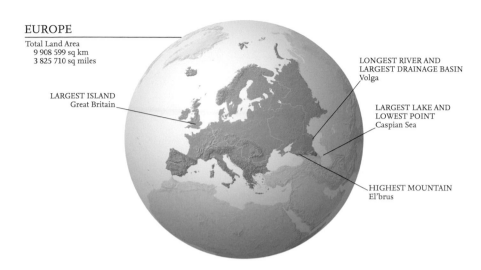

LARGEST ISLAND
Great Britain

LONGEST RIVER AND
LARGEST DRAINAGE BASIN
Volga

LARGEST LAKE AND
LOWEST POINT
Caspian Sea

HIGHEST MOUNTAIN
El'brus

HIGHEST MOUNTAINS	metres	feet	LARGEST ISLANDS	sq km	sq miles	LARGEST LAKES	sq km	sq miles	LONGEST RIVERS	km	miles
El'brus	5 642	18 510	Great Britain	218 476	84 354	Caspian Sea	371 000	143 243	Volga	3 688	2 292
Gora Dykh-Tau	5 204	17 073	Iceland	102 820	39 699	Lake Ladoga	18 390	7 100	Danube	2 850	1 771
Shkhara	5 201	17 063	Ireland	83 045	32 064	Lake Onega	9 600	3 707	Dnieper	2 285	1 420
Kazbek	5 047	16 558	Ostrov Severnyy (part of Novaya Zemlya)	47 079	18 177	Vänern	5 585	2 156	Kama	2 028	1 260
Mont Blanc	4 810	15 781				Rybinskoye Vodokhranilishche	5 180	2 000	Don	1 931	1 200
Dufourspitze	4 634	15 203	Spitsbergen	37 814	14 600				Pechora	1 802	1 120

AFRICA

Total Land Area
30 343 578 sq km
11 715 655 sq miles

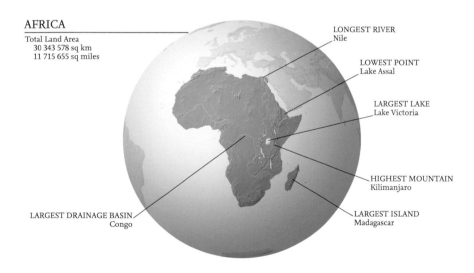

LONGEST RIVER
Nile

LOWEST POINT
Lake Assal

LARGEST LAKE
Lake Victoria

HIGHEST MOUNTAIN
Kilimanjaro

LARGEST ISLAND
Madagascar

LARGEST DRAINAGE BASIN
Congo

HIGHEST MOUNTAINS	metres	feet	LARGEST ISLANDS	sq km	sq miles	LARGEST LAKES	sq km	sq miles	LONGEST RIVERS	km	miles
Kilimanjaro	5 892	19 330	Madagascar	587 040	226 656	Lake Victoria	68 870	26 591	Nile	6 695	4 160
Mt Kenya	5 199	17 057				Lake Tanganyika	32 600	12 587	Congo	4 667	2 900
Margherita Peak	5 110	16 765				Lake Nyasa	29 500	11 390	Niger	4 184	2 600
Meru	4 565	14 977				Lake Volta	8 482	3 275	Zambezi	2 736	1 700
Ras Dejen	4 533	14 872				Lake Turkana	6 500	2 510	Wabē Shebelē Wenz	2 490	1 547
Mt Karisimbi	4 510	14 796				Lake Albert	5 600	2 162	Ubangi	2 250	1 398

© Collins Bartholomew Ltd

NORTH AMERICA

Total Land Area
24 680 331 sq km
9 529 076 sq miles
(includes Hawaiian Islands)

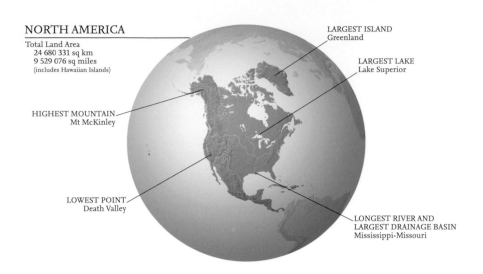

HIGHEST MOUNTAIN
Mt McKinley

LOWEST POINT
Death Valley

LARGEST ISLAND
Greenland

LARGEST LAKE
Lake Superior

LONGEST RIVER AND
LARGEST DRAINAGE BASIN
Mississippi-Missouri

HIGHEST MOUNTAINS	metres	feet	LARGEST ISLANDS	sq km	sq miles	LARGEST LAKES	sq km	sq miles	LONGEST RIVERS	km	miles
Mt McKinley	6 194	20 321	Greenland	2 175 600	839 999	Lake Superior	82 100	31 699	Mississippi-Missouri	5 969	3 709
Mt Logan	5 959	19 550	Baffin Island	507 451	195 927	Lake Huron	59 600	23 012	Mackenzie-Peace-Finlay	4 241	2 635
Pico de Orizaba	5 610	18 405	Victoria Island	217 291	83 896	Lake Michigan	57 800	22 317			
Mt St Elias	5 489	18 008	Ellesmere Island	196 236	75 767	Great Bear Lake	31 328	12 096	Missouri	4 086	2 539
Volcán Popocatépetl	5 452	17 887	Cuba	110 860	42 803	Great Slave Lake	28 568	11 030	Mississippi	3 765	2 340
			Newfoundland	108 860	42 031	Lake Erie	25 700	9 923	Yukon	3 185	1 979

SOUTH AMERICA

Total Land Area
17 815 420 sq km
6 878 534 sq miles

LONGEST RIVER AND
LARGEST DRAINAGE BASIN
Amazon

LARGEST LAKE
Lake Titicaca

HIGHEST MOUNTAIN
Cerro Aconcagua

LARGEST ISLAND
Isla Grande de Tierra del Fuego

LOWEST POINT
Laguna del Carbón

HIGHEST MOUNTAINS	metres	feet	LARGEST ISLANDS	sq km	sq miles	LARGEST LAKES	sq km	sq miles	LONGEST RIVERS	km	miles
Cerro Aconcagua	6 959	22 831	Isla Grande de Tierra del Fuego	47 000	18 147	Lake Titicaca	8 340	3 220	Amazon	6 516	4 049
Nevado Ojos del Salado	6 908	22 664	Isla de Chiloé	8 394	3 241				Río de la Plata-Paraná	4 500	2 796
Cerro Bonete	6 872	22 546	East Falkland	6 760	2 610				Purus	3 218	2 000
Cerro Pissis	6 858	22 500	West Falkland	5 413	2 090				Madeira	3 200	1 988
Cerro Tupungato	6 800	22 309							São Francisco	2 900	1 802

ATLANTIC OCEAN
Total Area
86 557 000 sq km
33 420 000 sq miles

Arctic Ocean
Hudson Bay
Baltic Sea
North Sea
Black Sea
Gulf of Mexico
Mediterranean Sea
Caribbean Sea
Deepest Point
Milwaukee Deep

ATLANTIC OCEAN	Area		Deepest Point	
	square km	square miles	metres	feet
Extent	86 557 000	33 420 000	8 605	28 231
Arctic Ocean	9 485 000	3 662 000	5 450	17 880
Caribbean Sea	2 512 000	970 000	7 680	25 197
Mediterranean Sea	2 510 000	969 000	5 121	16 801
Gulf of Mexico	1 544 000	596 000	3 504	11 496
Hudson Bay	1 233 000	476 000	259	850
North Sea	575 000	222 000	661	2 169
Black Sea	508 000	196 000	2 245	7 365
Baltic Sea	382 000	148 000	460	1 509

PACIFIC OCEAN
Total Area
166 241 000 sq km
64 186 000 sq miles

Bering Sea
Sea of Okhotsk
Sea of Japan (East Sea)
East China Sea and Yellow Sea
South China Sea
Deepest Point
Challenger Deep

PACIFIC OCEAN	Area		Deepest Point	
	square km	square miles	metres	feet
Extent	166 241 000	64 186 000	10 920	35 826
South China Sea	2 590 000	1 000 000	5 514	18 090
Bering Sea	2 261 000	873 000	4 150	13 615
Sea of Okhotsk	1 392 000	538 000	3 363	11 033
East China Sea and Yellow Sea	1 202 000	464 000	2 717	8 914
Sea of Japan (East Sea)	1 013 000	391 000	3 743	12 280

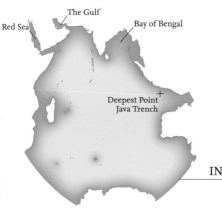

The Gulf
Red Sea
Bay of Bengal
Deepest Point
Java Trench

INDIAN OCEAN	Area		Deepest Point	
	square km	square miles	metres	feet
Extent	73 427 000	28 350 000	7 125	23 376
Bay of Bengal	2 172 000	839 000	4 500	14 764
Red Sea	453 000	175 000	3 040	9 974
The Gulf	238 000	92 000	73	239

INDIAN OCEAN
Total Area
73 427 000 sq km
28 350 000 sq miles

© Collins Bartholomew Ltd

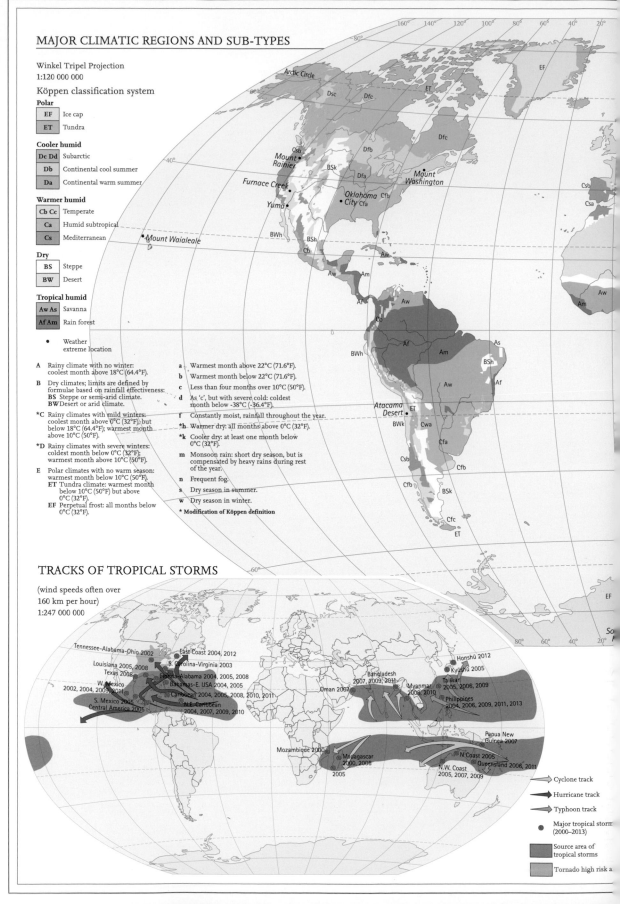

MAJOR CLIMATIC REGIONS AND SUB-TYPES

Winkel Tripel Projection
1:120 000 000

Köppen classification system

Polar

| EF | Ice cap |
| ET | Tundra |

Cooler humid

Dc Dd	Subarctic
Db	Continental cool summer
Da	Continental warm summer

Warmer humid

Cb Cc	Temperate
Ca	Humid subtropical
Cs	Mediterranean

Dry

| BS | Steppe |
| BW | Desert |

Tropical humid

| Aw As | Savanna |
| Af Am | Rain forest |

• Weather extreme location

A Rainy climate with no winter: coolest month above 18°C (64.4°F).

B Dry climates; limits are defined by formulae based on rainfall effectiveness:
BS Steppe or semi-arid climate.
BW Desert or arid climate.

*C Rainy climates with mild winters: coolest month above 0°C (32°F), but below 18°C (64.4°F); warmest month above 10°C (50°F).

*D Rainy climates with severe winters: coldest month below 0°C (32°F); warmest month above 10°C (50°F).

E Polar climates with no warm season: warmest month below 10°C (50°F).
ET Tundra climate: warmest month below 10°C (50°F) but above 0°C (32°F).
EF Perpetual frost: all months below 0°C (32°F).

a Warmest month above 22°C (71.6°F).

b Warmest month below 22°C (71.6°F).

c Less than four months over 10°C (50°F).

d As 'c', but with severe cold: coldest month below -38°C (-36.4°F).

f Constantly moist, rainfall throughout the year.

*h Warmer dry: all months above 0°C (32°F).

*k Cooler dry: at least one month below 0°C (32°F).

m Monsoon rain: short dry season, but is compensated by heavy rains during rest of the year.

n Frequent fog.

s Dry season in summer.

w Dry season in winter.

* Modification of Köppen definition

TRACKS OF TROPICAL STORMS

(wind speeds often over 160 km per hour)
1:247 000 000

Tennessee-Alabama-Ohio 2002
Louisiana 2005, 2008
Texas 2008
W. Mexico 2002, 2004, 2009, 2011
S. Mexico 2005
Central America 2005
East Coast 2004, 2012
S. Carolina-Virginia 2003
Florida-Alabama 2004, 2005, 2008
Bahamas-E. USA 2004, 2005
Caribbean 2004, 2005, 2008, 2010, 2011
N.E. Caribbean 2004, 2007, 2009, 2010
Oman 2007
Bangladesh 2007, 2009, 2011
Myanmar 2008, 2010
Honshū 2012
Kyūshū 2005
Taiwan 2005, 2006, 2009
Philippines 2004, 2006, 2009, 2011, 2013
Papua New Guinea 2007
Mozambique 2000
Madagascar 2000, 2008
2005
N Coast 2005
N.W. Coast 2005, 2007, 2009
Queensland 2006, 2011

→ Cyclone track
→ Hurricane track
→ Typhoon track
• Major tropical storm (2000–2013)
Source area of tropical storms
Tornado high risk a...

Arctic Circle

Dsc · Dfc · ET · EF · Dfb · Dfc · Csb · Mount Rainier · BSk · Dfa · Mount Washington · Furnace Creek · Oklahoma City · Cfb · Cfa · Yuma · BWh · BSh · Mount Waialeale · Cb · Aw · Am · Aw · Af · Aw · Cb · Af · BWh · Am · As · BSh · Af · Aw · Atacama Desert · ET · BWk · Cwa · Cfa · Csb · Cfb · BSk · Cfc · ET · Csb · Csa · Aw · Am

WORLD WEATHER EXTREMES

	Location		Location
Highest shade temperature	56.7°C/134°F Furnace Creek, Death Valley, California, USA (10 July 1913)	Highest surface wind speed	
Hottest place — Annual mean	34.4°C/93.9°F Dalol, Ethiopia	High altitude	372 km per hour/231 miles per hour Mount Washington, New Hampshire, USA (12 April 1934)
Driest place — Annual mean	0.1 mm/0.004 inches Atacama Desert, Chile	Low altitude	408 km per hour/254 miles per hour Barrow Island, Australia (10 April 1996)
Most sunshine — Annual mean	90% Yuma, Arizona, USA (over 4 000 hours)	Tornado	512 km per hour/318 miles per hour Oklahoma City, Oklahoma, USA (3 May 1999)
Least sunshine	Nil for 182 days each year, South Pole	Greatest snowfall	31 102 mm/1 224.5 inches Mount Rainier, Washington, USA (19 February 1971 — 18 February 1972)
Lowest screen temperature	-89.2°C/-128.6°F Vostok Station, Antarctica (21 July 1983)	Heaviest hailstones	1 kg/2.21 lb Gopalganj, Bangladesh (14 April 1986)
Coldest place — Annual mean	-56.6°C/-69.9°F Plateau Station, Antarctica	Thunder-days average	251 days per year Tororo, Uganda
Wettest place — Annual mean	11 873 mm/467.4 inches Meghalaya, India	Highest barometric pressure	1 083.8 mb Agata, Siberia, Russia (31 December 1968)
Most rainy days	Up to 350 per year Mount Waialeale, Hawaii, USA	Lowest barometric pressure	870 mb 483 km/300 miles west of Guam, Pacific Ocean (12 October 1979)
Windiest place	322 km per hour/200 miles per hour in gales, Commonwealth Bay, Antarctica		

© Collins Bartholomew Ltd

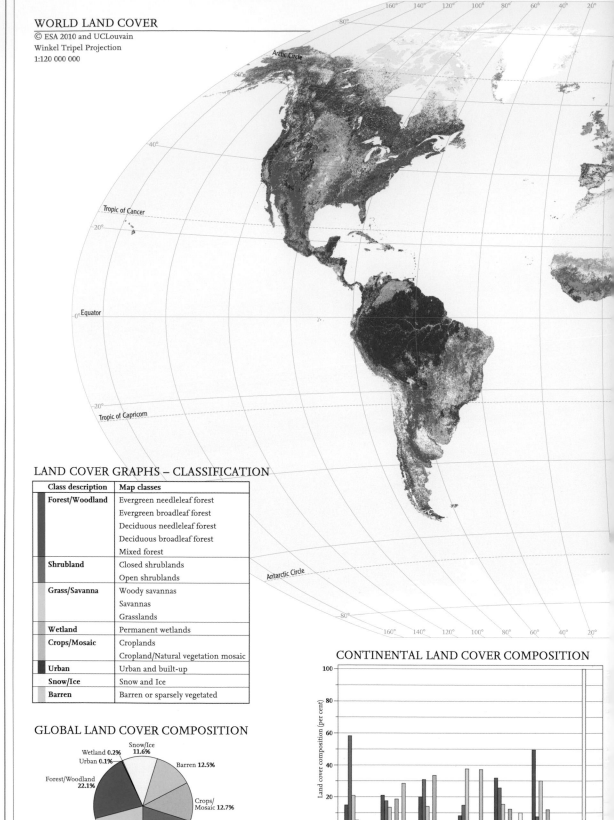

WORLD LAND COVER

© ESA 2010 and UCLouvain
Winkel Tripel Projection
1:120 000 000

LAND COVER GRAPHS – CLASSIFICATION

Class description	Map classes
Forest/Woodland	Evergreen needleleaf forest
	Evergreen broadleaf forest
	Deciduous needleleaf forest
	Deciduous broadleaf forest
	Mixed forest
Shrubland	Closed shrublands
	Open shrublands
Grass/Savanna	Woody savannas
	Savannas
	Grasslands
Wetland	Permanent wetlands
Crops/Mosaic	Croplands
	Cropland/Natural vegetation mosaic
Urban	Urban and built-up
Snow/Ice	Snow and Ice
Barren	Barren or sparsely vegetated

GLOBAL LAND COVER COMPOSITION

Wetland 0.2%
Snow/Ice 11.6%
Urban 0.1%
Barren 12.5%
Forest/Woodland 22.1%
Crops/Mosaic 12.7%
Grass/Savanna 20.9%
Shrubland 19.9%

CONTINENTAL LAND COVER COMPOSITION

Land cover composition (per cent)

Oceania   Asia   Europe   Africa   North America   South America   Antarctica

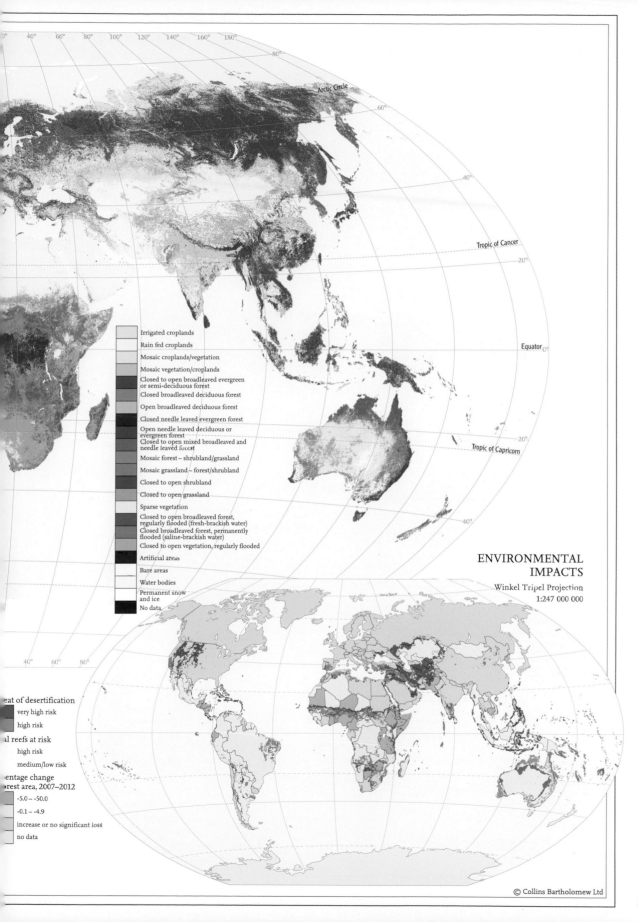

Irrigated croplands
Rain fed croplands
Mosaic croplands/vegetation
Mosaic vegetation/croplands
Closed to open broadleaved evergreen or semi-deciduous forest
Closed broadleaved deciduous forest
Open broadleaved deciduous forest
Closed needle leaved evergreen forest
Open needle leaved deciduous or evergreen forest
Closed to open mixed broadleaved and needle leaved forest
Mosaic forest – shrubland/grassland
Mosaic grassland – forest/shrubland
Closed to open shrubland
Closed to open grassland
Sparse vegetation
Closed to open broadleaved forest, regularly flooded (fresh-brackish water)
Closed broadleaved forest, permanently flooded (saline-brackish water)
Closed to open vegetation, regularly flooded
Artificial areas
Bare areas
Water bodies
Permanent snow and ice
No data

Arctic Circle
Tropic of Cancer
Equator
Tropic of Capricorn

ENVIRONMENTAL IMPACTS

Winkel Tripel Projection
1:247 000 000

eat of desertification
very high risk
high risk

al reefs at risk
high risk
medium/low risk

entage change
orest area, 2007–2012
-5.0 – -50.0
-0.1 – -4.9
increase or no significant loss
no data

© Collins Bartholomew Ltd

WORLD POPULATION DISTRIBUTION AND THE WORLD'S MAJOR CITIES

Winkel Tripel Projection
1:120 000 000

Major Urban Agglomerations

- over 20 million
- 10 million – 20 million
- 5 million – 10 million

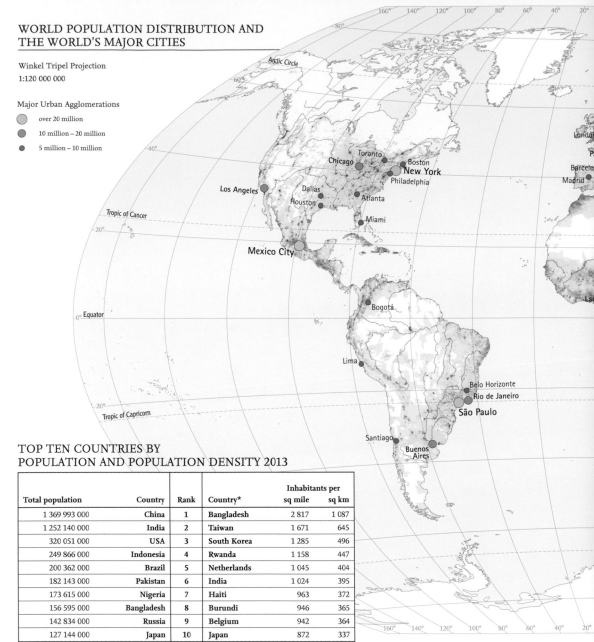

TOP TEN COUNTRIES BY POPULATION AND POPULATION DENSITY 2013

Total population	Country	Rank	Country*	Inhabitants per sq mile	Inhabitants per sq km
1 369 993 000	China	1	Bangladesh	2 817	1 087
1 252 140 000	India	2	Taiwan	1 671	645
320 051 000	USA	3	South Korea	1 285	496
249 866 000	Indonesia	4	Rwanda	1 158	447
200 362 000	Brazil	5	Netherlands	1 045	404
182 143 000	Pakistan	6	India	1 024	395
173 615 000	Nigeria	7	Haiti	963	372
156 595 000	Bangladesh	8	Burundi	946	365
142 834 000	Russia	9	Belgium	942	364
127 144 000	Japan	10	Japan	872	337

* Only countries with a population of over 10 million are considered.

KEY POPULATION STATISTICS FOR MAJOR REGIONS

	Population 2013 (millions)	Growth (per cent)	Infant mortality rate	Total fertility rate	Life expectancy (years)	% aged 60 and over 2010	% aged 60 and over 2050
World	7 162	1.1	37	2.5	70	11	22
More developed regions	1 253	0.3	6	1.7	78	22	32
Less developed regions	5 909	1.3	40	2.6	67	9	19
Africa	1 111	2.5	64	4.7	58	5	9
Asia	4 299	1.0	31	2.2	71	10	24
Europe	742	0.1	6	1.6	76	22	34
Latin America and the Caribbean	617	1.1	18	2.2	75	10	25
North America	355	0.9	6	1.9	79	19	27
Oceania	38	1.4	20	2.4	78	15	23

Except for population and % aged 60 and over figures, the data are annual averages projected for the period 2010–2015.

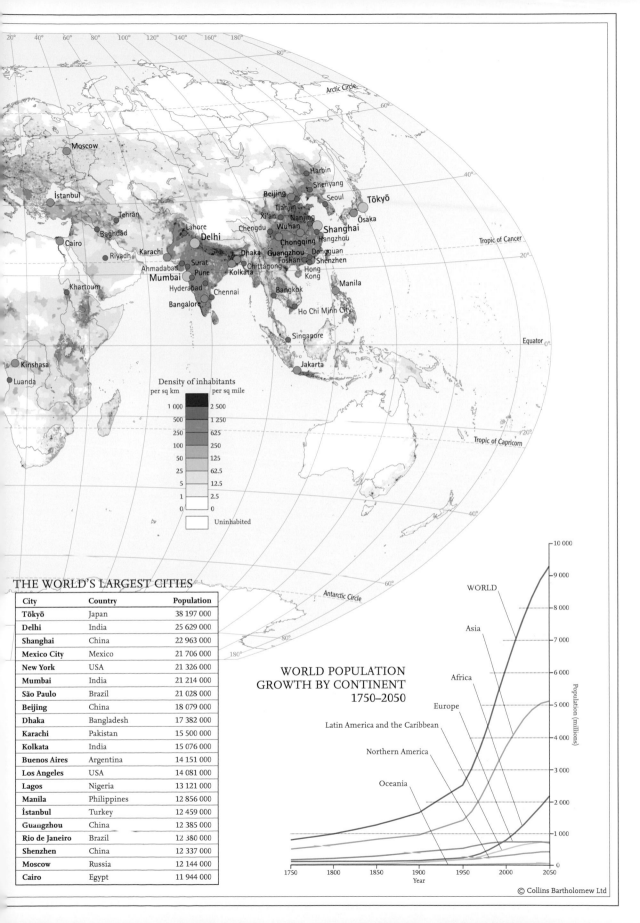

Density of inhabitants

per sq km	per sq mile
1 000	2 500
500	1 250
250	625
100	250
50	125
25	62.5
5	12.5
1	2.5
0	0
Uninhabited	

THE WORLD'S LARGEST CITIES

City	Country	Population
Tōkyō	Japan	38 197 000
Delhi	India	25 629 000
Shanghai	China	22 963 000
Mexico City	Mexico	21 706 000
New York	USA	21 326 000
Mumbai	India	21 214 000
São Paulo	Brazil	21 028 000
Beijing	China	18 079 000
Dhaka	Bangladesh	17 382 000
Karachi	Pakistan	15 500 000
Kolkata	India	15 076 000
Buenos Aires	Argentina	14 151 000
Los Angeles	USA	14 081 000
Lagos	Nigeria	13 121 000
Manila	Philippines	12 856 000
İstanbul	Turkey	12 459 000
Guangzhou	China	12 385 000
Rio de Janeiro	Brazil	12 380 000
Shenzhen	China	12 337 000
Moscow	Russia	12 144 000
Cairo	Egypt	11 944 000

WORLD POPULATION GROWTH BY CONTINENT 1750–2050

WORLD
Asia
Africa
Europe
Latin America and the Caribbean
Northern America
Oceania

© Collins Bartholomew Ltd

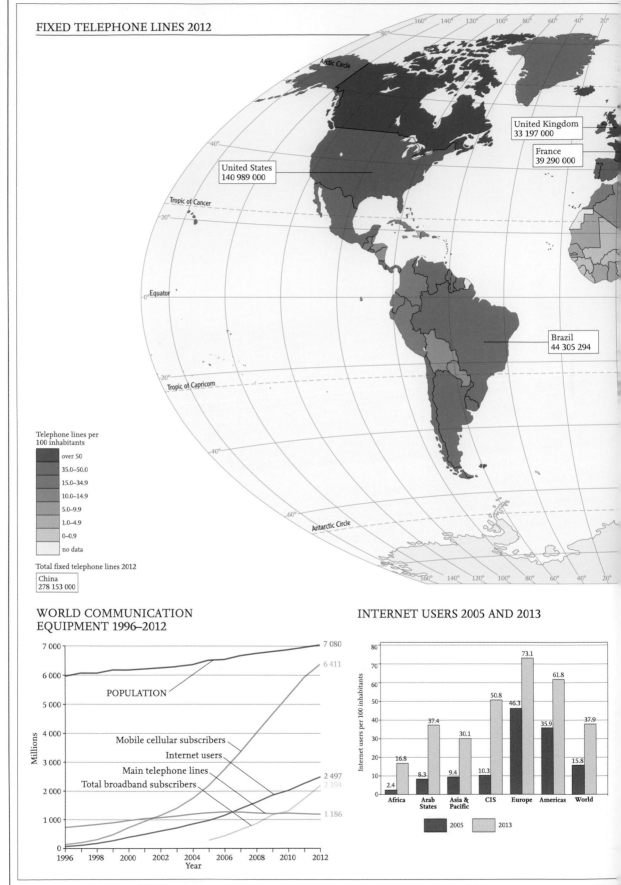

FIXED TELEPHONE LINES 2012

United Kingdom
33 197 000

France
39 290 000

United States
140 989 000

Brazil
44 305 294

Telephone lines per
100 inhabitants

- over 50
- 35.0–50.0
- 15.0–34.9
- 10.0–14.9
- 5.0–9.9
- 1.0–4.9
- 0–0.9
- no data

Total fixed telephone lines 2012

China
278 153 000

WORLD COMMUNICATION EQUIPMENT 1996–2012

POPULATION

Mobile cellular subscribers

Internet users

Main telephone lines

Total broadband subscribers

Millions

7 080
6 411
2 497
2 194
1 186

Year

INTERNET USERS 2005 AND 2013

Internet users per 100 inhabitants

	Africa	Arab States	Asia & Pacific	CIS	Europe	Americas	World
2005	2.4	8.3	9.4	10.3	46.3	35.9	15.8
2013	16.8	37.4	30.1	50.8	73.1	61.8	37.9

2005   2013

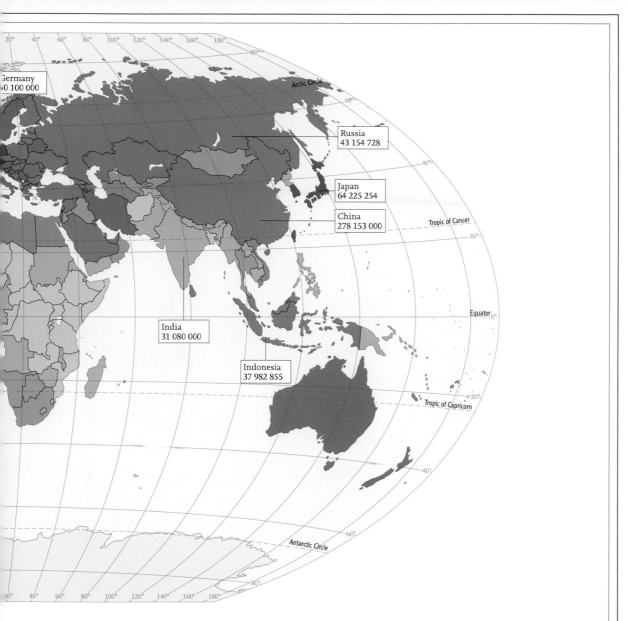

Germany
0 100 000

Russia
43 154 728

Japan
64 225 254

China
278 153 000

India
31 080 000

Indonesia
37 982 855

Arctic Circle

Tropic of Cancer

Equator 0°

Tropic of Capricorn

Antarctic Circle

TOP BROADBAND ECONOMIES 2012

Countries with highest broadband penetration rate – subscribers per 100 inhabitants

	Top Economies – Fixed Broadband	Rate
1	Switzerland	41.9
2	Netherlands	39.4
3	Denmark	38.2
4	France	37.8
5	South Korea	37.6
6	Norway	36.9

	Top Economies – Mobile Broadband	Rate
1	Singapore	123.3
2	Japan	113.1
3	Finland	106.5
4	South Korea	106.0
5	Sweden	101.3
6	Australia	96.2

INTERNET USERS

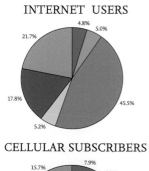

4.8%
5.0%
21.7%
45.5%
17.8%
5.2%

CELLULAR SUBSCRIBERS

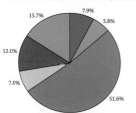

15.7%
7.9%
5.8%
12.0%
51.6%
7.0%

TELEPHONE MAIN LINES

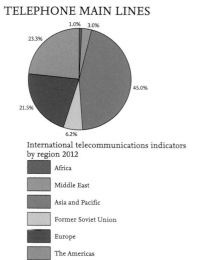

1.0%
3.0%
23.3%
45.0%
21.5%
6.2%

International telecommunications indicators by region 2012

- Africa
- Middle East
- Asia and Pacific
- Former Soviet Union
- Europe
- The Americas

© Collins Bartholomew Ltd

MAP POLICIES

PLACE NAMES

The spelling of place names on maps has always been a matter of great complexity, because of the variety of the world's languages and the systems used to write them down. There is no standard way of spelling names or of converting them from one alphabet, or symbol set, to another. Instead, conventional ways of spelling have evolved in each of the world's major languages, and the results often differ significantly from the name as it is spelled in the original language. Familiar examples of English conventional names include Munich (München), Florence (Firenze) and Moscow (from the transliterated form, Moskva).

In this atlas, local name forms are used where these are in the Roman alphabet, though for major cities, and main physical features, conventional English names are given first. The local forms are those which are officially recognized by the government of the country concerned, usually as represented by its official mapping agency. This is a basic principle laid down by the United Kingdom government's Permanent Committee on Geographical Names (PCGN) and the equivalent United States Board on Geographic Names, (BGN). Prominent English-language and historic names are not neglected, however. These, and significant superseded names and alternate spellings, are included in brackets on the maps where space permits, and are cross-referenced in the index.

Country names are shown in conventional English form and include any recent changes promulgated by national governments and adopted by the United Nations. The names of continents, oceans, seas and under-water features in international waters also appear in English throughout the atlas, as do those

of other international features where such an English form exists and is in common use. International features are defined as features crossing one or more international boundary.

BOUNDARIES

The status of nations, their names and their boundaries, are shown in this atlas as they are at the time of going to press, as far as can be ascertained. Where an international boundary symbol appears in the sea or ocean it does not necessarily infer a legal maritime boundary, but shows which offshore islands belong to which country. The extent of island nations is shown by a short boundary symbol at the extreme limits of the area of sea or ocean within which all land is part of that nation.

Where international boundaries are the subject of dispute it may be that no portrayal of them will meet with the approval of any of the countries involved, but it is not seen as the function of this atlas to try to adjudicate between the rights and wrongs of political issues. Although reference mapping at atlas scales is not the ideal medium for indicating the claims of many separatist and irredentist movements, every reasonable attempt is made to show where an active territorial dispute exists, and where there is an important difference between 'de facto' (existing in fact, on the ground) and 'de jure' (according to law) boundaries. This is done by the use of a different symbol where international boundaries are disputed, or where the alignment is unconfirmed, to that used for settled international boundaries. Ceasefire lines are also shown by a separate symbol. For clarity, disputed boundaries and areas are annotated where this is considered necessary. The atlas aims to take a strictly neutral viewpoint of all such cases, based on advice from expert consultants.

MAP PROJECTIONS

Map projections have been selected specifically for the area and scale of each map, or suite of maps. As the only way to show the Earth with absolute accuracy is on a globe, all map projections are compromises. Some projections seek to maintain correct area relationships (equal area projections), true distances and bearings from a point (equidistant projections) or correct angles and shapes (conformal projections); others attempt to achieve a balance between these properties. The choice of projections used in this atlas has been made on an individual continental and regional basis. Projections used, and their individual parameters, have been defined to minimize distortion and to reduce scale errors as much as possible. The projection used is indicated at the bottom left of each map page.

SCALE

In order to directly compare like with like throughout the world it would be necessary to maintain a single scale throughout the atlas. However, the desirability of mapping the more densely populated areas of the world at larger scales, and other geographical considerations, such as the need to fit a homogeneous physical region within a uniform rectangular page format, mean that a range of scales have been used. Scales for continental maps range between 1:20 000 000 and 1:44 000 000, depending on the size of the continental land mass being covered. Scales for regional maps are typically in the range 1:12 000 000 to 1:20 000 000. Mapping for most countries is at scales between 1:4 800 000 and 1:12 000 000, although for the more densely populated areas of Europe the scale increases to 1:2 400 000.

ABBREVIATIONS

Arch.	Archipelago			L.	Lake			Ra.	Range			mountain range
B.	Bay				Loch	(Scotland)	lake	S.	South, Southern			
	Bahia, Baía	Portuguese	bay		Lough	(Ireland)	lake		Salar, Salina,			
	Bahía	Spanish	bay		Lac	French	lake		Salinas	Spanish	salt pan, salt pans
	Baie	French	bay		Lago	Portuguese, Spanish	lake	Sa	Serra	Portuguese,	mountain range
C.	Cape			M.	Mys	Russian	cape, point		Sierra	Spanish	mountain range
	Cabo	Portuguese,		Mt	Mount			Sd	Sound			
		Spanish	cape, headland		Mont	French	hill, mountain	S.E.	Southeast,			
	Cap	French	cape, headland	Mt.	Mountain				Southeastern			
Co	Cerro	Spanish	hill, peak, summit	Mte	Monte	Portuguese, Spanish	hill, mountain	St	Saint			
E.	East, Eastern			Mts	Mountains				Sankt	German	Saint
Est.	Estrecho	Spanish	strait		Monts	French	hills, mountains		Sint	Dutch	Saint
G.	Gebel	Arabic	hill, mountain	N.	North, Northern			Sta	Santa	Italian, Portuguese,	
Gt	Great			O.	Ostrov	Russian	island			Spanish	Saint
I.	Island, Isle			Pk	Puncak	Indonesian, Malay	hill, mountain	Ste	Sainte	French	Saint
	Ilha	Portuguese	island	Pt	Point			Str.	Strait			
	Islas	Spanish	island	Pta	Punta	Italian, Spanish	cape, point	Tk	Teluk	Indonesian, Malay	bay, gulf
Is	Islands, Isles			R.	River			Tg	Tanjong, Tanjung	Indonesian, Malay	cape, point
	Islas	Spanish	islands		Rio	Portuguese	river	Vdkhr.	Vodokhranilishche	Russian	reservoir
Kep.	Kepulauan	Indonesian	islands		Río	Spanish	river	W.	West, Western			strait
Khr.	Khrebet	Russian	mountain range		Rivière	French	river		Wadi, Wâdi, Wādī	Arabic	watercourse

MAP SYMBOLS

LAND AND WATER FEATURES

Lake

Impermanent lake

Salt lake or lagoon

Impermanent salt lake

Dry salt lake or salt pan

—— River

- - - - Impermanent river

Ice cap / Glacier

_123_ Pass
Height in metres

∴ Site of special interest

⌄ Oasis

Wall

TRANSPORT

═══ Motorway

—— Main road

– – – Track

—— Main railway

Canal

✈ Main airport

BOUNDARIES

International boundary

Disputed international boundary
or alignment unconfirmed

Disputed territory boundary

Undefined international
boundary in the sea.
All land within this boundary is part
of state or territory named.

Administrative boundary
Shown for selected countries only.

Ceasefire line or other
boundary described on
the map

RELIEF

Contour intervals used in layer-colouring,
for land height and sea depth

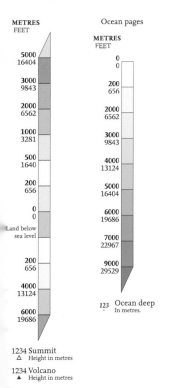

METRES
FEET

5000	16404
3000	9843
2000	6562
1000	3281
500	1640
200	656
0	0

Land below
sea level

200	656
4000	13124
6000	19686

Ocean pages

METRES
FEET

0	0
200	656
2000	6562
3000	9843
4000	13124
5000	16404
6000	19686
7000	22967
9000	29529

_123_ Ocean deep
In metres.

1234 Summit
△ Height in metres

1234 Volcano
▲ Height in metres

STYLES OF LETTERING

Cities and towns are explained separately

		Physical features	
Country	**FRANCE**	Island	*Gran Canaria*
Overseas Territory/Dependency	**Guadeloupe**	Lake	*Lake Erie*
Disputed Territory	WESTERN SAHARA	Mountain	*Mt Blanc*
Administrative name
Shown for selected countries only.	**SCOTLAND**	River	*Thames*
Area name	PATAGONIA	Region	*LAPPLAND*

CITIES AND TOWNS

| Population | National Capital | Administrative Capital
Shown for selected countries only | Other City or Town |
|---|---|---|---|
| over 10 million | **DHAKA** ⊡ | **Karachi** ⊙ | **New York** ⊙ |
| 5 million to 10 million | **MADRID** ⊡ | **Toronto** ⊙ | **Philadelphia** ⊙ |
| 1 million to 5 million | **KĀBUL** ☐ | **Sydney** ○ | **Gaoxiong** ○ |
| 500 000 to 1 million | **BANGUI** ☐ | Winnipeg ○ | Warangal ○ |
| 100 000 to 500 000 | WELLINGTON ☐ | Edinburgh ○ | Apucarana ○ |
| 50 000 to 100 000 | PORT OF SPAIN ☐ | Bismarck ○ | Invercargill ○ |
| under 50 000 | MALABO ▫ | Charlottetown ○ | Ceres ○ |

CONTINENTAL MAPS

BOUNDARIES —— International boundary   - - - - - - Disputed international boundary   •••••••• Ceasefire line

CITIES AND TOWNS   National Capital   **Beijing** ☐   Other City or Town   **New York** ○

© Collins Bartholomew Ltd

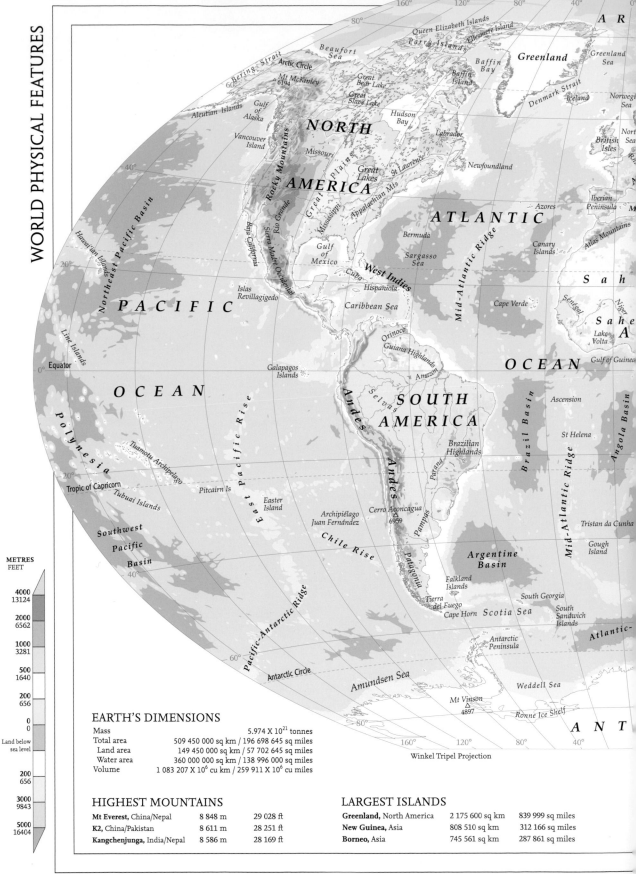

METRES / FEET

METRES	FEET
4000	13124
2000	6562
1000	3281
500	1640
200	656
0	0
Land below sea level	
200	656
3000	9843
5000	16404

Map labels

A R

Queen Elizabeth Islands
Parry Islands
Ellesmere Island
Baffin Bay
Greenland
Greenland Sea
Beaufort Sea
Bering Strait
Arctic Circle
Mt McKinley 6194
Great Bear Lake
Great Slave Lake
Baffin Island
Denmark Strait
Iceland
Norwegi
Aleutian Islands
Gulf of Alaska
NORTH
Hudson Bay
Labrador
Nort
British Isles
Sea
Vancouver Island
Missouri
Great Plain
Great Lakes
St Lawrence
Newfoundland
AMERICA
Rocky Mountains
Appalachian Mts
ATLANTIC
Azores
Iberian Peninsula
M
Baja California
Sierra Madre Occidental
Rio Grande
Mississippi
Bermuda
Mid-Atlantic Ridge
Canary Islands
Atlas Mountains
Gulf of Mexico
Sargasso Sea
Cape Verde
S a h
Hawaiian Islands
Northeast Pacific Basin
West Indies
Cuba
Hispaniola
Sénégal
Niger
S a h e l
Islas Revillagigedo
Caribbean Sea
A
Lake Volta
PACIFIC
Orinoco
Guiana Highlands
Gulf of Guinea
Line Islands
Galapagos Islands
Amazon
OCEAN
Equator
OCEAN
Polynesia
East Pacific Rise
Selvas
SOUTH
Brazil Basin
Ascension
Andes
AMERICA
Brazilian Highlands
Tuamotu Archipelago
St Helena
Angola Basin
Tropic of Capricorn
Pitcairn Is
Parana
Mid-Atlantic Ridge
Tubuai Islands
Easter Island
Andes
Southwest
Archipiélago Juan Fernández
Cerro Aconcagua 6959
Pampas
Tristan da Cunha
Pacific
Chile Rise
Patagonia
Argentine Basin
Gough Island
Basin
Falkland Islands
South Georgia
Pacific-Antarctic Ridge
Tierra del Fuego
Cape Horn
Scotia Sea
South Sandwich Islands
Atlantic-
Antarctic Circle
Antarctic Peninsula
Mt Vinson 4897
Amundsen Sea
Weddell Sea
Ronne Ice Shelf
A N T
Winkel Tripel Projection

EARTH'S DIMENSIONS

Mass	5.974×10^{21} tonnes
Total area	509 450 000 sq km / 196 698 645 sq miles
Land area	149 450 000 sq km / 57 702 645 sq miles
Water area	360 000 000 sq km / 138 996 000 sq miles
Volume	$1\,083\,207 \times 10^{6}$ cu km / $259\,911 \times 10^{6}$ cu miles

HIGHEST MOUNTAINS

Mt Everest, China/Nepal	8 848 m	29 028 ft
K2, China/Pakistan	8 611 m	28 251 ft
Kangchenjunga, India/Nepal	8 586 m	28 169 ft

LARGEST ISLANDS

Greenland, North America	2 175 600 sq km	839 999 sq miles
New Guinea, Asia	808 510 sq km	312 166 sq miles
Borneo, Asia	745 561 sq km	287 861 sq miles

Lambert Azimuthal Equal Area Projection

1: 100 800 000

Equatorial diameter	12 756 km / 7 927 miles	
Polar diameter	12 714 km / 7 900 miles	
Equatorial circumference	40 075 km / 24 903 miles	
Meridional circumference	40 008 km / 24 861 miles	

LARGEST LAKES

Caspian Sea, Asia/Europe	371 000 sq km	143 243 sq miles
Lake Superior, North America	82 100 sq km	31 699 sq miles
Lake Victoria, Africa	68 870 sq km	26 591 sq miles

LONGEST RIVERS

Nile, Africa	6 695 km	4 160 miles
Amazon, South America	6 516 km	4 049 miles
Yangtze, Asia	6 380 km	3 965 miles

© Collins Bartholomew Ltd

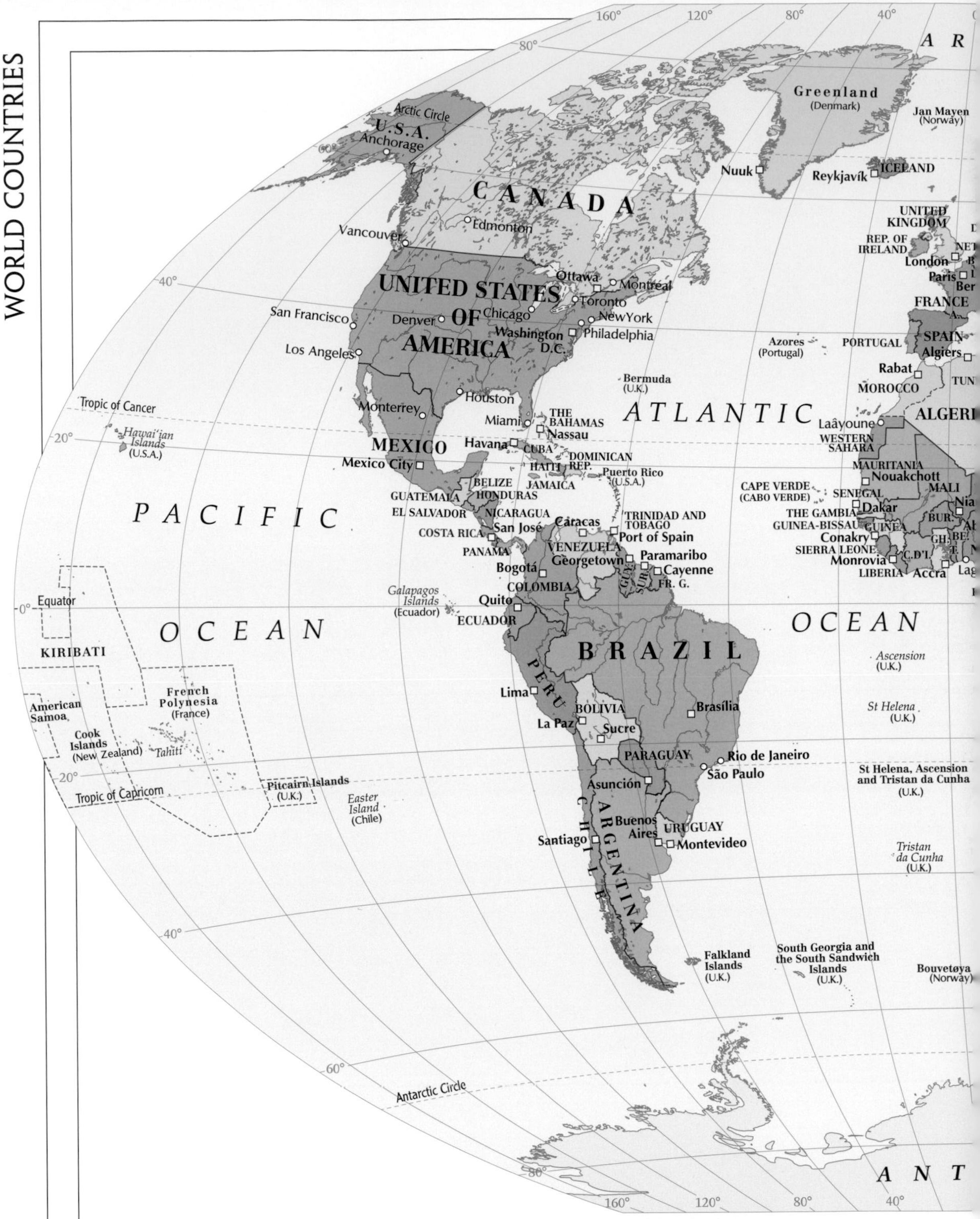

A R

Greenland
(Denmark)

Jan Mayen
(Norway)

Arctic Circle
U.S.A.
Anchorage

Nuuk

Reykjavík ICELAND

C A N A D A

Edmonton

UNITED
KINGDOM

REP. OF
IRELAND

London
Paris

NET
B
I

Vancouver

FRANCE

Ottawa
Montreal
Toronto
New York

UNITED STATES
OF
AMERICA

San Francisco

Denver
Chicago

SPAIN

PORTUGAL

Azores
(Portugal)

Algiers

Washington
D.C.
Philadelphia

Rabat
MOROCCO

TUN

Los Angeles

Bermuda
(U.K.)

A T L A N T I C

Laâyoune

ALGERI

Tropic of Cancer

Monterrey

Houston

THE
BAHAMAS
Nassau

WESTERN
SAHARA

Hawai'ian
Islands
(U.S.A.)

Miami

MAURITANIA

Nouakchott

MALI

Havana
CUBA

MEXICO

Mexico City

HAITI
DOMINICAN
REP.
Puerto Rico
(U.S.A.)

CAPE VERDE
(CABO VERDE)

SENEGAL
Dakar
THE GAMBIA
GUINEA-BISSAU
Conakry
SIERRA LEONE
Monrovia
LIBERIA

Nia

BELIZE
GUATEMALA HONDURAS
EL SALVADOR NICARAGUA
COSTA RICA San José
PANAMA

JAMAICA

Caracas

TRINIDAD AND
TOBAGO
Port of Spain

BUR.

GUINEA

C.D'I.

Accra

Lag

P A C I F I C

VENEZUELA
Bogotá
Georgetown
Paramaribo
Cayenne
FR. G.

KIRIBATI

O C E A N

Galapagos
Islands
(Ecuador)

COLOMBIA
Quito
ECUADOR

O C E A N

B R A Z I L

Equator

Ascension
(U.K.)

American
Samoa

French
Polynesia
(France)

Lima
PERU

St Helena
(U.K.)

BOLIVIA

Brasília

Cook
Islands
(New Zealand)

Tahiti

La Paz

Sucre

Rio de Janeiro
São Paulo

St Helena, Ascension
and Tristan da Cunha
(U.K.)

PARAGUAY

Pitcairn Islands
(U.K.)

Tropic of Capricorn

Easter
Island
(Chile)

Asunción

URUGUAY

Tristan
da Cunha
(U.K.)

Santiago

Buenos
Aires
Montevideo

ARGENTINA

Falkland
Islands
(U.K.)

South Georgia and
the South Sandwich
Islands
(U.K.)

Bouvetoya
(Norway)

Antarctic Circle

Winkel Tripel Projection

A N T

ABBREVIATIONS

A.	ANDORRA	BE.	BENIN	C.A.R.	CENTRAL AFRICAN	CZ.R.	CZECH REPUBLIC
AL.	ALBANIA	BEL.	BELGIUM		REPUBLIC	DEN.	DENMARK
ARM.	ARMENIA	B.H.	BOSNIA AND HERZEGOVINA	C.D'I.	CÔTE D'IVOIRE	EQ.G.	EQUATORIAL GUINEA
AUS.	AUSTRIA	BN.	BAHRAIN		(IVORY COAST)	FR.G.	FRENCH GUIANA
AZ.	AZERBAIJAN	BUR.	BURKINA FASO	CR.	CROATIA	GEOR.	GEORGIA
B.	BURUNDI	CAM.	CAMEROON	CYP.	CYPRUS	GER.	GERMANY

IC OCEAN

Svalbard
(Norway)

NORDEN

FINLAND

ESTONIA
LATVIA
LITH.
BELARUS
POLAND
HUN.
MO.
CRS
ROMANIA
BULGARIA
AL
ITALY
GREECE
Tripoli

BYA

EGYPT

LBYA

Moscow

Yekaterinburg

Omsk
Novosibirsk

Astana

KAZAKHSTAN

Kiev
UKRAINE

GEOR.
Tbilisi
Ankara
ARM. AZ.
TURKEY
SYRIA
LEB.
ISR.
Amman JOR.
Baghdad
Tehran
IRAQ
IRAN

Cairo
Riyadh

SAUDI
ARABIA

U.A.E.
Muscat
OMAN

RUSSIA

Magadan

Arctic Circle

Ulan-Bator

MONGOLIA

Harbin

Beijing

CHINA

Lanzhou
Tianjin

Xi'an
Chengdu
Wuhan

Chongqing

Shanghai

UZBEK.
Dushanbe
TURKM.
KYR.
TAJIK.
Kabul
AFGHAN.
ISTAN
Islamabad
New
Delhi
PAKISTAN

NEPAL
Kathmandu
BANGLA-
DESH
BHUTAN

Karachi
Dhaka
INDIA
MYANMAR
(BURMA)

N.KOREA
P'yongyang
Seoul
S.KOREA
Osaka

JAPAN
Tokyo

Taibei
TAIWAN

Hong Kong

PACIFIC

OCEAN

Tropic of Cancer

60°

40°

20°

Mumbai
Nay Pyi Taw
Ha Noi
Vientiane

Khartoum
ERITREA
YEMEN
Asmara
San'a'
DJIBOUTI

Addis
Ababa
SOUTH
SUDAN
ETHIOPIA

SUDAN

CHAD
Ndjamena

C.A.R.
Bangui

DEM.
REP.
OF THE
CONGO
Kinshasa

UGANDA
KENYA
Nairobi

Dodoma
TANZANIA

Rangoon

THAILAND
Bangkok

Chennai

SRI
LANKA

MALDIVES

Kuala Lumpur
Putrajaya

SINGAPORE

Manila

PHILIPPINES

BRUNEI

MALAYSIA

Northern
Mariana
Islands
(U.S.A.)

MARSHALL
ISLANDS

FEDERATED STATES
OF MICRONESIA

PALAU

Equator

0°

Mogadishu
SOMALIA

SEYCHELLES

British Indian
Ocean Territory
(U.K.)

COMOROS

INDIAN

Cocos
(Keeling)
Islands
(Australia)

Christmas
Island
(Australia)

Jakarta

INDONESIA

EAST TIMOR
(TIMOR-LESTE)

Port
Moresby

PAPUA
NEW
GUINEA

SOLOMON
ISLANDS

NAURU

TUVALU

KIRIBATI

ANGOLA
ZAMBIA
Lilongwe

MOZAMBIQUE

MADAGASCAR

Antananarivo

MAURITIUS

Réunion
(France)

Harare
ZIMBABWE
BOTS-
WANA
NAMIBIA
Windhoek

Pretoria
Maputo
SWAZILAND
LESOTHO
Maseru

SOUTH AFRICA

Perth

French Southern
and Antarctic Lands

Îles Kerguélen
(France)

OCEAN

AUSTRALIA

Brisbane

Sydney
Canberra

Wellington

NEW
ZEALAND

Coral Sea
Islands
Territory
(Aust.)

VANUATU

New
Caledonia
(France)

Norfolk
Island
(Australia)

SAMOA

FIJI

TONGA

Tropic of Capricorn

40°

CTICA

60°

Antarctic Circle

80°

1: 100 800 000

40°
80°
120°
160°

International boundaries in the sea shown on this
map indicate ownership of islands and island groups only.
They do not infer the alignments of legal maritime boundaries.

GH.	GHANA	KYR.	KYRGYZSTAN	NETH.	NETHERLANDS	SUR.	SURINAME
GUY.	GUYANA	LEB.	LEBANON	NI.	NIGERIA	SW.	SWITZERLAND
HUN.	HUNGARY	LITH.	LITHUANIA	Q.	QATAR	T.	TOGO
ISR.	ISRAEL	LUX.	LUXEMBOURG	R.	RWANDA	TAJIK.	TAJIKISTAN
JOR.	JORDAN	M.	MONTENEGRO	S.	SERBIA	TURKM.	TURKMENISTAN
K.	KOSOVO	MA.	MACEDONIA	SLA.	SLOVAKIA	U.A.E.	UNITED ARAB EMIRATES
KU.	KUWAIT	MO.	MOLDOVA	SL.	SLOVENIA	UZBEK.	UZBEKISTAN

© Collins Bartholomew Ltd

C 120° D 130° E 140° F 150° G 160° H 170°

Tropic of Cancer
TAIWAN

1

20°
Luzon Strait
Wake Island (U.S.A.)

Luzon
Pagan

Northern Mariana Islands (U.S.A.)

PHILIPPINES
Capitol Hill □ Saipan

2
MARSHALL ISLANDS

Hagåtña □ Guam (U.S.A.)

Ratak Chain

Palawan
Sulu Sea
Mindanao
Palau Islands
Hall Islands
Ralik Chain
Delap-Uliga-Dja

10°
Yap
Chuuk
Pohnpei
Palikir

Celebes Sea
Kepulauan Talaud
Caroline Islands
Bairiki Tarawa

Halmahera
Mortlock Islands
Kosrae
Gil Isla

FEDERATED STATES OF MICRONESIA

3
Admiralty Islands
New Ireland
Yaren
NAURU

Equator
Moluccas (Maluku)
Puncak Jaya 4884
Wewak
Bismarck Sea
Rabaul
Nukumanu Islands

Celebes (Sulawesi)
Seram
Mount Wilhelm 4509
Madang
New Britain
Choiseul
Santa Isabel
SOLOMON ISLANDS

INDONESIA
Laut Banda (Banda Sea)
Kep. Aru
New Guinea
PAPUA NEW GUINEA
Bougainville Island
Solomon Sea
New Georgia Islands
Malaita
Honiara
Duff Islands

Pulau Dolak
Daru
G. of Papua
Port Moresby
Guadalcanal
San Cristobal
Santa Cruz Islands

Laut Flores (Flores Sea)
Wetar
Kep. Tanimbar
Arafura Sea
Louisiade Arch.
Rennell
Banks Islands

4
Flores
Timor
Melville Island
Cape Arnhem
Torres Strait
Espiritu Santo
VANUATU

EAST TIMOR (TIMOR-LESTE)
Malakula
Port Vila
Erromango

Sumba
Cape Londonderry
Darwin
Gulf of Carpentaria
Coral Sea Islands Territory (Aust.)

Timor Sea
Cairns
Coral Sea
Tanna

Cape Lévêque
Wyndham
Normanton
New Caledonia (Fr.)
Îles Loyauté

5
Broome
Halls Creek
Townsville
Nouméa
Hunter Island

NORTHERN TERRITORY
Mount Isa
Great Barrier Reef
Île des Pins

Port Hedland
Great Sandy Desert
Rockhampton

Alice Springs
QUEENSLAND
Great Dividing Range

Newman
AUSTRALIA
Charleville
Brisbane
Norfolk Island (Aust.)

Tropic of Capricorn
WESTERN AUSTRALIA
Oodnadatta
Toowoomba
Gold Coast

Mt Magnet
SOUTH AUSTRALIA
Broken Hill
Tamworth
Lord Howe Island (N.S.W.) (Aust.)

6
Geraldton
Kalgoorlie
Port Augusta
Darling
NEW SOUTH WALES
Newcastle
Sydney
Wollongong

Great Australian Bight
Port Lincoln
Murray
Canberra
A.C.T.

Perth
Fremantle
Kangaroo Island
Adelaide
VICTORIA
Geelong
Melbourne
TASMAN SEA

Cape Leeuwin
Bass Strait
Flinders Island
Christch

30°
King Island
Launceston
South Island (Te Waipounamu)
Du

Hobart
TASMANIA
Invercargill

Cape Leeuwin
South East Cape
Stewart Island

7
Auckland Islands

Campbell Islan (N.Z.)

Macquarie Island (Aust.)

A 100° B 110° C 120° D 130° E 140° Longitude 150° east of Greenwich H

Lambert Azimuthal Equal Area Projection

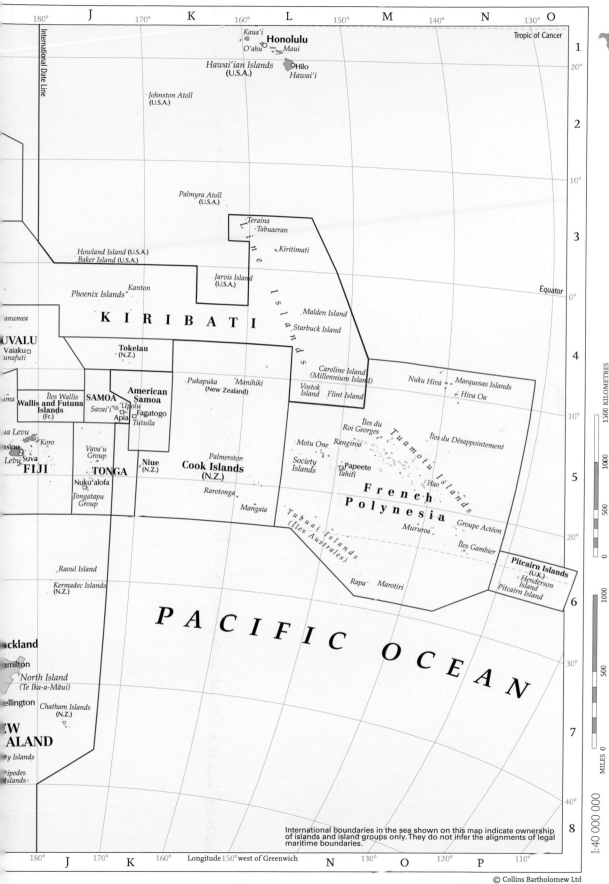

© Collins Bartholomew Ltd

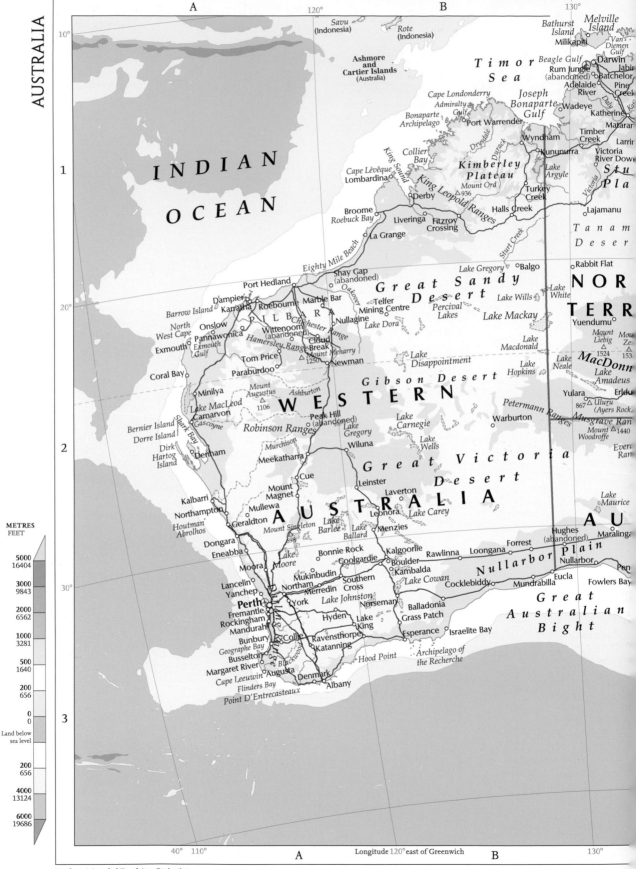

INDIAN

OCEAN

Savu
(Indonesia)

Rote
(Indonesia)

Ashmore
and
Cartier Islands
(Australia)

T i m o r
S e a

Bathurst
Island

Melville
Island

Milikapiti

Van
Diemen
Gulf

Beagle Gulf
Rum Jungle
(abandoned)
Darwin
Jabi

Batchelor

Adelaide
River

Pine
Creek

Cape Londonderry

Admiralty
Gulf

Joseph
Bonaparte
Gulf

Wadeye

Katherine

Matarar

Bonaparte
Archipelago

Port Warrender

Wyndham

Kununurra

Timber
Creek

Victoria
River Dow

Larrir

St u
Pla

Cape Lévêque
Lombardina

Mount Ord
△936

Halls Creek

Lajamanu

Collier
Bay

Derby

Turkey
Creek

Kimberley
Plateau

King Leopold Ranges

Broome

Roebuck Bay

Liveringa

Fitzroy
Crossing

Sturt Creek

T a n a m
D e s e r

La Grange

Eighty Mile Beach

Shay Gap
(abandoned)

Lake Gregory

Balgo

Rabbit Flat

N O R

Port Hedland

Dampier
Karratha
Roebourne
Marble Bar

Telfer
Mining Centre

G r e a t   S a n d y
D e s e r t

Lake Wills

Lake
White

T E R R

Barrow Island

Nullagine

P I L B A R A

Chichester Range

Percival
Lakes

Lake Mackay

Yuendumu

North
West Cape

Onslow
Pannawonica

Wittenoom
(abandoned)

Lake Dora

Mount
Liebig
△
1524

Mou
Ze
153

Hamersley Range

Cloud
Break

Exmouth

Exmouth
Gulf

Tom Price

Mount Meharry
△
1250

Newman

Lake
Disappointment

Lake
Macdonald

MacDonn

Coral Bay

Paraburdoo

G i b s o n   D e s e r t

Lake
Hopkins

Lake
Neale

Lake
Amadeus

Minilya

Lake MacLeod

Mount
Augustus
△
1106

Ashburton

Yulara

Erldu

W E S T E R N

Uluru
△867 (Ayers Rock,

Carnarvon

Gascoyne

Bernier Island
Dorre Island

Robinson Ranges

Peak Hill
(abandoned)

Lake
Gregory

Lake
Carnegie

Warburton

Petermann Ranges

Musgrave Ran

Mount
Woodroffe
△1440

Ever
Ran

Dirk
Hartog
Island

Denham

Murchison

Wiluna

Lake
Wells

G r e a t   V i c t o r i a

Lake
Maurice

Shark Bay

Meekatharra

D e s e r t

A U S T R A L I A

Cue

Leinster

A U

Kalbarri

Mount
Magnet

Laverton

Hughes
(abandoned)

Maralinga

Northampton

Mullewa

Leonora

Lake Carey

Forrest

Houtman
Abrolhos

Geraldton

Mount Singleton
△
698

Lake
Barlee

Lake
Ballard

Menzies

Rawlinna
Loongana

N u l l a r b o r   P l a i n

Nullarbor

Dongara

Bonnie Rock

Kalgoorlie

G r e a t

Eneabba

Coolgardie
Boulder
Kambalda

Cocklebiddy

Mundrabilla
Eucla

Pen

Fowlers Bay

Moora

Lake
Moore

Mukinbudin

Lake Cowan

A u s t r a l i a n

Lancelin
Yanchep

Northam

Southern
Cross

B i g h t

Perth

Merredin

Lake Johnston

Norseman

Balladonia

Fremantle
Rockingham
Mandurah

York

Hyden

Lake
King

Grass Patch

Esperance

Israelite Bay

Bunbury
Collie

Ravensthorpe

Geographe Bay

Katanning

Hood Point

Archipelago of
the Recherche

Busselton

Blackwood

Margaret River
Augusta

Denmark

Cape Leeuwin

Flinders Bay

Albany

Point D'Entrecasteaux

METRES
FEET

5000
16404

3000
9843

2000
6562

1000
3281

500
1640

200
656

0
0

Land below
sea level

200
656

4000
13124

6000
19686

10°

120°

130°

1

20°

2

30°

3

40° 110°

Longitude 120° east of Greenwich

120°

130°

A

B

A

B

Lambert Azimuthal Equal Area Projection

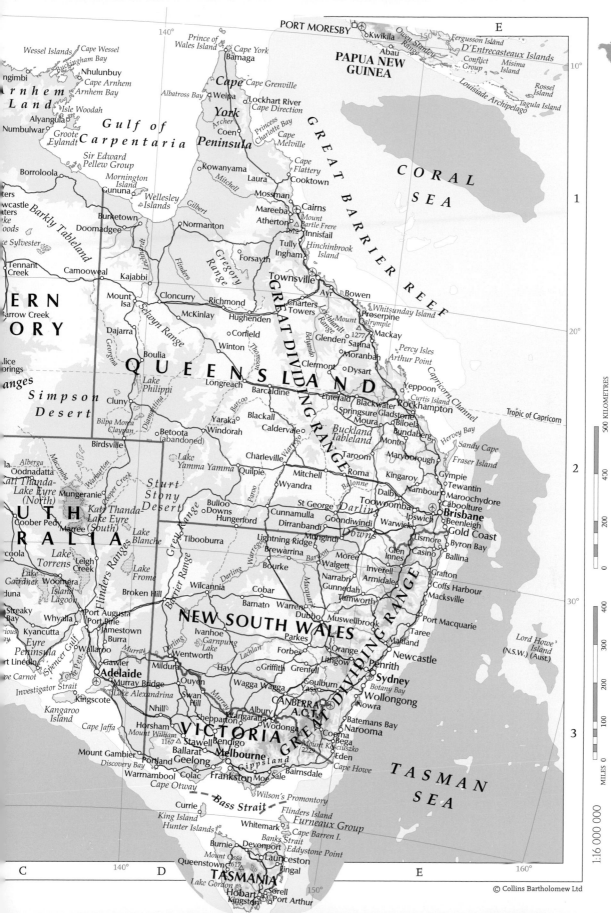

© Collins Bartholomew Ltd

A 140° B

1

Macumba

Warburton

Tirari Desert

Cooper Creek

Noccundra

Grey Range

Thargomindah

Kati Thanda-
Lake Eyre
(North)

Mungeranie

Cooper Creek

Innamincka

Moomba

Sturt Stony
Desert

Bulloo

QUE

Bulloo
Downs

Etadunna

Kati Thanda-
Lake Eyre
(South)

Lake
Blanche

*Caryapundy
Swamp*

Hungerford

William Creek

Tibooburra

Tilcha
(abandoned)

Mount Sturt
427 △

Milparinka

Wanaaring

Marree

Lake Callabonna

Parry

Moolawatana

Hawkers Gate

Tongo

Millers Creek

30°

S O U T H

Lyndhurst

Leigh
Creek

Balcanoona

Packsaddle

White Cliffs

Momba

Tilpa

Darling

Parakylia

Roxby
Downs

Lake
Frome

A U S T R A L I A

Mootwingee

Beltana

Parachilna

Wirraminna

Woomera

Lake
Torrens

Frome Downs

Flinders Ranges

Mount Robe
486 △

Euriowie

Wilcannia

N E W

Island
Lagoon

Pernatty
Lagoon

Curnamona

Broken
Hill

Stephens Creek

Woocalla

Hawker

Cockburn

Mingary

Mount Mana

Lake
Gairdner

Lake
Macfarlane

Cradock

Barrier Range

Menindee Lake

Menindee

Nonning

Quorn

Mannahill

Olary

Tandou Lake

2

Gawler Ranges

Port Augusta

Stirling North
Wilmington

Yunta

Paratoo

Coombah

Darnick

Ivanhoe

Buckleboo

Iron Knob

Mount
Ramarkable △ 969

Orroroo

Oakbank

Popiltah

Pooncarie

Mossgiel

Whyalla

Wirrabara

Peterborough

Terowie

Canopus

*Garnpung
Lake*

Kimba

Balumbah

Jamestown

Bool

Kyancutta

Port Pirie

Gladstone

Burra

Lake
Victoria

Burtundy

Hatfield

Oxley

Lock

Cleve

Cowell

Crystal
Brook

Snowtown

Clare

Darling

Wentworth

Sheringa *Eyre
Peninsula*

Arno
Bay

Wallaroo

Kadina

Blyth

Morgan

Murray

Renmark

Merbein

Mildura

Ungarra

Moonta

Port Wakefield

Waikerie

Barmera

Berri

Werrimull

Red
Cliffs

Robinvale

Murrumbidgee

Cockaleechie

Maitland

Balaklava

Kapunda

Nuriootpa

Loxton

Robinvale

Balranald

R I

Tumby
Bay

Minlaton

York Peninsula

Ardrossan

Gawler

Mannum

Alawoona

Mindarie

Hattah

Booroorba

Moulam

Port
Lincoln

Mount Lofty Range

Adelaide

Ouyen

Swan
Hill

Denil

Coffin
Bay

Cape
Carnot

Gambier
Islands

Yorketown

Mount Barker

Murray Bridge

Pinnaroo

Murrayville

Underbool

Lake
Tyrrell

Barham

Cohu

35°

Marion
Bay

Willunga

Tailem Bend

Lameroo

Sea Lake

Ultima

Murray

Investigator Strait

Backstairs Passage

Goolwa

Victor
Harbor

Lake
Alexandrina

Coonalpyn

Hopetoun

Birchip

Kerang

Echuc

Cape Borda

Kingscote

Penneshaw

Meningie

Tintinara

*Younghusband
Peninsula*

Keith

Lake
Hindmarsh

Wycheproof

Charlton

Rochester

Cape
du Couedic

*Kangaroo
Island*

Bordertown

Kaniva

Warracknabeal
Nhill

Donald

Dimboola

St Arnaud

Bendigo

Padthaway

Goroke

Horsham

V I C

Lacepede Bay

Kingston South East
Cape Jaffa

Naracoorte

Edenhope

Stawell

*Mount William
△
1167*

Avoca

Castlemaine

Kyneton

3

Robe

Lake
George

Penola

Glenelg

Balmoral

The Grampians

Ararat

Daylesford

Ma

Sunbu

Beachport

Millicent

Casterton

Coleraine

Ballarat

Melton

Mount Gambier

Hamilton

Skipton

Beaufort

Bacchus Marsh

Wyndham

Werribe

Mortlake

*Lake
Corangamite*

Geelong

Queenscli

Port MacDonnell

Heywood

Camperdown

Terang

Colac

Toro

Angles

*Discovery
Bay*

Portland

Port
Fairy

Warrnambool

Cape Nelson

Port Campbell

Lorne

Apollo Bay
Cape Otway

135°

A

Longitude 140° east of Greenwich

B

METRES
FEET

5000
16404

3000
9843

2000
6562

1000
3281

500
1640

200
656

0
0

Land below
sea level

200
656

4000
13124

6000
19686

52

Conic Equidistant Projection

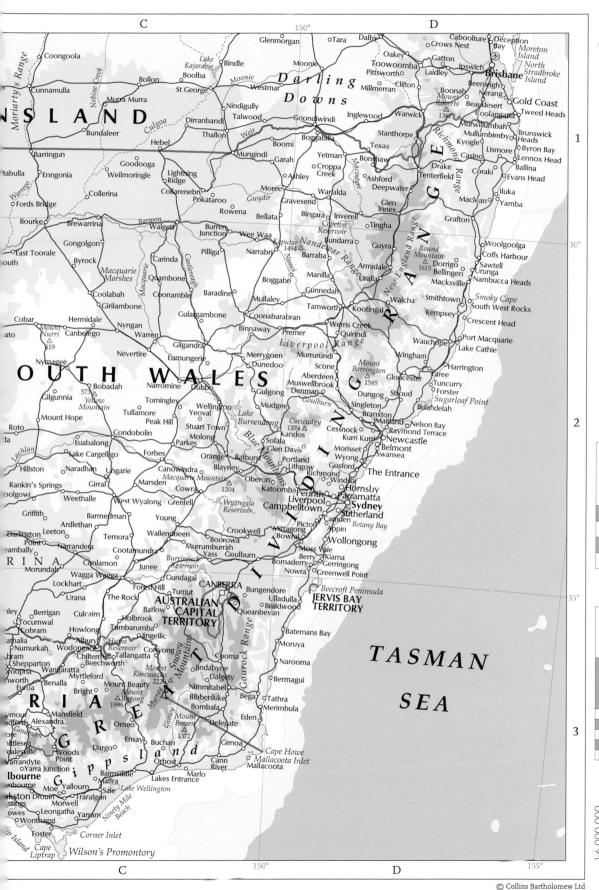

© Collins Bartholomew Ltd

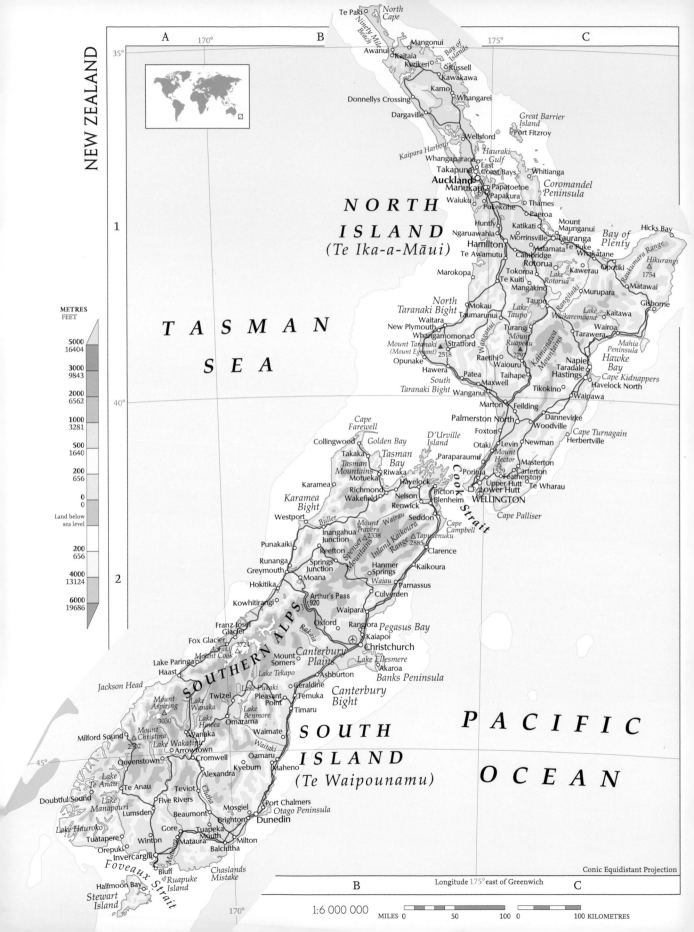

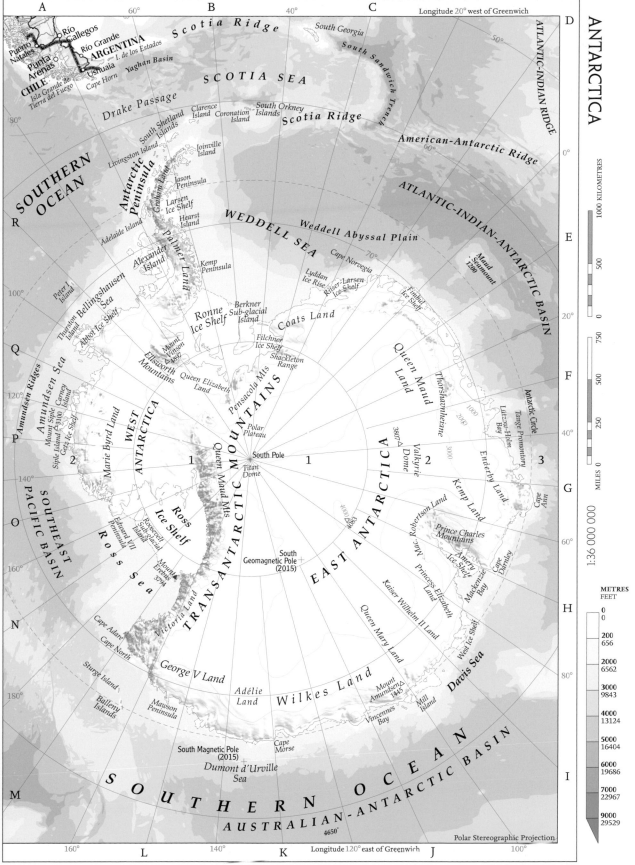

ANTARCTICA

1:36 000 000

1000 KILOMETRES

500

0

750

500

250

0

MILES 0

METRES FEET	
0	0
200	656
2000	6562
3000	9843
4000	13124
5000	16404
6000	19686
7000	22967
9000	29529

Polar Stereographic Projection

© Collins Bartholomew Ltd

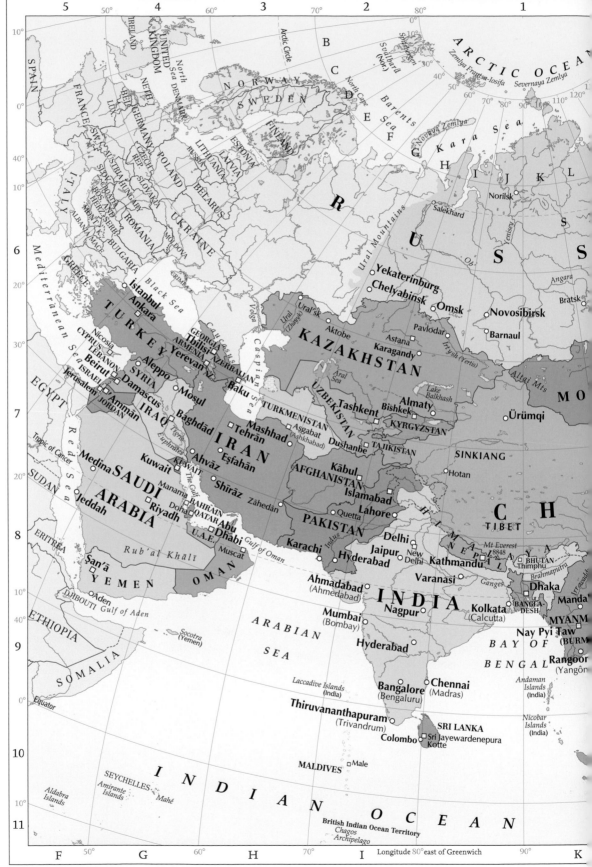

Two Point Equidistant Projection

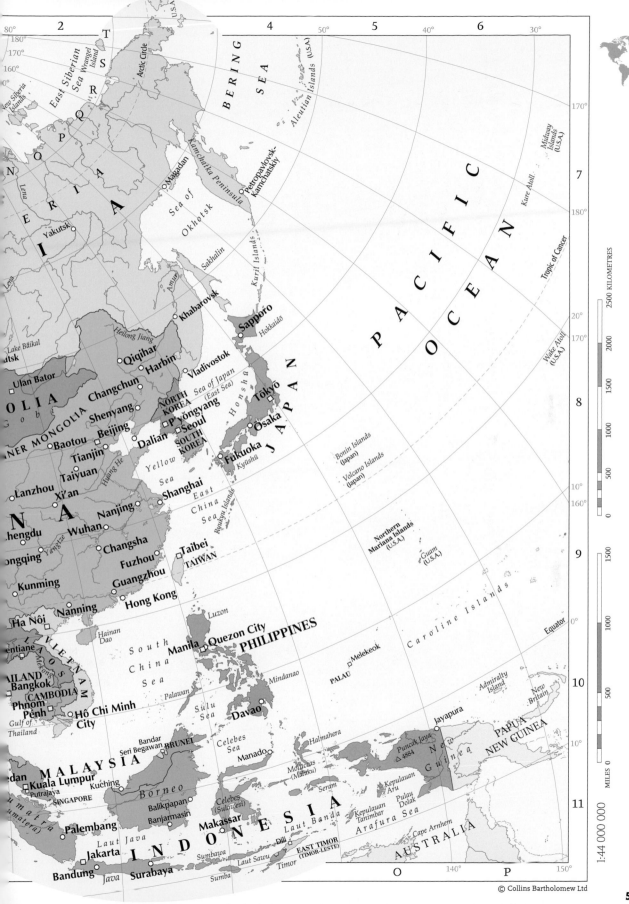

80°
180°
170°
160°
0°

T
S
R
Q
P
O
N

Arctic Circle

New Siberia Islands
East Siberian Sea
Wrangel Island (U.S.A.)
U.S.A.

BERING SEA

Aleutian Islands (U.S.A.)

170°

I B E R I A

Lena

Yakutsk

Magadan

Kamchatka Peninsula

Petropavlovsk-Kamchatskiy

Sea of Okhotsk

Sakhalin

Kuril Islands

Midway Islands (U.S.A.)

Kure Atoll

7

180°

Tropic of Cancer

20°
170°

Khabarovsk

Sapporo
Hokkaidō

Wake Atoll (U.S.A.)

P A C I F I C

Lake Baikal

utsk

Ulan Bator

Qiqihar   Harbin

Vladivostok

Sea of Japan (East Sea)

O C E A N

MONGOLIA   Changchun
Shenyang

NORTH KOREA
P'yŏngyang

Honshū

Tōkyō

J A P A N

8

INNER MONGOLIA
Baotou   Beijing   Dalian

SOUTH KOREA
Seoul

Osaka

Tianjin
Taiyuan

Huang He

Yellow
Sea

Fukuoka
Kyūshū

Bonin Islands (Japan)

Lanzhou   Xi'an

East
China
Sea

Volcano Islands (Japan)

10°
160°

hengdu

Nanjing

Shanghai

Ryukyu Islands

A   A

Wuhan

ongqing

Changsha

Yangtze

Taibei
TAIWAN

Northern Mariana Islands (U.S.A.)

9

Fuzhou

Kunming

Guangzhou

Nanning   Hong Kong

Hainan Dao

Guam (U.S.A.)

Luzon

Ha Nôi

entiane

LAOS

South
China
Sea

Quezon City

Manila   PHILIPPINES

Caroline Islands

Equator

0°

AILAND
Bangkok

VIETNAM

Palawan

Mindanao

Melekeok

PALAU

CAMBODIA
Phnom
Pênh

Hô Chi Minh
City

Sulu
Sea

Davao

Admiralty Islands

New Britain

10

Gulf of
Thailand

Bandar
Seri Begawan

BRUNEI

Celebes
Sea

Manado

Halmahera

Jayapura

PAPUA
NEW GUINEA

Puncak Jaya
△ 4884

N e w   G u i n e a

dan
Kuala Lumpur
Putrajaya

MALAYSIA

Borneo

Moluccas
(Maluku)

Kepulauan
Aru

Pulau
Dolak

SINGAPORE

Kuching

Balikpapan

Seram

11

umatra)

Banjarmasin

Makassar

Celebes
(Sulawesi)

Laut Banda

Kepulauan
Tanimbar

Arafura Sea

Cape Arnhem

Palembang

Jakarta

Bandung   Surabaya

Java

I N D O N E S I A

Sumbawa

Dili

EAST TIMOR
(TIMOR-LESTE)

Laut Sawu   Timor

AUSTRALIA

Sumba

2500 KILOMETRES
2000
1500
1000
500
0

1500
1000
500
0
MILES

1:44 000 000

90° A 105° B 120°

CHINA

Khulna
Chittagong
Cox's Bazar
Wuntho · Lunglei · Lincang · Yuxi
Namtu · Kaiyuan · Hechi · Liuzhou · Longyan · Xiamen (Amoy) · TAIBEI
Monywa · Mandalay · Lashio · Jinghong · Wenshan · Baise · Jingxi · Wuzhou · Meizhou · Xinzi · Hualian
Myingyan · Mogok · Gejiu · Nanning · Guangzhou (Canton) · Chaozhou
MYANMAR
Meiktila · Taunggyi · Keng Tung · Lao Cai · Cao Bằng · Pingxiang · Yulin · Qinzhou · Beihai · Maoming · Shantou (Swatow)
Minbu · Louangnamtha · Thai Nguyên · Hai · Duong · Zhanjiang · Shenzhen · Gaoxiong
Sittwe · **NAY PYI TAW** · Pyinmana · Chiang Rai · Louangphabang · **Hai Phong** · Xuwen · Hong Kong
Kyaukpyu · Phôngsali · Nam Đinh · Haikou · Wencheng
Thandwe · (BURMA) · Chiang Mai · Phayao · Xaignabouli · Nan · Phônsavan · **HA NOI (Hanoi)** · Nam Định · Chengmai · Qionghai
Pye · Taung-ngu · Lampang · Phrae · Thanh Hoa · **Hainan Dao** · Wanning
Hinthada · Pegu · Uttaradit · **VIENTIANE (Viangchan)** · Ha Tinh · Dongfang
Bassein · Thaton · Udon Thani · Sakon Nakhon · Savannakhét · Vinh · Đông Hoi
Rangoon (Yangon) · Mawlamyaing · Phitsanulok · Khon Kaen · Salavan · **Huê** · Quang Ngai
THAILAND · Lop Buri · Surin · Ubon Ratchathani · Pakxé · **Đa Năng**
Ye (Yai) · Ayutthaya · Nakhon Ratchasima · Sâmraông · Plây Ku
Dawei · **BANGKOK (Krung Thep)** · Chon Buri · **CAMBODIA** · Bátdâmbâng · Quy Nhon · **SOUTH CHINA SEA**
Myeik · Pailin · Pattaya · Pouthisăt · Buôn Ma Thuột
Tenasserim · Chanthaburi · **PHNOM PENH** · Nha Trang
Prachuap Khiri Khan · Kampóng · Spœ · Đa Lat
Myeik Kyunzu (Mergui Archipelago) · Sihanoukville · Đôn Kêv · Biên Hoa · Phan Rang-Thap Cham
Chumphon · **Ho Chi Minh City** · Phan Thiêt
Ranong · Long Xuyên · My Tho
Takua Pa · Surat Thani · Rach Gia · Cân Tho
Phangnga · Nakhon Si Thammarat · Ca Mau · Bac Liêu
Phuket · Krabi · Phatthalung · Mui Ca Mau
Hat Yai · Songkhla
Kangar · Yala · Kota Bharu
Alor Star · Pasir Putih
Sungai Petani · Kuala Terengganu
George Town · Taiping · Gunung Tahan 2189
Ipoh · Kuala Nipis · **MALAYSIA**
KUALA LUMPUR · Kuantan · Natuna Besar · Kepulauan Natuna (Indonesia)
PUTRAJAYA · Seremban · Keluang · Kepulauan Anambas
Melaka · Muar · **Johor Bahru** · Kepulauan Tambelan
Dumai · Minas · **SINGAPORE** · Kepulauan Riau
Pekanbaru · Kuching · Singkawang · Mempawah
Pontianak

BRUNEI · BANDAR SERI BEGAWAN · Labuan
Kota Kinabalu · Gunung Kinabalu · Sandakan
SABAH · Lahad Datu · Semporna · Tawau
Miri · Bintulu · Mukah · Sibu · Tanjungselor · Tarakan
SARAWAK · Sri Aman · Datadian · Tanjungredeb
Debak · Lubok Antu · Sangkulirang · Tolitoli
BORNEO · Muaralaung · **Samarinda** · Donggala
Nangahpinoh · Sukadana · Balikpapan · Palu
Ketapang · Pangkalanbuun · Palangkaraya · Sampit · Amuntai · Kotabaru · Mamuju
Banjarmasin · Martapura · Parepare
Tanjung Selatan · Watampone
INDONESIA · **CELEBES (SULAWESI)** · **Makassar (Ujung Pandang)** · Bontosunggu

Puerto Princesa · Palawan · Brooke's Point · Balabac Strait · Sulu Sea · Banggi · Kudat
MANILA · **Quezon City** · Lucena · Batangas · Mindoro · Calamian Group
San Fernando · Dagupan · Tarlac · Laoag City · Tuguegarao · Vigan · Bontoc · Luzon · Batan Islands · Babuyan Islands
Taytay · Roxas

North Andaman · Middle Andaman · **Andaman Islands (India)** · South Andaman · Port Blair · Nachuge · Little Andaman · Car Nicobar · Nicobar Islands (India) · Dakoank · Great Nicobar

Banda Aceh · Sigli · Bireun · Langsa · Gunung Abongabong 2985 · Pangkalansusu · Simeulue · Medan · Prapat · Sinabang · Gunungsitoli · Sibolga · Nias
SUMATERA · Pangkalpinang · Belinyu · Sungailiat · Manggar
Payakumbuh · Bukittinggi · Bangko · Sungaipenuh · Pulau-pulau Batu · Siberut
Padang · Sijunjung · Jambi · Belitung · Toboali
Sipura · Gunung Kerinci 3805 · Pagai Utara · Pagai Selatan · Sekayu · **Palembang** · Kepulauan Lingga · Selat Karimata
Bengkulu · Lahat · Menggala · Gunung Dempo 3159 · Bintuhan · Kotabumi · Krui · Enggano · **Bandar Lampung**
Teluk Palabuhanratu · Selat Sunda · **JAKARTA** · Serang · Cirebon · Sukabumi · Bandung · **Semarang** · **Surabaya**
Cilacap · **Bandung** · Yogyakarta · Surakarta · Malang · Jember
JAVA (JAWA) · Denpasar · Bali · Mataram · Lombok · **Sumbawa**
Madura · Kepulauan Kangean · Laut Bali (Bali Sea) · Praya · Dompu · Raba · Flo
Pulau Selayar · Tanahjampea · Bonerate · Waikabubak · Sumba

INDIAN OCEAN
Cocos (Keeling) Islands (Australia)
Christmas Island (Australia)
Lesser Sunda Islands

Gulf of Tongking
Gulf of Thailand
Strait of Malacca
Andaman Sea
Irrawaddy · Mouths of the Irrawaddy
Arakan Yoma
Tenasserim
Salween
Mekong · Mouths of the Mekong
Greater Sunda Islands
Natuna Besar
Selat Makassar (Makassar Strait)
Laut Jawa (Java Sea)
Sulu Sea
Celebes Sea
Pegunungan Barisan

15° · 1 · 2 · 3 · 15°
0° Equator
Longitude 105° east of Greenwich

METRES / FEET

METRES	FEET
5000	16404
3000	9843
2000	6562
1000	3281
500	1640
200	656
0	0
Land below sea level	
200	656
4000	13124
6000	19686

Albers Equal Area Conic Projection

Tropic of Cancer

Ryukyu Islands
(Nansei-shotō)
(Japan)

AIWAN

The People's Republic
of China claims Taiwan
as its 23rd Province.

1

Philippine

Sea

Northern
Mariana
Islands
Pagan   **(U.S.A.)**

15°

arri

uzon

CAPITOL HILL ○ □ Saipan
                    ● Tinian

P A C I F I C

O C E A N

llos
nds

PHILIPPINES

Catanduanes

Legazpi
○ Sorsogon
nblon ○ Catarman
Irosin   *Samar*
Masbate ○ Catbalogan
nay   Tacloban
Bacolod
○ *Cebu*
tos   *Bohol*   Surigao
biliaran Butuan
Cagayan de Oro ○
quieta ○ Iligan *Mindanao*
adian ○ Cotabato **Davao**
amboanga   ○ Mati
bela
Moro ● General Santos
Gulf

Rota ○
HAGÅTÑA ○ ⊕
Guam
(U.S.A.)

Mariana Trench

2

Ulithi   *Fais*

Yap ○ ⊕
Colonia   *Faraulep*

● *Ngulu*   *Sorol*

FEDERATED STATES
OF MICRONESIA

C a r o l i n e
I s l a n d s

Eauripik

b e s

a

Kepulauan
Talaud

○ *Sangir*
Kepulauan
Sangir

PALAU
MELEKEOK ⊕ *Babeldaob*

East Caroline
Basin

1:20 000 000

0°

Morotai

0°

Equator

enanjung ○ **Manado**
inahasa
kwandang ○ Tondano ○ *Ternate* ● *Biak*
Gorontalo   Sao-Siu ○
Laut Maluku
(Molucca Sea)
uwuk   *Labuna*
○ *Peleng* *Bacan*
ba ○ **Banggai** *Obi*
pulauan Talaba *Dofa*
nggai *Kepulauan*
Sula

○ *Daruba*
○ *Tobelo*

Halmahera

Waigeo
Selat Dampir   *Kwoka*
Salawati ○ Sorong   ● 3000 *Jazirah*
Misoöl ○ Fafanlap   *Doberai*
      ○ Ransiki
SIA

Pelleluhu
Islands

St Matthias
Group
Mussau Island

Admiralty
Islands ○ Lorengau
Hermit Islands
Wuvulu *Manus Island*
Island

Sabel Channel
New Hanover
○ Kavieng

Umbukul ○
Bismarck Archipelago *Ireland*
Rabaul
○

Bismarck Sea

Manado
Numfoor
Selat Yapen *Yapen*
Manokwari ○ *Biak*
Inanwatan ○ Serui ○ Sarmi ○ Jayapura
Nabire   *Pegunungan Van Rees*
Babo   *Taritatu*
Maprik ○ Sepik
Wewak ○ Bogia ○
Aitape ○ Schouten Islands *Manam Island*
Vanimo

Moluccas
(Maluku)
Seram *Laut Seram (Seram Sea)*
Namlea ○ Piru ○ Bula
Ambon ○ *619* Saparua Fakfak ○
Buru Ambon Adi ○ Kaimana ○
S I A Enarotali ○
Puncak
Jaya   *4884*
Puncak
Trikora
4730
Pegunungan Maoke

Madang ○
Umbon ○

Long
Island

Kimbe ○

New Britain

Lau ○

2438

Ulamona ○

Gasmata ○

PAPUA
GUINEA
Mendi ○

Central Ra. Mt
Wilhelm ○ 4509
Goroka ○
Hagen ○

Huon
Peninsula

Lae ○

Morobe ○

Trobriand
Islands
l'Osuia ○

○ *Manui*
Wowoni
ka ○
○ *Buton*

Kendari

Mangole
Taliabu *Dofa*

Kepulauan
Banggai

Kepulauan
Watubela *Kepulauan*
Kai
○ *Buru* Ambon
Kepulauan Kai Kecil
Kai Tual ○ Dobo ○ Wokam
○ *Sula* *Kepulauan*
Kai Besar Benjina ○
4700 *Puncak*
Mandala
Digul
Lorentz

NEW GUINEA

Kiunga ○

Lake
Murray

Balimo ○

Morehead ○

Kikori ○

Kerema ○

Wau ○

Mount
Victoria
4073 △

Gulf
of Papua

Bereina ○

Kwikila ○

D'Entrecasteaux Is.

Goschen Strait

Bolubolu ○

Alotau ○

Samarai ○

u Island

ores
Sea)

AUSTRALIA

Timor Sea

Laut Banda
(Banda Sea)

Kepulauan
Barat Daya
Kepulauan
Alor *Pulau*
Damar *Romang* ○ *Wuliaru*
Kalabahi ○ Alor *Wetar* *Babar*
antuka ○ Huaki ○ Kaiwatu ○
Maliana ○ Manatuto ○ *Leti*
DILI ○
OCUSSI ○ *Kepulauan*
Sermata *Selaru*
wu   Kefamenanu ○ 2960
(Sea) **EAST**
Timor ○ **TIMOR**
Kupang ○ **(TIMOR-LESTE)**
Rote

○ Tepa
Larat ○
Kepulauan
Saumlakki *Kepulauan Tanimbar*

Tanjung Deyong

Pulau
Dolok

Tanjung Vals

Digul

Merauke ○

Morehead ○

Daru ○

PORT
MORESBY ○ Abau ○

Kwikila ○

A r a f u r a   S e a

Thursday
Island ○
Prince of Wales
Island
Bamaga ○
Cape York

Melville
Island
Bathurst Island *Beagle Gulf*
Van Diemen
Gulf
Milikapiti ○
Batchelor ○ ⊕ Darwin ○ Jabiru ○
Adelaide River *Pine Creek*
Croker Island
Cape Wessel
Wessel Islands
Nhulunbuy ○
Cape Arnhem
Milingimbi ○
Arnhem
Alyangula ○
Land

Cape Wessel

Gulf
of
Carpentaria

Cape Grenville

Weipa ○ Lockhart River ○

Coen ○

Cape York
Peninsula

Cape Melville ○
Cape
Flattery
Cooktown ○
Laura ○

15°

© Collins Bartholomew Ltd

59

500 KILOMETRES
250
0

500
250
MILES 0

THAILAND

Phangnga
Ban Khok Kloi
Thalang
Phuket
Krabi
Thung Song
Khao Chum Thong
Nakhon Si Thammarat

Mui Ca Mau
Nam Căn
Đảo Côn Sơn

VIETNAM

SOUTH CHI

Trang
Phatthalung
Thale Luang
Hat Yai
Songkhla
Satun
Sadao
Pattani
Yala
Kangar
Langkawi
Rangae
Narathiwat
Kota Bharu
Pasir Putih
Alor Star
Sungai Petani
Pinang
Butterworth
George Town
Taiping
Kuala Kerai
Kuala Terengganu

Andaman Sea

Pulau We
Sabang
Banda Aceh
Sigli
Bireun
Calang
Lhokseumawe
Peureula
Takengon
Gunung Abongabong △2985
Langsa
Pangkalansusu
Blangkejeren
Gunung Leuser △3145
Binjai
Belawan
Tapaktuan

Strait of Malacca

MALAYSIA

Kuala Kangsar
Ipoh
Gunung Tahan △2189
Tasik Kenyir
Dungun
Kampar
Kuala Lipis
Cukai
Teluk Intan
Bagan Datuk
Kuantan

PENINSULAR MALAYSIA

KUALA LUMPUR
Klang
PUTRAJAYA
Temerluh
Pekan

Medan
Tebingtinggi
Pematangsiantar
Sidikalang
Nisaran
Tanjungbalai
Prapat
Danau Toba
Labuhanbilik
Balige
Rantauprapat
Bagansiapiapi
Dumai
Muar
Batu Pahat
Bengkalis
Mersing
Keluang

Sibolga
Gunungsitoli
Padangsidimpuan
Gunungtua
Rokan
Duri
Daludalu
Padang Endau

SINGAPORE
Johor Bahru
Batam

Simeulue

Pulau-pulau Banyak

Nias
Sirombu
Telukdalam

Minas
Pekanbaru
Bangkinang
Kampar

Tanjungpinang
Kepulauan Riau
Bintan

Kepulauan Tambelan (Indonesia)

Natuna Besar
Panarik

Kepulauan Natuna (Indonesia)

Subi Besar

Selat Serasan

Liku
Sema
Sambas
Pemangkat
Kuchi
Siluas
Singkawang
Bengkayan
Mempawah
Ngabang
Pontianak

Hutanopan
Natal
Airbangis
Talu

Tembilahan
Rengat

Singkep
Daik
Kepulauan Lingga

Lingga

Balaiberk
Kubu
Telukbatang

Telo
Pulau-pulau Batu
Payakumbuh
Padangpanjang
Bukittinggi
Padang
Solok
Sijunjung
Kualatungal
Simpang

Kagologolo
Siberut
Painan
Muarabungo
3805
Gunung Kerinci
Batanghari
Jambi

Belinyu
Sungailiat
Mentok
Pangkalpinang
Bangka
Koba

Pulau-pulau Karimata
Ketapang
Suka
Kendawang

Muarasiberut
Sipura
Kaliet
Sungaipenuh
Bangko
Sarolangun
Muaratembesi

Pagai Utara
Mukomuko
Pagai Selatan
Buriai
Surulangun
Rajik
Tanjungpandan
Manggar
Dendang
Belitung

S U M A T E R A

Pegunungan Barisan

Lubuklinggau
Tebingtinggi
Sekayu
Plaju
Palembang
Kayuagung
Prabumulih

Selat Karimata

Tan Sari

Cusup
Bengkulu
Gunung Dempo △3159
Lahat
Musi
Menggala
Muaradua
Martapura
Kotabumi

Mega
Bintuhan
Gunung Resagi △2232
Metro
Kotaagung
Krui
Bandar Lampung

Enggano

Tanjung Cina
Deli
Teluk Semangka
Panaitan
Krakatau
Selat Sunda
Rangkasbitung
Sukabumi
Serang
Karawang
JAKARTA
Bogor 3019
Bandung
Garut
Ciamis

Tanjung Indramayu
Cirebon
Tegal
Pekalon
Gunung Slamet △3428
Temanggu

I N D I A N   O C E A N

Teluk Palabuhanratu
Sindangbarang
Cilacap
Kebum

J A V A (J A W A)

I N D
L A U T
(J A V
(J A W A

G r e a t e r   S u n d a   I s l

Kepulauan Mentawai

Equator 0°

2

10°

A Longitude 100° east of Greenwich B
Albers Equal Area Conic Projection

METRES FEET
5000 16404
3000 9843
2000 6562
1000 3281
500 1640
200 656
0 0
Land below sea level
200 656
4000 13124
6000 19686

MALAYSIA AND INDONESIA WEST

Palawan Rio Tuba
Bugsuk
Balabac
Balabac
SULU
Balabac Strait Banggi
SEA
Presidente Manuel A Roxas Oroquieta
Liloy Ozamis Iligan
Siocon Pagadian
Zamboanga Peninsula
Zamboanga Cotabato
Sulu Moro Datu Piang
Kudat Isabela *Gulf*
Kanibongan *Cagayan de Tawi-Tawi* *Basilan* Lebak
Kota Belud *Archipelago*
Kota *Turtle Islands* Jolo Jolo **PHILIPPINES**
Kinabalu *Gunung* *(Philippines)* Siasi
Beaufort *Kinabalu* Sandakan
△4095 Ranau Tambisan
Labuan *Gunung Trus Madi* Lamag Panglima
BANDAR SERI *△2649* Lahad Sugala *Tawi-Tawi*
BEGAWAN Tenom Kuamut Datu
BRUNEI Lawas *Sibutu*
Kuala Belait **SABAH** Semporna
Lutong Tomani Pensiangan
Seria *Lumbis* Tawau
Miri *Bukit Harden* Mensalong **CELEBES**
2136 **SEA**
Bintulu Labang Long Kubuang
Igan Mukah Akah
Tanjung Sirik Belaga Tarakan
Sibu **SARAWAK** Tanjungselor
Sarikei Kapit
Saratok Datadian
Debak *Rajang* Tanjungredeb
Sri Aman *△2988*
Lubok **BORNEO** Gunung Menyapa
Antu *△2000* Sepinang
Semitau Putusibau *Telen* Sangkulirang *Tanjung* Tolitoli Kwandang
Sintang *Mahakam* *Mangkalihat* *Semenanjung Minahasa*
Longiram Bontang Gorontalo
Tenggarong Sidoan *Kepulauan Togian*
Muaralaung Moutong
Samarinda *Teluk* *Togian* *Tanjung*
Tewah Muarateweh Tomali *Tomini* *Pangkalsiang*
Samboja *Batudaka* Luwuk
Rantaupanjang Balikpapan Donggala Peleng
KALIMANTAN Palu Tataba
Palangkaraya Tanahgrogot Mapane Poso Banggai
Sampit Tanjung Tenteno Uekuli *Kepulauan*
Amuntai *Bukit △3074* Kolonedale *Banggai*
Kualapembuang Kandangan Babana Masamba Rantepang
Kotabaru Mamuju Wotu
Martapura *Sebuku* Somba *Gandawata* Palopo
Banjarmasin Pagatan Majene Makale Malili Manui
Laut Polewali Malamala
Tanjung Parepare Anabanua Kendari
Selatan **CELEBES** Kolaka Wowoni
Tanjung **(SULAWESI)** Sengkang
Puting Watampone Raha
Kepulauan Maros Muna Buton
I N D O N E S I A *Laut Kecil* Sinjai
Makassar *Gunung Lompobattang* Baubau
(Ujung Pandang) Bulukumba
J A W A Bontosunggu
(S E A) *Bawean* *Masalembu* Benteng *Pulau Selayar*
Besar *Batuata*
Tanjung
Bugel *Kepulauan*
Kangean
Tuban *Madura* Arjasa Sabalana *Tanahjampea* Kalao *Kalaotoa*
Bangkalan Sumenep *Kepulauan* *Kepulauan Solor*
Surabaya Raas *Tengah* *Kepulauan Bonerate*
Jombang *Genteng* Situbondo *Laut Bali* Reo **Flores** Larantuka
Pasuruan *(Bali Sea)* *Sumbawa* *Laut Flores* Maumere Labala
Malang Banyuwangi *Gunung* *(Flores Sea)*
△3676 Singaraja *Tambora △2821* Raba Labuhanbajo
Lumajang Jember Gianyar *3142* Alas Dompu Bajawa Ende
Denpasar Mataram Sumbawabesar Ruteng
Bali Praya Taliwang Plampang *Selat Sumba* *Laut Sawu*
Sumbawa *Lombok* *Selat Lombok* Memboro *(Savu Sea)*
s **Sumba** Waikabubak Waingapu

METRES 400 00 KILOMETRES
200
0
300
200
100
0
1:9 000 000

© Collins Bartholomew Ltd

METRES
FEET

METRES	FEET
5000	16404
3000	9843
2000	6562
1000	3281
500	1640
200	656
0	0
Land below sea level	
200	656
4000	13124
6000	19686

Major regions and countries:

HUNAN · GUIZHOU · GUANGXI · ZHUANGZU ZIZHIQU · HAINAN · CHINA · SICHUAN · YUNNAN · MYANMAR (BURMA) · INDIA · ARUNACHAL PRADESH · NAGALAND · MANIPUR · ASSAM · MEGHALAYA · TRIPURA · MIZORAM · BANGLADESH · BHUTAN · LAOS · VIETNAM · Gulf of Tongking · BAY OF BENGAL

Selected cities and features:

Changde · Kunming · Guiyang · Zunyi · Nanning · Beihai · Zhanjiang · Haikou · Sanya · HA NOI · Hai Phong · VIENTIANE · Mandalay · NAY PYI TAW · Dhaka (Dacca) · Chittagong · Guilin · Liuzhou · Wuzhou · Luangphabang

Hengduan Shan · Nu Shan (Lancang) Jiang · Mekong (Lancang) Jiang · Salween (Nu Jiang) · Wuliang Shan · Shan Plateau · The Triangle · Kumon Ra. · Irrawaddy · Arakan Yoma · Pegu Yoma

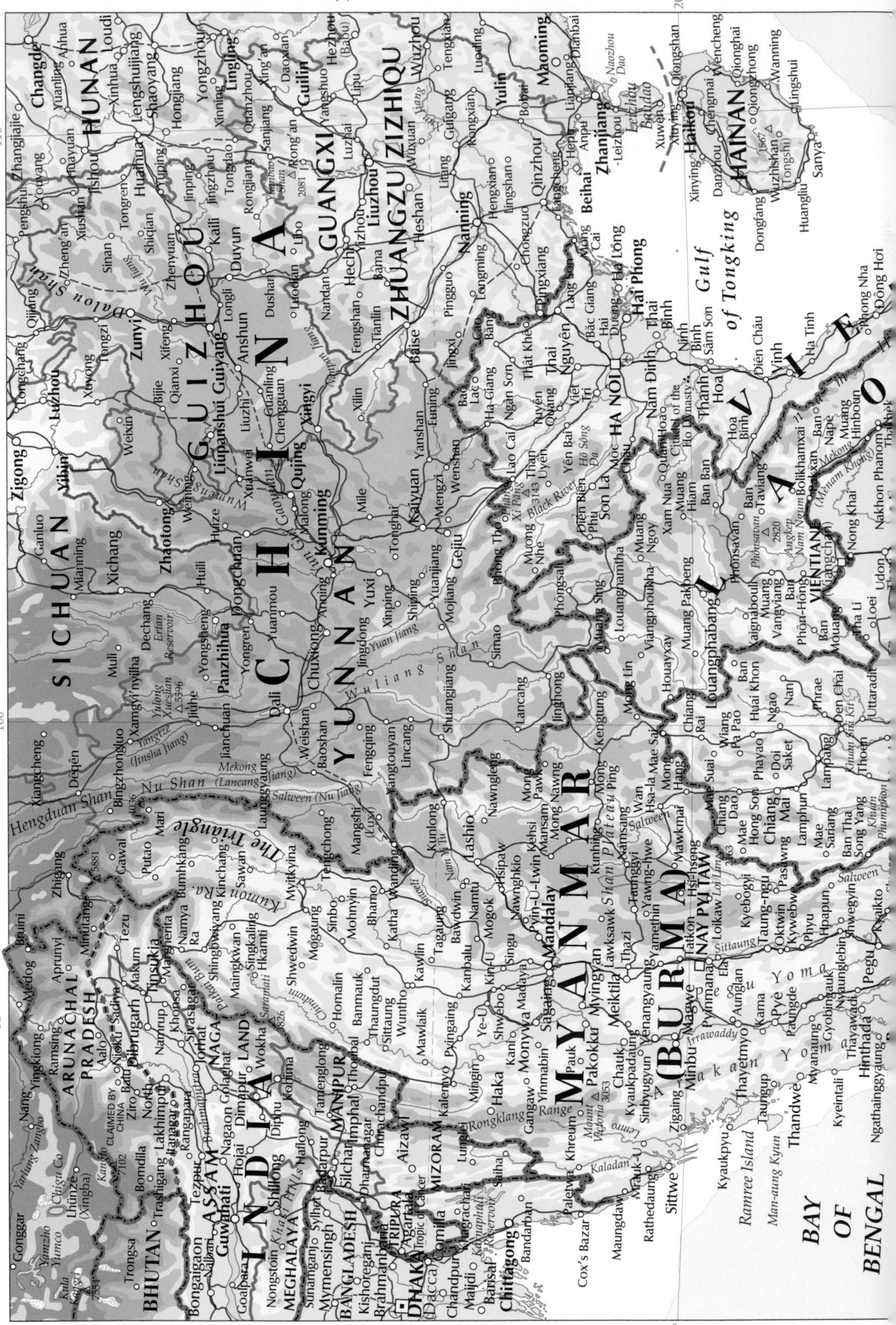

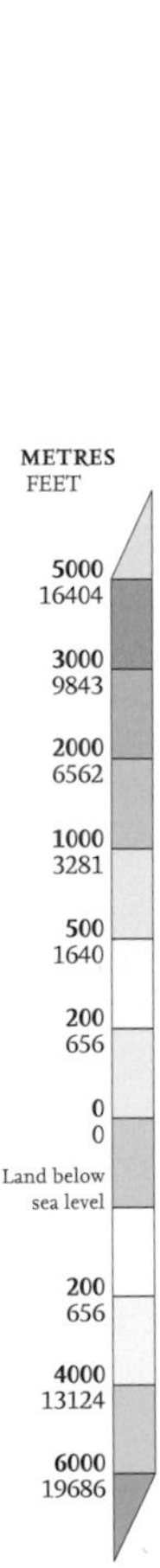

Albers Equal Area Conic Projection

VIÊT NAM

Da Nang
Hôi An
Nouei
Hue

Quang Ngai
Bông Son
Quy Nhon
Sông Cau
Tuy Hoa
Ninh Hoa
Nha Trang
Ba Ngoi (Cam Ranh)
Vinh Cam Ranh
Phan Rang-Thap Cham
Phan Thiêt
Da Lat

S O U T H

C H I N A

S E A

Laut (Indonesia)

INDO-CHINA

Salavan
Muang
Khôngxêdôn
Pakxé
Phôiphisai
Bolavên

Ban Tôp
Khemmarat
Muang Phin
Mangsahan
Ubon Ratchathani
Khu Khan
Kantaralak

Ngok Linh 2598
Kon Tum
Pláy Ku
Chu Sê
Buôn Ma Thuôt
Chu Yang Sin 2405
Kon Chro'
Krông Nô
Duc Xuyên
Đức Bôn
Nghia
Bao Lôc
Biên Hoa
Vung Tau
Ho Chi Minh City (Saigon)

Attapu
Xékong
Sekong
Stoeng Trêng
Sênmônourôm
Viroche
Phumi Kâmpóng Trâbêk
Prêk Kak
Krâchéh

CAMBODIA

PHNOM PENH

Temple of Preah Vihear
Preah Vihear
Samraong
Rôviĕng
Siĕm Réab
Thong
Stung
Tônle Sap
Kâmpóng Cham
Prey Vêng
Ta Khmau

Sisophôn
Bâtdâmbâng
Pouthisăt
Kâmpóng Chhnang
Kâmpóng Spœ
Don Kêv

Tây Ninh
Tân Châu
Môc Hoa
Châu Dôc
Long Xuyên
Sa Đéc
Rach Gia
Mui Ca Mau
Ca Mau
Nam Căn
Bac Liĕu
Soc Trang
Can Tho
Vinh Long
Bên Tre
My Tho
Long An
Tân An

THAILAND

Savannakhet
Maha Sarakham
Yasôthon
Phimun Mangsahan
Sisaket
Surin
Buriram
Nakhon Ratchasima
Phon
Chaiyaphum

Ban Phai
Pak Thong Chai
Khon Kaen
Nang Rong
Khorat

Roi Et

Ko Kong
Krông Kaôh Kong
Sihanoukville
Kâmpôt
Don Kêv

Gulf
of
Thailand

Khlung
Trat
Ko Chang

Khao Phit (Dong Phraya Yen)
San Khao Phit

Phichit
Nakhon Sawan
Sing Buri
Lop Buri
Ayutthaya
Nakhon Pathom
Thon Buri
Sara Buri
Nakhon Nayok
Chanthaburi
Rayong
Pattaya
Phet Buri
Prachuap Khiri Khan

BANGKOK (Krung Thep)

Suphan Buri
Nam Tok
Kanchanaburi
Three Pagodas Pass
RatBuri
Samut Songkhram
Bang Saphan Yai
Chumphon

Kadonkani
Bogale
Labutta
Pyapon
Thanbyuzayat
Thayetchaung
Thagyettaw
Onbingwin
Dawei
Migyaunglaung
Ye (Yai)
Winkana
Kyaiti
Seikvi
Kyaui
Mudon
Mawlamyaing

Mouths of the Irrawaddy
Gulf of Martaban

Cape Negrais
Preparis North Channel
Preparis South Channel
Preparis Island
Great Coco Island
Narcondam Island
Barren Island

Andaman
Sea

Mergui Archipelago
(Myeik Kyunzu)

Palaw
Myeik
Kawthaung
Tenasserim
Bokpyin
Kawthaung
Kra Buri
Ranong

Letsok-aw Kyun
Lanbi Kyun
Zadetkyi Kyun

Isthmus of Kra

Chaiya
Surat Thani
Ban Na San
Ban Chiang
Krabi
Thung Song
Trang
Phatthalung
Thale Luang
Songkhla
Hat Yai
Sadao
Satun
Langkawi

Nakhon Si Thammarat
Sichon
Khao Chum Thong
Phangnga
Thalang
Phuket
Takua Pa
Ban Khok Kloi

Ko Samui
Ko Phangan

MALAYSIA

Pattani
Narathiwat
Yala
Rangae
Kota Bharu
Pasir Putih
Kuala Terengganu
Dungun
Kuala Kerai
Tasik Kenyir
Gunung Tahan 2189
Kuala Kangsar
Taiping
Butterworth
George Town
Pinang
Ipoh
Alor Star
Sungai Petani
Kangar

INDONESIA

Banda Aceh
Sabang
Pulau We
Sigli
Sangsa
Bireun
Peureula
Lhokseumawe
Takengon
Calang
Gunung Abongabong 2985

North Andaman
Interview Island
Middle Andaman
North Sentinel Island
South Andaman
Ritchie's Archipelago
Wrightmyo
Port Blair
Nachuge
Little Andaman

Andaman Islands (India)

Ten Degree Channel

Car Nicobar
Tilanchong Island
Camorta
Nancowry
Teressa Island
Katchall

Nicobar Islands (India)

Dakoank
Little Nicobar
Great Nicobar

INDIAN
OCEAN

10°
10°
110°
100°

Longitude 100° east of Greenwich

A **B** **C**

1: 9 600 000

MILES 0   100   200   300

0   200   400 KILOMETRES

© Collins Bartholomew Ltd

A 120° **B**

1

20°

Pratas
Islands
*Administered by Taiwan,
claimed by China*

*Batan
Islands*
Itbayat
Basco
Batan

*Luzon
Strait*

Balintang Channel

Babuyan

Calayan
*Babuyan
Islands*
Camiguin

Bangui
Fuga
Camiguin

**Laoag
City**
Babuyan Channel
San Vicente

Bangued
Aparri

Vigan
Tuguegarao
*Mount
Sapocoy*
Ilagan

Tagudin
Bontoc
Palanan

*Mount
Pulog*
Santiago

San Fernando
2929

La Trinidad
Bayombong

Dagupan
Baguio
LUZON

Lingayen
San Carlos

Tarlac
San Jose

Mount Pinatubo
Cabanatuan

Iba 1660
Gapan

Angeles
San Fernando

Olongapo
Valenzuela
Polillo Islands

Balanga
Quezon City

MANILA
Pasig

Tagaytay City
Santa Cruz
Labo

San Pablo
Lucena
Daet
Pandan

*Lubang
Islands*
Batangas
Lopez
Libmanan
Catanduanes

Calapan
Boac
Naga
Virac

*Mount
Halcon*
Naujan
Oas
Tabaco

Mamburao
2585
Mayon
Legazpi
2421

Mindoro
Roxas
Sibuyan
Sorsogon

Mindoro Strait
San Jose
Romblon
Irosin

Catarman

*New
Busuanga*
Tablas
Masbate
Calbayog

*Calamian
Group*
Pandan
Samar

Culion
Coron
*Sibuyan
Sea*
Masbate
Catbalogan

Roxas
*Visayan
Sea*
Tacloban

El Nido
Limpacan
Culasi
Panay
Ormoc
Guiuan

Taytay
*Cuyo
Islands*
*Cordillera
Range*
Pototan
Cadiz
Leyte
Dinagat

*Dalanganem
Islands*
Bacolod
Siargao

**San Jose de
Buenavista**
Iloilo
2450
Cebu
Maasin
Dapa

Dumaran
Negros
Cebu

Bohol
Surigao

Palawan
Roxas
Talisay

Cauayan
*Bohol
Sea*
Tagbilaran
Tandag

Puerto Princesa
Tanjay
Siquijor
Camiguin
Mambajao

Apurahuan
Bayawan
Butuan

Aborlan
Dumaguete
**Cagayan
de Oro**
Gingoog

*Mount
Mantalingajan*
**Presidente
Manuel A Roxas**
Dapitan

2054
Brooke's Point
Oroquieta
Iligan
Malaybalay
Bislig

Rio Tuba
Liloy
MINDANAO
Baganga

Bugsuk
Ozamis
Mount Ragang
2815
Tagum

Balabac
Pagadian
*Mount
Apo*
Davao

Balabac
Siocon
Cotabato
2954
Digos
Mati

Balabac Strait
*Zamboanga
Peninsula*
Datu Piang
*Davao
Gulf*

Banggi
*Cagayan de
Tawi-Tawi*
Zamboanga
*Moro
Gulf*
Lebak
Banga

Kudat
Isabela
General Santos

Kota Belud
*Turtle Islands
(Philippines)*
Basilan
Kiamba
Batulaki

**Kota
Kinabalu**
*Gunung
Kinabalu*
Jolo
Kanibongan

Gunung
4095
Jolo
Sarangani Islands
*Kepulauan
Nanusa*

Ranau
Sandakan
Siasi

Trus Madi
Lamag
Tambisan
Karakelong
*Kepulauan
Talaud*

Tenom
2649
**Panglima
Sugala**
Sulu Archipelago
Pulutan

Lawas
Sibutu
Tawi-Tawi
MALAYSIA

Tomani
SABAH
**Lahad
Datu**
*CELEBES
SEA*
INDONESIA

Kuamut
Pensiangan
Semporna
Sangir
Tahuna

Lumbis
Tawau
Kaburuang

INDONESIA
Mensalong

Kubuang
Tarakan

A Longitude 120° east of Greenwich **B**

*SOUTH
CHINA
SEA*

SULU SEA

*Palawan
Passage*

Albers Equal Area Conic Projection

Scarborough
Reef
*Claimed by China,
Taiwan and Philippines*

200 KILOMETRES
100
0

150
100
0
MILES

1:9 600 000

**METRES
FEET**

5000
16404

3000
9843

2000
6562

1000
3281

500
1640

200
656

0
0

Land below
sea level

200
656

4000
13124

6000
19686

*PHILIPPINE
SEA*

PHILIPPINES

64

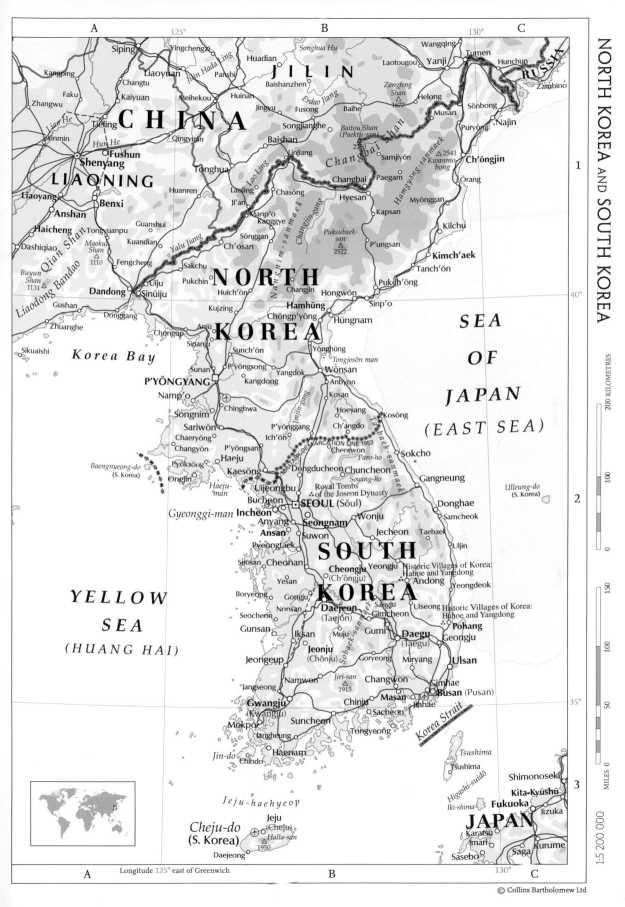

A   125°   B   130°   C

JILIN

Siping
Yingchengxi
Huadian
Songhua Hu
Wangqing
Tumen
Hunchun
RUSSIA
Kangping
Liaoyuan
Jilin Hada Ling
Baishanzhen
Laotougou
Yanji
Zarubinó
Faku
Changtu
Meihekou
Huinan
Zengfeng
Shan
1677
Helong
Musan
Sönbong
Zhangwu
Kaiyuan
Jingyu
Fusong
Baihe
Puryöng
Najin
Xinmin
Liao He
Tieling
Qingyuan
Baishan
Songjianghe
Baitou Shan
(Paektu-san)
2750
Samjiyön
Örang
Hun He
Fushun
Tonghua
Linjiang
Changbai Shan
2541
Kwanmo
bong
Ch'öngjin
Shenyang
LIAONING
Huanren
Laoling
Ji'an
Changbai
Paegam
Myönggan
Kilchu
Liaoyang
Benxi
Guanshui
Chasöng
Hyesan
Kapsan
Anshan
Kuandian
Manp'o
Kanggye
Puksubaek-
san
2522
P'ungsan
Tanch'ön
Kimch'aek
Haicheng
Tongyuanpu
Sönggan
Ch'osan
NORTH
Dashiqiao
Maokun
Shan
1110
Fengcheng
Yalu Jiang
Sakchu
Pukchin
Huich'ön
Changjin
Hongwön
Pukch'ong
Qian Shan
Buyun
Shan
1131
Dandong
Uiju
Sinüiju
Kujzing
KOREA
Sinp'o
Gushan
Donggang
Huich'ön
Anju
Chöngp'yöng
Hamhüng
Hüngnam
Sikuaishi
Korea Bay
Sinanju
Sunch'ön
Yönghüng
Tongjosön man
Liaodong Bandao
Zhuanghe
Chöngup
Yangdok
Wönsan
SEA
OF
JAPAN
(EAST SEA)
Sunan
P'yöngsong
Kangdong
Anbyön
P'YÖNGYANG
Chinghwa
Kosan
Namp'o
Hoeyang
Songnim
Sariwön
P'yönggang
Ch'ando
Kosöng
Chaeryöng
Ich'ön
MILITARY DEMARCATION LINE 1953
Sokcho
Changyön
P'yöngsan
Cheorwon
Paro-ho
Pyöksöng
Haeju
Kaesöng
Dongducheon
Chuncheon
Gangneung
Baengnyeong-do
(S. Korea)
Ongjin
Haeju-
man
Uijeongbu
Soyang-ho
Donghae
SEOUL (Söul)
Wonju
Samcheok
Gyeonggi-man
Bucheon
Royal Tombs
of the Joseon Dynasty
Taebaek
Ulleung-do
(S. Korea)
Incheon
Anyang
Seongnam
Jecheon
Uljin
Ansan
Suwon
SOUTH
Yeongju
Historic Villages of Korea:
Hahoe and Yangdong
Andong
YELLOW
Seosan
Cheonan
Cheongju
(Ch'öngju)
Yeongdeok
SEA
Boryeong
Gongju
KOREA
Sangju
Uiseong Historic Villages of Korea:
Hahoe and Yangdong
(HUANG HAI)
Yesan
Nonsan
Daejeon
(Taejön)
Gimcheon
Gumi
Pohang
Seocheon
Gunsan
Iksan
Muju
Goryeong
Geongju
Jeonju
(Chönju)
Daegu
(Taegu)
Jeongeup
Namwon
Jiri-san
1915
Changwon
Miryang
Ulsan
Jangseong
Chinju
Gimhae
Busan (Pusan)
Gwangju
(Kwangju)
Suncheon
Masan
Jinhae
Mokpo
Sacheon
Korea Strait
Tsushima
Haenam
Tongyeong
Tsushima
Jin-do
Chindo
Higashi-suidö
Shimonoseki
Kita-Kyūshū
Jeju-haehyeop
Iki-shima
Fukuoka
Iizuka
Cheju-do
(S. Korea)
Jeju
(Cheju)
Halla-san
1950
JAPAN
Karatsu
Daejeong
Imari
Kurume
Saga
Sasebo

A   Longitude 125° east of Greenwich   B   130°   C

200 KILOMETRES

1:5 000 000

© Collins Bartholomew Ltd

65

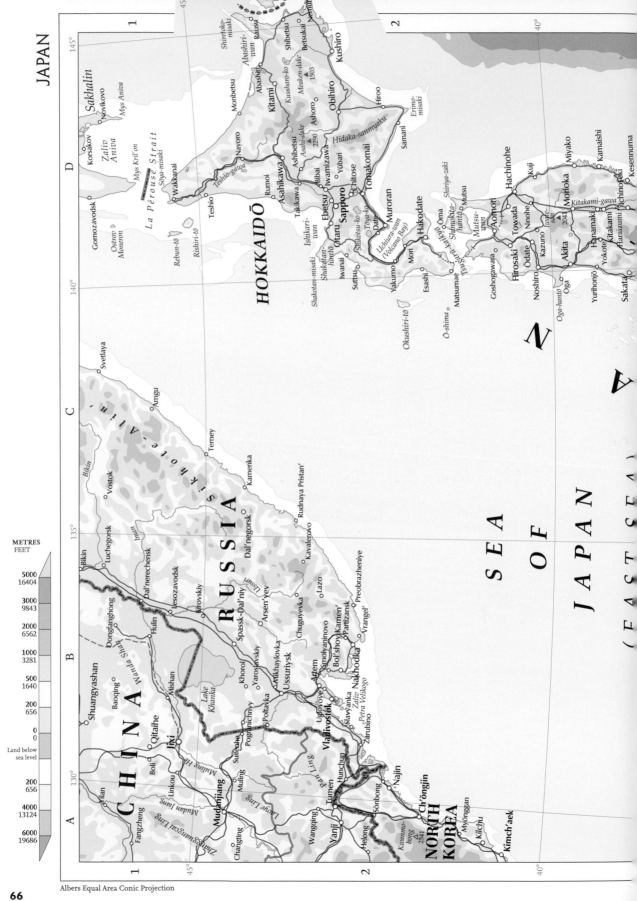

JAPAN

45°

145°

Sakhalin

Shiretoko-misaki
Rausu
Abashiri-wan
Shibetsu
Betsukai
Nemuro

Korsakov
Novikovo

Mys Aniva

Kussharo-ko
Meaken-dake
1503
Kushiro

Monbetsu

Zaliv
Aniva

Gomozavodsk

Mys Kril'on

Abashiri
Kitami
Ashoro
Obihiro
Hiroo

Sōya-misaki

Ostrov
Moneron

La Pérouse Strait

Wakkanai

Nayoro
Asahi-dake
2290
Erimo-misaki

Teshio-gawa
Shibetsu
Asahi-dake
Hidaka-sammyaku
Samani

Rebun-tō

Teshio

Rumoi
Asahikawa
Ashibetsu
Iwamizawa
Yūbari

Rishiri-tō

Takikawa
Ebetsu
Bibai
Chitose
Tomakomai
Shiriya-zaki

HOKKAIDŌ

Ishikari-wan
Otaru
Sapporo
Muroran

Ōma
Mutsu
Shimokita-hantō

Shakotan-misaki
Date
Shikotsu-ko
Tōya-ko
Uchiura-wan
(Volcano Bay)
Hakodate
Ōhata
Shimokita-
wan
Mutsu-wan

Iwanai

Hachinohe
Kuji

Shakotan-
hantō
Suttsu
Yakumo
Mori
Tsugaru-kaikyō

Hirosaki
Aomori
Towada
Ninohe

Morioka
Miyako

Esashi
Matsumae

Odate
Kazuno
Iwate-
san
204

Kitakami-gawa
Ichinoseki

Okushiri-tō
Ō-shima

Noshiro
Oga
Yurihonjō
Akita
Yokote
Hanamaki
Kitakami
Haraizumi
Kamaishi
Kesennuma

Oga-hantō

Sakata

N

A

Svetlaya

Amgu

Terney

Kamenka

Rudnaya Pristan'

Sikhote-Alin'

Bikin

Vostok

Dal'negorsk

Kavalerovo

Preobrazheniye

SEA

Luchegorsk

Iman

Krasnyy Yar

Bikin

Dal'nerechensk

Kirovsky

Ussuri

Spassk-Dal'niy

Arsen'yev

Chuguyevka

Lazo

Vrangel'

OF

RUSSIA

Khorol

Yaroslavskiy

Mikhaylovka

Smolyaninovo

Kamen

Bol'shoy

Partizansk

JAPAN

Dongfanghong

Hulin

Mishan

Lake
Khanka

Ussuriysk

Artem

Nakhodka

(EAST SEA)

CHINA

Shuangyashan

Baoqing

Qitaihe

Jixi

Wanda Shan

Suifenhe
Pogranichnyy

Poltavka

Ugolovye
Vladivostok

Slavyanka
Zaliv
Petra Velikogo
Zarubino

Boli

Linkou
Muling

Laoye Ling

Hunchun

Najin

Fangzheng

Mudanjiang

Muling

Mudan Jiang

Tumen

Sŏnbong
Chʻŏngjin

Changting

Wangqing

Yanji

Helong

Sŏngbong

Myŏnggan

NORTH
KOREA

Xilan

Zhangguangcai Ling

Kwanmo-
bong
2541

Kilchu

Kimchʻaek

Albers Equal Area Conic Projection

D

C

B

A

40°

2

1

130°

45°

40°

135°

140°

METRES
FEET

5000
16404

3000
9843

2000
6562

1000
3281

500
1640

200
656

0
0

Land below
sea level

200
656

4000
13124

6000
19686

JAPAN

SOUTH KOREA

PACIFIC OCEAN

HONSHŪ

SHIKOKU

KYŪSHŪ

Ulju

Tsushima

Liancourt Rocks
Claimed and administered
by South Korea as Dok-do;
claimed by Japan as Take-shima

Ulleung-do
(S. Korea)

Oki-shotō
Dōgo
Dōzen
Okinoshima

Gōtsu
Hamada
Masuda
Hagi
Nagato
Yamaguchi
Shimonoseki
Kita-Kyūshū
Fukuoka
Saga
Karatsu
Imari
Iki-saima
Sasébo
Amakusa-Shimo-shima
Nagasaki
Ōmuta
Isahaya
Koshikijima-rettō
Satsuma-Sendai
Makurazaki

Tsushima
Tsushima
Hirado-shima

Matsue
Izumo
Kurayoshi
Tottori
Niimi
Shobara
Hiroshima
Iwakuni
Ube
Hōfu
Iizuka
Kurume
Yatsushiro
Kumamoto
Nishino-omote
Kanoya
Kagoshima
Ōsumi-shotō
Yaku-shima
Tanega-shima

Chūgoku-sanchi
Hyōno-san △ 1510
Tsuyama
Akō
Himeji
Okayama
Kurashiki
Mihara
Kure
Matsuyama
Imabari
Iyo
Uwajima
Ōita
Beppu
Saiki
Nobeoka
Hyūga
Miyazaki
Miyakonojō
Usa
Usuki
Kuju-san △ 1788
Bungo-suidō
Nankoku
Kōchi
Tosashimizu
Ashizuri-misaki
Tosa-wan

Suō-nada

Niihama
Sakaide
Takamatsu
Naruto
Tokushima
Ishizuchi-san △ 1981
Anan
Kainan
Kii-suidō
Muroto
Muroto-zaki

Kurayoshi
Akashi
Kōbe
Ōsaka
Sakai
Kishiwada
Wakayama
Tanabe
Gobō
Shingū
Kumano

Kii-sanchi
Ise-wan
Ise
Tsu
Iwase
Matsusaka
Owase
Shio-no-misaki

Maizuru
Wakasa-wan
Tsuruga
Biwa-ko
Kyōto
Kameoka

Fukui
Sabae
Echizen (takefu)
Ono
Komatsu
Kaga
Kanazawa
Takaoka
Toyama
Takayama
Noto-hantō
Wajima
Suzu-misaki
Suzu
Nanao
Himi

CHŪBU

Matsumoto
Ueda
Nagano
Myōkō
Jōetsu
Kashiwazaki
Nagaoka
Sanjō
Niitsu
Niigata
Shibata
Ryōtsu
Sado-shima
Murakami

Takasaki
Maebashi
Kiryū
Ōta
Kumagaya
Kawagoe
Sagamihara
Hachiōji
Chino
Kōfu
Fuji-san △ 3776
Shirane-san △ 3192
Shiranesugidaira
Mt Shirane-san △ 2578

Nasushiobara
Utsunomiya
Ōtawara
Kōriyama
Aizuwakamatsu
Yamagata
Yonezawa
Natori
Sendai

Kinka-san

Murakami

Uonuma

Nagoya
Komaki
Gifu
Nakatsugawa
Toyota
Okazaki
Ina
Okaya
Suwa
Iida

Hamamatsu
Toyohashi
Matsusaka
Toyokawa
Shizuoka
Fuji
Yaizu
Shimizu

Numazu
Odawara
Atami
Itō
Shimoda
Iro-zaki

TOKYO
Kawasaki
Yokohama
Ōmiya
Tsuchiura
Narita
Chōshi
Tateyama
Nojima-zaki

Mito
Hitachi
Hitachinaka
Kitaibaraki
Iwaki
Fukushima
Kōriyama

Kashima-nada

O-shima
Nii-jima
Miyake-jima
Mikura-jima
Izu-shotō
Hachijō-jima
Aoga-shima
Sumisu-jima
Tori-shima

PACIFIC OCEAN

SOUTH KOREA

1:6 000 000

Longitude 135° east of Greenwich

MILES 0   50   100   150

0   100   200 KILOMETRES

© Collins Bartholomew Ltd

67

3   35°   4

130°   35°   140°

B   C   D

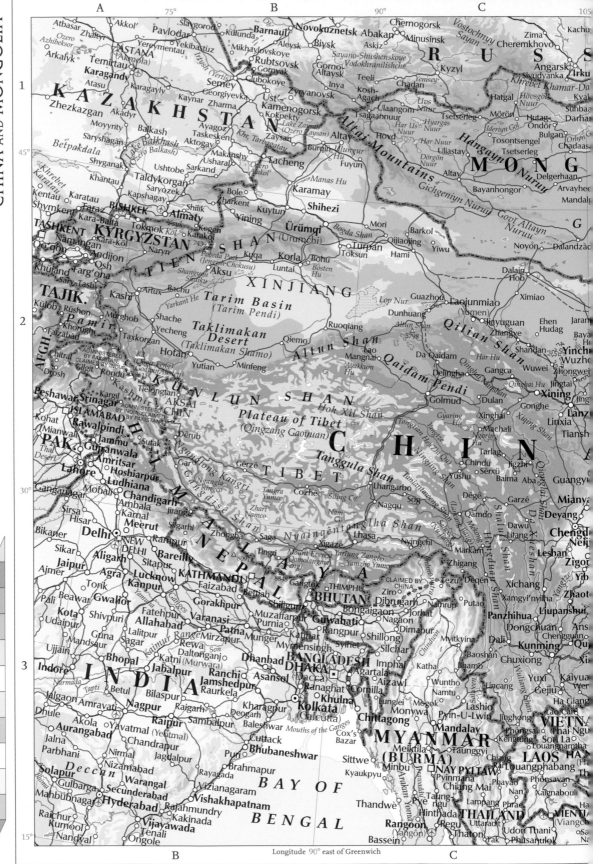

METRES
FEET

5000
16404

3000
9843

2000
6562

1000
3281

500
1640

200
656

0
0

Land below
sea level

200
656

4000
13124

6000
19686

Albers Equal Area Conic Projection

Longitude 90° east of Greenwich

© Collins Bartholomew Ltd

C

1

2

Korea
Bay

Bo
Hai

Liaodong
Wan

Yellow Sea
(Huang Hai)

Shandong
Bandao

Laizhou
Wan

Bohai
Wan

Hangzhou
Wan

MONGOLIA

NEI MONGOL ZIZHIQU
(INNER MONGOLIA)

Yin Shan

LIAONING

HEBEI

SHANXI

SHANDONG

HENAN

JIANGSU

ANHUI

NINGXIA HUIZU ZIZHIQU

GANSU

SHAANXI

HUBEI

CHONGQING

SICHUAN

Helan Shan

Tengger
Shamo

Badain
Jaran Shamo

Ordos
(Mu Us
Shadi)

Huangtu Gaoyuan

Lüliang Shan

Taihang Shan

Qin Ling

Qionglai
Shan

Yellow River
(Huang He)

Great Wall

Yangtze
(Chang Jiang)

Grand Canal

BEIJING
TIANJIN (Tientsin)
SHANGHAI

Shenyang
Fushun
Benxi
Anshan
Haicheng
Liaoyang
Dandong
Dalian
Qinhuangdao
Tangshan
Cangzhou
Shijiazhuang
Baoding
Datong
Taiyuan
Hohhot
Baotou
Yinchuan
Lanzhou
Xining
Xi'an
Zhengzhou
Luoyang
Kaifeng
Qingdao (Tsingtao)
Yantai
Weihai
Weifang
Jinan
Zibo
Linyi
Xuzhou
Nanjing
Wuxi
Suzhou
Changzhou
Hangzhou
Ningbo
Hefei
Wuhan
Chongqing
Chengdu
Mianyang
Deyang
Nanchong

METRES / FEET

METRES	FEET
5000	16404
3000	9843
2000	6562
1000	3281
500	1640
200	656
0	0
Land below sea level	
200	656
4000	13124
6000	19686

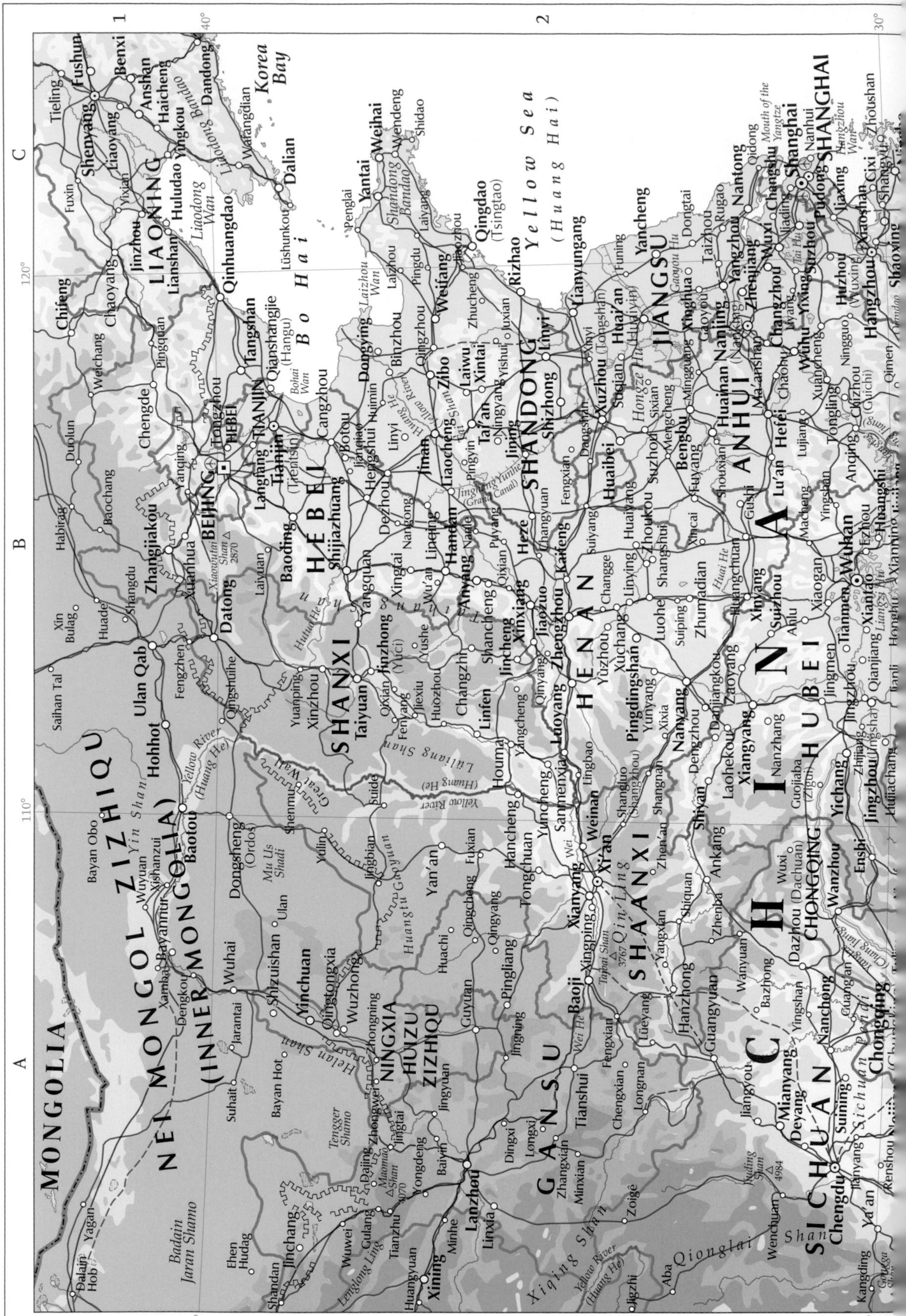

A

B

40°

120°

110°

40°

30°

30°

1

2

Albers Equal Area Conic Projection

© Collins Bartholomew Ltd

71

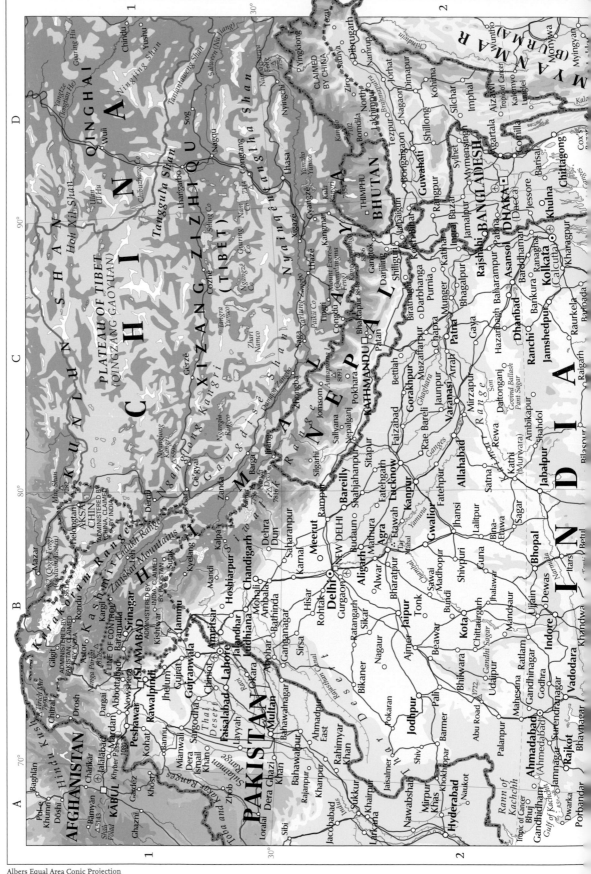

METRES
FEET

5000
16404

3000
9843

2000
6562

1000
3281

500
1640

200
656

0
0

Land below
sea level

200
656

4000
13124

6000
19686

72

Albers Equal Area Conic Projection

© Collins Bartholomew Ltd

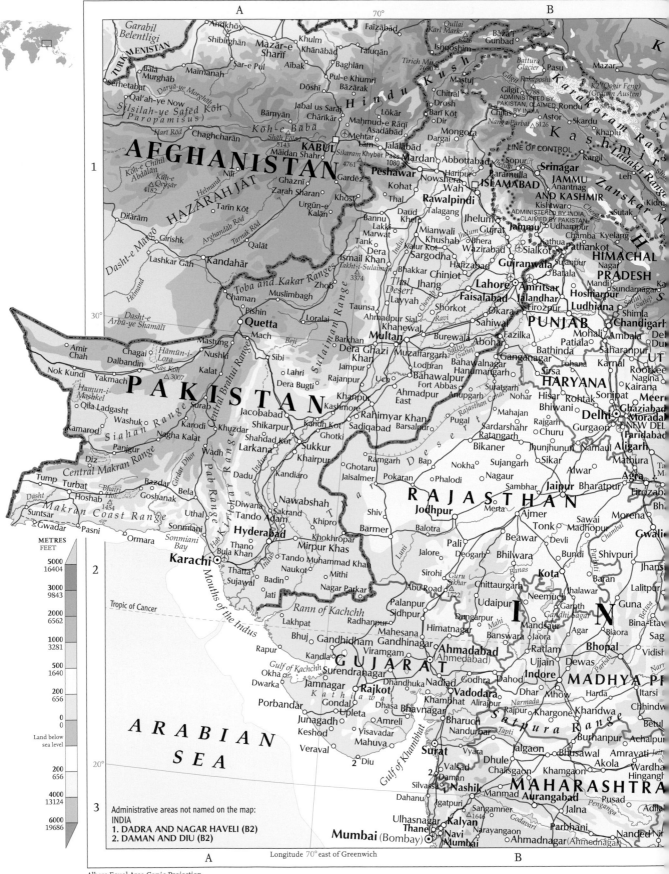

Albers Equal Area Conic Projection

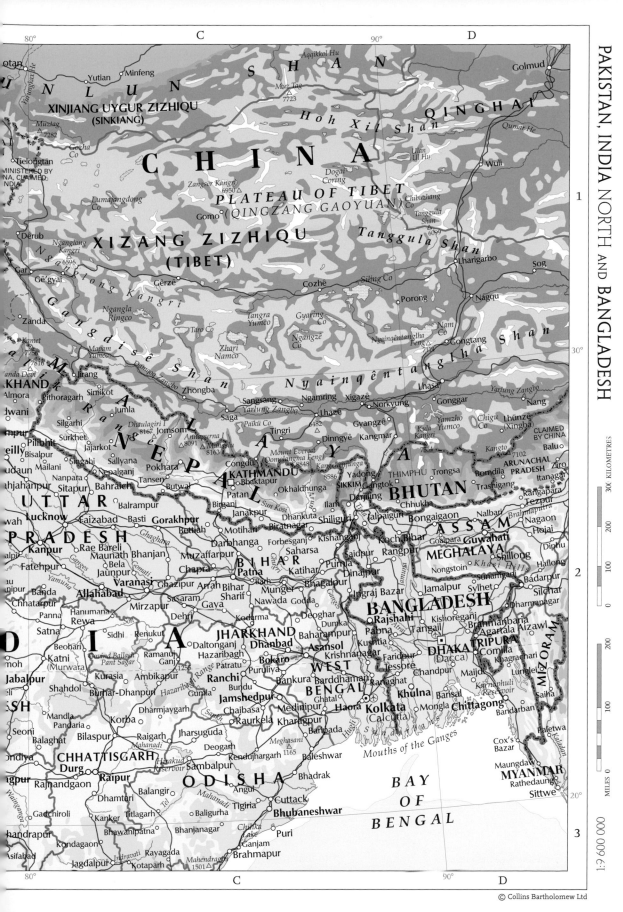

Map labels

Grid references (top): C 80° · 90° D

Grid references (right): 1 · 30° · 2 · 3

China / Tibet region:

KUNLUN SHAN
Yutian · Minfeng
Aqqikkol Hu
Golmud
XINJIANG UYGUR ZIZHIQU (SINKIANG)
Muz Tag 7723
QINGHAI
Muztag △ 7282
Hoh Xil Shan
Qumar He
Gozha Co
Tieliongtan
MINISTERED BY NA, CLAIMED NDIA
C H I N A
Dogai Coring
Ulan Ul Hu
Wuli
Zangsêr Kangri 6950 △
PLATEAU OF TIBET
Chibuzhang
Dêrub
Nganglong Kangri
5596 △
Lumajangdong Co
Gomo
(QINGZANG GAOYUAN)
Tanggula Shan
Tharigarbo
Sog
XIZANG ZIZHIQU (TIBET)
Gar · Ge'gyai
Gêrzê
Cozhê
Siling Co
Porong
Nagqu
Tanggula Shan 6099
Zanda
Ngangla Ringco
Taro Co
Tangra Yumco
Gyaring Co
Ngangzê Co
Nam Co
Gongtang 7111 △
Gongtang
Nyainqêntanglha Feng 7162
Nyainqêntanglha Shan
Kamet 7756 △
7816 △
Mapam Yumco
Zhari Namco
△ anda Devi
Jirang
Gangdisê Shan
Damqog Zangbo
Zhongba
Nyainqêntanglha
Lhasa
Yarlung Zangbo
Nang
30°

Nepal / Himalaya:

KHAND
Almora · Pithoragarh · Simikot
Jumla
Sangsang · Ngamring · Xigazê · Norkyung
Gonggar
wani · Silgarhi
Surkhet · Jajarkot
Saga
Yarlung Zangbo
Lhazê
Gyangzê
Yamzho Yumco
Chigu Co
Lhünzê (Xingba)
CLAIMED BY CHINA
eilly · Pilibhit
Bisalpur
Dhaulagiri I 8167 △
Jomsom
Paikü Co
6482
Tingri
Dinngyê
Kangmar
Kula Kangri 7554
Kangto △ 7102
Balu
ikshanpur · Mailani
Singahi
Saliyana
Annapurna I 8091 △
Manaslu 8163 △
Congdü
Mount Everest (Qomolangma Feng) 8848
8586
Yadong
ARUNACHAL PRADESH
Bomdila · Itanagar
Ziro
udaun · Nanpara · Nepalganj
Pokhara · Tansen
KATHMANDU
Bhaktapur
Kangchenjunga
THIMPHU · Trongsa
Trashigang
Rangapara · Tezpur
Sitapur · Bahraich
Patan
Okhaldhunga
SIKKIM · Gangtok
Darjiling
BHUTAN
Chhukha
Nalbari
Brahmaputra
Nagaon · Hojai

India plains / Bangladesh:

UTTAR
Lucknow · Faizabad · Basti · **Gorakhpur**
Bettiah · Motihari
Birganj · Janakpur
Biratnagar
Ilam
Shiliguri · Jalpaiguri
Bongaigaon
ASSAM
wah
Ghaghara
PRADESH
Rae Bareli · Maunath Bhanjan
Bela · Jaunpur
Darbhanga · Forbesganj · Kishanganj
Koch Bihar
Goalpara · **Guwahati**
MEGHALAYA
Shillong
Diphu
Kanpur
Muzaffarpur
Gandak
Saharsa
Saidpur
Rangpur
Khasi Hills
Haflong
Fatehpur · Maunath · Bela
Ganges
Chapra
BIHAR
Patna
Katihar · Purnia
Dinajpur
Jamalpur · Sylhet
Badarpur
alpi · Banda
Varanasi · Ghazipur · Arrah · Bihar
Munger · Bhagalpur
Ingraj Bazar
Nongstoin
Sunamganj
Silchar
Allahabad
Sasaram · Sharif
Nawada · Godda
Deoghar
Rajshahi
BANGLADESH
Dharmanagar
Mirzapur
Gaya
Kodarma · Dumka
Pabna · Tangail
Brahmanbaria
Agartala · Aizawl
Chhatarpur
Rewa · Sidhi · Renukoot
Dumka
Rajshahi
Kishoreganj
Panna · Satna
Daltonganj
Hazaribagh
Baharampur
Kushtia
DHAKA (Dacca)
TRIPURA
Comilla
Khagrachari
Lunglei
Beohari
Gobind Ballash Pant Sagar
Ramanuj Ganj
JHARKHAND
Dhanbad
Asansol
Krishnanagar
Faridpur
Jessore
Chandpur
Maijdi
MIZORAM
moh
Katni (Murwara)
Patratu 1255
Purulia
Bokaro
WEST
Ranaghat
Khulna · Barisal
Karnaphuli Reservoir
Saiha
Jabalpur
Kurasia · Ambikapur
Ranchi
Bankura · Barddhaman
Bandarban
SH
Burhar-Dhanpuri
Dharmjaygarh
Hazaribagh Range
Gumla
Jamshedpur
BENGAL
Chittagong
Mandla · Pandaria · Korba
Chaibasa
Medinipur
Haora
Kolkata (Calcutta)
Mongla
Cox's Bazar
Pulatwa
Seoni · Balaghat · Bilaspur
Raigarh
Mahanadi
Kharagpur
Bamada
Hugli
Patalganga
Kuakata
Mouths of the Ganges
Maungdaw
MYANMAR
ondiya
CHHATTISGARH
Hirakud Reservoir
Deogarh
Rendujhargarh
Baleshwar
Rathedaung
ngpur
Durg · **Raipur**
Sambalpur
Bhadrak
BAY
Sittwe
Rajnandgaon
Dhamtari
Balangir
Angul
Cuttack
OF
handrapur
Gadchiroli · Kanker · Titlagarh
Bhawanipatna
Bhanjanagar · Chilka Lake · Puri
Bhubaneshwar
BENGAL
asifabad
Jagdalpur
Kotaparh
Rayagada
Mahendragiri 1501 △
Ganjam
Brahmapur

Scale bars (right margin):
300 KILOMETRES · 200 · 100 · 0
200 · 100 · 0 MILES

1:9 000 000

Bottom grid: C 80° · 90° D

© Collins Bartholomew Ltd

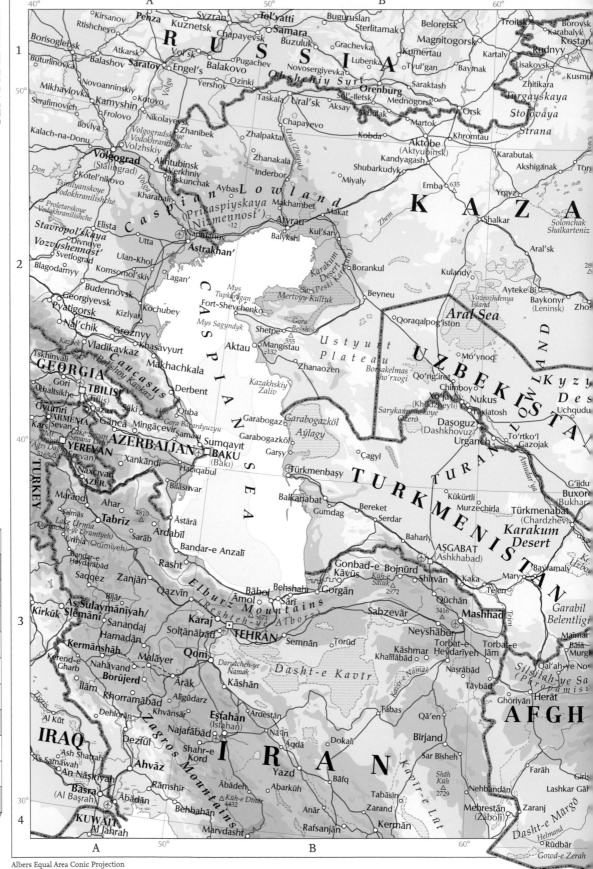

RUSSIA

Kirsanov · Penza · Syzran · Tol'yatti · Bugüruslan · Sterlitamak · Beloretsk · Troitsk · Borovoy
Rtishchevo · Kuznetsk · Samara · Buzuluk · Grachevka · Magnitogorsk · Karabalyk · Kostan
Borisoglebsk · Atkarsk · Chapayevsk · Obshchiy Syrt · Kümertau · Kartaly · Lisakovsk · Kusmu
Buturlinovka · Balashov · Saratov · Engel's · Balakovo · Novosergiyevka · Orenburg · Saraktash · Zhitikara · Tobyl
Mikhaylovka · Kamyshin · Kotovo · Yershov · Ozinki · Sal'-Iletsk · Mednogorsk · Orsk · Turgayskaya
Novoanninskiy · Frolovo · Nikolayevsk · Zhanibek · Zhalpaktal · Taskala · Ural'sk · Aksay · Akbulak · Martok · Khromtau · Stolovaya
Serafimovich · Ilovlya · Volga · Zhanakala · Inderbor · Kobda · Aktobe (Aktyubinsk) · Karabutak · Strana
Kalach-na-Donu · Volgograd (Stalingrad) · Akhtubinsk · Verkhniy · Baskunchak · Aybas · Lowland · Makhambet · Makat · Zhem · Emba · 635 · Yrgyz · Karabutak · Akshiganak · Torg
Don · Kotel'nikovo · Volga · Kharabali · Prikaspiyskaya (Nizmennost') · Atyrau · Kul'sary · Shalkar · KAZA
Tsimlyanskoye Vodokhranilishche · Elista · Utta · Narimanov · Balykshi · Borankul · Kulandy · Ayteke Bi · Baykonyr (Leninsk) · Zhos
Proletarskoye Vodokhranilishche · Stavropol'skaya Vozvyshennost' · Divnoye · Komsomol'skiy · Lagan' · Mys Tupkaragan · Fort-Shevchenko · Mertvyy Kultuk · Beyneu · Qoraqalpog'iston · Mo'ynoq · Vozrozhdenya Island · 28
Blagodarnyy · Budennovsk · Georgiyevsk · Pyatigorsk · Kizlyar · Kochubey · Mys Sagyndyk · Shetpe · Aktau · Mangistau · 132 · Zhanaozen · Borsakelmas sho'rxogi · Aral Sea · UZBEKISTA
Nal'chik · Groznyy · Khasavyurt · Kazbek · Vladikavkaz · Makhachkala · Derbent · Kazakhskiy Zaliv · Ustyurt Plateau · Qong'irot · Chimboy · Nukus · Mo'ynoq · Xo'jayli (Khodzheyli) · Taxiatosh · To'rtko'l · Kungrad · Des
GEORGIA · Gori · TBILISI (Tiflis) · Şäki · Quba · Gora Bazardyuzyu · Garabogaz · Garabogazköl Aylagy · Daşoguz (Dashkhovuz) · Urganch · Gazojak · Uchqudu
TURKEY · ARMENIA · Gyumri · Lake Sevan · GäncÄ · Mingäçevir · Şamaxi · Sumqayit · Garabogaz · Sarykamyshskoye Ozero · Urganch · G'ijdu
Kars · YEREVAN · AZERBAIJAN · BAKU (Baki) · Garsy · Çagyl · To'rtko'l · Gazojak · Buxoro (Bukhara)
Naxçivan · AZER. · Xankändi · Haçiqabul · TURKMENISTAN · Kükürtli · Murzechirla · Türkmenabat (Chardzhev)
Marand · Ahar · Biläsuvar · Türkmenbaşy · Balkanabat · Bereket · Serdar · Kaka · Bayramaly · Karakum
Salmäs · Täbriz · Ästärä · Gumdag · Baharly · AŞGABAT (Ashkhabad) · Mary · Tejen · Karakum Desert
Lake Urmia (Daryächeh-ye Orümiyeh) · Saräb · Ardabil · Bandar-e Anzali · Gonbad-e Kävus · Bojnürd · Shirvän · Quchän · Mashhad · Garabil Belentligi
Urmia (Orümiyeh) · Bäbol · Behshahr · Gorgän · Kuh-e Saluk · 2972 · Sabzevär · 3416 · Mashhad · Silsilah-ye Ko
Saqqez · Zanjän · Qazvin · Ämol · Säri · Elburz Mountains (Reshteh-ye Alborz) · Neyshäbür · Torbat-e Heydariyeh · Torbat-e Jäm · Maima
As Sulaymäniyah/Slëmäni · Bijär · Soltänäbäd · Karaj · TEHRÄN · Semnän · Torüd · Käshmar · Khalilabad · Naşräbäd · Täybäd · Herät
Kirkük · Sanandaj · Hamadän · Qom · Daryächeh-ye Namak · Dasht-e Kavir · Tábas · Qä'en · Ghöriyän · AFGH
Kermänshäh · Kerend-e Gharb · Näväand · Maläyer · Äräk · Käshän · Kavir-e Namak · Birjand · Sar Bisheh · Farah
Iläm · Khorramäbäd · Borüjerd · Aligüdarz · Khvänsär · Nä'in · Äqdä · Naşräbäd · Shäh Kuh · 2729 · Nehbandän · Lashkar Gäh · Giris
Dehlorän · Najaräbäd · Eşfahän (Isfahan) · Ardestän · Tabäsin · Zarand · Mehrestän (Zäboli) · Zaranj
IRAQ · Al Küt · Dezfül · Shahr-e Kord · Zagros Mountains · Yazd · Abädeh · Dokali · Qä'en · Dasht-e Märgo
Ash Shatrah · Ähväz · Najafäbäd · IRAN · Abarküh · Bäfq · Anär · Rafsanjän · Kermän · Helmand · Rüdbär
An Näşiriyah · As Samäwah · Rämshir · Küh-e Dinär · 4432 · Marvdasht · Behbahän · Kuh-e Dinär · Gowd-e Zerah
Basra (Al Başrah) · Äbädän · KUWAIT · Al Jahrah · Marvdasht

Albers Equal Area Conic Projection

METRES
FEET

5000 / 16404
3000 / 9843
2000 / 6562
1000 / 3281
500 / 1640
200 / 656
0 / 0
Land below sea level
200 / 656
4000 / 13124
6000 / 19686

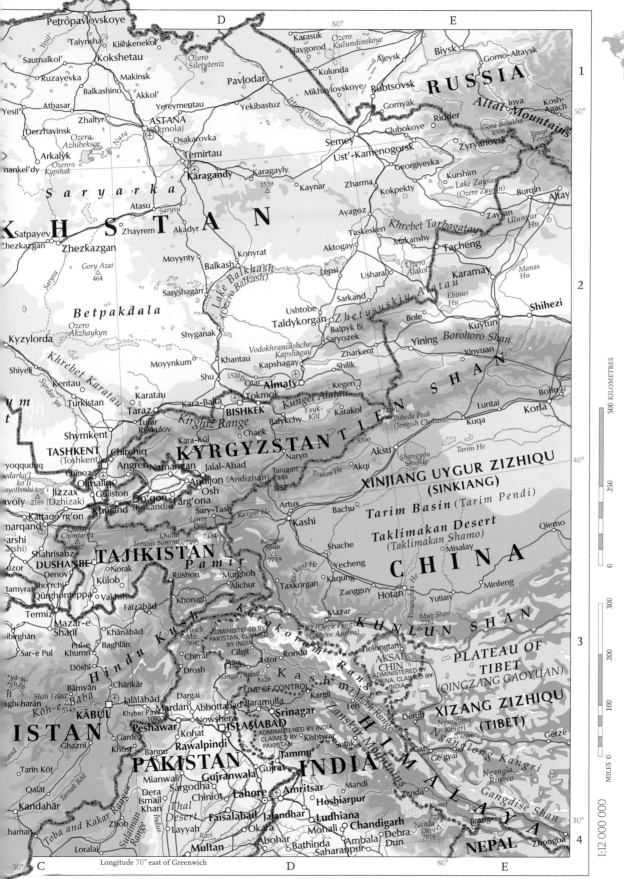

ISRAEL
JORDAN
IRAQ
EGYPT
SUDAN
ERITREA
ETHIOPIA
SAUDI ARABIA
YEMEN

Port Said (Būr Saʻīd)
Gaza
Al ʻArīsh
Beersheba
Al Ḩayy
An Najaf
Ad Dīwānīyah
ʻAmārah
Ash Shatrah
Al Ismāʻīlīyah
At Karak
Ţurayf
40°
Suez (As Suways)
Dead Sea
As Samāwah
An Nāṣirīyah
Sūq ash Shuyūkh
Basra (Al Baṣrah)
Petra
Maʻān
Al ʻĪsāwīyah
ʻArʻar
30°
Ash Shabakah
Hawr al Ḩammār
Raudhatain
Eilat
Al ʻAqabah
Wadi Rum Protected Area
Al Mudawwarah
Dawmat al Jandal
Sakākah
Ḩālat ʻAmmār
KU
Ḩawr
Al Jahra
Nuwaybiʻ al Muzayyinah
Jabal Katrine (Mount Catherine) 2637
Haql
Jabal al Lawz 2579
Al Biʻr
Rafḩaʻ
An Nafūd
Ash Shuʻbah
Aş Şubayḩ
Zaʻfarānah
Raʻs Ghārib
Jabal Gharib 1751
At Ṭūr
Ar Raʼ 979
Hafar al Bāṭin
Wādī al Bāṭin
Jamsah
Sharm ash Shaykh
Al Muwaylih
Jabal ad Dubbāgh 2350
QaḰat al Muʻaẓẓam
Taymāʼ
Jubbah
Hāʼil
Ash Shuʻaybah
Jabal al Kū 325
Qary al Ul
Al Ghurdaqah (Hurghada)
Dubā
Tabūk
Ghazzālah
Ṭābah
Al Kahfah
Al Quwārah
Buraydah
Al Arṭāwīyah
Asharat
Ru
Ash Shum
Būr Safājah
Qalʻat al Azlam
Ad Dār al Ḩamrāʼ
Al ʻUla
Jabal az Zalmā 1258
Samīrah
Jabal al Tin
ʻUnayzah
Az Zilfī
Al Majmaʻah
Al Quṣayr
Al Badāʼiʻ
Khaybar
As Sulaymī
Ar Rass
Nafy
Ad Dirʻīyah
Al Jubaylah
Al Wajh
Ḩulayfah
Hujr
Nuqrah
Wādī al Ḩamd
Shubaykīyah
Jabal Shiʻr
ʻAriah
Ad Dawādimī
RIYADH (Ar Riyāḑ)
2
Marsá al ʻAlam
Hanak
Umm Lajj
Buwāṭ
Sūq Suwayq
ʻAfīf
Al Qāʻīyah
As Salamīyah
Jabal Ḩamāṭah 1977
Jabal Raḍwá 1814
Medina (Al Madīnah)
Al Musayjid
Maḥd adh Dhahab
Al Quwayʻīyah
Ad Dilam
Al Hillah
Baranīs
Yanbuʻ al Bahr
Rayyis
Badr Ḩunayn
Halabān
Ar Ruwaydah
Jabal Tuwayq
Tropic of Cancer
Biʼr Shalatayn
Mastūrah
Umm Birak
Ad Dafīnah
Ẕalim
Jabal Kufsh
Jabal Ḩasan
Khashm Māwān 1025
Layla
ARABIA
HALAIB TRIANGLE ADMINISTERED BY EGYPT, CLAIMED BY SUDAN
Rābigh
Jabal Umm Mukhbir
Ḩādhah
As Sūq
Jabal Tuwayq
Al Badī
Wadi al Allaqi
Halāʼib
King Abdullah Economic City
Tuwwal
Khulays
Madrakah
Jebel Asoteriba 2215
Marsa Delwein
Jeddah (Jiddah)
Al Hawīyah
Turabah
Amaʼir
Ranyah
As Sulayyil
Salāla
Dungunab
Muhammad Qol
Mecca (Makkah)
At Ṭāʼif
Al ʻAqīq
PENINSULA
Nubian Desert
Jebel Oda 2259
Mastābah
Jabal Abū Šādi
Al Līth
Al Junaynah
Qalʻat Bīshah
Al Kumdah
Banī Maʻārid
20°
SUDAN
Port Sudan
Dawqah
Baljurshi
Al ʻAlāyah
Khamāsīn
Wādī Tathlīth
ʻUrūq al Awārik
Wadi Amur
Kamob Sanha
Al Mindak
Qam Hadil
Tathlīth
RUB (EM
Sinkat
Suakin
Al Qunfidhah
An Nimāṣ
Ḩamdah
Musmar
Erheib
Tokar
Al Birk
Abhā
Dirs
Khamis Mushayt
Haiya
Derudeb
Karora
Algena
2780
Ash Shuqayq
Ad Darb
Harajā
Zahrān
Najrān
Ash Sharawrat
3
Hagar Nish Plateau
Nakfa
Suara 2603
Afabet
ʻAsīr
Sabyā
Jāzān
Abū ʻArīsh
Saʻdah
Ramlat Dahm
Aroma
Kassala
Keren
Massawa
Jazāʼir Farasān
Mīdī
Khamr
Hazm al Jawf
Husn Āl
New Halfa
Teseney
Barentu
ASMARA
Dekemharé
Dahlak Archipelago
Aş Şalif
Kamarān
Al Mahwīt
Amrān
Maʼrib
Khashm el Girba Dam
Showak
Mendefera
Adi Keyih
Mersa Fatma
Az Zaydīyah
Bājil
3760
ŞANʻĀ
Maʼbar
YEM
Bayḩān al Qiṣāb
ʻAtaq
Gedaref
Om Hajer
Inda Silase
Ādigrat
Koluli
Hodeidah (Al Ḩudaydah)
Manākhah
Dhamār
Radāʻ
Hal
Reihad
Aksum
329
Aḍwa
Āṣalē
Bayt al Faqīh
Zabīd
Ibb
Yarīm
Al Baydāʼ
Lawdar
Gallabat
Mekʻelē
Ras Dejen Simēn 4533
2131
Al Khawkhah
Hays
Qaftabah
Jabal Thamar 2512
Shuqrah
Ādī Ārkʻay
Danakil
Ed Az Zuqur
Ta'izz
3267
Musaymir
Zinjibār
Reloula
Atbara
Mocha (Al Mukhā)
Mawza
Dhubāb
Lahij
An Nābiyah
At Turbah
Bab al Mandab
Ash Shaykh ʻUthman
Aden (ʻAdan)

Longitude 40° east of Greenwich

Albers Equal Area Conic Projection

METRES FEET
5000 16404
3000 9843
2000 6562
1000 3281
500 1640
200 656
0 0
Land below sea level
200 656
4000 13124
6000 19686

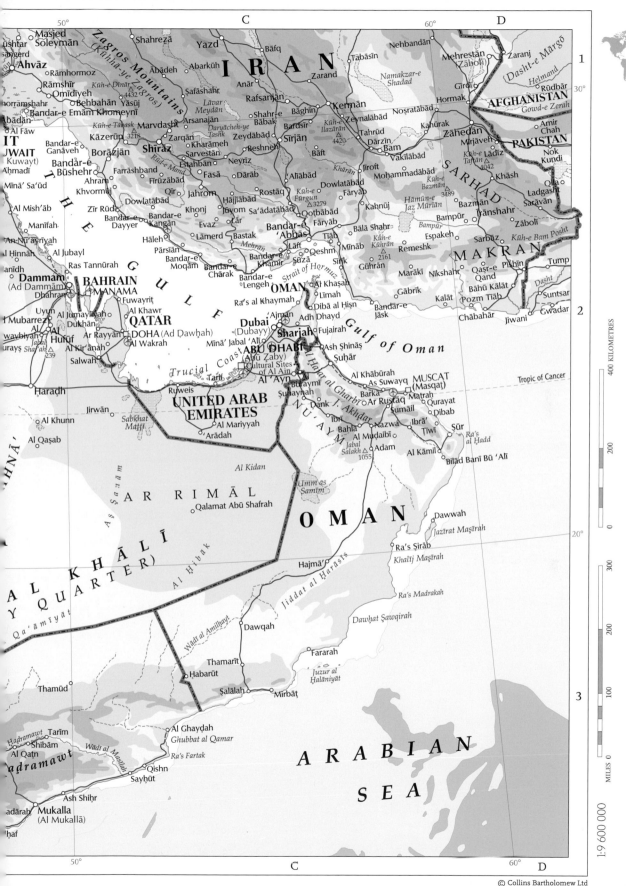

© Collins Bartholomew Ltd

1:9 600 000

A | 30° | B

UKRAINE

CHIȘINĂU
Târgu Mureș
Miercurea-Ciuc
Bacău
Vaslui
Tiraspol
Berezivka
Mykolaiv
Tokmak
Nova
Kakhovka
Mariupol'
Taganrog
Sighișoara
Sebeș
Sfântu Gheorghe
Tecuci
Comrat
Bender
Tighina
Kherson
Melitopol'
Berdyans'k
Rosto na-Do
Lugoj
Deva
Sibiu
Focșani
Artsyz
Odessa (Odesa)
Skadovs'k
Novooleksiyivka
Armyans'k
Yeysk
Starominskaya
Primorsko Akhtarsk
Pavlovsk
Reșița
Caransebeș
Fǎgǎraș
Brașov
Galați
Buzău
Izmayil
Kiliya
Bilhorod-Dnistrovs'kyy
Karkinits'ka Zatoka
Krasnoperekops'k
Nyzhn'ohirs'kyy
Gulf of Taganrog
Timashevsk
Slavyansk na-Kubani
ROMANIA
Drobeta-Turnu Severin
Pitești
Ploiești
Brăila
Babadag
Chornomors'ke
CRIMEA ADMINISTERED BY RUSSIA
Dzhankoy
Temryuk
Kerch
Krymsk
Krasnod
Craiova
Slatina
Rosiori de Vede
BUCHAREST (București)
Danube (Dunărea)
Yevpatoriya
Nyzhn'ohirs'kyy
Feodosiya
Khadyzhensk
Zaicar
Calafat
Caracal
Corabia
Ruse
Călărași
Constanța
Simferopol'
Sudak
Novorossiysk
Tuapse
Montana
Vratsa
Lovech
Razgrad
Dobrich
Mangalia
Sevastopol'
Roman-Kosh
Soc
Botevgrad
Shumen
Kavarna
BLACK SEA
Pernik
SOFIA (Sofiya)
Kazanlak
Sliven
Burgas
Varna
Kyustendil
Blagoevgrad
Plovdiv
Stara Zagora
BULGARIA
Musala △ 2925
Kočani
Sandanski
Smolyan
Dimitrovgrad
Haskovo
Kârdzhali
Babaeski
Kırklareli
Saray
Cide
İnebolu
İnce Burun
Sinop
Strumica
Drama
Xanthi
Komotini
Edirne
Çorlu
Silivri
Zonguldak
Bartın
Boyabat
Bafra
Samsun
Serres
Kavala
Tekirdağ
Keşan
İstanbul
Kadıköy
Karabük
Kastamonu
Vezirköprü
Merzifon
Terme
Ordu
Trabz
Thessaloniki
Polygyros
Thasos
Gallipoli (Gelibolu)
Körfez
Adapazarı (Sakarya)
Düzce
Gerede
Tosya
Osmancık
Amasya
Giresun
Aegean Sea
Gökçeada
İmroz
Çanakkale
Gemlik
Bursa
İnegöl
Bolu
Göynük
Mudurnu
Çankırı
Çorum
Anadolu Dağla
Volos
Limnos
Ezine
Can
Susurluk
Uludağ △ 2543
Bilecik
Sakarya
ANKARA
Kalecik
Sungurlu
Turhal
Tokat
Sivas
Zara
Erzin
GREECE
Evvoia
Lesbos
Mytilini
Bergama
Demirci
Simav
Kütahya
Eskişehir
Sivrihisar
Polatlı
Kırıkkale
Yozgat
Yıldızeli
Suşehri
Chalkida
Aliağa
Bornova
Akhisar
Manisa
Banaz
Afyon
Emirdağ
Cihanbeyli
Kaman
Kırşehir
Kayseri
Erciyes Dağı △ 3917
Pınarbaşı
Elazığ
Tur
ATHENS (Athína)
Chios
İzmir
Salihli
Uşak
Sandıklı
Akşehir
Yunak
Nevşehir
Aksaray
Yahyalı
Elbistan
Kahramanmaraş
Piraeus
Andros
Tinos
Kuşadası
Aydın
Denizli
Çivril
Dinar
Lake Tuz (Tuz Gölü)
Niğde
Hasan Dağı △ 3268
Adıyaman
Malatya
Siver
Ermoupoli
Syros
Samos
İkaria
Söke
Nazilli
Isparta
Eğirdir △ 2799
Eğirdir Gölü
Karapınar
Konya
Adana
Gaziantep
TURKEY
Milos
Naxos
Ios
Paros
Milas
Yatağan
Muğla
Burdur
Bucak
Beyşehir Gölü
Ereğli
Karaman
Tarsus
Osmaniye
Kilis
Şanlı
Birecik
Euphrates (Fırât)
Viranş
Santorini (Thira)
Karpathos (Scarpanto)
Marmaris
Fethiye
Dalaman
Elmalı
Korkuteli
Serik
Manavgat
Mersin (İçel)
İskenderun (Alexandretta)
Chania
Irakleio
Agios Nikolaos △ 2456
Siteia
Ierapetra
Rhodes (Rodos)
Kaş
Megisti
Antalya
Antalya Körfezi
Alanya
Anamur
Ermenek
Erdemli
Silifke
Taurus Mountains (Toros Dağları)
Antakya (Antioch)
Aleppo (Halab)
Ar Raqqah
Eup
Rethymno
CRETE (KRITI)
Lindos
△ 3073
Cape Apostolos Andreas
Cape Arnauti
İdlib
Latakia
Ma'arrat an Nu'mān
Kyrenia (Keryneia)
NICOSIA (Lefkosia)
ADMINISTERED AS NORTHERN CYPRUS
Aigialousa
Famagusta
Ḥamāh
SYRI
Paphos
Polis
Evrychou
Larnaca
Tartūs
Bāniyās
Chania
MEDITERRANEAN SEA
Limassol (Lemesos)
Tripoli (Trâblous)
Homs
Tadmur
CYPRUS
LEBANON
BEIRUT (Beyrouth)
Al Qaryatayn
An Nabk
Sidon
Zahle
Sab' Ābār
Tyre
Al Qunayṭirah (abandoned)
△ 2814
DAMASCUS (Dimashq)
Sea of Galilee (Yam Kinneret)
Haifa (Hefa)
Nazerat
Dar'ā
As Suwaydā'
Syrian Des (Bādiyat ash Sh
Tel Aviv-Yafo
Nabulus
Irbid
Al Mafraq
Az Zarqā'
ISRAEL
Rehovot
JERUSALEM
Gaza
WEST BANK
AMMAN
EGYPT
Al Burdī
Umm Sa'ad
Marsá Maṭrūḥ
Kafr ash Shaykh
Baltīm
Dumyāṭ
Al Mansūrah
Port Said (Būr Sa'īd)
Al 'Arīsh
Beersheba
Al Karak
At Tafilah
Al 'Isāwīyah
Turayf
Alexandria (Al Iskandarīya)
Al 'Āmirīyah
Al Hammām
Damanhūr
Banhā
Tanţā
Az Zaqāzīq
Al Ismā'īlīyah
Suez Canal
Dead Sea
Petra
Ma'ān
JORDAN
Wadi an Sirhān
Libyan Plateau (Ad Diffah)
Qattara Depression
Wāḥāt Sīwah (Siwa Oasis)
Qārah
Shubrā al Khaymah
Giza (Al Jīzah)
CAIRO (Al Qāhirah)
Pyramids of Giza
Memphis
Suez (As Suways)
Gulf of Suez
Wadi Rum Protected Area
Al 'Aqabah
Ḥālat 'Ammār
Dawmat al Jand
SAUD
Siwah
Al Fayyūm
Za'farānah
Sinai
Elat
Al Mudawwarah
Al Bawītī
Banī Suwayf (Beni Suef)
Nuwaybi' al Muzayyinah
Jabal Katrina (Mount Catherine) △ 2637
Haql
Jabal al Lawz △ 2579
Al Bi'r
Al Bawītī
Banī Mazār
Maghāghah

LIBYA

SERBIA

METRES / FEET
5000 / 16404
3000 / 9843
2000 / 6562
1000 / 3281
500 / 1640
200 / 656
0 / 0
Land below sea level
200 / 656
4000 / 13124
6000 / 19686

1
2
3

Longitude 30° east of Greenwich

Albers Equal Area Conic Projection

C · 50° · D

Novocherkassk
Sor
Donyztau

Balykshi

Karakul'
Desert
(Peski Karakum)

Borankul

Proletarskoye
Vodokhranilishche
Volga
Utta
Astrakhan'

RUSSIA

Sal'sk
Elista

Beyneu

UZBEKISTAN

choretsk
Ulan-
Khol
Lagan'

Ipatovo
Divnoye

KAZAKHSTAN

Kropotkin
Stavropol'skaya
Komsomol'skiy

Mys
Tupkaragan
Fort-Shevchenko

Ustyurt
Plateau

1

Kuban'
Stavropol'
Budennovsk
Kochubey
Gora
Besshoky
555

chir
Labinsk
Vozvyshennost'
Nevinnomyssk
Kizlyar

kop
Cherkessk
Georgiyevsk
Makhachkala
Mangistau
132

Borsakelmas
sho'rxogi

Psebay
Pyatigorsk
Prokhladnyy
Khasavyurt
Aktau
Zhanaozen

Udil Karabaur

Karachayevsk
Kislovodsk
Mozdok
Buynaksk
Kuryk

Elbrus
5642
Nal'chik
Alagir
Vladikavkaz
Izberbash
Kazakhskiy
Zaliv

gra
Grozny
Sarykamyshskoye
Ozero

KHAZIA
Sokhumi
T'q'varcheli
Derbent

Zugdidi

GEORGIA
SOUTH
OSETIA
Telavi

Garabogaz

TURKMENISTAN

Kutaisi
Khashuri
Garabogazköl
Aýlagy

Poti
Samt'redia
Gori
TBILISI
(Tiflis)
Zaqatala
Gora
Bazardyuzyu
Garabogazköl
Çagyl

Bat'umi
Akhaltsikhe
Şäki
1466
Garşy

Artvin
Ardahan
Rustavi
Quba

Akhalkalaki
Qazax
Mingäçevir
Göyçay
Şamaxı
Sumqayıt

Kaçkar
3932
Yusufeli
Vanadzor
Gäncä
Abşeron
Yarımadası
Jaňňa

Oltu
Gyumri
Sevan
AZERBAIJAN
BAKU
(Bakı)

40°

Kars
ARMENIA
YEREVAN
(Erevan)
Hacıqabul
Türkmenbaşy

Horasan
Sarıkamış
Ararat
abandoned
Ağdam
Xankändi
Şirvan
Hazar

Erzurum
Ağrı
Doğubayazıt
Mt Ararat
Ağrı Dağı
NAGORNO-
KARABAKH
Sälyan
Jebel
Balkanabat

Hınıs
Tutak
Siśtan
Biläsuvar
Cälilabad
Gumdag
Bereket

Patnos
Ercis
Mākū
Naxçıvan
AZER.
Länkäran
Serdar

Malazgirt
Ahlat
Stephan Dağı
Khvoy
Āstārā

Ogurjaly
Adasy
Magtymguly

Muş
1058
Marand
Ahar
4810

Bitlis
Lake Van
(Van Gölü)
Salmas
Sarāb
Ardabil

Van
Urmia
Tabrīz

Başkale
Lake Urmia
(Daryācheh-ye Orūmīyeh)
Mīāneh

Hakkâri
Marāgheh

Şırnak
Semdinli
Oshnavīyeh
Bandar-e Heydarābād
Fūman
Bandar-e Anzalī
Gomīsh
Tappeh
Gorgān
Gonbad-e
Kāvūs

Zākhū
Dahūk
'Amādīyah
Amēdī
Mahābād
Mīāndoāb
Zanjān
Rasht
Lāhījān
Behshahr
Māyāmey

Al Mayādīn
Mosul
Saqqez
Qazvīn
Tonekābon
Nowshahr
Bābol
Amol
Sārī
Shāhrūd

Tall 'Afar
Elburz Mountains
(Reshteh-ye Alborz)
5671
Dāmghān

Arbīl/Hewlêr
Bījār
Abhar
Karaj
TEHRĀN
Semnān
Torūd

As Sulaymānīyah/Slêmānī
Sanandaj
Soltānābād

2

Ash
Sharqāt
Kirkūk
Qorveh
Hamadān
Qom
Daryācheh-ye
Namak
Dasht-e Kavir

Ar Mayādīn
Tuz Khurmātū
Halabjah
Helebce
Ravānsar
Kangāvar
Malāyer
Arāk
Jandaq

'Ānah
Bayjī
Tikrīt
Qasr-e
Shīrīn
Kermānshāh
Nahāvand
Ardestān

Al Hadīthah
Sāmarrā'
Kerend-e Gharb
Eslāmābād-e
Gharb
Borūjerd
Golpāyegān
Khvānsār
Nā'īn
Dokali

Hīt
Al Muqdadīyah
 Īlām
Dorūd
Dārān
Kūh-e
Karbūsh
Najafābād
Esfahān
(Isfahan)
Āqdā
Meybod

Ar Ramādī
Al Kāzimīyah
Khorramābād
Aligūdarz
4294

IRAN

Hawr al Habbānīyah
BAGHDAD
Dehlorān
Dezfūl
Shahr-e
Kord
Shahreżā
Yazd

Buhayrat ar
Razāzah
Karbalā
Hillah
Al Kūt
Shūshtar
Zagros Mountains
(Kūhhā-ye Zāgros)
Kūh-e
Dīnār
4432
Ābādeh
Bāfq

IRAQ

Al Hayy
Masjed
Soleymān
Ābarkūh

An Najaf
Ad Dīwānīyah
Al 'Amārah
Rāmhormoz
Yāsūj
Şafāshahr
Anār

Ash Shatrah
Susangerd
Ahvāz
 Omīdiyeh
Behbahān
Kūh-e
Tāshk
Lāvar
Meydān
Shahr-e
Bābak

As Samāwah
Sūq ash
Shuyūkh
Rāmshīr
Khorramshahr
Marvdasht
Arsanjān
Ābādeh Tashk

An Nāşirīyah
Hawr al
Hammār
Basra
(Al Başrah)
Ābādān
Bandar-e
Emām Khomeynī
Kūh-e
Tāshk
3215
Zarqān
Kharāneh
Daryācheh-ye
Tashk
Beshneh

30°

Ash Shabakah
Raudhatain
Al Fāw
Kāzerūn
Shīrāz
Salvestān
Neyrīz

kah
KUWAIT
Hawalli
KUWAIT
(Al Kuwayt)
Bandar-e
Ganāveh
Borāzjān
Fasā
Estahbān
Dārāb

3

IRAQ
Rafhā'
Al Jahrah
Bandar-e
Būshehr
Farrāshband
Fīrūzābād
Qīr
Jahrom
Hājjīābād

RABIA
An Nafūd
Ash
Shu'bah
Al Ahmadī
Mīnā' Sa'ūd
Ahram
Khvormūj
Dowlatābād
Jūyom
Rostāq

C · 50° · D

1:9 600 000

400 KILOMETRES
200
0

MILES 0
100
200
300

METRES
FEET

5000
16404

3000
9843

2000
6562

1000
3281

500
1640

200
656

0
0

Land below
sea level

200
656

4000
13124

6000
19686

ARCTIC

BARENTS SEA

Svalbard
(Norway)

Spitsbergen

Longyearbyen

Jan Mayen
(Norway)

Greenland Sea

A

B

C

D

E

F Kara Sea G
(Karskoye More)

Zemlya
Frantsa-Iosifa

Novaya Zemlya

NORWAY

SWEDEN

FINLAND

Gulf of Bothnia

Baltic Sea

ESTONIA

LATVIA

LITHUANIA

BELARUS

UKRAINE

MOSCOW

St Petersburg

White Sea
(Beloye More)

Archangel

Ural Mountains
(Uralskiy Khrebet)

West Siberian Plain

R S F S U B

Yamal Peninsula
(Poluostrov Yamal)

Gydan Peninsula
(Gydanskiy
Poluostrov)

Murmansk

Yekaterinburg
(Sverdlovsk)

Chelyabinsk

Omsk

Novosibirsk

Tomsk

Barnaul

GEORGIA

ARM. AZER.

TURKEY

IRAN

Black Sea

Sea of
Azov

Caspian Sea

Caspian Lowland
(Prikaspiyskaya Nizmennost')

KAZAKHSTAN

Aral Sea

UZBEKISTAN

TURKMEN.

Kyzylkum
Desert

ASTANA
(Akmola)

Lake Balkhash
(Ozero Balkash)

CHIN

Altai
Mountains

Black
Sea

Conic Equidistant Projection

Longitude 75° east of Greenwich

© Collins Bartholomew Ltd

A 40° B 30° C 20° D 10° E 0° F 10° 20°

Greenland
(Denmark)

Bjørnøya
(Nor.)

Arctic Circle

Denmark Strait

2
60°

Jan Mayen
(Nor.)

Tromsø

ICELAND
Reykjavík

NORWEGIAN
SEA

L

3

Faroe
Islands
(Den.)
Tórshavn

Trondheim

N
O
R
W
A
Y

S
W
E
D
E
N

Gulf of Both

Shetland
Islands

Bergen

Orkney
Islands

Oslo

Stockho

ATLANTIC

OCEAN

SCOTLAND

NORTH
SEA

Vänern

Gothenburg

Gotland

Baltic Sea

50°

Glasgow
Edinburgh

N. Belfast
IRELAND

DENMARK
Copenhagen

Malmö

L

IRELAND
Dublin

Manchester

UNITED
KINGDOM

Hamburg

Berlin

POLAN

RUSS

Manchester

NETHERLANDS
Amsterdam

Hannover

Hamburg

Poznań
Warsaw

WALES ENGLAND
Cardiff Birmingham
London

The Hague
Essen

GERMANY

Berlin

Łódź

English Channel

Brussels

BELGIUM

Rhine

Frankfurt

Prague

Katow

4

K.   KOSOVO
LIE.   LIECHTENSTEIN
MACE.   MACEDONIA
MONT.   MONTENEGRO

Channel Is
(U.K.)

Seine

Luxembourg
LUXEMBOURG

CZECH
REPUBLIC

Paris

Munich

Vienna

SLOVAKI

Bay of
Biscay

Loire

FRANCE

Danube

Bratislava

Budap

Cape Finisterre

SWITZERLAND
Lyon

Bern LIE.

AUSTRIA

S

HUNGARY

Mont Blanc
4808 m

A

SLOVENIA

Ljubljana

Zagreb

Bordeaux

Rhône

Turin

Milan

CROATIA

Belgrad

40°

Bilbao

Pyrenees

SAN
MARINO

BOS. AND
HERZ.

SER

Oporto

PORTUGAL

SPAIN

Andorra
la Vella ANDORRA

Marseille

MONACO

I

Sarajevo

MONT.

Pris

Madrid

Barcelona

Corsica

VATICAN
CITY
Rome

T

Podgorica

K.

Skopj

Lisbon

Cabo de
São Vicente

Valencia

Balearic Islands

Sardinia

A

Adriatic Sea

Tirana

M

ALBANIA

5

Seville

Gibraltar
(U.K.)

Naples

L

Y

Palermo

Sicily

Ionian
Sea

MEDITERRANEAN
SEA

A

MOROCCO

ALGERIA

TUNISIA

MALTA
Valletta

10° D   Greenwich 0° meridian   E   10°   F   20°

Chamberlin Trimetric Projection

© Collins Bartholomew Ltd

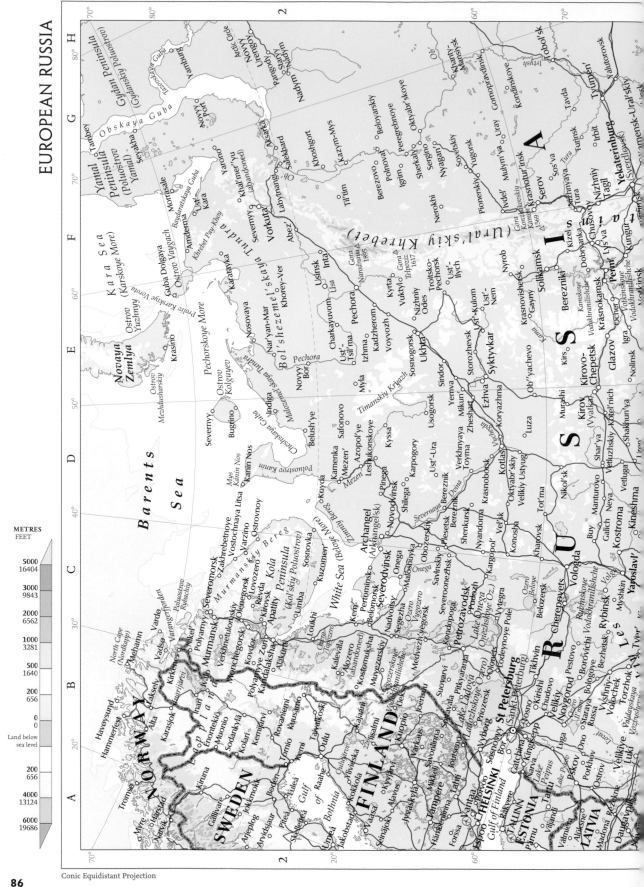

METRES
FEET

5000
16404

3000
9843

2000
6562

1000
3281

500
1640

200
656

0
0

Land below
sea level

200
656

4000
13124

6000
19686

NORWAY

SWEDEN

FINLAND

ESTONIA

LATVIA

Barents
Sea

Kara Sea
(Karskoye More)

Novaya
Zemlya

Pechorskoye More

B a r e n t s

S e a

White Sea
(Beloye More)

R U S S I A

Conic Equidistant Projection

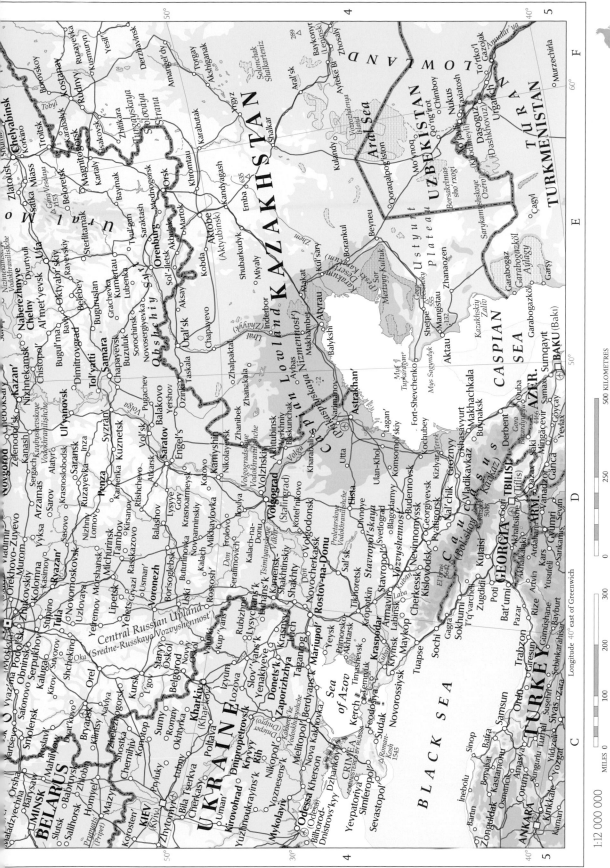

© Collins Bartholomew Ltd

1:12 000 000

MILES 0 100 200 300

0 250 500 KILOMETRES

Longitude 40° east of Greenwich

BELARUS
MINSK
Slutsk
Sallihorsk
Babruysk
Homyel'
Mazyr
Korosten'
Zhytomyr
KIEV
(Kyiv)

UKRAINE
Yuzhnoukrayins'k
Mykolayiv
Kirovohrad
Kropyvnyts'kyy
Odessa
(Odesa)
Bilhorod-
Dnistrovs'kyy
Voznesens'k
Kherson
Nova Kakhovka
Nikopol'
Melitopol'
Simferopol'
Sevastopol'
Yevpatoriya

CRIMEA
ADMINISTERED
BY RUSSIA

Kerch
Feodosiya
Sudak

Dnipropetrovs'k
Dnipro
Kamyans'ke
Kryvyy
Rih
Zaporizhzhya
Berdyans'k
Mariupol'

Kharkiv
(Kharkov)
Poltava
Cherkasy
Kremenchuk
Uman'
Bila Tserkva

Sea of Azov

BLACK SEA

Novorossiysk
Tuapse
Sochi
Gagra
Sokhumi
Zugdidi
Poti
Bat'umi

GEORGIA
Kutaisi
TBILISI
(Tiflis)

TURKEY
ANKARA
Zonguldak
Bartın
Kastamonu
Çorum
Amasya
Samsun
Sinop
İnebolu
Bafra
Yozgat
Sivas
Tokat
Trabzon
Rize
Artvin
Erzurum

ARM.
YEREVAN
Gyumri
Vanadzor

AZER.
BAKU
(Baki)
Sumqayıt
Gäncä
Mingäçevir

CASPIAN SEA

KAZAKHSTAN

UZBEKISTAN

TURKMENISTAN

Aral Sea

TURAN LOWLAND

Ustyurt Plateau

Caspian Lowland
(Prikaspiyskaya Nizmennost')

MOSCOW
(Moskva)

Central Russian Upland
(Sredne-Russkaya Vozvyshennost')

U r a l
Chelyabinsk
Zlatoust
Miass
Troitsk
Magnitogorsk

Samara
Ulyanovsk
Kazan'
Nizhnekamsk
Naberezhnye Chelny
Ufa
Orenburg
Orsk
Aktobe

Saratov
Volgograd
(Stalingrad)
Astrakhan'

Voronezh
Rostov-na-Donu
Krasnodar
Maykop

C a u c a s u s
(Bol'shoy Kavkaz)
Vladikavkaz
Groznyy
Makhachkala
Derbent

NORTHEAST EUROPE

METRES / FEET

5000	16404
3000	9843
2000	6562
1000	3281
500	1640
200	656
0	0

Land below sea level

200	656
4000	13124
6000	19686

Longitude 25° east of Greenwich

Conic Equidistant Projection

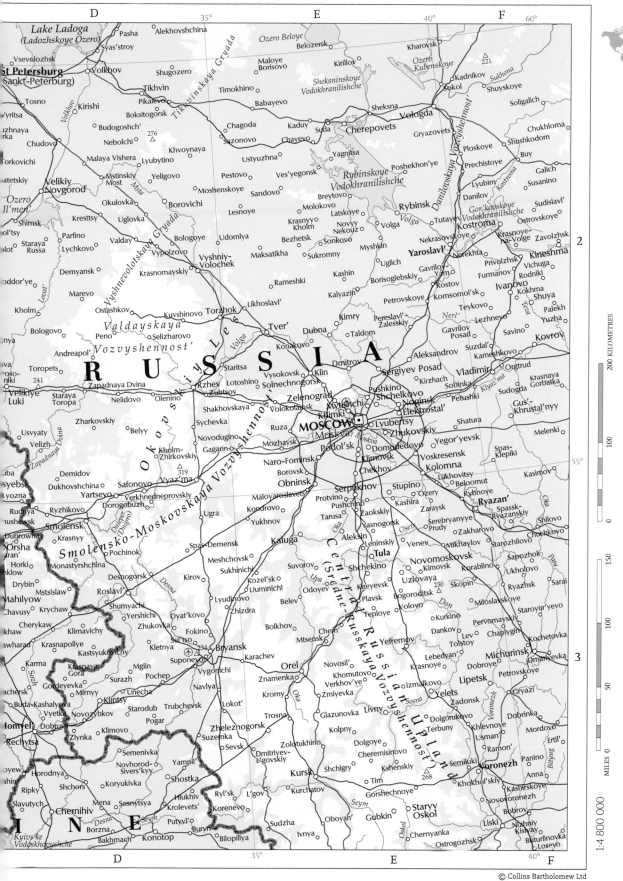

© Collins Bartholomew Ltd

A 25° B 30°

BELARUS

POLAND

1

Ciechanów · Zambrów · Vawkavysk · Zel'va · Baranavichy · Nyasvizh · Asipovichy · Babruysk · Karm
Ostrów · Białystok · Svislach · Slonim · Lyakhavichy · Klyetsk · Kapyl' · Staryya · Rahachow · Chachersk
Mazowiecka · Hajnówka · Ivatsevichy · Hantsavichy · Salihorsk · Lyuban' · Hlusk · Zhlobin · Buda-
Wyszków · Pruzhany · Kamyanyets · Byaroza · Tsyelyakhany · Mal'kavichy · Svyetlahorsk · Kashalyova
Legionowo · WARSAW · Kobryn · Drahichyn · Luninyets · Dzyatlavichy · Zhytkavichy · Kapatkyevichy · Mazyr · Loyew · Homyel'
(Warszawa) · Siedlce · Zhabinka · Brest · Ivanava · Pina · Pinsk · Davyd- · Zarichne · Stolin · Lyel'chytsy · Khoyniki · Brahin · Slavut
Pruszków · Biała · Malaryta · Lyubeshiv · Pripet · Marshes · Yel'sk · Narowlya · Chornobyl'

50°

UKRA

SLOVAKIA

2

HUNGARY

MOLDOVA

ROMANIA

45°

3

SERBIA · **BULGARIA**

Longitude 25° east of Greenwich

A B 30°

Conic Equidistant Projection

METRES / FEET

| 5000 / 16404 |
| 3000 / 9843 |
| 2000 / 6562 |
| 1000 / 3281 |
| 500 / 1640 |
| 200 / 656 |
| 0 / 0 |
| Land below sea level |
| 200 / 656 |
| 4000 / 13124 |
| 6000 / 19686 |

Map grid references: C, D, E (top and bottom)

Longitude markers: 35°, 40° (top); 35°, 40° (bottom)

RUSSIA

UKRAINE (shown as "I N E")

Central Russian Upland (Sredne-Russkaya Vozvyshennost')

Selected place names:

Krasnaya Gora, Mglin, Suponevo, 234, Bryansk, Karachev, Orel, Lebedyan', Dobroye, Dmitriyevka, Tambov, Kotovsk

Surazh, Gordeyevka, Vygonichi, Novosil', Khomutovo, Verkhov'ye, Krasnoye, Izmalkovo, Yelets, Lipetsk, Gryazi, Petrovskoye, Znamenka

Unecha, Klintsy, Pochep, Navlya, Znamenka, Kromy, Zmiyevka, Glazunovka, Kolpny, Dolgorukovo, Zadonsk, Usman', Ramon', Mordovo, Tokareyka

Mirnyy, Novozybkov, Starodub, Trubchevsk, Lokot', Trosna, Oka, Dolgoye, Terbuny, Khlevnoye, Semiluki, Voronezh, Panino, Anna, Ertil', Zherdevka

Zlynka, Klimovo, Pogar, Zheleznogorsk, Suzemka, Sevsk, Zolotukhino, Cheremisinovo, Shchigry, Tim, Khokhol'skiy, Novovoronezh, Talovaya, Novokhopersk

brush, Semenivka, Novhorod-Sivers'kyy, Yampil', Dmitriyev-L'govskiy, Kursk, Gorshechnoye, Kshenskiy, Voronezh, Bobrov, Buturlinovka

Horodnya, Koryukivka, Shostka, Hlukhiv, Ryl'sk, L'gov, Kurchatov, Gubkin, Staryy Oskol, Liski, Nizhniy Kislyay, Vorontsovka

Chnors, Mena, Sosnytsya, Krolevets', Korenevo, Oboyan', Ivnya, Rakitnoye, Chernyanka, Novyy Oskol, Kamenka, Iosevo, Pavlovsk, Kalach

Nizhyn, Bakhmach, Konotop, Buryn', Bilopillya, Sumy, Borisovka, Belgorod, Alekseyevka, Korocha, Biryuch, Alekseyevka, Podgorenskiy, Rossosh', Verkhniy Mamon, Boguchar

Nosivka, Ichnya, Romny, Lebedyn, Hadyach, Okhtyrka, Zolochiv, Derhachi, Vovchans'k, Velykyy Burluk, Veydelevka, Roven'ki, Kantemirovka, Meshkovskaya

obrovytsya, Pryluky, Pyryatyn, Lokhvytsya, Zin'kiv, Bohodukhiv, Kharkiv (Khar'kov), Chuhuyiv, Kup"yans'k, Novopskov, Markivka, Milove, Chertkovo

Pereyaslav-Khmel'nyts'kyy, Lubny, Myrhorod, Lyubotyn, Merefa, Shevchenkove, Kivsharivka, Svatove, Starobil's'k, Bilovods'k, Degtevo, Millerovo

Zolotonosha, Khorol, Reshetylivka, Poltava, Karlivka, Zmiyiv, Balakliya, Izyum, Kreminna, Rubizhne, Syeverodonets'k, Lysychans'k, Tarasovskiy

Chornobay, Cherkasy, Hlobyne, Krasnohrad, Pervomays'kyy, Northern Donets, Slov"yans'k, Stakhanov, Luhans'k, Kamensk-Shakhtinskiy

Kremenchuts'ke Vodoskhovyshche, Kremenchuk, Komsomol's'k, Lozova, Barvinkove, Kramators'k, Artemiv's'k, Alchevs'k, Krasnodon, Gukovo

Svitlovods'k, Dnieper (Dnipro), Novomoskovs'k, Pavlohrad, Druzhkivka, Dzerzhyns'k, Donets'kyy Kryazh, 367, Ust'-Donetskiy

Oleksandrivka, Oleksandriya, Dniprodzerzhyns'k, Dnipropetrovs'k, Kostyantynivka, Horlivka, Makiyivka, Krasnyy Luch, Roven'ki, Krasnyy Sulin, Shakhty

vomyrhorod, Znam"yanka, P"yatykhatky, Vil'nohirs'k, Pershotravens'k, Krasnoarmiys'k, Avdiyivka, Donets'k, Novoshakhtinsk, Kamenolomni

Kirovohrad, Zhovti Vody, Synel'nykove, Vasyl'kivka, Vil'nyans'k, Makiyivka, Novocherkassk, Semikarakorsk

Novoukrainka, Novyy Buh, Ordzhonikidze, Marhanets', Zaporizhzhya, Hulyaypole, Dokuchayevs'k, Matveyev Kurgan, Pokrovskoye, Rostov-na-Donu, Aksay, Bagayevskiy

Bobrynets', Kryvyy Rih, Shyroke, Apostolove, Enerhodar, Vasylivka, Orikhiv, Volnovakha, Novoazovs'k, Taganrog, Azov, Batay'sk, Kugey, Kagal'nitskaya, Veselyy

oznesens'k, Bashtanka, Kakhovs'ke Vodoskhovyshche, Nikopol', Dniprorudne, Tokmak, Smyrnove, Volodars'k, Kuybysheve, Mariupol', Gulf of Taganrog, Yeysk, Zernograd, Yegorlykskaya

Nova Odesa, Snihurivka, Beryslav, Nyzhni Sirohozy, Vesele, Chernivka, Berdyans'k, Prymors'k, Staroshcherbinovskaya, Starominskaya, Leningradskaya, Kushchevskaya

Mykolayiv, Kakhovka, Melitopol', Molochna, Dolgaya Kosa, Starominskaya, Pavlovskaya, Sosyka

nnyy Buh, Nova Kakhovka, Yakymivka, Pryazovs'ke, Kamyshevatskaya, Beysugskiy Liman, Primorsko-Akhtarsk, Brin'kovskaya, Krylovskaya, Novopokrovskaya

hakiv, Kherson, Tsyurupyns'k, Kalanchak, Novotroyits'ke, Syvas'ke, Timashevsk, Bryukhovetskaya, Beysug, Berezanskaya, Vyselki, Tikhoretsk

Hola Prystan', Chaplynka, Novooleksiyivka, Henichesk, Kosa Biryuchyy Ostriv, Sea of Azov, Achuyevo, Korenovsk, Dinskaya, Ust'-Labinsk, Kropotkin

Skadovs'k, Armyans'k, Krasnoperekops'k, Rozdol'ne, Zatoka Syvash, Dzhankoy, Primorsko, Kalininskaya, Poltavskaya, Slavyansk-na-Kubani, Tbilisskaya, Kavkazskaya

Karkinits'ka Zatoka, Pervomays'ke, Nyzhn'ohirs'kyy, Sovyets'kyy, Kerch, Temryukskiy Zaliv, Temryuk, Krasnodar, Adygeysk, Giaginskaya

Mys Prubiynyy, Chornomors'ke, Krasnohvardiys'ke, Kirovs'ke, Lenine, Prymors'kyy, Starotitarovskaya, Krymsk, Abinsk, Severskaya, Belorechensk, Maykop

CRIMEA, Yevpatoriya, Saky, Crimea (Kryms'kyy Pivostriv), Bilohirs'k, Feodosiya, Anapa, Verkhnebakanskiy, Goryachiy Klyuch, Apsheronsk, Kamennomostskiy

CRIMEA: ADMINISTERED BY RUSSIA, Simferopol', Demerdzhi, 1239, Sudak, Novorossiysk, Gelendzhik, Khadyzhensk, Dzhubga

Bakhchysaray, Roman-Kosh 1545, Alushta, Kuban', Apsheronsk

Sevastopol', Balaklava, Yalta, Mys Ayya, Novomikhaylovskiy, Tuapse, Sochi

B L A C K   S E A

Scale bars:
200 KILOMETRES, 100, 0
150, 100, 50, 0 MILES

1:4 800 000

© Collins Bartholomew Ltd

91

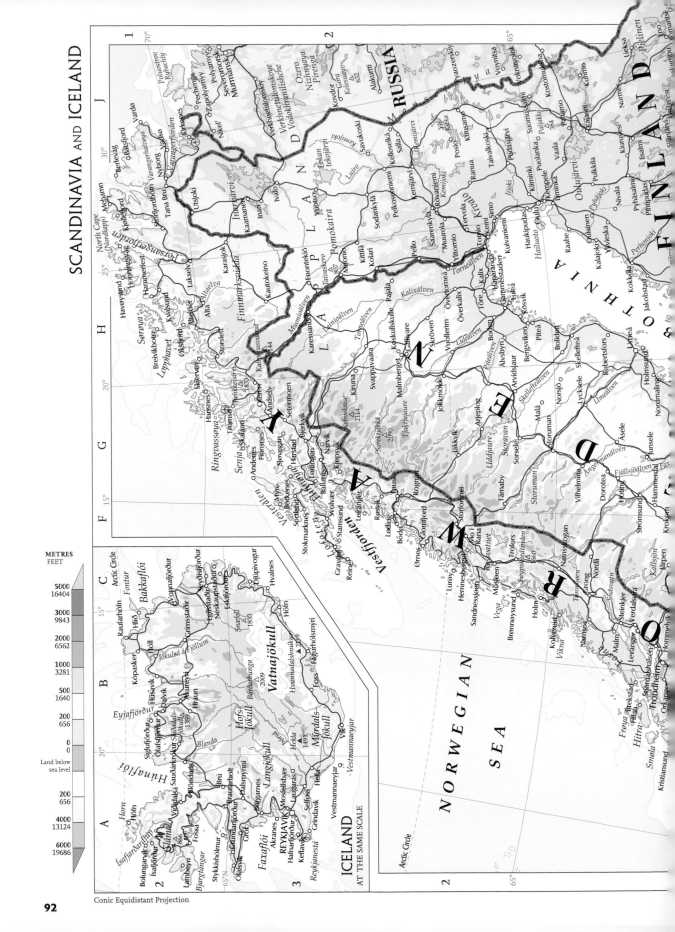

RUSSIA

FINLAND

OF BOTHNIA

GULF

NORWEGIAN SEA

SWEDEN

NORWAY

ICELAND
AT THE SAME SCALE

Conic Equidistant Projection

METRES
FEET

5000
16404

3000
9843

2000
6562

1000
3281

500
1640

200
656

0
0

Land below
sea level

200
656

4000
13124

6000
19686

Arctic Circle

Vatnajökull

Bakkaflói

Hofsjökull

Langjökull

Mýrdalsjökull

Faxaflói

Húnaflói

REYKJAVIK

Hekla

Eyjafjörður

North Cape
(Nordkapp)

Finnmarksvidda

Lofoten

Vestfjorden

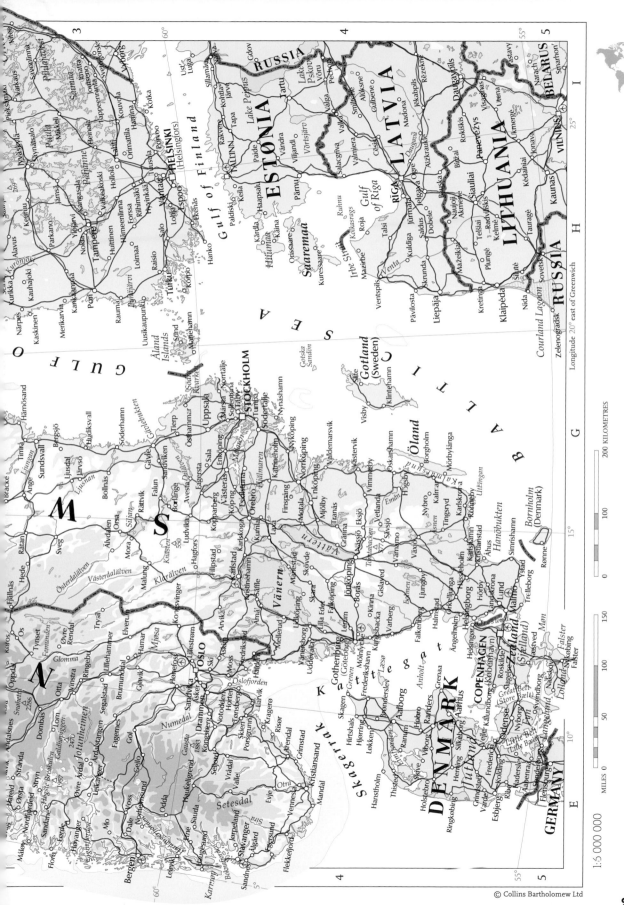

© Collins Bartholomew Ltd

1:5 000 000

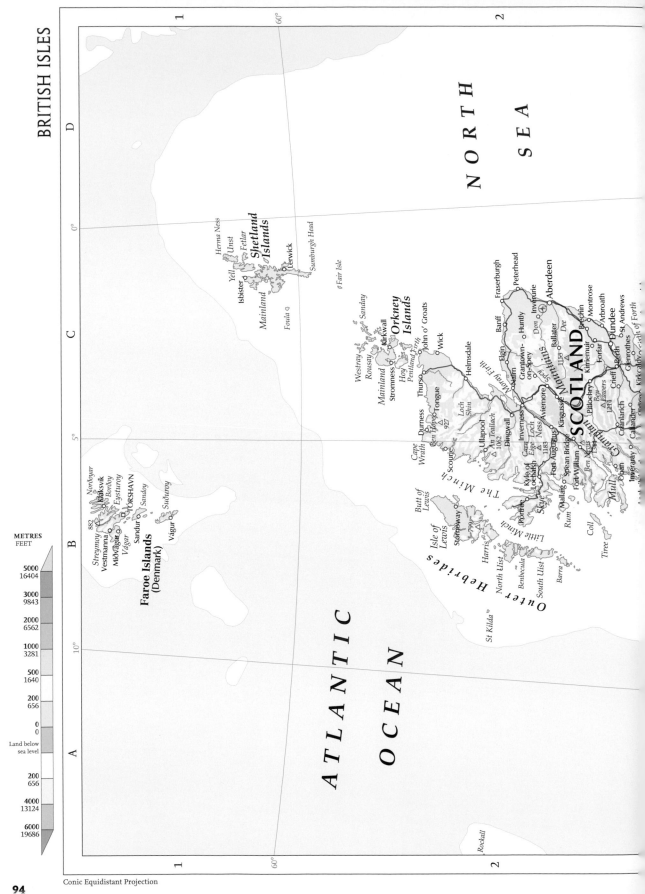

METRES
FEET

5000
16404

3000
9843

2000
6562

1000
3281

500
1640

200
656

0
0

Land below
sea level

200
656

4000
13124

6000
19686

ATLANTIC

OCEAN

NORTH

SEA

Faroe Islands
(Denmark)

SCOTLAND

Shetland
Islands

Orkney
Islands

Outer Hebrides

The Minch

Isle of
Lewis

Rockall

Conic Equidistant Projection

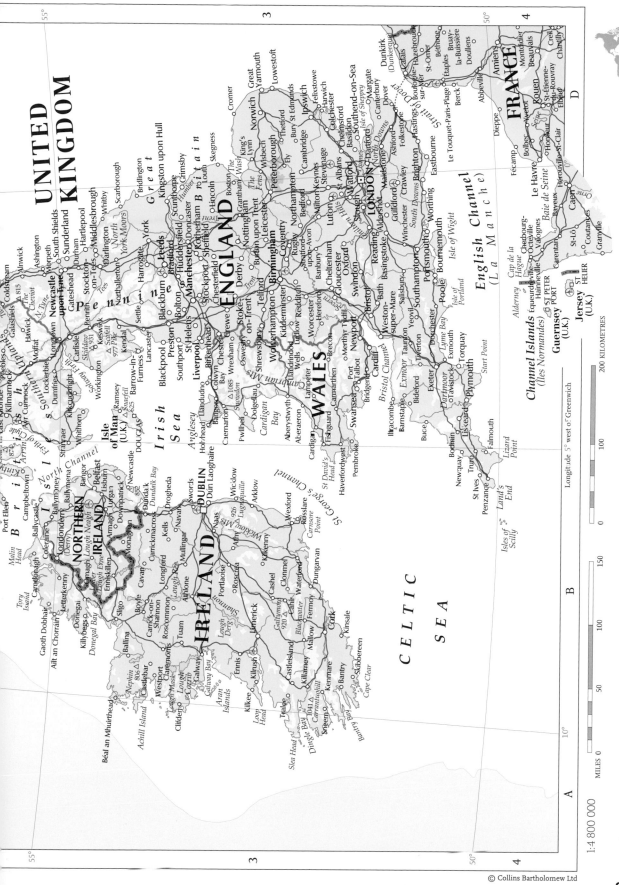

UNITED KINGDOM

Great Britain

ENGLAND

WALES

NORTHERN IRELAND

IRELAND

Isle of Man (U.K.)

British Isles

Southern Uplands

Pennines

Cambrian Mts

FRANCE

Irish Sea

North Channel

St George's Channel

CELTIC SEA

English Channel (La Manche)

Channel Islands (Îles Normandes)

Guernsey (U.K.)

Jersey (U.K.)

ST PETER PORT

ST HELIER

DUBLIN

LONDON

Belfast

Cardiff

Baie de Seine

Strait of Dover

Bristol Channel

Lyme Bay

Cardigan Bay

Galway Bay

Dingle Bay

Bantry Bay

Donegal Bay

Dundalk Bay

Firth of Clyde

Cromer
Great Yarmouth
Lowestoft
Norwich
King's Lynn
Wisbech
Peterborough
Thetford
Bury St Edmunds
Ipswich
Harwich
Felixstowe
Colchester
Chelmsford
Southend-on-Sea
Margate
Canterbury
Dover
Folkestone
Hastings
Eastbourne
Brighton
Worthing
Crawley
Maidstone
Dartford
Watford
St Albans
Basildon
Ashford

Dunkirk (Dunkerque)
Calais
Boulogne-sur-Mer
St-Omer
Hazebrouck
Béthune
Bruay-la-Buissière
Douliens
Le Touquet-Paris-Plage
Berck
Abbeville
Amiens
Dieppe
Fécamp
Le Havre
Bolbec
Honfleur
Caen
Bayeux
St-Lô
Coutances
Granville
Carentan
Valognes
Octeville
Cherbourg-Octeville
Montdidier
Beauvais
Rouen
St-Étienne-du-Rouvray
Elbeuf
Héricourt-St-Clair
Montivilliers
Creil
Chantilly

Alnwick
Ashington
Morpeth
Newcastle upon Tyne
South Shields
Sunderland
Gateshead
Durham
Hartlepool
Stockton-on-Tees
Middlesbrough
Darlington
Whitby
Scarborough
Bridlington
York
Harrogate
Leeds
Bradford
Huddersfield
Halifax
Wakefield
Doncaster
Scunthorpe
Grimsby
Kingston upon Hull
Louth
Skegness
Boston
Lincoln
Grantham
Nottingham
Derby
Chesterfield
Sheffield
Rotherham
Barnsley
Manchester
Stockport
Bolton
Blackburn
Preston
Blackpool
Southport
St Helens
Liverpool
Birkenhead
Chester
Crewe
Stoke-on-Trent
Telford
Stafford
Wolverhampton
Kidderminster
Birmingham
Coventry
Leicester
Rugby
Northampton
Milton Keynes
Bedford
Luton
Stevenage
Cambridge
Ely

Newquay
Truro
St Ives
Penzance
Land's End
Isles of Scilly
Lizard Point
Falmouth
Bodmin
Liskeard
Plymouth
Tavistock
Dartmoor
Okehampton
Exeter
Exmouth
Tiverton
Torquay
Start Point
Bideford
Barnstaple
Ilfracombe
Taunton
Yeovil
Dorchester
Bridport
Weymouth
Isle of Portland
Poole
Bournemouth
Isle of Wight
Southampton
Portsmouth
Winchester
Salisbury
Basingstoke
Reading
Newbury
Oxford
Swindon
Gloucester
Cheltenham
Worcester
Hereford
Ludlow
Shrewsbury
Oswestry
Wrexham
Bristol
Bath
Weston-super-Mare
Bridgwater

St David's Head
Fishguard
Haverfordwest
Pembroke
Cardigan
Aberaeron
Aberystwyth
Lampeter
Carmarthen
Swansea
Port Talbot
Bridgend
Newport
Brecon
Merthyr Tydfil
Llandrindod Wells
Dolgellau
Pwllheli
Caernarfon
Bangor
Holyhead
Anglesey
Colwyn Bay
Llandudno
Rhyl

Douglas
Ramsey
Snaefell 621

Carlisle
Workington
Whitehaven
Keswick
Kendal
Lancaster
Morecambe
Barrow-in-Furness
Skiddaw 931
Scafell Pike 977
Settle

Dumfries
Lockerbie
Moffat
Dalkeith
Kilmarnock
Ayr
Cumnock
Galashiels
Hawick
Jedburgh
Peebles
Coldstream
Berwick-upon-Tweed
Stranraer
Girvan
Newton Stewart
Kirkcudbright
Whithorn
The Cheviot 815
Gretna

Campbeltown
Port Ellen
Arran

Malin Head
Carndonagh
Gaoth Dobhair
Ailt an Chorráin
Letterkenny
Killybegs
Donegal
Ballybofey
Carrickmacross
Sligo
Ballina
Westport
Castlebar
Claremorris
Roscommon
Carrick-on-Shannon
Boyle
Tuam
Galway
Ennis
Kilrush
Kilkee
Tralee
Loop Head
Limerick
Nenagh
Thurles
Portlaoise
Athlone
Mullingar
Longford
Cavan
Monaghan
Clones
Enniskillen
Armagh
Dungannon
Cookstown
Coleraine
Londonderry (Derry)
Strabane
Omagh
Lurgan
Portadown
Lisburn
Ballymena
Larne
Newtownabbey
Bangor
Downpatrick
Newcastle
Newry
Dundalk
Drogheda
Swords
Dún Laoghaire
Bray
Wicklow
Arklow
Gorey
Naas
Kells
Navan
Kildare
Portarlington
Tullamore
Birr
Mountmellick
Kilkenny
Carlow
New Ross
Wexford
Rosslare
Carnsore Point
Waterford
Dungarvan
Clonmel
Cashel
Tipperary
Mitchelstown
Fermoy
Mallow
Castleisland
Killarney
Kenmare
Sneem
Bantry
Skibbereen
Cape Clear
Kinsale
Cork
Youghal

Lough Neagh
Lough Erne
Lough Ree
Lough Derg
Lough Corrib
Lough Mask
Shannon
Blackwater
Barrow
Nore
Suir

Wicklow Mts
Lugnaquillia 926

Carrantuohill 1041
Nephin 806

Tory Island
Achill Island
Aran Islands
Clifden
Loop Head
Slea Head
Dingle

Béal an Mhuirthead

Beachy Head
North Downs
South Downs
The Fens
The Wash
Humber
Trent
Severn
Thames
Tees
North York Moors

Cap de la Hague
Alderney
Écréhou
Équeurdreville-Hainneville

CELTIC SEA

Start Point

Exmoor

Longitude 5° west of Greenwich

1:4 800 000

MILES 0   50   100   150

KILOMETRES 0   100   200

© Collins Bartholomew Ltd

95

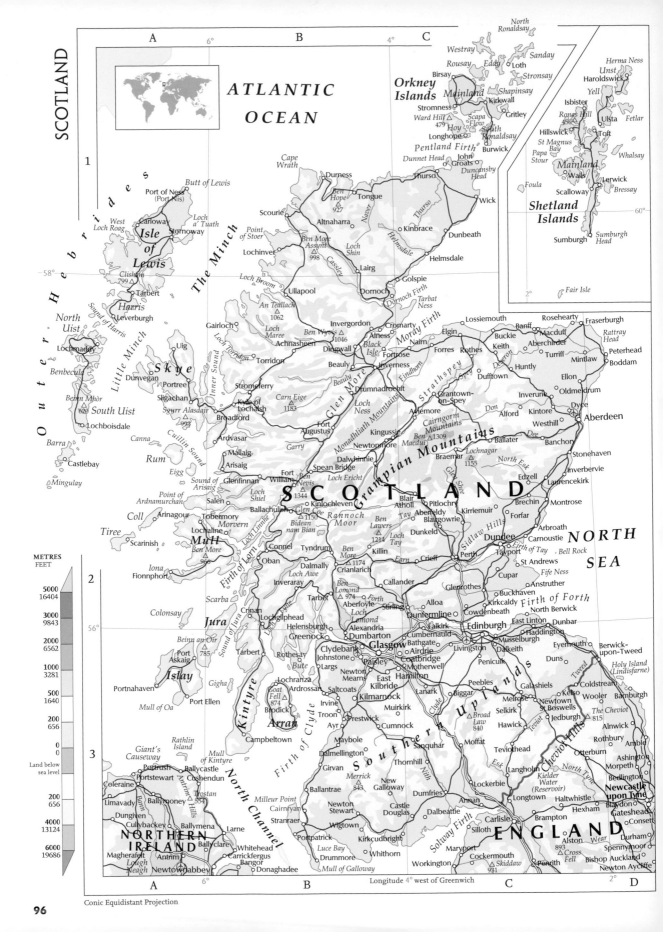

ATLANTIC
OCEAN

1

Outer Hebrides

The Minch

Orkney Islands

North Ronaldsay
Westray
Rousay
Eday
Loth
Sanday
Birsay
Shapinsay
Stronsay
Mainland
Stromness
Kirkwall
Gritley
Ward Hill △
479
Scapa
Flow
Hoy
South
Ronaldsay
Longhope
Burwick
Pentland Firth
Dunnet Head
John o'
Groats
Duncansby
Head

Herma Ness
Unst
Haroldswick
Yell
Isbister
Ronas Hill △
458
Ulsta
Fetlar
Hillswick
St Magnus
Bay
Papa
Stour
Toft
Mainland
Foula
Walls
Lerwick
Whalsay
Bressay
Scalloway
Sumburgh
Sumburgh
Head
Shetland
Islands
60°
2°
Fair Isle

Cape
Wrath
Durness
Tongue
Thurso
Wick
Ben
Hope △
927
Scourie
Altnaharra
Naver
Kinbrace
Dunbeath
Point
of Stoer
Lochinver
Ben More
Assynt △
998
Loch
Shin
Lairg
Helmsdale
Helmsdale
Golspie
Ullapool
Loch Broom
An Teallach △
1062
Dornoch
Dornoch Firth
Tarbat
Ness

Butt of Lewis
Port of Ness
(Port Nis)
Loch
a' Tuath
West
Loch Roag
Carloway
Stornoway
Isle
of
Lewis
Clishim △
799
Tarbert
Harris
Leverburgh
Sound of Harris

North
Uist
Lochmaddy
Benbecula
Beinn Mhòr △
620
South Uist
Lochboisdale
Barra
Castlebay
Mingulay

Little Minch

Gairloch
Loch
Maree
Achnasheen
Ben Wyvis △
1046
Alness
Invergordon
Cromarty
Moray Firth
Lossiemouth
Rosehearty
Fraserburgh
Elgin
Banff
Macduff
Buckie
Aberchirder
Rattray
Head
Peterhead
Boddam
Torridon
Loch Torridon
Inner
Sound
Dingwall
Black
Isle
Fortrose
Nairn
Forres
Rothes
Keith
Turriff
Mintlaw
Ellon
Oldmeldrum
Skye
Stromeferry
Carn Eige △
1183
Beauly
Glen More
Inverness
Drumnadrochit
Findhorn
Dufftown
Huntly
Spey
Inverurie
Westhill
Dyce
Aberdeen
Dunvegan
Portree
Kyle of
Lochalsh
Broadford
Sligachan
Sgurr Alasdair △
993
Loch
Ness
Strathspey
Deveron
Don
Alford
Kintore
Banchory
Cuillin Sound
Ardvasar
Fort
Augustus
Monadhliath Mountains
Aviemore
Grantown-
on-Spey
Cairngorm
Mountains
Ballater
Stonehaven
Inverbervie
Mallaig
Garry
Kingussie
Newtonmore
Ben
Macdui △
1309
Dee
Braemar
Lochnagar △
1155
North Esk
Laurencekirk
Arisaig
Spean Bridge
Dalwhinnie
Glen Shee
Edzell
Brechin
Montrose

Point of
Ardnamurchan
Sound of
Arisaig
Glenfinnan
Loch
Shiel
Fort
William
Ben
Nevis △
1344
Loch Eil
Blair
Atholl
Pitlochry
Grampian Mountains
SCOTLAND
Forfar
Arbroath
Carnoustie
NORTH

Eigg
Rum
Salen
Ballachulish
Kinlochleven
Glen Coe △
1150
Bidean
nam Bian
Rannoch
Moor
Ben
Lawers △
1214
Loch
Tay
Aberfeldy
Blairgowrie
Dunkeld
Kirriemuir
Sidlaw Hills
Dundee
SEA

Coll
Arinagour
Tobermory
Morvern
Lochaline
Connel
Oban
Tyndrum
Ben
More △
1174
Killin
Earn
Crieff
Perth
Tayport
Firth of Tay
Bell Rock

Tiree
Scarinish
Mull
Ben More △
966
Loch Linnhe
Loch Etive
Dalmally
Crianlarich
Loch Awe
Callander
St Andrews
Fife Ness

Iona
Fionnphort
Loch Lomond
Inveraray
Ben
Lomond △
974
Forth
Aberfoyle
Stirling
Cupar
Glenrothes
Anstruther

Colonsay
Scarba
Jura
Crinan
Lochgilphead
Tarbet
Helensburgh
Greenock
Alexandria
Dumbarton
Alloa
Clackmannan
Kirkcaldy
Buckhaven
North Berwick
Firth of Forth
Beinn an Òir △
785
Sound of Jura
Tarbert
Rothesay
Bute
Largs
Clydebank
Glasgow
Airdrie
Coatbridge
Motherwell
Cumbernauld
Falkirk
Livingston
Edinburgh
Dalkeith
Musselburgh
Haddington
East Linton
Dunbar

Islay
Port
Askaig
Gigha
Johnstone
Paisley
Newton
Mearns
East
Kilbride
Hamilton
Bathgate
Penicuik
Peebles
Duns
Eyemouth
Berwick-
upon-Tweed

Portnahaven
Port Ellen
Mull of Oa
Lochranza
Ardrossan
Saltcoats
Kilmarnock
Lanark
Biggar
Galashiels
Melrose
Newtown
St Boswells
Kelso
Coldstream
Holy Island
(Lindisfarne)
Tweed
Wooler
Bamburgh

Kintyre
Goat
Fell △
874
Brodick
Arran
Irvine
Troon
Prestwick
Ayr
Muirkirk
Clyde
Broad
Law △
840
Selkirk
Jedburgh
The Cheviot △
815
Alnwick
Amble

Giant's
Causeway
Rathlin
Island
Mull of
Kintyre
Campbeltown
Firth of Clyde
Maybole
Cumnock
Sanquhar
Moffat
Southern Uplands
Hawick
Teviot
Cheviot Hills
North Tyne
Rothbury
Otterburn
Ashington
Morpeth
Bedlington

Portrush
Portstewart
Ballycastle
Cushendun
Antrim Hills
Trostan △
554
North Channel
Milleur Point
Cairnryan
Stranraer
Dalmellington
Thornhill
Merrick △
843
New
Galloway
Dumfries
Teviothead
Langholm
Esk
Kielder
Water
(Reservoir)
Longtown
Haltwhistle
Hexham
Blaydon
Newcastle
upon Tyne
Gateshead

Coleraine
Limavady
Ballymoney
Ballymena
Larne
Ballyclare
Whitehead
Carrickfergus
Bangor
Ballantrae
Newton
Stewart
Wigtown
Portpatrick
Kirkcudbright
Castle
Douglas
Dalbeattie
Annan
Gretna
Carlisle
Brampton
Silloth
Maryport
Solway Firth
Alston
Wear
Consett
Durham
Spennymoor
Bishop Auckland
Newton Aycliffe

ENGLAND
Cross
Fell △
893
Skiddaw △
931
Penrith
Cockermouth
Workington

Magherafelt
Cullybackey
Dungiven
NORTHERN
IRELAND
Lough
Neagh
Antrim
Newtownabbey
Donaghadee
Luce Bay
Drummore
Mull of Galloway
Whithorn

Longitude 4° west of Greenwich

METRES
FEET

5000
16404

3000
9843

2000
6562

1000
3281

500
1640

200
656

0
0

Land below
sea level

200
656

4000
13124

6000
19686

A
6°
B
4°
C
2°
D

96

Conic Equidistant Projection

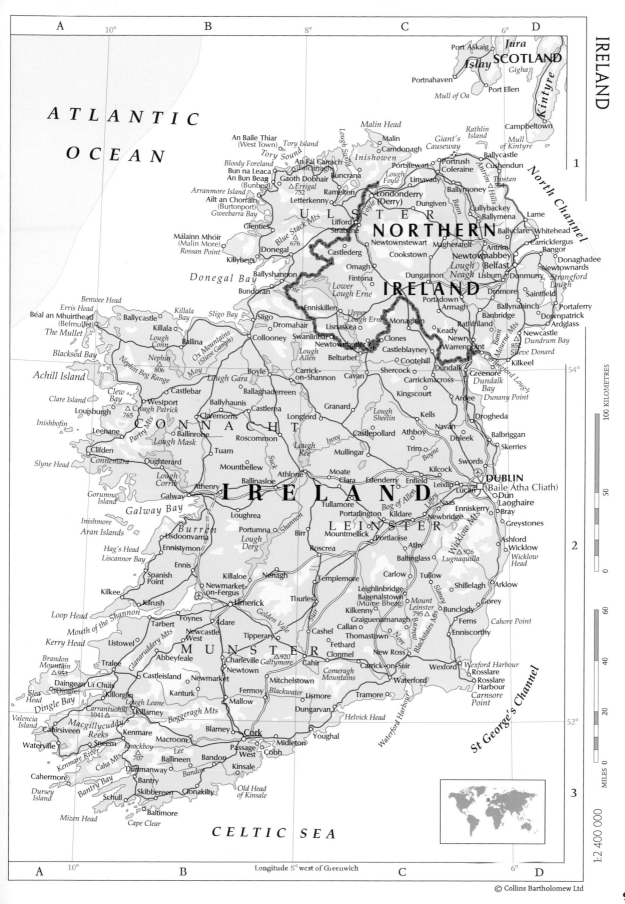

ATLANTIC

OCEAN

SCOTLAND
Port Askaig
Jura
Islay
Gigha
Portnahaven
Port Ellen
Mull of Oa
Campbeltown
Mull of Kintyre

North Channel

Malin Head
An Baile Thiar (West Town)
Tory Island
Tory Sound
Malin
Carndonagh
Giant's Causeway
Rathlin Island
Ballycastle
Cushendun
Bloody Foreland
An Fál Carrach (Falcarragh)
Inishowen
Portstewart
Portrush
Coleraine
Antrim Hills
Bun na Leaca
An Bun Beag (Bunbeg)
Gaoth Dobhair
Lough Foyle
Limavady
Ballymoney
Trostan △ 554
Arranmore Island
Errigal △ 752
Letterkenny
Ramelton
Londonderry (Derry)
Dungiven
Cullybackey
Larne
Ailt an Chorráin (Burtonport)
Gweebarra Bay
Lifford
Strabane
Ballymena
Ballyclare
Whitehead
Carrickfergus
Rossan Point
Glenties
Blue Stack Mts △ 676
Castlederg
Newtownstewart
Magherafelt
Newtownabbey
Bangor
Donaghadee
Málainn Mhóir (Malin More)
Donegal
Cookstown
Antrim
Belfast
Newtownards
Killybegs
Ballyshannon
Omagh
Dungannon
Lough Neagh
Lisburn
Dunmurry
Strangford Lough
Saintfield
Donegal Bay
Bundoran
Fintona
Erne
Lower Lough Erne
Enniskillen
Upper Lough Erne
Monaghan
Portadown
Armagh
Dromore
Ballynahinch
Downpatrick
Ardglass
Portaferry
Benwee Head
Erris Head
Béal an Mhuirthead (Belmullet)
Ballycastle
Killala Bay
Killala
Sligo Bay
Sligo
Dromahair
Lisnaskea
Clones
Keady
Newry
Warrenpoint
Mourne Mts
Newcastle
Sleve Donard △ 853
Dundrum Bay
Kilkeel
The Mullet
Lough Conn
Ballina
Ox Mountains (Sliebe Gamph)
Collooney
Swanlinbar
Newtownbutler
Castleblayney
Cootehill
Dundalk
Carlingford Lough
Greenore
Blacksod Bay
Nephin △ 806
Moy
Belturbet
Shercock
Dundalk Bay
Dunany Point
Achill Island
Nephin Beg Range
Lough Gara
Boyle
Carrick-on-Shannon
Cavan
Carrickmacross
Ardee
Clare Island
Clew Bay
Westport
Castlebar
Lough Allen
Kingscourt
Louisburgh
Croagh Patrick △ 765
Ballaghaderreen
Granard
Kells
Drogheda
Inishbofin
Leenane
Claremorris
Castlerea
Longford
Lough Sheelin
Navan
Duleek
Balbriggan
Clifden
Ballinrobe
CONNACHT
Roscommon
Castlepollard
Athboy
Trim
Skerries
Slyne Head
Partry Mts
Lough Mask
Tuam
Lough Ree
Mullingar
Boyne
Swords
Connemara
Lough Corrib
Oughterard
Mountbellew
Suck
Athlone
Moate
Kilcock
Gorumna Island
Athenry
Ballinasloe
Clara
Edenderry
Enfield
Leixlip
DUBLIN
Inishmore
Aran Islands
Galway
Loughrea
IRELAND
Tullamore
Bog of Allen
Naas
(Baile Átha Cliath)
Lucan
Dún Laoghaire
Galway Bay
Burren
Portumna
Birr
Portarlington
Kildare
Newbridge
Enniskerry
Bray
Lisdoonvarna
Lough Derg
Shannon
LEINSTER
Mountmellick
Portlaoise
Wicklow Mts △ 926
Greystones
Hag's Head
Ennistymon
Roscrea
Athy
Ashford
Wicklow
Liscannor Bay
Ennis
Slaney
Baltinglass
Lugnaquilla
Wicklow Head
Spanish Point
Killaloe
Nenagh
Templemore
Carlow
Tullow
Shillelagh
Arklow
Kilkee
Newmarket-on-Fergus
Thurles
Leighlinbridge
Gorey
Kilrush
Limerick
Bagenalstown (Muine Bheag)
Mount Leinster △ 795
Bunclody
Ferns
Cahore Point
Loop Head
Tarbert
Foynes
Adare
Golden Vale
Suir
Cashel
Kilkenny
Graiguenamanagh
Enniscorthy
Mouth of the Shannon
Newcastle West
Tipperary
Callan
Thomastown
New Ross
Kerry Head
Listowel
MUNSTER
Abbeyfeale
Charleville
Fethard
Clonmel
Carrick-on-Suir
Wexford
Wexford Harbour
Rosslare
Brandon Mountain △ 953
Tralee
Galtymore △ 920
Cahir
Comeragh Mountains
Rosslare Harbour
Slea Head
Daingean Uí Chúis (Dingle)
Castleisland
Newtown
Mitchelstown
Waterford
Tramore
Carnsore Point
Killorglin
Blackwater
Fermoy
Lismore
Dingle Bay
Newmarket
Kanturk
Mallow
Dungarvan
Helvick Head
Valencia Island
Carrantuohill △ 1041
Lough Leane
Killarney
Boggeragh Mts
Waterford Harbour
Waterville
Macgillycuddy's Reeks
Kenmare
Macroom
Blarney
Cork
Midleton
Youghal
St George's Channel
Cahirsiveen
Sneem
Knockboy △ 707
Lee
Passage West
Cobh
Caha Mts
Ballineen
Bandon
Kinsale
Cahermore
Kenmare River
Dunmanway
Bantry
Bandon
Dursey Island
Bantry Bay
Skibbereen
Clonakilty
Old Head of Kinsale
Schull
Baltimore
Mizen Head
Cape Clear

CELTIC SEA

Longitude 8° west of Greenwich

100 KILOMETRES
50
0

MILES
60
40
20
0

1 : 2 400 000

© Collins Bartholomew Ltd

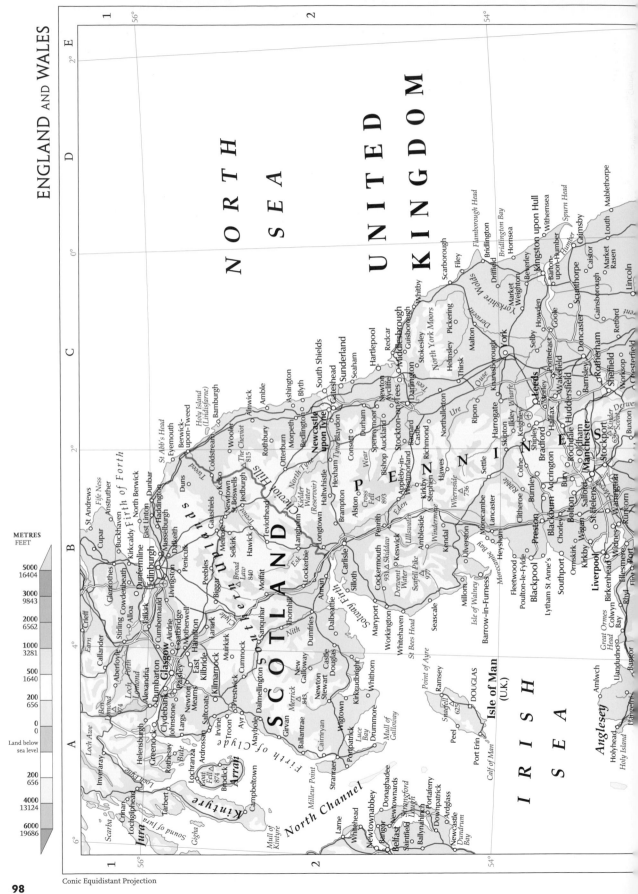

Conic Equidistant Projection

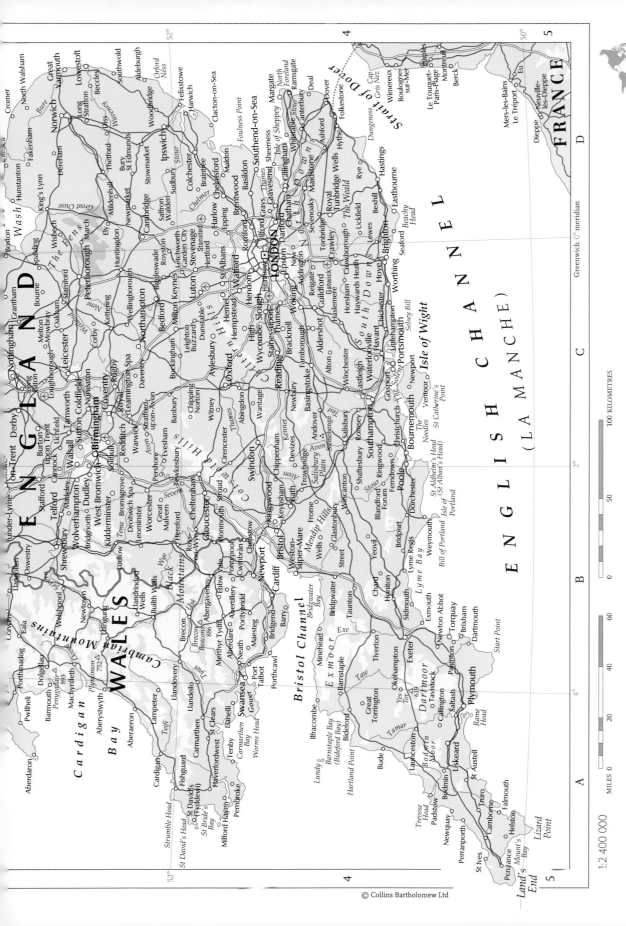

© Collins Bartholomew Ltd

1:2 400 000

A   4°   B   6°   C

NORTH SEA

East Frisian Islands
Spiekeroog
Norderney Langeoog Langeoog
Juist Norderney
Borkum Norden Westerham
Borkum Wittm
OSTFRIESLAN
West Frisian Islands Schiermonnikoog The Wadden Sea Hinte Aurich Wiesn
Terschelling Ameland Uithuizen Delfzijl Emden
West- Hollum Eemnum Appingedam Leer Westers
Terschelling Ferwerd Dokkum Bedum (Ostfriesland) Strücklingen (Saterland)
Oost- Burdaard Oenkerk Kollum Hoogezand- Groningen Winschoten Papenbur
Vlieland Harlingen Franeker Leeuwarden Sappemeer Veendam Friesoy
Vlieland Witmarsum Reduzum Drachten Assen Stadskanaal Walchum
Texel Den Burg Bolsward Sneek Heerenveen Beilen Emmen Haren (Ems)
Marsdiep Wieringerwerf Sloten Wolvega Steenwijk Hoogeveen Coevorden Löningen
Den Helder IJsselmeer Meppel Hardenberg Groß Hesepe Meppen
Schagen Enkhuizen Creil Emmeloord Kloosterhaar Lingen Fürste
Nieuwe Niedorp Urk Kraggenburg Ommen Vriezenveen Nordhorn
Heerhugowaard Hoorn Markermeer Kampen Zwolle Almelo Rheine
Bergen Berkhout Lelystad Dronten Raalte Borne Oldenzaal Ibben
Alkmaar Purmerend NETHERLANDS Heerde Nijverdal Gronau
Castricum Zaandam Driemond Harderwijk Deventer Hengelo (Westfalen)
Beverwijk AMSTERDAM Naarden Nijkerk Torenberg Apeldoorn Enschede Emsdetten
IJmuiden Haarlem Amstelveen 107 Eibergen Ahaus Steinfurt
Zandvoort Hillegom Hilversum Amersfoort Zutphen Hoog- Winterswijk Greven
Noordwijk Leiden Maarssen Utrecht Veenendaal Ede Doesburg Keppel Borken Coesfeld Mür
Katwijk aan Zee Alphen aan den Rijn Waddinxveen Neder Rijn Arnhem Doetinchem Velen Dülmen Asche
THE HAGUE Delft Gouda Nieuwegein Wageningen Zevenaar Bocholt MÜNSTERLAND
('s-Gravenhage) Nieuwegein Culemborg Waal Andelst Borken Dorsten Recklinghausen Hamm
(Den Haag) Rotterdam Schoonhoven Tiel Nijmegen Kleve Rhein Dorsten Marl Lüne Dort
Hook of Holland Vlaardingen Capelle aan Gorinchem Oss Wijchen Goch Wesel Gelsenkirchen Herne
(Hoek van Holland) den IJssel Maas Kevelaer Dinslaken Bottrop Tröndenberg Hagen
Helvoetsluis Spijkenisse Dordrecht Hertogenbosch Wanroij Venray Duisburg Essen Bochum Iserl
Scharendijke Middelharnis Oosterhout Uden Erp St Anthonis Moers Mülheim an der Ruhr Hagen
Burgh- Zierikzee Zevenbergen Waalwijk Best Deurne Venlo Krefeld Ratingen Lüdenscheid
Haamstede Oosterschelde Roosendaal Breda Tilburg Boxtel Helmond Asten Mönchengladbach Düsseldorf Remscheid
Westkapelle Middelburg Halsteren Soes Bergen op Zoom Eindhoven Roermond Viersen Hilden Wuppertal Attend
Koudekerke Vlissingen Zandvliet Brecht Veldhoven Neuss Solingen Gummersb
Knokke- Breskens Sluis Kapellen Westmalle Lille Valkenswaard Weert Herkenbosch Wegberg Grevenbroich Dormagen Leverkusen
Zeebrugge Heist Maldegem St-Niklaas Schilde Turnhout Lommel Kessel Sittard Hückelhoven Bergisch Gladbach
Blankenberge Temeuzen Philippine Antwerp Lier Geel Bocholt Steyl Genk Heerlen Cologne (Köln)
Ostend Meetkerke St-Laureins Antwerpen Hechtel Maaseik Kerkrade Hürth Wieh
(Oostende) Zedelgem Eeklo (Anvers) Willebroek Mechelen Aarschot Beringen Eschweiler Kerpen Troisdorf
Nieuwpoort Brugge Evergem Dendermonde Diest Hasselt Maastricht Aachen Düren Bonn St Augustin
Veurne (Bruges) Wingene Ghent Wichelen Vilvoorde Leuven Tienen Mechelen Stolberg Kreuzau Königswinter Altenk
Diksmuide Tielt (Gent) Schaerbeek Borgloon Tongeren (Rheinland) Zülpich Meckenheim Neuwied Monta
Roeselare Deinze Anderlecht BRUSSELS Oupeye Raeren Bad Neuenahr- Koblenz
Ieper Kortrijk Zulte Brussel/Bruxelles Liège Mechernich Ahrweiler Rhine (Rhein)
Menen Leie Oudenaarde Uccle Waterloo Halle Verviers Kallo Malmedy Blankenheim Adenau Mayen Lahnstein
Roubaix Mouscron Ronse Nivelles Spa Dahlem Hilleshm Wes
Lille Ath Ottignies Eghezée Seraing St-Vith Gerolstein Cochem Emmelsha
Villeneuve- Soignies Fleurus Braives Durbuy Prüm Daun Mosel
d'Ascq Tournai Lens La Louvière Namur Andenne Huy Vielsalm Arzfeld Manderscheid Blankenrath Simme
Péruwelz Mons Charleroi Assesse Marche- La Roche- Neuerburg Wittlich Bernkastel- Hunsrü
Lens Boussu Frameries Thuin Châtelet Ciney en-Famenne en-Ardenne Bastogne Kues am R
Douai Valenciennes Montignies- Houffalize Clervaux Bitburg Erbeskopf Bir
le-Tilleul Beaumont Dinant Rochefort St-Hubert Salmtal 818 Idar-Oberstein
Maubeuge Aulnoye- Hastière- Philippeville Wiltz Echternach Morbach Bad Kreuzna
Cambrai Aymeries Lavaux Beauraing Libin Ettelbruck Mersch Trier Bad Bad
Caudry Avesnes- Couvin Fumay Bièvre Neufchâteau LUXEMBOURG Konz Reinsfeld Sobernheim Donner
Bohain-en- sur-Helpe Hirson Rocroi Montherné Paliseul 567 Redange Nohfelden Wolfstein
Vermandois La Capelle Bouillon Vresse-sur- Bastogne Virton Arlon Mettlach St Wendel
Péronne Vervins Bogny-sur-Meuse Semois Sedan Carignan LUXEMBOURG Esch- Merzig Neunkirchen Kaisersla
St-Quentin Marle Charleville- Pétange sur-Alzette Saarlouis Homburg
Chauny Serre Rozoy- Mézières Mouzon Pétange Thionville
Tergnier sur-Serre Signy- 316 Stenay Hayange Florange
Noyon Montcornet l'Abbaye Longuyon Rombas
Laon FRANCE Vouziers Dun-sur- Spincourt
Attichy Rethel Meuse Consenvoye
Soissons Aisne Guignicourt Longuyon
Courmelles Guise Béthen Oise Mézières
Villers- Fismes Tinqueux Reims
Cotterêts Oise

METRES FEET

5000 / 16404
3000 / 9843
2000 / 6562
1000 / 3281
500 / 1640
200 / 656
0 / 0
Land below sea level
200 / 656
4000 / 13124
6000 / 19686

Longitude 6° east of Greenwich

A   4°   B   C

Conic Equidistant Projection

A   B   C

5°

10°

DENMARK

Rør

NORTH

SEA

Helgoland

Helgoländer
Bucht

West Frisian Islands

East Frisian Islands

The Wadden Sea

NETHERLANDS

AMSTERDAM

Haarlem

THE HAGUE
('s-Gravenhage)
(Den Haag)

Breda

Tilburg

Antwerp
(Antwerpen)

Mechelen

BRUSSELS
(Brussel)
(Bruxelles)

BELGIUM

Namur

Dinant

50°

LUXEMBOURG

LUXEMBOURG

FRANCE

LORRAINE

G E R M A N Y

BAYERN

Munich
(München)

Stuttgart

Black Forest
(Schwarzwald)

AU

SWITZERLAND

BERN

ZÜRICH

**LIECHTEN-
STEIN**

VADUZ

A L P S

ITALY

Dolomites

SL

LJUBLJAN

1

2

METRES
FEET

5000
16404

3000
9843

2000
6562

1000
3281

500
1640

200
656

0
0

Land below
sea level

200
656

4000
13124

6000
19686

B

C

Longitude 10° east of Greenwich

Conic Equidistant Projection

© Collins Bartholomew Ltd

A · 5° · B · 0°

UNITED KINGDOM

Bristol Channel
Ilfracombe
Bideford
Barnstaple
Exmoor
Bude
Tiverton
Newquay
Tavistock
St Ives
Truro
Bodmin
Liskeard
Penzance
Plymouth
Land's End
Falmouth
Lizard Point
Isles of Scilly

Bath · Reading · **LONDON** · Dartford · Isle of Sheppey
Weston-super-Mare
Basingstoke · Aldershot · Gillingham · Margate
Taunton · Salisbury · Winchester · Guildford · Maidstone · Canterbury
Yeovil · Southampton · Crawley · Dover · Dunki (Dunkerque)
Dorchester · Exmouth · Worthing · Brighton · Hastings · Folkestone
Poole · Bournemouth · Portsmouth · Eastbourne
Isle of Wight

Start Point
Lyme Bay
Dartmoor
Exeter
Torquay

St-Omer
Calais
Boulogne-sur-Mer · Hazebr
Étaples · Bruay · Buissi
Le Touquet-Paris-Plage · Berck
Strait of Dover

ARTO

English Channel
(La Manche)

Dieppe · Doulle
Fécamp · Amiens
Neufchâtel-en-Bray · PICARDY
Cap de la Hague
Alderney
Tourlaville · Yvetot · Monti
Équeurdreville · Cherbourg-Octeville · Le Havre · Bolbec · Rouen · Beauvai
Guernsey (U.K.) · Hainneville · Honfleur · Elbeu · St-Étienne-du-Rouvray
ST PETER PORT · Valognes · Deauville · Hérouville-St-Clair · Cre
Baie de Seine · Caen · Lisieux · Chanti · Pontoi
Channel Islands · Carentan · Bayeux · Évreux · Mantes-la-Jolie · St-Denis
(Îles Normandes) · Jersey (U.K.) · St-Lô · Coutances · L'Aigle · Dreux · Boulogne-Billancourt · PAI
ST HELIER · Golfe de St-Malo · Granville · Vire · NORMANDY · Argentan · Versailles
Roscoff · Lannion · Cap Fréhel · Avranches · Flers · Sées · Rambouillet · Ev
Lesneven · Guingamp · Dinard · St-Malo · Dol-de-Bretagne · Mayenne · Alençon · Chartres · Mennée
Île d'Ouessant · Guipavas · Morlaix · St-Brieuc · Fougères · Nogent-le-Rotrou · Étampes
Plouzané · Châteaulin · Rostrenen · Lamballe · Dinan · BRITTANY · Mayenne · Nogent-le-Rotrou · Fleury
Brest · Montagnes Noires · Loudéac · Cesson-Sévigné · Château-du-Loir · Artenay · Aubra
Douarnenez · Quimperlé · Pontivy · Rennes · Vitré · Laval · Le Mans · Vendôme · Orléans · Château
Pointe du Raz · Quimper · Redon · Château-Gontier · La Flèche · Châteaudun · sur-L
Concarneau · Lorient · Vannes · Châteaubriant · Blois · Château
Ploemeur · Auray · Angers · Baugé-en-Anjou · Romorantin-Lanthenay · Colline
Île de Groix · Carnac · Ancenis · Tours · Joué-lès-Tours · St-Avertin · Salb
Quiberon · St-Nazaire · Nantes · Saumur · Loire · Sance
Belle-Île · Guérande · Orvault · St-Sébastien-sur-Loire · Vienne · Loches · Vier
La Baule-Escoublac · Vertou · Cholet · Thouars · Châtellerault · Indre · Bourges
Pornic · Les Herbiers · Chinon · Vatan · Château
Noirmoutier-en-l'Île · Challans · Bressuire · Le Blanc · Châteauroux · FRA
Île de Noirmoutier · La Roche-sur-Yon · Parthenay · Poitiers · Argenton-sur-Creuse
St-Jean-de-Monts · Fontenay-le-Comte · Niort · Plaines et · Montmorillon · Montlu
Île d'Yeu · Les Sables-d'Olonne · Seuilau Poitou · Bellac · Le Dorat · Guéret
Talmont-St-Hilaire · Civray · Bourganeuf · Ahur
Île de Ré · La Rochelle · Confolens · Limoges · Aubuss
Pointe de Chassiron · St-Jean-d'Angély · St-Junien · AU
St-Pierre-d'Oléron · Rochefort · Charente · Angoulême · St-Yrieix-la-Perche · Plateaux
Pointe de la Coubre · Saintes · Cognac · Soyaux · au Limousin · Égle
Pointe de Grave · Royan · Barbezieux-St-Hilaire · Uzerche · Tulle
Soulac-sur-Mer · Montendre · Riberac · Brive-la-Gaillarde · Pleau
Gironde · Périgueux · Montignac · Souillac
Pauillac · Ambares · Contras · Le Bugue · Sarlat-la-Canéda
Vet-Lagrave · Libourne · Dordogne
Mérignac · Bordeaux · Montignac · Gourdon · Figeac
Arcachon · Pessac · Gradignan · Bergerac · Lot · Cahors
La Teste-de-Buch · Cestas · Garonne · Marmande · Villeneuve-sur-Lot · Ro
Gujan-Mestras · Langon · Aveyro
Mimizan · Bazas · AQUITAINE · Casteljaloux · Agen · Villefranche-de-Rouergue
Labouheyre · Nérac · Moissac · Carmaux
Morcenx · Roquefort · Castelsarrasin · All
Mont-de-Marsan · Condom · Montauban · Gaillac
Soustons · Grenade · Union · Carmaux
Mar Cantábrico · Dax · Aire-sur-l'Adour · Auch · Toulouse · Puylaurens · Ca
Cabo de Peñas · Tartas · Colomiers · Maza
Gijón/Xixón · Santander · Algorta · San Sebastián · Biarritz · Orthez · Maubourguet · Cugnaux
Avilés · Ribadesella/Ribeseya · (Donostia) · Bayonne · Pau · Muret
Luarca · Salas · Llanes · Laredo · Barakaldo · Irún · St-Jean-de-Luz · Billère · Tarbes · Carcassonne
ASTURIAS · Oviedo · Llangreu/Llanes · Santillana · Arizgoiti · Lourdes · GASCON · Pamiers
Mieres · Langreo · Torrelavega · Bilbao · Eibar · Oloron-Ste-Marie · (GASCOGNE) · Limoux
del Camín · Cabanaquinta/ · Durango · Etxarri-Aranatz · Soulom · St-Gaudens · Foix · Quillan
Peña Ubiña 2417 · Cabañaquinta · Reinosa · Laudio · Arrasate · Tolosa · Bagnères-de-Luchon · ANDORR
Villablino · Torrecerredo 2648 · Vitoria-Gasteiz · NAVARRA · Lourdes · PYRENEES · ANDORRA
Cordillera Cantábrica · Guardo · Aguilar de Campoo · Briviesca · Pamplona · Monte Perdido 3348 · Vielha · LA VELLA · Les Escaldes
San Andres del Rabanedo · León · Saldaña · Estella · Aragón · Aneto 3404 · La Seu d'Urgell
Astorga · Osorno · Nájera · Logroño · Tafalla · Jaca · Berga · Ripoll
Valencia de Don Juan · Sahagún · Burgos · Calahorra · Arguis · La Seu d'Urgell · Tremp
Benavente · Medina de Rioseco · Palencia · Lerma · Sierra de la Demanda · Alfaro · Sadaba · Ejea de los Caballeros · Huesca · Graus

SPAIN

BAY OF BISCAY

Gulf of Gascony

Île d'Ouessant

Pointe du Raz

METRES / FEET

METRES	FEET
5000	16404
3000	9843
2000	6562
1000	3281
500	1640
200	656
0	0
Land below sea level	
200	656
4000	13124
6000	19686

A · 5° · B · Greenwich 0° meridian

Conic Equidistant Projection

© Collins Bartholomew Ltd

1:4 800 000

A 10° B 5°

ATLANTIC
OCEAN

Cabo Ortegal
Punta de Estaca de Bares
Ortigueira
Cervo
Mar Cantábrico
Ferrol
Viveiro
Ribadeo
Luarca
Avilés
Cabo de Peñas
Gijón/Xixón
Ribadesella/Ribeseya
Santander
Laredo
Algorta
A Coruña
Gándara
Salas
La Pola
Siero
Oviedo
Llangreu/
Langreo
Cabanaquinta
Cabañaquinta
Torrecerredo △ 2648
Torrelavega
Santillana
Reinosa
Barakaldo
Bilbao
Laudio
Betanzos
Vilalba
Cangas del Narcea
Mieres
△ 2417 Peña Ubiña
Santiago de Compostela
Ordes
Melide
Lugo
ASTURIAS
Villablino
Cordillerá Cantábrica
Vitoria-Gasteiz
GALICIA
Muros
Becerreá
San Andres del Rabanedo
León
Guardo
Aguilar de Campoo
Miranda de Ebro
Briviesca
Logr
Cape Finisterre (Cabo Fisterra)
Vilagarcía de Arousa
Santa Uxía de Ribeira
Estrada
Lalín
Sarria
Chantada
Ponferrada
Astorga
Saldaña
Osorno
Burgos
Nájera
Pontevedra
Monforte de Lemos
Sierra de Eo
Barco
△ El Teleno 2188
Valencia de Don Juan
Sahagún
Ebro
Sierra de la Dema
Cangas
Marín
Redondela
Ourense
Xinzo de Limia
Truchas
Sierra de la Cabrera
Benavente
Medina de Rioseco
Palencia
Lerma
Aranda de Duero
Vigo
Miño
Cañiza
Fondevila
Verín
Chaves
CASTILLA Y LEÓN
Valladolid
Duero
Ayllón
Tui
Viana do Castelo
Braga
Bragança
Macedo de Cavaleiros
Zamora
Toro
Tordesillas
Cuéllar
Póvoa de Varzim
Guimarães
Vila Real
Mirandela
Tuela
Hermeselle
Medina del Campo
Olmedo
Cerezo de Abajo
Maia
Oporto
Torre de Moncorvo
Serra de Mogadouro
Embalse de Almendra
Salamanca
Arévalo
Segovia
Sierra de Guadarrama
Sigüenz
Guadalajara
Matosinhos
(Porto)
Lamego
Douro
Ledesma
Peñaranda de Bracamonte
△ Peñalara 2430
Alcalá de Henares
Vila Nova de Gaia
Pedroso
São João da Madeira
Meda
Vilar Formoso
Ciudad Rodrigo
Lumbrales
Tormes
Ávila
MADRID
Ovar
Aveiro
Ílhavo
Viseu
Mangualde
Guarda
Nuñomoral
Sierra de Gredos
Móstoles
Fuenlabrada
Getafe
Buei
Mealhada
Águeda
Mondego
Torre 1993
Coimbra
Serra da Estrela
Covilhã
Sabugal
Béjar
Valle del Tiétar
Parla
Emb
Figueira da Foz
Lousã
Fundão
Plasencia
Navalmoral de la Mata
Talavera de la Reina
Torrijos
Aranjuez
Ocaña
Taran
Marinha Grande
Pombal
Castelo Branco
Coria
Toledo
CASTILLA-LA MANCH
Leiria
Batalha
Tomar
Alcántara
Embalse de Valdecañas
Tagus (Tajo)
Madridejos
Alcázar
San Juan
Caldas da Rainha
Torres Novas
Abrantes
Cáceres
Sierra de San Pedro
Trujillo
Miajadas
Sierra de Guadalupe
Embalse de Cíjara
Socuéllamos
Villarrobl
Peniche
Entroncamento
Santarém
Ponte de Sor
Portalegre
EXTREMADURA
Navalvillar de Pela
Guadiana
Ciudad Real
Daimiel
Tome
Torres Vedras
Coruche
Campo Maior
Elvas
Montijo
Mérida
Don Benito
Villanueva de la Serena
Almadén
Manzanares
Vila Franca de Xira
Amadora
LISBON
Cacém
(Lisboa)
Estremoz
Badajoz
Cabeza del Buey
Puertollano
Valdepeñas
Villanueva de los Infantes
Cascais
Almada
Montijo
Redondo
Olivenza
Zafra
Almendralejo
Hinojosa del Duque
Pozoblanco
Cabo Espichel
Setúbal
Alcácer do Sal
Évora
Barragem de Alqueva
Fregenal de la Sierra
Peñarroya-Pueblonuevo
Los Pedroches
Baía de Setúbal
Grândola
Torrão
Amareleja
Azuaga
Sierra Morena
Andújar
Linares
Sines
Cabo de Sines
Beja
Moura
Rosal de la Frontera
Serpa
Cortegana
Constantina
Córdoba
Baeza
Úbeda
Odemira
Aljustrel
Castro Verde
Guadiana
Valverde del Camino
Palma del Río
Guadalquivir
Jaén
Sierra de
Baza
Mértola
Huelva
Almonte
Coria del Río
Lora del Río
Écija
Montilla
Alcaudete
Alcalá la Real
ALGARVE
Portimão
Loulé
Ayamonte
Playa de Castilla
Utrera
Morón de la Frontera
Carmona
Genil
Lucena
Cabra
Priego de Córdoba
Guadix
Granada
Sierra Nevada
△ Mulhacén 3482
Aljezur
Almodôvar
Cabo de Lagos
São Vicente
Sagres
Albufeira
Olhão
Tavira
Cabo de Faro
Santa Maria
Sanlúcar de Barrameda
Lebrija
Seville
(Sevilla)
Marchena
Osuna
Antequera
Loja
Vélez-Málaga
Motril
Adra
El
Las Marismas
Arcos de la Frontera
ANDALUCÍA
Ronda
Málaga
Gol Alm
Costa de la Luz
El Puerto de Santa María
Cádiz
Jerez de la Frontera
Chiclana de la Frontera
Torremolinos
Costa del Sol
Golfo de Cádiz
San Fernando
Vejer de la Frontera
Barbate
Algeciras
La Línea de la Concepción
Estepona
Marbella
Alboran Sea
I. de Albor
Cabo Trafalgar
Pta Almina
Gibraltar (U.K.)
Strait of Gibraltar
Ceuta (Spain)
Tangier
(Tanger)
Cabo Negro
Asilah
Tétouan
Cap des Trois Fourches
MOROCCO

PORTUGAL

SPAIN

METRES
FEET

5000 / 16404
3000 / 9843
2000 / 6562
1000 / 3281
500 / 1640
200 / 656
0 / 0
Land below sea level
200 / 656
4000 / 13124
6000 / 19686

A 10° B 5°

Conic Equidistant Projection

© Collins Bartholomew Ltd

1:4 800 000

200 KILOMETRES

100

0

MILES 0

50

100

150

Greenwich 0° meridian

A   L   P   S

3738

Merano
Bolzano
Dolomites
Cortina d'Ampezzo
Tarvisio
Ortles 3905
Tirano
Laives
Trento
Belluno
Feltre
Vittorio Veneto
Udine
Gorizia
SLO
LJUBLJANA
Logatec

Chiavenna
Sondrio
Riva del Garda
Rovereto
Schio
Conegliano
Pordenone
Monfalcone
Trieste
Koper

Bonneville
Cluses
Martigny
Matterhorn
Chamonix
Mont-Blanc 4478
Bellinzona
Lugano
Lake Como
Lecco
Lake Garda
Valdagno
Treviso
Portogruaro
Rijeka

Annecy 4810
Albertville
Mont Blanc
Aosta
Verbania
Lake Maggiore
Como
Varese
Busto Arsizio
Monza
Bergamo
Brescia
Verona
Vicenza
Padua (Padova)
Venice (Venezia)
Gulf of Venice
Poreč
Istria
Crikve

Chambéry
Aix-les-Bains
St-Egreve
Grenoble
Vercors
Ivrea
Biella
Novara
Milan (Milano)
Rho
Treviglio
Crema
Cremona
Mantua (Mantova)
Legnago
Rovigo
Chioggia
Rovinj
Pula
Cres
Veli Lošin
Lošin

45°

Voiron
Modane
Barre des Ecrins 4102
Cuorgnè
Cirie
Rivoli
Turin (Torino)
Vigevano
Pavia
Lodi
Piacenza
Po
Ferrara
Codigoro
Porto Tolle
Rt Kamenjak

La Mure
Gap
Briançon
Giaveno
Moncalieri
Asti
Casale Monferrato
Alessandria
Tortona
Novi Ligure
Parma
Carpi
Modena
Reno
Argenta
Comacchio

Barcelonnette
Digne-les-Bains
Sisteron
St-Bonnet-en-Champsaur
Cottian Alps
Pinerolo
Saluzzo
Fossano
Alba
Acqui Terme
Reggio nell'Emilia
Bologna
Imola
Faenza
Forlì
Ravenna
Cesenatico

Manosque
Castellane
Verdon
1871 Col de Tende
Tende
Cuneo
Mondovì
Savona
Sestri Levante
Fivizzano
Monte Cimone 2165
Rimini

Grasse
MONTE-CARLO
Ventimiglia
Imperia
Sanremo
Maritime Alps
Albenga
Rapallo
Gulf of Genoa
Carrara
Massa
Barga
Pistoia
Prato
San Marino
SAN MARINO
Pesaro
Fano
Senigallia
Ancona

Draguignan
Cannes
Antibes
Nice
MONACO
Capo Mele
Genoa (Genova)
La Spezia
Viareggio
Lucca
Florence (Firenze)
Scandicci
Arno
Sansepolcro
Cagli
Jesi
Osimo
Civitanova Marche

Brignoles
Fréjus
St-Raphaël
St-Tropez
Côte d'Azur
Cap de St-Tropez
Pisa
Livorno
Empoli
Siena
Arezzo
Cortona
Gubbio
Perugia
Fabriano
Macerata
Fermo
San Bened del Tronto

Toulon
Hyères
Cap Sicié
Îles d'Hyères
Ligurian Sea
Cap Corse
Isola di Capraia
Cecina
San Vincenzo
Piombino
Isola d'Elba
Montepulciano
Marsciano
Todi
Foligno
Ascoli Piceno
Giuliano
Teramo

2

Monte Stello 1307
Isola Pianosa
Castiglione della Pescaia
Grosseto
Lago di Bolsena
Orvieto
Narni
Terni
Rieti
Corno Grande 2912
Penne
Pesc
L'Aquila
Chie
Monte Amaro 2793

St-Florent
L'Île-Rousse
Bastia
Isola di Montecristo
Arcipelago Toscano
Viterbo
Tarquinia
Tivoli
Guidonia Montecelio

Corsica (Corse) (France)
Calvi
Vescovato
Cervione
Monte Rotondo 2622
Corte
Ghisonaccia
Prunelli-di-Fiumorbo
Civitavecchia
VATICAN CITY
ROME (Roma)
Sora
Triven

Capo Rosso
Capo di Feno
Ajaccio
Olmeto
Sartène
Zonza
Punta d'Ovace 1410
Porto-Vecchio
Pomezia
Velletri
Sezze
Frosinone
Cassino
Campoba
Venafr

Capo Pertusato
Bonifacio
Strait of Bonifacio
Anzio
Aprilia
Latina
Sessa
Aurunca

Punta Caprara
Isola Asinara
Golfo dell'Asinara
Arzachena
La Maddalena
Capo Ferro
Sabaudia
Fondi
Gaeta
Golfo di Gaeta
Case
Naples (Napoli)

Porto Torres
Punta Balestrieri 1359
Olbia
Isole Ponziane
Pozzuoli
Pompe
Sorrento

Capo Caccia
Sassari
Oschiri
Budoni
Isola d'Ischia
Isola di Capri

Alghero
Ploaghe
Bonorva
Buddusò
Siniscola
Capo Comino

Sardinia (Sardegna) (Italy)
Macomer
Nuoro
Orosei
Golfo di Orosei

40°

Abbasanta
Oristano
Laconi 1834
Punta La Marmora
Capo di Monte Santu

Capo della Frasca
Mandas
Tertenia
Tortolì
T Y R R H E N I A N   S E A

Guspini
San Gavino Monreale
Serramanna
Villaputzu
Isola di Ustica

Monte Linas 1236
Iglesias
Assemini
Quartu Sant'Elena
Isol Lipa

Portoscuso
Isola di San Pietro
Punta Maxia 1017
Cagliari
Capo Carbonara
Isola Filic

Sant'Antioco
Isola di Sant'Antioco
Pula
Golfo di Cagliari

3

Sicily (Sicilia)
Isola di Ustica
Capo San Vito
Monte Sparagio
Partinico
Palermo
Cefalù

Isola Marettimo
Trapani
Alcamo
Rocca Busambra 1613
Termini Imerese

Marsala
Partanna
M E D I T E R R A N E A N   S E A

La Galite
Mazara del Vallo
Castelvetrano
Caltanissetta
Leonforte

Capo Granitola
Sciacca
Canicattì
Caltagiro
Niscemi

Cap de Fer
Chetaïbi
Sicilian Channel
Agrigento
Licata
Gela

Collo
Skikda
Annaba
Menzel Bourguiba
Bizerte
Rass Jebel
Cap Bon
Di
Golfo di Gela
Vitt

ALGERIA
TUNISIA
Nefza
Mateur
El Kala
Tabarka
Golfe de Tunis

Azzaba
El Hadjar
Cap de Garde
El Tarf
Jedeida

Conic Equidistant Projection

Longitude 10° east of Greenwich

METRES
FEET

5000 / 16404
3000 / 9843
2000 / 6562
1000 / 3281
500 / 1640
200 / 656
0 / 0
Land below sea level
200 / 656
4000 / 13124
6000 / 19686

© Collins Bartholomew Ltd

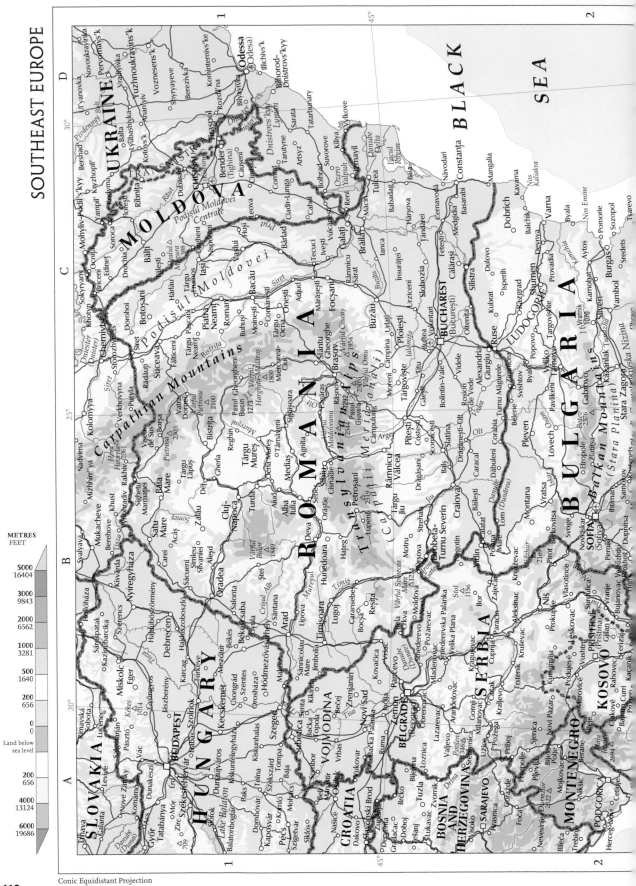

METRES
FEET

5000
16404

3000
9843

2000
6562

1000
3281

500
1640

200
656

0
0

Land below
sea level

200
656

4000
13124

6000
19686

Conic Equidistant Projection

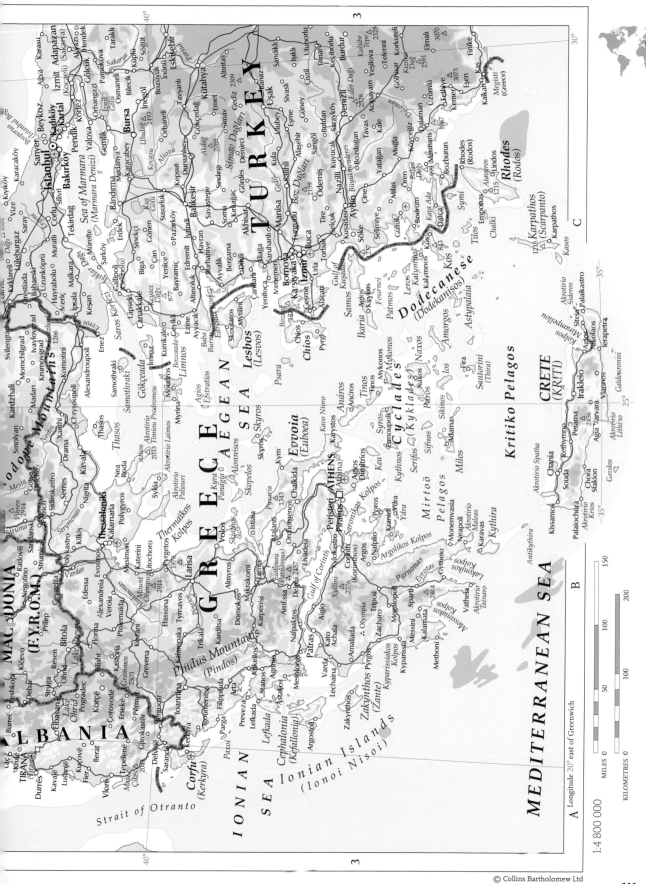

© Collins Bartholomew Ltd

1:4 800 000

© Collins Bartholomew Ltd

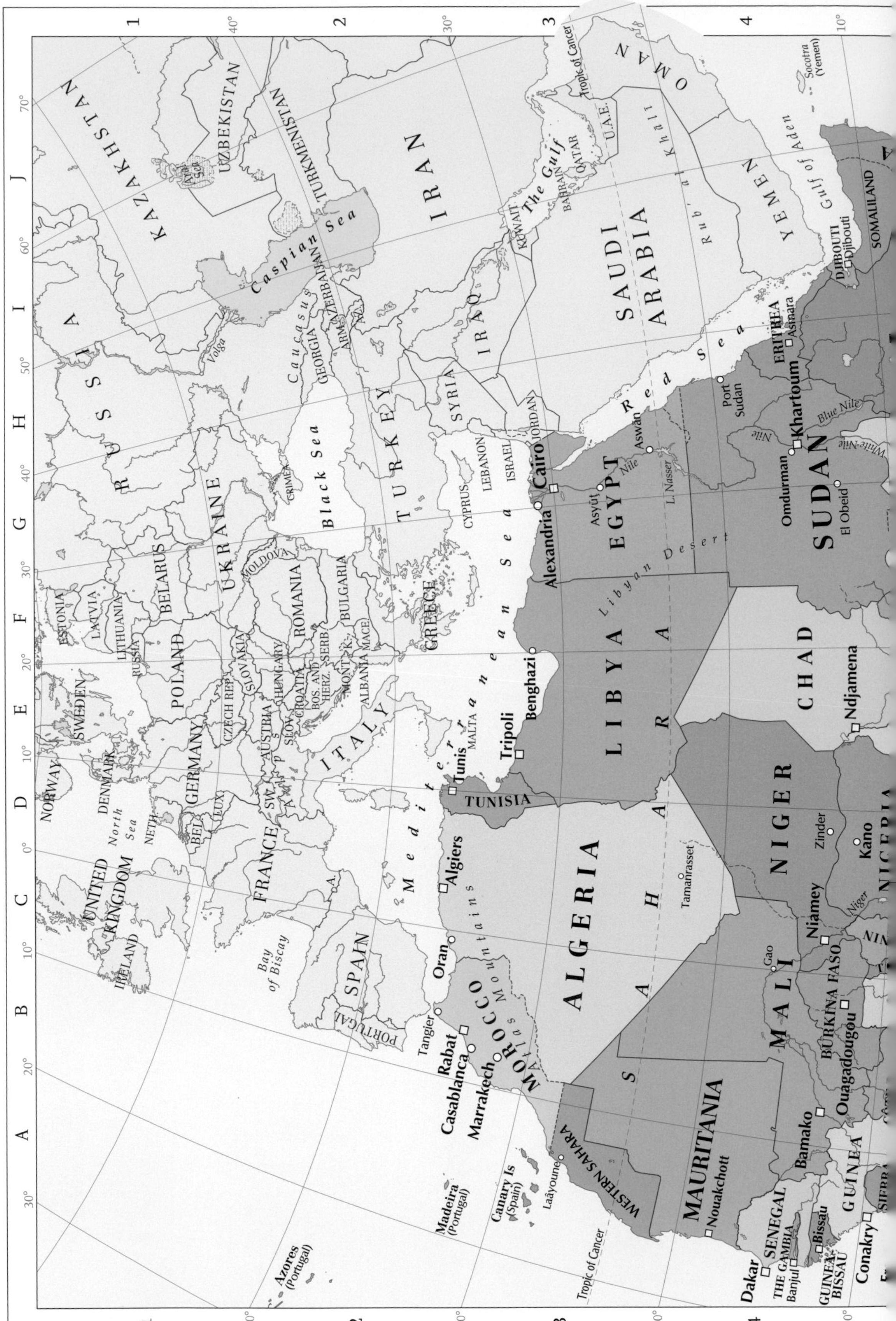

1

2

3

4

70°

60°

50°

40°

30°

20°

10°

0°

10°

20°

30°

J

I

H

G

F

E

D

C

B

A

KAZAKHSTAN

UZBEKISTAN

TURKMENISTAN

Aral Sea

Caspian Sea

Volga

RUSSIA

AZERBAIJAN

ARMENIA

GEORGIA

Caucasus

Black Sea

CRIMEA

IRAN

TURKEY

SYRIA

IRAQ

KUWAIT

The Gulf

BAHRAIN

QATAR

U.A.E.

OMAN

Rub' al Khali

YEMEN

Gulf of Aden

— Socotra
    (Yemen)

SAUDI
ARABIA

JORDAN

ISRAEL
LEBANON

CYPRUS

GREECE

Mediterranean Sea

MALTA

Tropic of Cancer

Red Sea

Port
Sudan

Aswān

L. Nasser

ERITREA
Asmara

DJIBOUTI
Djibouti

SOMALILAND

Khartoum

Omdurman

El Obeid

SUDAN

Blue Nile

White Nile

Nile

Nile

Cairo

Alexandria

Asyūṭ

EGYPT

LIBYA

Libyan Desert

Benghazi

Tripoli

Tunis

TUNISIA

CHAD

Ndjamena

NIGER

Zinder

Kano

NIGERIA

Niger

Niamey

BURKINA FASO

Ouagadougou

BENIN

Gao

MALI

Bamako

GUINEA

Conakry

SIERRA

GUINEA-
BISSAU

Bissau

THE GAMBIA

Banjul

SENEGAL

Dakar

Nouakchott

MAURITANIA

WESTERN SAHARA

Laâyoune

Tropic of Cancer

Canary Is
(Spain)

Madeira
(Portugal)

Azores
(Portugal)

MOROCCO

Atlas Mountains

Marrakech

Casablanca

Rabat

Tangier

Oran

Algiers

ALGERIA

SAHARA

Tamanrasset

SPAIN

PORTUGAL

*Bay
of Biscay*

FRANCE

UNITED
KINGDOM

IRELAND

NORWAY

SWEDEN

DENMARK

*North
Sea*

NETH.
BEL.
LUX.

GERMANY

SWITZ.
AUST.
CZECH REP.
SLOVAKIA

POLAND

BELARUS

UKRAINE

MOLDOVA

ROMANIA

HUNGARY

SLOV.

CROATIA

BOS.
&HERZ.

SERB.

MONT.

ALBANIA

MACE.

BULGARIA

ITALY

ESTONIA

LATVIA

LITHUANIA

RUSSIA

Oblated Stereographic Projection

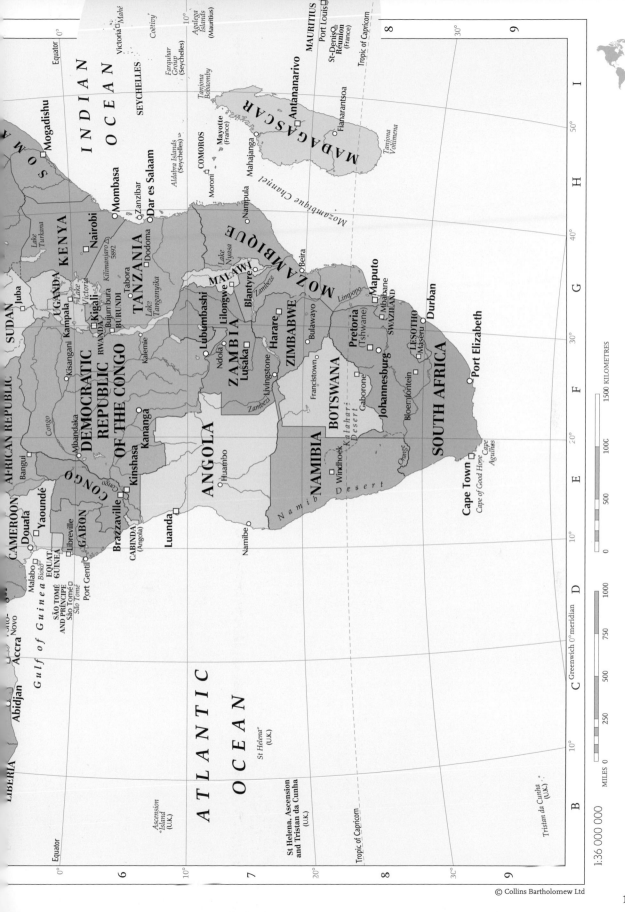

A 20° A 10°

SPAIN
Gibraltar Cartage
Gibraltar (UK) Málaga Almería
Strait Ceuta (Spain) Mostaganem Chl
Tangier Gibraltar Tétouan Oran
(Tanger) Melilla Sidi Bel
Larache (Spain) Tlemcen Saïda
Ksar el Kebir Sidi Taounate Oujda
Ben Slimane Kacem Taourirt
RABAT Kenitra Faza Taourirt Baya
Casablanca Meknès Fès (Fez) Hauts Plateaux Atla
El Jadida Oued
Settat Khouribga Er
Safi Beni Mellal Rachidia Aïn Sefra Sahar
El Kelaâ des Srahna
Marrakech Jebel Ouarzazate Béchar Grand Erg
Essaouira Toubkal Abadla Occidenta
Haut Atlas (High Atlas) 4167 Beni
Taroudannt Zagora Abbès
Agadir Anti-Atlas Tabelbala El Homr
Tiznit Hamada du Drâa Timimou
Sidi Ifni Guelmim Plateau
Tan- Ksabi Adrar Sbaa
Tan Erg Iabès **ALGI**
Laâyoune El-Mahbas Tindouf Bordj Flye Aoulef In Sa
Ste-Marie
Boujdour Reggane
Es-Smara El Eglab Chenachane Poste
Bir Lahlou Sebkha Azzel Weygand
Galtat-Zemmour Bir Matti Sebkh
Skaymat Mogrein Aïn Ben Tili Chegga Mekerr
WESTERN Iguidi
Tropic of Cancer **SAHARA** S
Dakhla ADMINISTERED El Hammâmi Taoudenni Oued Ilafert
BY MOROCCO Maqteïr A
Zouérat Erg Chech Bordj
Aoussard Fdérik Chech Mokhtar
Tichla Choûm Tanezrouft
Nouâdhibou Atâr Guelb er Richât OURÂNE Aoukâr AZAWAD Adrar d
485 Araouane Aguelhok
Nouâmghâr Akchâr Akjoujt Ifôghas
MAURITANIA Azawâd Kidal
Sebkhet Azaouâd **MALI** Anéfis
NOUAKCHOTT Te-n-Dghâmcha Tidjikja Dahr Tîchît IRÎGUI Gourma-
Boutilimit Tîchît Oualâta Rharous Bourem
Tiguent Moudjéria Dahr Lac Timbuktu
Rosso Magta Lahjar HÔD Oualâta Faguibine (Tombouctou) Gao Ménaka
St-Louis Aleg Ayoûn el Néma Goundam Doro Ansongo
Louga Bogué Atroûs Lac Hombori
Dagana Kaédi Kiffa Bassikounou Youvarou Niangay
Linguère Mbout Timbedgha Nara Nampala Mopti Douentza Gorom Tillabéri
DAKAR Dara Matam Sélibabi Nioro Kogoni Bandiagara Gorom NIAM
Thiès Diourbel Bakel Yélimane Ballé S Koro Djibo Dori Filin
Mbour Kaffrine Sandaré A Bla Djenné Ouahigouya Gourcy Kaya Bogandé Dos
SENEGAL Kayes Diéma Niono Ten-N'Gouma Kantcha
Fatick Kaolack Goudiri Bafoulabé Koulikoro Ségou Tougan Yako **BURKINA FASO** Fada-N'Gourma Diapa
BANJUL Tambacounda Kita Koutiala Koudougou Zorgho Gayeri
Brikama **THE GAMBIA** Kédougou Satadougou Dioila **BAMAKO** Sikasso San **OUAGADOUGOU** Tenkodogo Bawku **BEN**
Ziguinchor Sédhiou Kolda Mali Kati Kangaba Bougouni Koutiala Bobo- Manga Pô Téo Bolgatanga Dapaong Kara Natiting
Cacheu Bafata Gabu Koundara Gaoual Siguiri Dioulasso Banfora Léo Bawku Bimbila Djougou Paral
GUINEA Bissau Fouta Pita Labé Dabola Kolondiéba Orodara Wa Yendi Bassila
BISSAU Buba Bolama Djallon Dinguiraye Kouroussa Mandiana Minignan Ferkessédougou Gaoua Bolgatanga Kara Bassa
Arquipélago Boké Fria Mamou Kankan Kadiolo Korhogo Bouna Tamale Sokodé
dos Bijagós Dubréka Kindia Faranah Odienné Boundiali Damongo **GHANA** Salaga Atakpamé Abeo
GUINEA Kissidougou Beyla Dianra Korhogo Bouna Techiman Kete Krachi
CONAKRY Lungi Falaba Kérouané Séguéla Katiola Sunyani Kintampo Mampong
Port Toula Lola Man Bouaké Wenchi La
Loko Makeni Nzérékoré **CÔTE D'IVOIRE** Abomey PORTO-NO
FREETOWN Kamakwie Koidu Zorzor Sanniquellie Daloa Bouaflé Bongouanou Kumasi LOMÉ Slave Co
SIERRA Magburaka Sefadu Danané Gagnoa Divo Bekwai Aného
LEONE Bo Kenema Zimmi Tapeta **YAMOUSSOUKRO** Obuasi **ACCRA**
Bonthe Ganta Zwedru Tiassalé Koforidua Winneba Tema
MONROVIA Harbel Lakota Aboisso Tarkwa
Buchanan River Cess **LIBERIA** Sassandra Abidjan Axim Sekondi Cape Coast Big
Greenville Bingerville Gold Coast of Ber
Barclayville Grand- San-Pédro Cape
Harper Lahou Three Points
Cape Palmas Tabou **GULF OF GUINEA**

A A

ATLANTIC OCEAN

30°

Madeira (Portugal) FUNCHAL

La Palma Lanzarote
Canary Islands SANTA CRUZ DE TENERIFE Fuerteventura
(Islas Canarias) Pico del Teide 3718 Tenerife Jandía
(Spain) La Gomera 807 Gran LAS PALMAS DE GRAN CANARIA
El Hierro Canaria

2

20°

3

10°

METRES / FEET

METRES	FEET
5000	16404
3000	9843
2000	6562
1000	3281
500	1640
200	656
0	0
Land below sea level	
200	656
4000	13124
6000	19686

4

114

Lambert Azimuthal Equal Area Projection

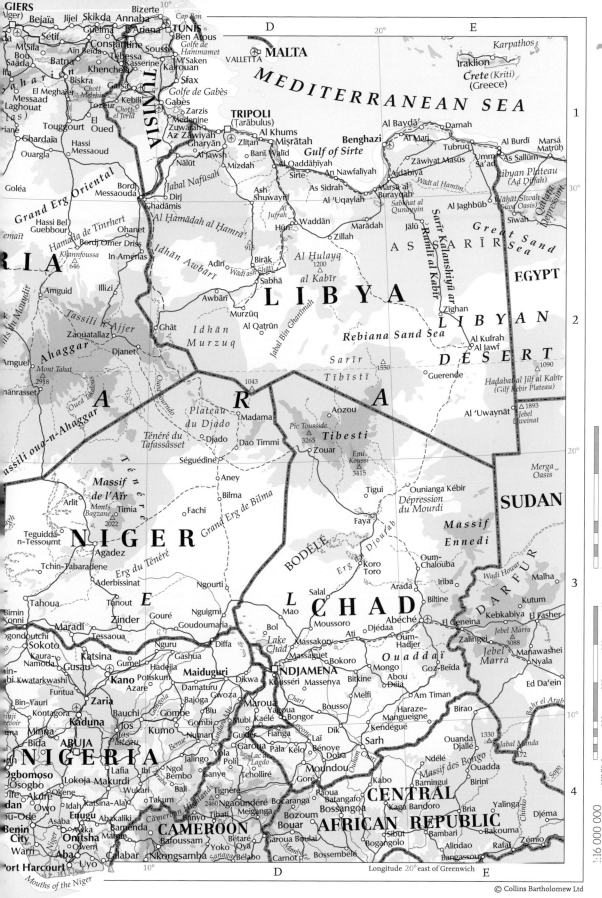

GIERS
(Alger) Bejaïa Jijel Skikda Annaba Bizerte 10°
Sétif Guelma E'Ariana TUNIS
M'Sila Constantine Ben Arous
Bou Aïn Beïda M'Saken
Saâda Batna Khenchela Kairouan Sousse
Ifa Biskra Tebessa Sfax Golfe de Hammamet
Messaad El Meghaïer Kasserine Gafsa Golfe de Gabès Cap Bon
Laghouat Chott Melrhir Tozeur Kebili Gabès VALLETTA MALTA
(as) El Oued Chott el Jerid Zarzis
Touggourt Medenine
Ghardaïa Hassi Zuwārah TRIPOLI (Tarābulus) MEDITERRANEAN SEA
Ouargla Messaoud Az Zāwiyah Al Khums Karpathos
Goléa Gharyān Zlitan Mişrātah Benghazi Al Bayḍā' Iraklion Darnah
Grand Erg Oriental Bordj Al Jawsh Banī Walīd Sirte Gulf of Sirte Al Marj Tubruq Crete (Kriti) As Sallūm
Messaouda Dirj Nālūt Mizdah Al Qaddāḥīyah An Nawfalīyah Ajdabiya Zāwiyat Masūs Umm Marsā (Greece)
Hassi Bel Ghadāmis Jabal Nafūsah Ash As Sidrah Marsá al Wādī al Ḥamīm Sa'ad Maţrūḥ
Guebbour Al Hamādah al Hamrā' Shuwayrif Al Uqaylah Burayqah Sabkhat al Al Jaghbūb Al Burdī 30°
Ohanet Al Jufrah Ḥūn Waddān Marādah Qunayyin Wāḥāt Sīwah Libyan Plateau
Hamada de Tinrhert Bordj Omer Driss Zillah Jālū (Siwa Oasis) (Ad Diffah)
Khannfoussa 646 In Amenas Adīrī 915 Al Ḥulayq AS Siwah Qattara
emaït Birāk 1200 SARĪR Great Sand Depression
Wādī ash Shāṭī' al Kabīr Ramlat al Kabīr Sea
ARIA Amguid Illizi Sabhā Zighān LIBYA Al Kufrah EGYPT
Tassili n'Ajjer Zaouatallaz Awbārī Murzūq Jabal Bin Ghanīmah Rebiana Sand Sea Al Jawf LIBYAN
Amguel Ghāt Idhān Rebiana Sand Sea DESERT 2
Mont Tahat Djanet Murzūq Sarīr Tībīstī 1550 Al Uwaynāt Hadabat al Jilf al Kabīr
Ahaggar 2918 Al Qaţrūn 1043 Guerende (Gilf Kebir Plateau)
anrasset A Oued Tamanrasset R Madama A Aozou 1893 Merga
assili oug-n-Ahaggar Plateau du Djado Pic Toussidé Jebel Oasis
Oued Tafassâsset Djado 3265 Tibesti Uweinat
Ténéré du Dao Timmi Zouar Emi SUDAN
Tafassâsset Séguédine Koussi DARFUR 20°
Massif Aney 3415 Tigui Ouninga Kébir
de l'Aïr Bilma Dépression Massif
Arlit Monts Timia Fachi Grand Erg de Bilma Faya du Mourdi Ennedi
Bagzane 2022 NIGER Erg Koro Oum- Iriba Malha
Teguidda- Agadez Djourab Toro Chalouba Wadi Howar Kebkabiya El Fasher 3
n-Tessoumt BODÉLÉ Erg Arada Biltine Kutum
Tchin-Tabaradene Aderbissinat E Ngourti Salal L CHAD El Geneina Zalingei
Tahoua Tânout Mao Moussoro Abéché DARFUR 3088
Birnin Zinder Gouré Nguigmi Ati Djédaa Oum- Jebel Marra
Konni Tessaoua Nguru Diffa Lake Massakory Hadjer Goz-Beïda Jebel Manawashei
gondoutchi Maradi Gashua Chad Massaguet Mongo Ouaddaï Marra Nyala
Sokoto Katsina Gumel Hadejia Bokoro Bitkine Abou Ed Da'ein
Kaura- Gusau Potiskum Dikwa NDJAMENA Massenya Déïa Am Timan
Namoda Kano Azare Damaturu Kousséri Melfi Haraze- Birao 10°
inbi Kwatarkwashi Funtua Gwoza Maroua Massif des Bongo
Bin-Yauri Zaria Bajoga Yagoua Bousso Mangueigne Ouanda 1330
Kontagora Bauchi Gombe Biu Mubi Kaélé Bongor Djallé Jabal Manda
na Minna Jos Kumo Numan Guider Laï Dik Kendégué Ndélé 172
Bida ABUJA Plateau Gombi Garoua Fianga Pala Kélo Sarh Kabo Bamingui Birini
NIGERIA Lafia Ibi Wukari Yola Poli Bénoye Doba Goré Paoua Batangafo Sopo
gbomoso Lokoja Makurdi Bali Takum Tignère Moundou Bossangoa CENTRAL
Osogbo Akure Idah Katsina-Ala Ganye 2460 Ngaoundéré Bocaranga Kaga Bandoro Bria
ife Oke Enugu Abakaliki Tibati Meiganga Bozoum Bambari AFRICAN REPUBLIC Yalinga
dan Owo Asaba Awka Banyo Bétaré Oya Bouar Sibut Djéma
u-Ode Benin Onitsha Owerri Cameroon Highlands Garoua Boulaï Bogangolo Bakouma
City Aba Bafoussam Yoko Bélabo Bossembélé Alindao Rafaï 4
Warri Uyo Nkongsamba CAMEROON Carnot Bangassou Zémio Chinko
Port Harcourt Calabar Bossangoa
Mouths of the Niger 10° Longitude 20° east of Greenwich

© Collins Bartholomew Ltd

1:16 000 000

600 KILOMETRES
400
200
0

400 MILES
300
200
100
0

115

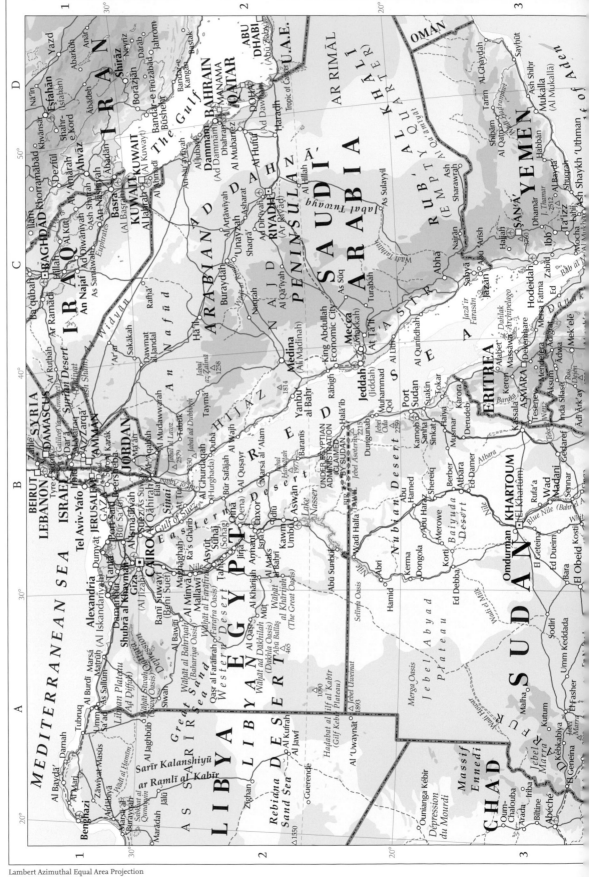

METRES
FEET

5000	16404
3000	9843
2000	6562
1000	3281
500	1640
200	656
0	0
Land below sea level	
200	656
4000	13124
6000	19686

Lambert Azimuthal Equal Area Projection

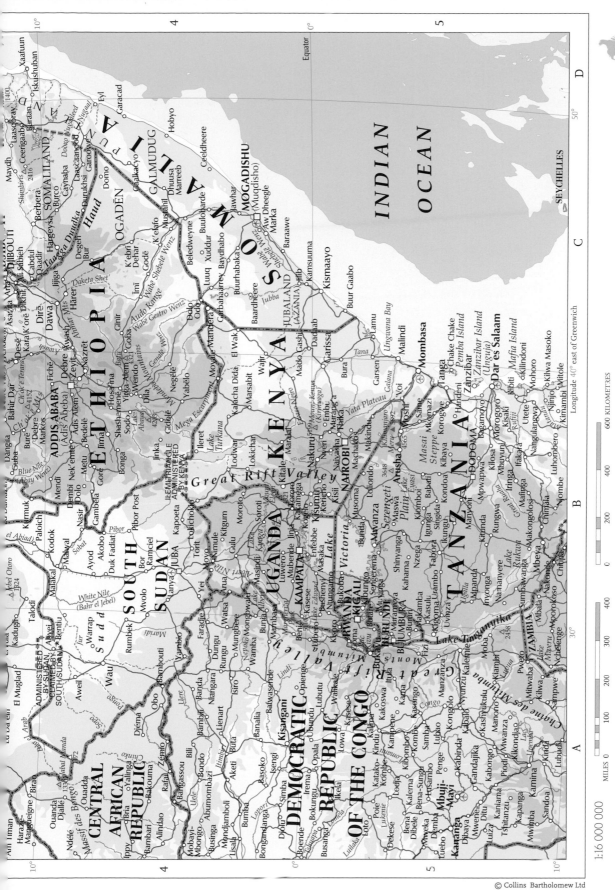

© Collins Bartholomew Ltd

1:16 000 000

Longitude 40° east of Greenwich

INDIAN OCEAN

SEYCHELLES

S O M A L I A

ETHIOPIA

KENYA

TANZANIA

UGANDA

RWANDA

BURUNDI

SOUTH SUDAN

DEMOCRATIC REPUBLIC OF THE CONGO

CENTRAL AFRICAN REPUBLIC

ZAMBIA

DJIBOUTI

SOMALILAND

PUNTLAND

GALMUDUG

Great Rift Valley

A 10° B 20°

CHAD

Tudun Wada · Kari · Damboa · Gwoza · Mora · Maroua · Massenya · Abou Déia · Bourtoutou
Bauchi · Bajoga · Biu · Mubi · Kaélé · Yagoua · Bongor · Bousso · Am Timan · Plaine de Garar
Gombe · Combi · Mokolo · Guider · Fianga · Dik · Kendégué · Haraze-Mangueigne · Birao
Jos · Dindima · Kumo · Kaltungo · Numan · Garoua · Laï · Kélo · Béinamar · Koumra · Sarh · Tiroungoulou · Ouanda Djallé · 1330 △ · Ja
Jos Plateau · Pankshin · Benue · Yola · Pala · Poli · Tcholliré · Koum · Mbé · Touboro · Doba · Maro · Kabo · Markounda · Bamingui · Ouadda · Birini

NIGERIA

Shendam · Jalingo · Ngol Bembo · Gashaka · Tignère · Ngaoundéré · Baïbokoum · Batangafo · Kaga Bandoro · Dékoa · Ippy · Bria · Yalinga · Chinko
Lafia · Ibi · Wukari · Ganye · Bali · Banyo · Meiganga · Bocaranga · Bozoum · Bossangoa · Massif des Bongo
Makurdi · Donga · Takum · 2460 △ · Tibati · Ngaoundal · Garoua Boulaï · Bouar · Baoro · Bogangolo · Sibut · Grimari · Bambari · Bakouma
Katsina-Ala · Gboko · Cameroun Highlands · Bamenda · Yoko · Bétaré Oya · Carnot · Gadzi · Damara · Alindao · Bangassou · Rafaï

CENTRAL AFRICAN REPUBLIC

Ikom · Nkambe · Wum · Lac de Mbakaou · Ngaoundal · Bossembélé · **BANGUI** · Zongo · Mobayi-Mbongo · Bondo · Abumombazi
Mamfe · Mbouda · Foumban · Bélabo · Nanga · Bertoua · Berbérati · Boda · Bimbo · Bosobolo · Businga · Gemena
Calabar · Kumba · Nkongsamba · Bafoussam · Bafia · Sanaga · Eboko · Mbandjok · Batouri · Bambio · Mbaïki · Libenge · Uele · Ebola
Mont Cameroun 4100 △ · Buea · Mbanga · Monatélé · Obala · **YAOUNDÉ** · Abong Mbang · Nola · Salo · Dongou · Libenge · Mondjamboli · Ake
Limbe · **Douala** · Edéa · Mbalmayo · Akonolinga · Yokadouma · Bambio · Kungu · Lisala · Bumba · Lolo · Basc

CAMEROON

Bioko · Kribi · Ebolowa · Sangmélima · Boumba · Moloundou · Makanza · Bongandanga · Simba · Iseng · Tshuapa
MALABO · Bata · Niefang · Djoum · Souanké · Quesso · Epéna · Impfondo · Losombo · Bolomba · Djolu · Langa
EQUATORIAL GUINEA · Ebebiyin · Oyem · Sembé · Mékambo · Mbomo · Makoua · Bikoro · Embondo · Boende · Bokungu · Opal
Cogo · Evinayong · Mitzic · Makokou · Mbandaka · Bokatola · Watsi · Busanga · Ikela
Ntoum · Alembé · Booué · Makoua · Owando · Bolia · Bokele · Eyangu
LIBREVILLE · Bifoun · Lambaréné · Okondja · Obouya · Loukoléla · Lac Tumba · Inongo · Boleko · Ifumo · Loto · Lomela
Cap Lopez · Port-Gentil · Lastoursville · Akiéni · Koulamoutou · Okoyo · Ntandembele · Kutu · Lac Mai-Ndombe · Poie · Lodja · Kot

GABON · **CONGO** · **DEMOCRATIC REPUBLIC OF THE CONGO**

Iguéla · Lagune Nkomi · Fougamou · Mimongo · Moanda · Franceville · Gamboma · Ngo · Bolobo · Mushie · Dekese
Mouila · Ndendé · Mayoko · Bouanza · Lékana · Djambala · Bagata · Domiongo · Bena Dibele
Lagune Ndogo · Tchibanga · Nyanga · Mossendjo · Komono · Ngabé · Bandundu · Ilebo · Bena-Sung
Mayumba · Nzambi · Sibiti · Makabana · Loudima · **BRAZZAVILLE** · **KINSHASA** · Bulungu · Mangai · Lusan · Pe
Pointe-Noire · Loubomo · Madingou · Mindouli · Belize · Kasangulu · Kenge · Masi-Manimba · Mweka · Luebo · Demba
CABINDA (Angola) · Tshela · Luozi · Kisantu · Kikwit · Idiofa · Kananga · **Mbuji-Ma**
Cabinda · Boma · Kimpese · Mbanza-Ngungu · Kingandu · Gungu · Kilembe · Tshikapa · Kazumba · Dibaya
Kitona · Matadi · Maquela do Zombo · Popokabaka · Feshi · Tshikapa · Kamonia · Mwene-Ditu · Gandajika
Muanda · M'banza Congo · Mawanga · Kasongo-Lunda · Bumba · Kamona · Luiza · Plateau du Kasaï · Kan

ATLANTIC OCEAN

Tomboco · Lucunga · Songo · Uíge · Quimbele · Tembo Aluma · Bindu · Kahemba · Chitato · Cambulo · Kimp
N'zeto · Ambriz · Muxaluando · Negage · Massango · Caungula · Cuilo · Lucapa · Sombo · Mwimba
Caxito · Camabatela · Calandula · Saurimo · Kapanga · Sandoa
LUANDA · Catete · Lucala · Malanje · Xá-Muteba · Capenda-Camulemba · Mona · Chiluage · Quimbundo · Muconda · Muriege · Kasaï
N'dalatando · Dondo · Calulo · Quitapa · Cacolo · Sombo · Saurimo
Gabela · Quibala · 1613 △ · **ANGOLA** · Dala · Luau · Luacano · Caianda · Mwhili
Waku-Kungo · Sumbe · Andulo · N'harea · Quirima · Camanongue · Calunda
Planalto do Bié · Camacupa · Luena · Sachanga · Cazombo · Calunda

Lambert Azimuthal Equal Area Projection

Longitude 20° east of Greenwich

METRES / FEET
5000 / 16404
3000 / 9843
2000 / 6562
1000 / 3281
500 / 1640
200 / 656
0 / 0
Land below sea level
200 / 656
4000 / 13124
6000 / 19686

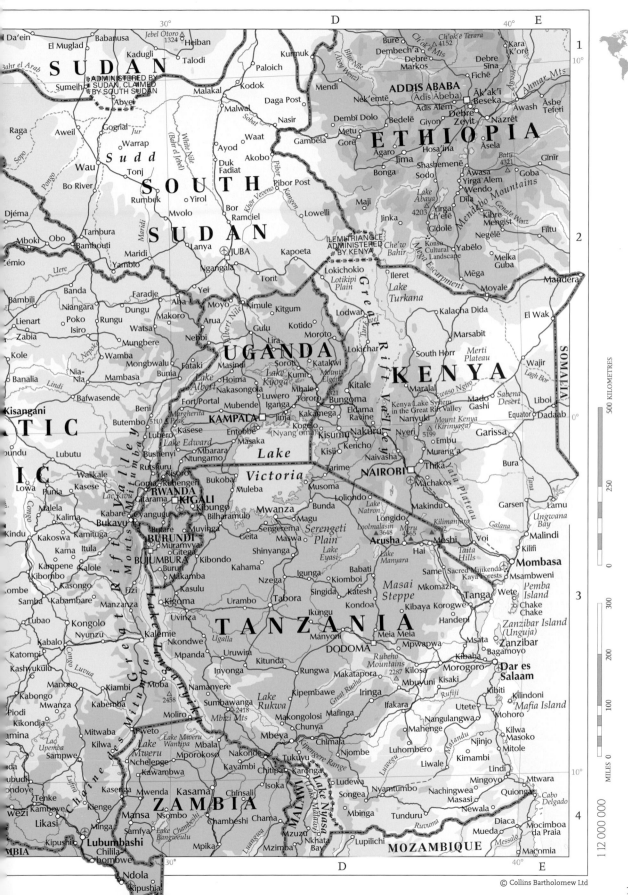

SUDAN

SOUTH

SUDAN

ETHIOPIA

UGANDA

KENYA

SOMALIA

RWANDA

BURUNDI

TANZANIA

ZAMBIA

MALAWI

MOZAMBIQUE

Da'ein
El Muglad
Babanusa
Kadugli
Jebel Otoro
1324
Heiban
Talodi
Kurmuk
Paloich
Bure
Dembech'a
Ch'ok'e Terara
4152
Kara
K'ore
Debre
Markos
Debre
Sina
Fiché
Āk'ak'ī
Beseka
Āwash
Ahmar Mts
Debre Teferi

ADMINISTERED BY
SUDAN, CLAIMED
BY SOUTH SUDAN
Abyei
Malakal
Kodok
Daga Post
Mendī
ADDIS ABABA
(Ādīs Ābeba)
Ādīs Alem
Nazrēt
Āwash

Sumeih
Aweil
Gogrial
Jur
Warrap
White Nile
(Bahr el Jebel)
Malwal
Nasir
Sobat
Waat
Dembī Dolo
Bedelē
Giyon
Debre
Zeyit
Metu
Gambela
Gorē
Hosa'ina
Āsela

Raga
Wau
Bo River
Tonj
Rumbek
Ayod
Duk
Fadiat
Akobo
Pibor
Gambela
Āgaro
Jīma
Shashemenē
Āwasa
Yīrga 'Alem
Goba
Batu
4321
Gīnīr

Djéma
Mboki
Obo
Tambura
Bambouti
Maridi
Yambio
Lanya
Kapoeta
Nimule
JUBA
ILEMI TRIANGLE
ADMINISTERED
BY KENYA
Lokichokio
Che'w
Bahir
Iléret
Lake
Turkana
Moyale
Mandera

Bo River
Tonj
Yirol
Mvolo
Bor
Ramciel
Lowelli
Majī
Jinka
Gīdolē
4203
Yīrga
Ch'efē
Negēlē
Filtu

NAIROBI

DODOMA

Dar es
Salaam

Dar es
Salaam

Lubumbashi

Ndola
Kipushi

© Collins Bartholomew Ltd

500 KILOMETRES

250

0

300

200

100

MILES 0

1 12 000 000

A 20° B

Pointe-Noire
CABINDA
(Angola)
Tshela · Luozi · Kisantu · Kenge · Masi-Manimba · Idiofa · Mweka · Bena-Sungu · Lusambo
Cabinda · Kimpese · Mbanza-Ngungu · Kingandu · Kikwit · Kilembe · Luebo · Demba · Mbuji-Mayi · Penge · Lubao · Kongolo
Boma · Muanda · Kitona · Matadi · Maquela do Zombo · Mawanga · Feshi · Gungu · Kananga · Kabinda · Kashyukulu · Kaba
M'banza Congo · Damba · Kasongo-Lunda · Bumba · Kamonia · Dibaya · Mwene-Ditu · Gandajika · Mwanza · Mano
Tomboco · Lucunga · Songo · Uíge · Tembo Aluma · Bindu · Kahemba · Chitato · Luiza · Tshitanzu · Piodi · Kikon
N'zeto · Muxaluando · Negage · Massango · Camabatela · Caungula · Cambulo · Plateau du Kasaï · Kaniama · Mwanza · Samp
Ambriz · Caxito · Calandula · Xá-Muteba · Capenda-Camulemba · Sombo · Mwimba · Kapanga · Kamina · Kinda · Uemba

DEMOCRATIC REPUBLIC OF THE CONGO

LUANDA · Catete · Lucala · Malanje · Saurimo · Chiluage · Muriege · Kafakumba · Lubudi · Kie
N'dalatando · Dondo · Mona Quimbundo · Muconda · Malonga · Kasaji · Tenke · Lik
10° Calulo · Quibala · Quitapa · Cacolo · Luau · Dilolo · Nasondoye · Kolwezi · Kambove · Lubumbas
Gabela · Waku-Kungo · Andulo · Quirima · Dala · Luacano · Caianda · Kipus
Sumbe · N'harea · Camanongue · Cazombo · Calunda · Mwinilunga · Solwezi · Ching
Lobito · Balombo · Planalto do Bié · Camacupa · Cuemba · Sachanga · Luena · Lucusse · Lumbala · Mufumbwe · Ingwe
Benguela · Caála · Huambo · Kuito · Chinguar · Cangamba · Kaquengue · Zambezi · Kasempa · Luns
Cubal · Caconda · Chipindo · Umpulo · Tempué · Lumbala N'guimbo · Kabompo · Mumbeji
Caluquembe · Quilengues · Kuvango · Menongue · Cuando · Cangombe · Kalabo · Lukulu

ANGOLA **ZAMI**

Bibala · Matala · Planalto da Huíla · Cuito · Cuanavale · Chiume · Mavinga · Neriquinha · Mongu · Namwala · Mumbwa
2 Namibe · Lubango · Cassinga · Caiundo · Baixo-Longa · Rivungo · Senanga · Mulobezi · Choma · Ka
Tombua · Chiange · Cuvelai · Uamanda · Acampamento de Caça do Mucusso · Kalomo
Tigres (abandoned) · Cahama · Mucope · Nankova · Katima Mulilo · Victoria Falls · Livingst
Foz do Cunene · Chitado · Xangongo · Ondjiva · Calai · Dirico · Bukalo · Victoria Falls · Hwan
Oshikango · Cuangar · Rundu · Bagani · CAPRIVI STRIP · Kasan · Shumba
Oshakati · Oshikati · Gumare · Okavango Delta · Phuduhudu · Nata · Maiten
Opuwo · Etosha Pan · Tsumeb · Grootfontein · Tsumkwe · Maun · Tutume
Sesfontein · Kamanjab · Otavi · Kombat · Sehithwa · Xhumo · Orapa · Letlhakane · Serowe · Francistown
20° Outjo · Otjiwarongo · Makgadikgadi
Khorixas · Okakarara · Eiseb · Ghanzi

NAMIBIA **BOTSWANA**

Kalkfeld · Omaruru · Steinhausen
Uis Mine · Usakos · Omitara · Buitepos · Takatshwaane · Serowe · Pala
Hentiesbaai · Okahandja · Witvlei · Gobabis · Ghanzi · Mahalapye
Swakopmund · WINDHOEK · Dordabis · Tshootsha
Walvis Bay · Rehoboth · Leonardville · Ncojane · Tsetseng · Molepolole · Mochudi
Tsumis Park · Kang · Hukuntsi · Tshane · Jwaneng · GABORONE · Thabazi
Tropic of Capricorn · Solitaire · Hoachanas · **Kalahari** · Khakhea · Kanye · Lobatse · Soshang
Nauchas · Aranos · Werda · Mabule · Mmabatho · Johannesb
3 Maltahöhe · Narib · Stampriet · Mariental · Gochas · **Desert** · Mahikeng · Sow
Helmeringhausen · Tses · Koës · Tshabong · Terra Firma · Sasol
NAMAQUALAND · Keetmanshoop · Severn · Vryburg · Delareyville
Lüderitz · Aus · Seeheim · Aroab · Bokspits · Kuruman · Tswelelang · Kgotsong · Magk
Ai-Ais · Grünau · Van Zylsrus · Olifantshoek · Lime Acres · Vaalspan · Thabor
Karasburg · Ariamsvlei · Upington · Rostmasburg · Phahameng · Masilo
Oranjemund · Alexander Bay · Keimoes · Galeshewe · Kimberley

SOUTH AFRIC

ATLANTIC OCEAN

Cuanza 1613
2620
1506
2050 Onjati Mountain
2202

Barragem do Gove
Etosha Pan
Okavango Delta
Victoria Falls

Longitude 20° east of Greenwich

METRES FEET
5000 · 16404
3000 · 9843
2000 · 6562
1000 · 3281
500 · 1640
200 · 656
0 · 0
Land below sea level
200 · 656
4000 · 13124
6000 · 19686

Lambert Azimuthal Equal Area Projection

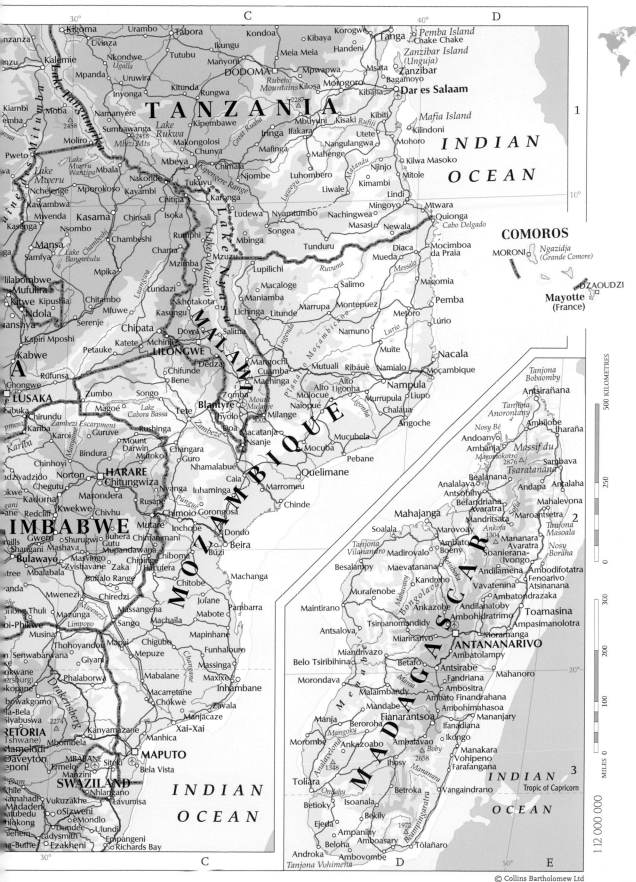

This is a map of southeastern Africa including Tanzania, Mozambique, Malawi, Zimbabwe, and Madagascar.

TANZANIA

Kigoma · Urambo · Tabora · Kondoa · Kibaya · Korogwe · Tanga · Pemba Island / Chake Chake
Uvinza · Ikungu · Meia Meia · Handeni · Zanzibar Island (Unguja)
Kalemie · Nkondwe · Ugalla · Tutubu · Manyoni · Msata · Zanzibar
Mpanda · Uruwira · DODOMA · Mpwapwa · Bagamoyo
Inyonga · Kitunda · Rungwa · Rubeho Mountains · Kilosa · Morogoro · Kibaha · Dar es Salaam
Namanyere · Kipembawe · 2287 · Mbuyuni · Kisaki · Kibiti · Mafia Island
Sumbawanga · Lake Rukwa · Mafinga · Iringa · Ifakara · Rufiji · Kilindoni
Mbeya · Chimala · Chunya · Nangulangwa · Utete · Mohoro
Nakonde · Njombe · Mahenge · Njinjo · Kilwa Masoko
Tukuyu · Luhombero · Kimambi · Mitole
Karonga · Liwale · Lindi

INDIAN OCEAN

Chitipa · Isoka · Ludewa · Nyamtumbo · Nachingwea · Mingoyo · Mtwara
Rumphi · Songea · Masasi · Newala · Quionga
Mbinga · Tunduru · Diaca · Cabo Delgado
Chama · Mzuzu · Lupilichi · Mueda · Mocimboa da Praia

MALAWI

COMOROS

MORONI · Ngazidja (Grande Comore)

DZAOUDZI
Mayotte (France)

Macaloge · Maçomia
Salimo · Pemba
Maniamba · Marrupa · Montepuez · Metoro · Lúrio
Lichinga · Litunde · Namuno
Kasungu · Nkhotakota · Namialo · Nacala
LILONGWE · Mangochi · Cuamba · Ribáué · Moçambique
Dedza · Machinga · Alto Ligonha · Nampula
Zomba · Molocuè · Murrupula · Liupo
Blantyre · Naiopuè · Chaláua · Angoche

MOZAMBIQUE

INDIAN OCEAN

© Collins Bartholomew Ltd

A B

20°

Brakwater
Witvlei
Gobabis
Takatshwaane
Khomas Highland
WINDHOEK
2489
Doreenville
Kule
Ncojane
Palamakoloi
Tsetseng
Bergland
Dordabis
Louwater-Suid
Gross Ums
One
Lehututu
Kang
Salajwe
Khudumela
Wortel
Rehoboth
Leonardville
K A L A H A R I
Hukuntsi
Tshane
Motokwe
Takatokwane
B O T S W A
Tropic of Capricorn
Heide
Aminuis
Lokgwabe
Kokong
Mabutsane
Nauchas
Tsumis Park
Hoachanas
D E S E R T
Khakhea
Jwaneng
Solitaire
Narib
Aranos
Bullsport
Kuis
Stampriet
Werda
Moselebe
Maltahöhe
Salzbrunn
Mariental
Makopong
Molopo
Bossiesvlei
Nananib Plateau
Gibeon
Witbooisvlei
25°
N A M I B I A
Omaweneno
Terra Firma
Senlac
Mal
Schwarzrand
Twee Rivier
Auob
Tshabong
Morokweng
Tosca
Tses
Koës
Nossob
Berseba
Wasser
Helmeringhausen
Kolonkwaneng
Molopo
Severn
Laxey
N O R T
Tiraz Mountains 2040
G R E A T
N A M A Q U A L A N D
Bokspits
Kuruman
Van Zylsrus
Lolwane
Vryburg
Huhu
Tsaukaib
Bethanie
Keetmanshoop
Aroab
Rietfontein
Hakseen Pan
Hotazel
Kuruman
Taun
Reivilo
Garub
Aus
Sandverhaar
Seeheim
Molopo
Dibeng
Kathu
Valspr
Gawachab
Little Karas Berg
2202
Groot Karas Berg
Gaiab
Sishen
Gakarosa 1855
Warren
Holoog
Klein Karas
Olifantshoek
Ghaap Plateau
2
Rosh Pinah
Grünau
Ariamsvlei
Postmasburg
Lime Acres
Ai-Ais
Karasburg
Kokerboom
Lutzputs
Langberg
Barkly West
Kimberley
Galeshewe
Warmbad
Onseepkans
Keimoes
Upington
Grootdrink
Griquatown
Campbell
Ritchie
Oranjemund
Alexander Bay
Kakamas
Kleinbegin
Groblershoop
G R I Q U A L A N D
W E S T
Modder
Bongani
Wreck Point
Eksteenfontein
Pella
Pofadder
Hartbees
Putsonderwater
Orange
Douglas
Koffiefo
Lekkersing
Kenhardt
Asbestos Mountains
Hopetown
Luck
Port Nolloth
Steinkopf
Concordia
Aggeneys
Marydale
E'Thembini
Prieska
N A M A Q U A L A N D
Nababeep
Carolusberg
Verneuk Pan
Strydenburg
Orange
Kleinsee
Springbok
N O R T H E R N   C A P E
Copperton
Petrusville
Vander
Komaggas
Kamieskroon
Grootvloer
Kalingvolei
De Naawte
Houwater
Philipstown
30°
Hondeklipbaai
Kamiesberge
Onderstedorings
Van Wyksvlei
Vosburg
Britstown
De Aar
Wallekraal
Swartkolkvloer
Brandvlei
S O U T H A
Kareeberge
Hanover
Nonzwa
Garies
Loeriesfontein
Sakrivier
Carnarvon
Noup
Bitterfontein
Kootjieskolk
Sterling
Ongers
Victoria West
Richmond
KwaNon
Nuwerus
Williston
Sak
Masinyusane
Sabelo
Hardeveld
Nieuwoudtville
Great Karoo
Murraysburg
Sneeuw
Lutzville
Calvinia
Fish
Fraserburg
Nuweveldberge
Graaff-Reinet
Vredendal
Vanrhynsdorp
Klawer
Doring
Roggeveldberg
Sutherland
Beaufort West
Sidesaviwa
A T L A N T I C
Lambert's Bay
Graafwater
Clanwilliam
Aberdeen
Baboon Point
St Helena Bay
Wuppertal
Komsberg
Merweville
Leeu-Gamka
KwaZamukucinga
Jansenville
O C E A N
Citrusdal
Prince Albert Road
Salt
Cape St Martin
St Helena Bay
Velddrif
Piketberg
Olifants
Laingsburg
Prince Albert
Steytlervi
Vredenburg
Porterville
Prince Alfred Hamlet
Ladismith
Zoar
Dysselsdorp
Willowmore
Saldanha
Moorreesburg
Ceres
2258
Touwsrivier
2325
Groot Swartberge
Calitzdorp
De Rust
Oudtshoorn
Kougaberge
Joubertina
Malmesbury
Wellington
W E S T E R N
Uniondale
Haarlem
Kruisfo
Atlantis
Paarl
Worcester
C A P E
Montagu
Little Karoo
George
Plettenberg Bay
Humansc
Durbanville
Stellenbosch
Robertson
Barrydale
Mossel Bay
Groot Brakrivier
Knysna
Cape Seal
Bellville
Khayelitsha
Swellendam
Riversdale
Mossel Bay
CAPE TOWN
Somerset West
Heidelberg
Caledon
Port Beaufort
St Sebastian Bay
Kanonpunt
False Bay
Strand
Stilbaai
Hermanus
Bredasdorp
Cape of Good Hope
Hawston
Gansbaai
Arniston
Cape Agulhas
Struis Bay

METRES
FEET
5000 | 16404
3000 | 9843
2000 | 6562
1000 | 3281
500 | 1640
200 | 656
0 | 0
Land below sea level
200 | 656
4000 | 13124
6000 | 19686

Lambert Azimuthal Equal Area Projection

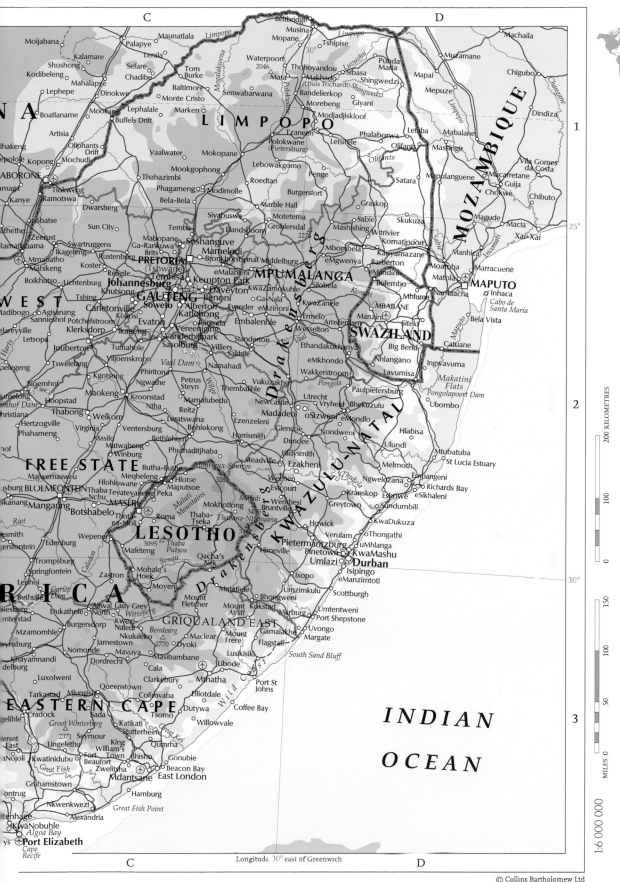

© Collins Bartholomew Ltd

Bi-Polar Oblique Projection

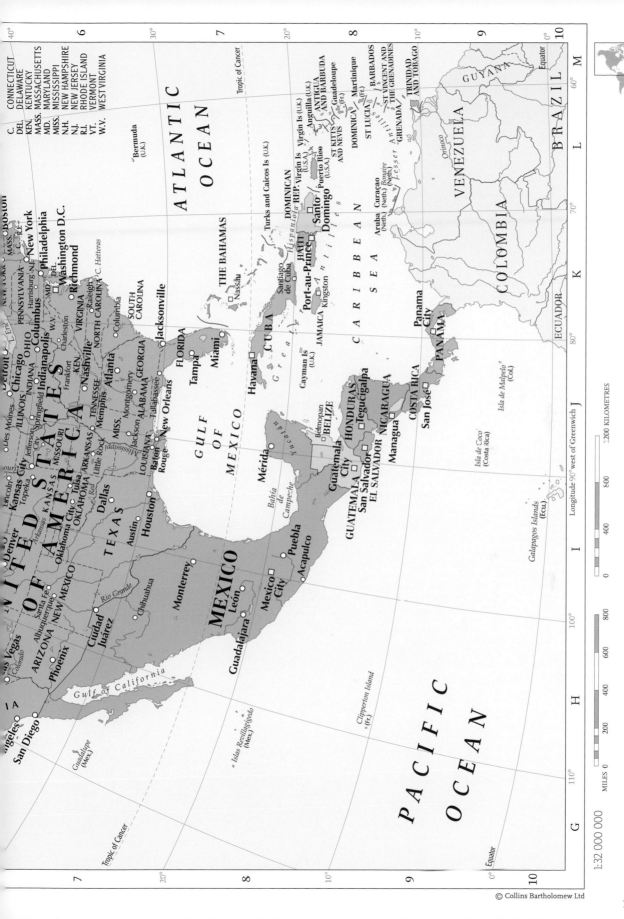

PACIFIC OCEAN

ATLANTIC OCEAN

UNITED STATES OF AMERICA

MEXICO

GULF OF MEXICO

CARIBBEAN SEA

Greater Antilles

Lesser Antilles

CUBA
THE BAHAMAS
JAMAICA
HAITI
DOMINICAN REP.
PUERTO RICO

GUATEMALA
BELIZE
EL SALVADOR
HONDURAS
NICARAGUA
COSTA RICA
PANAMA

COLOMBIA
VENEZUELA
GUYANA
BRAZIL
ECUADOR

© Collins Bartholomew Ltd

1:32 000 000

MILES 0   200   400   600   800

0   400   800   1200 KILOMETRES

Longitude 90° west of Greenwich

Tropic of Cancer

Equator

Lambert Azimuthal Equal Area Projection

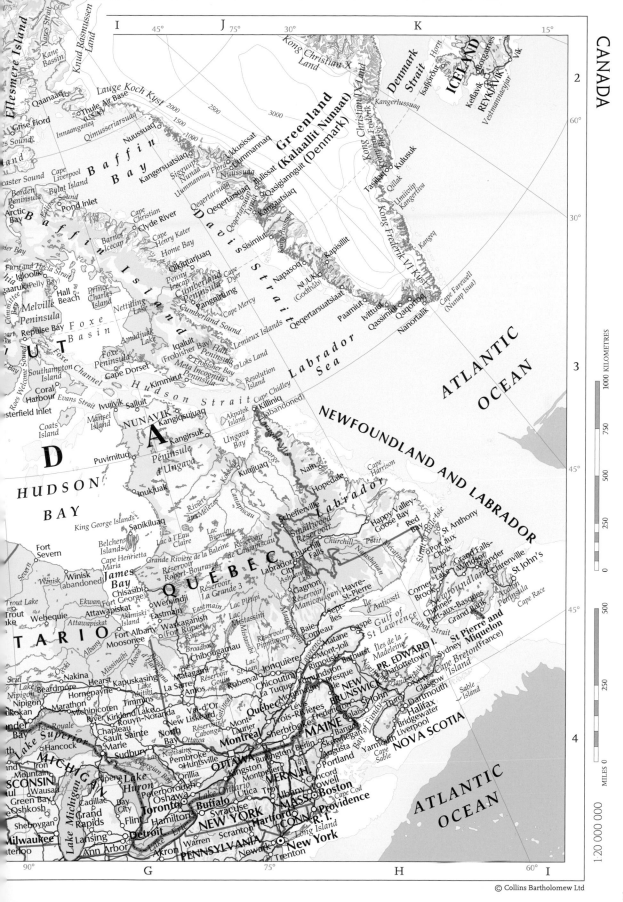

1 20 000 000

© Collins Bartholomew Ltd

Lambert Azimuthal Equal Area Projection

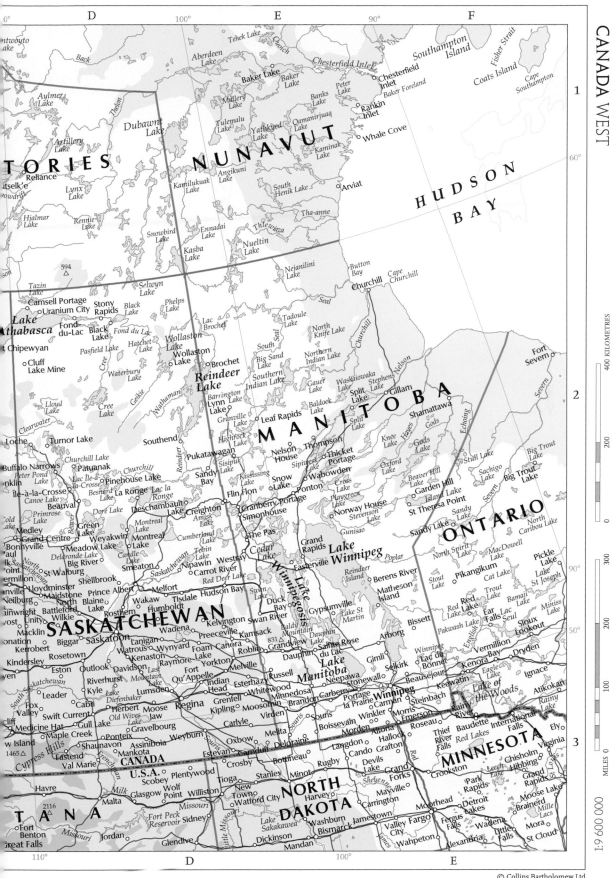

D · 100° · E · 90° · F

1

60°

2

90°

50°

3

HUDSON

BAY

NUNAVUT

MANITOBA

ONTARIO

SASKATCHEWAN

MINNESOTA

NORTH DAKOTA

CANADA
U.S.A.

400 KILOMETRES

200

0

300

200

01

MILES 0

1:9 000 000

© Collins Bartholomew Ltd

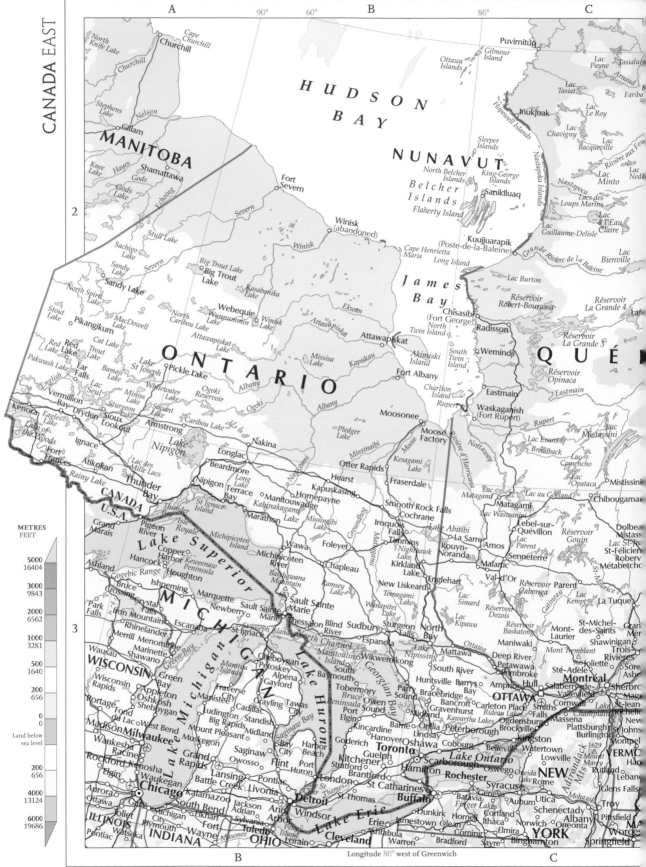

A　　　　90°　60°　　　　B　　　　80°　　　　C

H U D S O N

B A Y

North
Knife Lake

Cape
Churchill

Churchill

Churchill

Ottawa
Islands

Gilmour
Island

Puvirnituq

Lac
Payne

Tasialujj

Stephens
Lake

Nelson

MANITOBA

Gillam

Inukjuak

Lac
Le Roy

Lac
Chavigny

Fariba

Rivière aux Feu

2

Knee
Lake

Hayes

Shamattawa

Gods

Shoal

Gods
Lake

Echoing

Fort
Severn

Sleeper
Islands

King George
Islands

N U N A V U T

North Belcher
Islands

Sanikiluaq

Lac
Bacqueville

Lac
Ned

Lac
Minto

Lacs des
Loups Marins

Lac
à l'Eau
Claire

Stull Lake

Severn

**Belcher
Islands**

Flaherty Island

Nastapoka Islands

Nastapoca

Winisk
(abandoned)

Sachigo
Lake

Sandy
Lake

Severn

Big Trout Lake

Big Trout
Lake

Kasabonika

Winisk

J a m e s

B a y

Lac
Guillaume-Delisle

Stout
Lake

North Spirit
Lake

MacDowell
Lake

North
Caribou
Lake

Wunnummin
Lake

Webequie

Winisk
River

Ekwan

Cape Henrietta
Maria

Kuujjuarapik
(Poste-
de-la-Baleine)

Long Island

Lac
Burton

Réservoir
Robert-Bourassa

Réservoir
La Grande 4

Lac
Bienville

Pikangikum

Red
Lake

Cat Lake

Bamaji
Lake

Attawapiskat
Lake

Attawapiskat

Missisa
Lake

Kapiskau

Akimiski
Island

Chisasibi
(Fort George)

North
Twin Island

Radisson

Réservoir
La Grande 3

Red
Lake

Ear
Falls

Trout
Lake

Pakwash Lake

Lake
St Joseph

Pickle Lake

O N T A R I O

Albany

Fort Albany

South
Twin
Island

Wemindji

Eastmain

Eastmain

Réservoir
Opinaca

QUÉ

English

Vermillion
Bay

Miniss
Lake

Sioux
Lookout

Whitewater
Lake

Ogoki
Reservoir

Albany

Moosonee

Charlton
Island

Rupert

Waskaganish
(Fort Rupert)

Lac
Mistassini

Kenora

Dryden

Lac
Seul

Sturgeon
Lake

Caribou Lake

Armstrong

Ogoki

Pledger
Lake

Moose
Factory

Rupert

Lac
Evans

Lac
Comencho

Mistissini

Eagle
Lake of
the Woods

Ignace

Savant
Lake

**Moose
Factory**

Rivière d'Harricana

Broadback

Lac
Opataca

Fort
Frances

Atikokan

Lac des
Mille Lacs

Nakina

Missinaibi

Otter Rapids

Kesagami
Lake

Nottaway

Chibougama

Rainy Lake

Thunder
Bay

Nipigon

Longlac

Hearst

Fraserdale

Lac
Matagami

Lac au Goéland

Mistissini

CANADA

U.S.A.

Grand
Marais

Pigeon
River

Beardmore

Nipigon
Lake

Terrace
Bay

Kapuskasing

Hornepayne

Smooth Rock Falls

Cochrane

Matagami

Lac Waswanipi

Lebel-sur-
Quévillon

Réservoir
Gouin

Dolbea
Mistass

Ashland

Thunder
Bay

Isle
Royale

St Ignace
Island

Manitouwadge

Kabinakagami
Lake

Iroquois
Falls

Timmins

La Sarre

Amos

Lac St-
Roberv

St-Félicien

Métabetcho

Hancock

Harbor

Keweenaw
Peninsula

Michipicoten
River

Marathon

Missinaibi
Lake

Groundhog

Nighthawk
Lake

Rouyn-
Noranda

Malartic

Lac
Parent

Senneterre

Gogebic Range

Houghton

Lake Superior

Michipicoten
Island

Wawa

Foleyet

Chapleau

Kirkland
Lake

Englehart

Val-d'Or

Réservoir
Parent

La Tuque

Park
Falls

Crystal
Falls

Iron Mountain

Ishpeming

Marquette

Batchawana
Mountain

Sault Sainte
Marie

Ramsey
Lake

Sturgeon
Falls

New Liskeard

Lac
Simard

Réservoir
Dozois

Réservoir
Cabonga

Lac
Kempt

MICHIGAN

Newberry

St Ignace

St-Joseph Island

Thessalon

North Channel

Blind
River

Sudbury

Wanapitei
Lake

North
Bay

Mattawa

Maniwaki

Mont-
Laurier

St-Michel-
des-Saints

Grar
Mer

Rhinelander

Merrill

Menominee

Manitoulin
Island

Wikwemikong

Espanola

Lake
Nipissing

Ottawa

Deep River

Petawawa

Mont Tremblant

Shawinigan

Trois-
Rivières

Wausau

Shawano

Marinette

Cheboygan

Petoskey

Manitou
Islands

South
Baymouth

Tobermory

Owen
Sound

Parry
Sound

Huntsville

Bracebridge

Barrys
Bay

South River

Pembroke

Arnprior

Hull

OTTAWA

Sté-Adèle

Joliette

Montréal

Sore

Ashe

Wisconsin

Green
Bay

Traverse
City

Gaylord

Alpena

Bruce
Peninsula

Gravenhurst

Carleton Place

Smiths
Falls

Salaberry-
de-
Valleyfield

Sherbr

Mag

3

Wisconsin
Rapids

Appleton

Oshkosh

Sheboygan

Grayling

Tawas
City

Georgian Bay

Midland

Orillia

Peterborough

Kawartha Lakes

Rideau Lakes

Cornwall

Champlain

St-Jean

Richelie

New

WISCONSIN

Portage

Fond
du Lac

Ludington

Big Rapids

Cadillac

Saginaw Bay

Elgin

Kincardine

Barrie

Lindsay

Brockville

Massena

Plattsburgh

St-Johns

Montpel

Madison

Milwaukee

West Bend

Muskegon

Mount Pleasant

Midland

Bay
City

Hanover

Oshawa

Cobourg

Belleville

Kingston

Burlington

1629

Watertown

VERMC

Waukesha

Racine

**Grand
Rapids**

Saginaw

Flint

Port
Huron

Goderich

Toronto

Scarborough

Lake Ontario

Oswego

Lowville

Mount
Marty

St-Johns

Montpel

Rockford

Kenosha

Battle Creek

Lansing

Owosso

Pontiac

Stratford

Kitchener

Guelph

Hamilton

Rochester

Syracuse

Oneida
Lake

Rome

Utica

Adirondack
Mts

Rutland

Lebano

Aurora

Elgin

Waukegan

Kalamazoo

Jackson

Livonia

London

Brantford

St Catharines

Buffalo

Batavia

Geneva

Finger Lakes

Auburn

Schenectady

Albany

Glens Falls

Pittsfield

Ottawa

Joliet

Chicago

South Bend

Ann
Arbor

Windsor

Detroit

St Thomas

Dunkirk

Jamestown

Olean

Cortland

Ithaca

Norwich

Oneonta

Troy

MA

Wores

Pontiac

Gary

Michigan
City

Plymouth

Elkhart

Fort
Wayne

Sylvania

Toledo

Pelee
Island

Lake Erie

Erie

Ashtabula

Warren

Bradford

Corning

Elmira

Binghamton

**NEW
YORK**

Springfield

Watseka

INDIANA

OHIO

Maumee

Lorain

Cleveland

Sayre

Longitude 80° west of Greenwich

**METRES
FEET**

5000	16404
3000	9843
2000	6562
1000	3281
500	1640
200	656
0	0

Land below
sea level

200	656
4000	13124
6000	19686

B　　　　　　　　　　C

Lambert Azimuthal Equal Area Projection

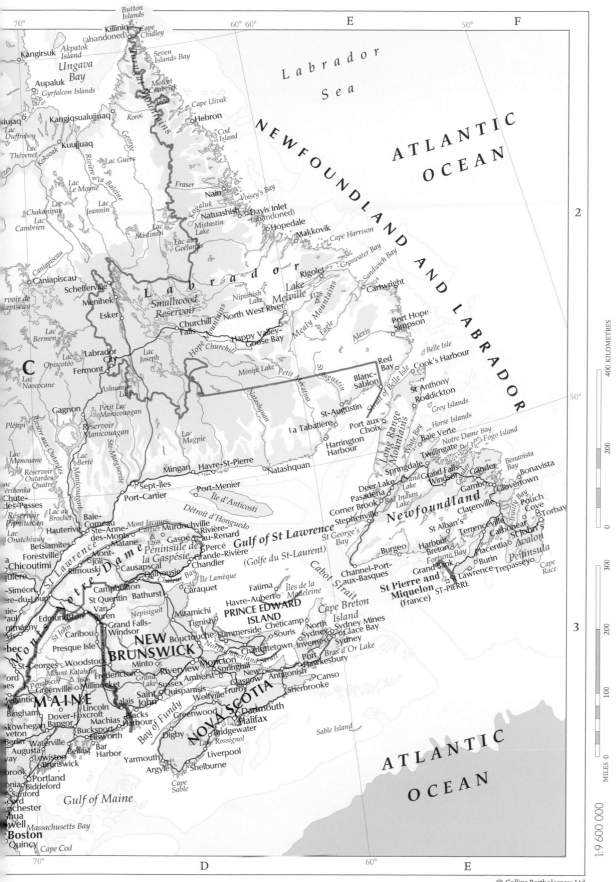

© Collins Bartholomew Ltd

50° 130° A 120° B 110° C 100°

Grid / coordinate labels: 2, 40°, 3, 30°, 4, 20°

PACIFIC OCEAN

CANADA

BRITISH COLUMBIA

ALBERTA

SASKATCHEWAN

MANITOBA

Port Hardy, Gold River, Nanaimo, Victoria, Vancouver Island, Campbell River, Powell River, Nanaimo, Bellingham, Everett, Cape Flattery, Cape Flattery

Mount Waddington 4042, 100 Mile House, Kamloops, Kelowna, Vernon, Nelson, Penticton, Okanagan, Jasper, Mount Columbia 3747, Wetaskiwin, Edmonton, Leduc, Vegreville, Lloydminster, Vermilion, Prince Albert, Nipawin, The Pas, Cedar Lake, Meadow Lake

Banff, Airdrie, Calgary, Okotoks, Brooks, Medicine Hat, Lethbridge, Red Deer, Hanna, Unity, Biggar, Saskatoon, Rosthern, Humboldt, Canora, Yorkton, Melville, Moose Jaw, Davidson, Regina, Weyburn, Virden, Estevan, Williston, Minot

Mount Olympus 2428, Tacoma, Olympia, Seattle, WASHINGTON, Spokane, Coeur d'Alene, Kalispell, Shelby, Havre, Glasgow, Glendive, Fort Peck Reservoir, Lake Sakakawea

Astoria, Portland, Salem, Oregon City, Eugene, Albany, Bend, Pendleton, La Grande, Richland, Yakima, Moscow, Lewiston, Missoula, Great Falls, Helena, MONTANA, Billings, Miles City, N. DAKOTA, Dickinson, Bismarck, Bowman, Mobridge, Lake Oahe, S. DAKOTA, Pierre, Aberdeen

Coos Bay, Crescent City, OREGON, Grants Pass, Burns, Caldwell, Nampa, Boise, IDAHO, Butte, Bozeman, Dillon, Idaho Falls, Cody, Sheridan, Gillette, Buffalo, Black Hills, Rapid City

Eureka, Klamath Falls, Lakeview, Alturas, Twin Falls, Jerome, Pocatello, Pinedale, Lander, WYOMING, Casper, Chadron, Scottsbluff, NEBRASKA, Ogallala

Redding, Red Bluff, Winnemucca, Elko, Wendover, Logan, Brigham City, Ogden, Salt Lake City, Provo, Evanston, Laramie, Cheyenne, Greeley, Sidney, North Platte, McCook

Ukiah, Reno, Sparks, Carson City, NEVADA, Lovelock, Great Salt Lake, Kings Peak, Green River, Craig, Boulder, Denver, Aurora, Burlington, KANSAS

Point Arena, Sacramento, Santa Rosa, Stockton, San Francisco, Oakland, Modesto, San Jose, Salinas, Fresno, Visalia, Tonopah, Ely, UTAH, Grand Junction, Moab, Richfield, Colorado Springs, Pueblo, Great Bend, Dodge City

Monterey Bay, Bakersfield, Santa Maria, Santa Barbara, Death Valley, Beatty, Las Vegas, Henderson, St George, Cedar City, Kanab, Page, Lake Powell, Durango, Alamosa, Trinidad, Ulysses, Liberal

Point Conception, Oxnard, Los Angeles, Pasadena, Long Beach, Riverside, Santa Ana, Oceanside, San Diego, Tijuana, Ensenada, Barstow, Kingman, Lake Havasu City, Prescott, Flagstaff, Winslow, Grand Canyon, Tuba City, Kayenta, Gallup, Santa Fe, Los Alamos, Taos, Clayton, Stratford, Dumas, Amarillo, Elk City

Mexicali, Yuma, San Luis Río Colorado, ARIZONA, Glendale, Phoenix, Mesa, Casa Grande, Tucson, Albuquerque, St Johns, Socorro, NEW MEXICO, Las Cruces, Alamogordo, Roswell, Tucumcari, Clovis, Portales, Lubbock, Vernon, Wichita Falls

Picacho del Diablo 3096, Lázaro Cárdenas, San Felipe, Puerto Peñasco, Nogales, Nogales, Douglas, Agua Prieta, Silver City, Deming, El Paso, Ciudad Juárez, Las Cruces, Artesia, Pecos, Hobbs, Lovington, Midland, Odessa, San Angelo, Snyder, Post, TEXAS

Guadalupe (Mexico), Cabo San Quintín, Caborca, Benjamin Hill, Magdalena, Nuevo Casas Grandes, Moctezuma, Van Horn, Fort Stockton, Alpine, Edwards Plateau

Isla Ángel de la Guarda, Bahía Sebastián Vizcaíno, Isla Cedros, Punta Eugenia, Rosarito, Hermosillo, Madera, Cuauhtémoc, Chihuahua, Ciudad Delicias, Ojinaga, Emory Peak, Ciudad Acuña, Del Rio, Piedras Negras, Sonora

Isla Tiburón, Guaymas, Bahía, Tortugas, Santa Rosalía, Sonora, Ciudad Obregón, Navojoa, Los Mochis, Madera, Ciudad Camargo, Jiménez, Hidalgo del Parral, Bolsón de Mapimí, Monclova, Sabinas, Nuevo Laredo

PACIFIC OCEAN, Villa Insurgentes, Isla Carmen, Isla San José, Isla Cerralvo, Guasave, Guamúchil, Culiacán, Costa Rica, Gómez Palacio, Torreón, Saltillo, Matamoros, Sabinas Hidalgo, Monterrey, Montemorelos

Tropic of Cancer, Santa Margarita, La Paz, San José del Cabo, Mazatlán, Durango, Río Grande, Matehuala, MEXICO, Sierra Madre Occidental

Gulf of California, Baja California

Scale:
METRES / FEET
5000 / 16404
3000 / 9843
2000 / 6562
1000 / 3281
500 / 1640
200 / 656
0 / 0
Land below sea level
200 / 656
4000 / 13124
6000 / 19686

UNITED STATES

Lambert Azimuthal Equal Area Projection

Longitude 110° west of Greenwich

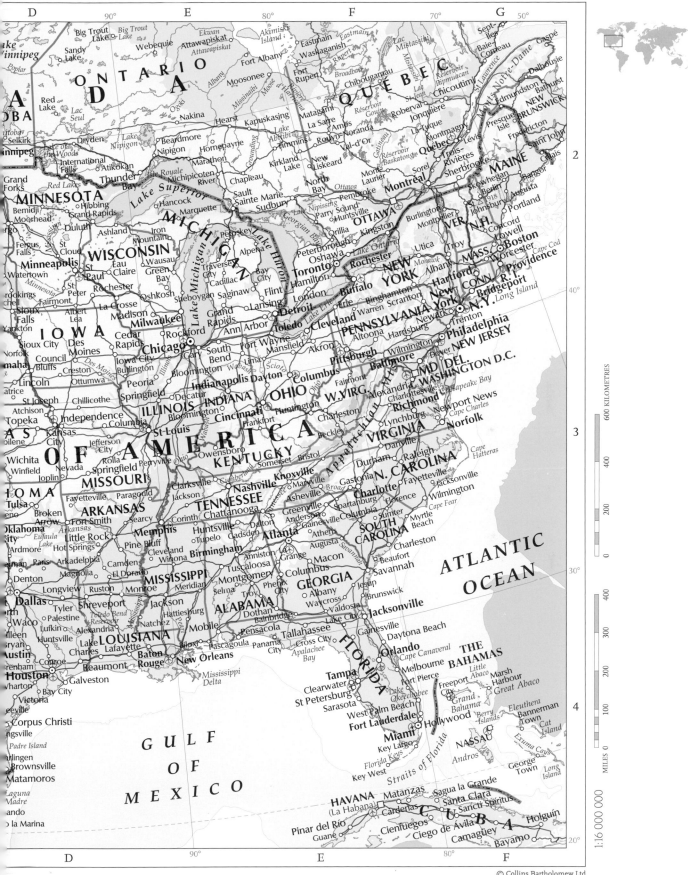

CANADA

ONTARIO

Big Trout Lake
Sandy Lake
Webequie
Attawapiskat
Attawapiskat
Ekwan
Akimiski Island
Eastmain
Waskaganish
Rupert
Lac Mistassini
Sept-Îles
Baie-Comeau
Gaspé

Poplar
Lake Winnipeg
Red Lake
Lac Seul
Nakina
Fort Albany
Moosonee
Broadback
Chibougamau
QUÉBEC
Bipinuacan
Edmundston
Moisie
Bathurst
NEW BRUNSWICK

Selkirk
Winnipeg
Lake of the Woods
Dryden
Nipigon
Beardmore
Hornepayne
Kapuskasing
Timmins
La Sarre
Amos
Rouyn-Noranda
Val-d'Or
Roberval
Lac St-Jean
Jonquière
La Tuque
Montmagny
Lévis
Trois-Rivières
Sorel
Presque Isle
Fredericton
Saint John
Calais

Grand Forks
Red Lakes
International Falls
Atikokan
Thunder Bay
Isle Royale
Michipicoten
River
Marathon
Chapleau
Kirkland Lake
New Liskeard
North Bay
Ottawa
Pembroke
Huntsville
Sherbrooke
Montpelier
MAINE
Skowhegan
Berlin
VER. **N.H.**
Augusta
Bangor
918

MINNESOTA
Bemidji
Moorhead
Fergus Falls
Duluth
Ashland
Iron Mountain
MICHIGAN
Petoskey
Sault Sainte Marie
Lake Huron
Alpena
Sudbury
Lake Nipissing
Parry Sound
Orillia
Peterborough
Oshawa
Lake Ontario
Kingston
Utica
Burlington
Concord
Portland
Lowell
MASS. Boston
Worcester
R.I. Providence
Cape Cod

Grand Rapids
Hibbing
WISCONSIN
Wausau
Eau Claire
Green Bay
Oshkosh
Sheboygan
Cadillac
Traverse City
Bay City
Saginaw
Flint
London
Hamilton
Buffalo
Erie
NEW YORK
Rochester
Syracuse
Albany
Troy
Mohawk
Hartford
CONN. Bridgeport
Long Island

Minneapolis
St Paul
Rochester
Madison
Milwaukee
Lake Michigan
Grand Rapids
Lansing
Detroit
Toledo
Cleveland
Warren
Scranton
New York
Newark
N.Y. Long Island

IOWA
Sioux City
Des Moines
Cedar Rapids
Rockford
Chicago
Gary
South Bend
Fort Wayne
Mansfield
Akron
PENNSYLVANIA
Altoona
Harrisburg
Trenton
Philadelphia
NEW JERSEY

Council Bluffs
Creston
Burlington
Peoria
Bloomington
Lafayette
Lima
Scioto
Fairmont
Pittsburgh
Wilmington
Dover
DEL.
MD.
WASHINGTON D.C.

Lincoln
St Joseph
Chillicothe
Springfield
Decatur
INDIANA
Indianapolis Dayton
Columbus
OHIO
Huntington
W.VIRG.
Charleston
Charlottesville
Alexandria
Chesapeake Bay
Cape Charles

Topeka
Kansas City
Jefferson City
Columbia
St Louis
Frankfort
Cincinnati
Beckley
Lynchburg
Richmond
Newport News
Norfolk

Wichita
Winfield
Joplin
Nevada
Springfield
Rolla
Perryville
Owensboro
Ohio
KENTUCKY
Somerset
Bristol
Danville
VIRGINIA
Cape Hatteras

OF AMERICA
MISSOURI
Fayetteville
Paragould
Jonesboro
Clarksville
Jackson
Nashville
Knoxville
Maryville
Durham
Raleigh
N. CAROLINA
Fayetteville
Jacksonville

OKLAHOMA
Tulsa
Broken Arrow
Fort Smith
ARKANSAS
Searcy
Corinth
TENNESSEE
Chattanooga
Asheville
Greenville
Spartanburg
Charlotte
Florence
Wilmington

Oklahoma City
Ardmore
Little Rock
Hot Springs
Memphis
Arkansas
Tupelo
Gadsden
Atlanta
Athens
Anderson
Columbia
SOUTH CAROLINA
Sumter
Myrtle Beach
Cape Fear

Paris
Arkadelphia
Magnolia
Camden
Cleveland
Winona
Birmingham
Anniston
La Grange
Macon
Charleston
Beaufort

Dallas
Tyler
Longview
Ruston
Monroe
MISSISSIPPI
Meridian
Tuscaloosa
Columbus
GEORGIA
Savannah
ATLANTIC OCEAN

Waco
Palestine
Lufkin
Shreveport
Jackson
Selma
Montgomery
Phenix City
Albany
Jesup
Brunswick

Austin
Huntsville
Alexandria
Natchez
Hattiesburg
ALABAMA
Dothan
Troy
Waycross
Valdosta
Jacksonville

Houston
Conroe
Beaumont
LOUISIANA
Lake Charles
Lafayette
Baton Rouge
Biloxi
Pascagoula
Mobile
Pensacola
Tallahassee
Bainbridge
Lake City
Gainesville
Daytona Beach

Galveston
Bay City
Mississippi Delta
New Orleans
Panama City
Apalachee Bay
Cross City
Cape Canaveral
Melbourne
Fort Pierce
THE BAHAMAS
Little Abaco
Marsh Harbour
Great Abaco

Victoria
Corpus Christi
Kingsville
Padre Island
Orlando
FLORIDA
Tampa
Clearwater
St Petersburg
Sarasota
West Palm Beach
Fort Lauderdale
Freeport
Grand Bahama
Berry Islands
Eleuthera
Bannerman Town
Cat Island

GULF
OF
MEXICO
Harlingen
Brownsville
Matamoros
Laguna Madre
Miami
Key Largo
Florida Keys
Key West
Straits of Florida
NASSAU
Andros
Exuma Cays
George Town
Long Island

HAVANA
(La Habana)
Matanzas
Cárdenas
Sagua la Grande
Santa Clara
Sancti Spíritus
Pinar del Río
Guane
Cienfuegos
CUBA
Ciego de Ávila
Camagüey
Bayamo
Holguín

1:16 000 000

600 KILOMETRES
400
200
0

400
300
200
100
0 MILES

© Collins Bartholomew Ltd

133

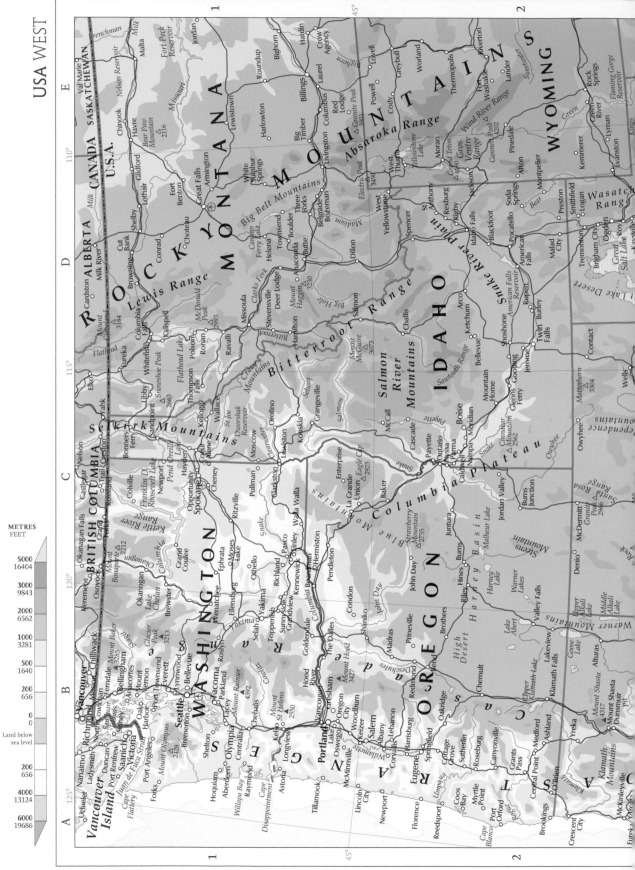

METRES
FEET

5000	16404
3000	9843
2000	6562
1000	3281
500	1640
200	656
0	0
Land below sea level	
200	656
4000	13124
6000	19686

134

Lambert Azimuthal Equal Area Projection

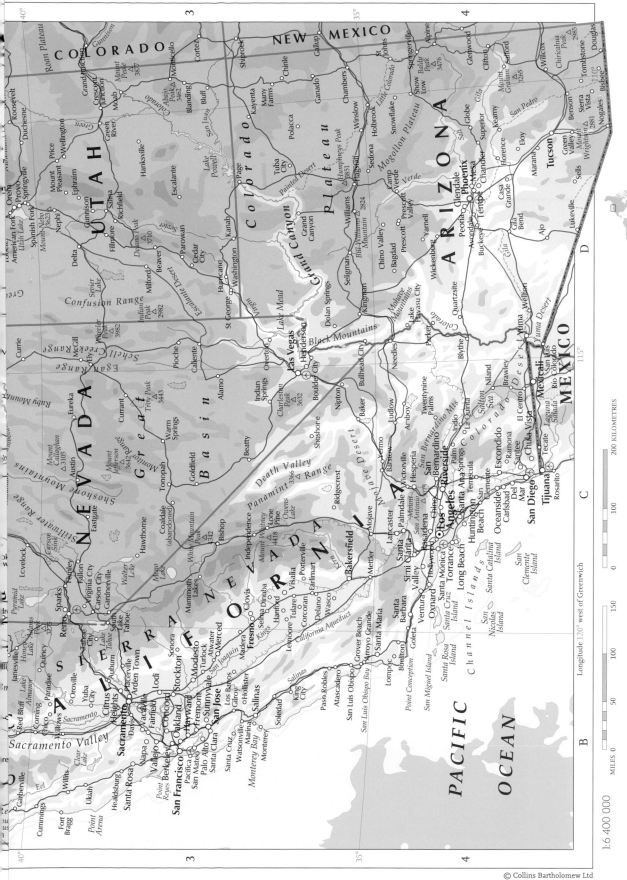

© Collins Bartholomew Ltd

1:6 400 000

Longitude 120° west of Greenwich

A 95° B 90° C

MISSOURI

1

Vinita
Owasso
Tulsa
Sapulpa
Muskogee
Okmulgee
Henryetta
Broken Arrow
Pryor
Siloam Springs
Bentonville
Rogers
Springdale
Fayetteville
Tahlequah
Sallisaw
Van Buren
Chicotah
Okmulgee

West Plains
Mountain Home
Harrison
Marshall
White
Boston Mountains
Clarksville
Heber Springs
Batesville
Newport

Alton
Poplar Bluff
Pocahontas
Hoxie
Jonesboro
Blytheville
Trumann

Charleston
Sikeston
Dexter
Paducah
Mayfield
Murray
Union City
Paris
Dyersburg
McKenzie
Humboldt
Jackson
Brownsville

Hopkinsville
Oak Grove
Russellville
Clarksville
Gallatin
Springfield
Franklin
Dickson
Nashville
Lebanon
Columbia
Shelbyville
McMinnville
Manchester
Lewisburg
Linden

KENTU
TENNES

35°
McAlester
Atoka
Poteau
742
Mansfield △ 839
Fort Smith
Magazine Mountain
Mansfield
Morrilton
Conway
Searcy
Jacksonville
Forrest City
West Memphis
Memphis
Millington
Bartlett
Bolivar
Savannah
Lawrenceburg
Fayetteville
Tullahoma
Huntsville
Scotts
Athens
Decatur
Wheeler Lake

OKLAHOMA
ARKANSAS

Hugo
Idabel
De Queen
Ashdown
Hope
Paris
Commerce
New Boston
Texarkana
Sulphur Springs
Mount Pleasant

Mena
Hot Springs
Little Rock
Arkadelphia
Malvern
Pine Bluff
Fordyce
Dumas
Warren
Camden
El-Dorado
Magnolia
Hamburg
Crossett
Homer
Minden

Stuttgart
Helena
Marianna
Southaven
Holly Springs
Oxford
Clarksdale
Batesville
Cleveland
Grenada
Indianola
Greenwood
Leland
Greenville
Yazoo City

Booneville
Russellville
Tupelo
Amory
Hamilton
Columbus
Starkville
Louisville
Macon
Eutaw
Demopolis

Corinth
Florence
Athens
Huntsville
Cullman
Jasper
Birmingham
Vestavia Hills
Bessemer
Alabaster
Tuscaloosa
Alexander City
Clanton
Prattville
Selma
Montgomery

Cheaha Mountain
Gadsden
Center Point Annis
Sylacauga

MISSISSIPPI
ALABAM

TEXAS
LOUISIANA

2
Longview
Gladewater
Tyler
Athens
Jacksonville
Palestine
Nacogdoches
Crockett
Lufkin
Shreveport
Kilgore
Henderson
Carthage
Bossier City
Gibsland
Mansfield
Jonesboro
Winnfield
Natchitoches
Many
Tenaha
Driskill Mountain △ 163
Ruston
Monroe
Tallulah
Winnsboro
Olla
Pineville
Vicksburg
Jackson
Brandon
Forest
Lake Providence
Canton
Crystal Springs
Natchez
Brookhaven
Meridian
York
Laurel
Jackson
Thomasville
Greenville
Monroeville
Evergreen
Troy
Andalusia
Enterpr

Corrigan
Livingston
Jasper
Alexandria
Marksville
Lecompte
DeRidder
Leesville
Oakdale
Opelousas
Ville Platte
Port Allen
Baton Rouge
Hattiesburg
Petal
McComb
Kentwood
Bogalusa
Picayune
Lumberton
Mobile
Prichard
Atmore
Century
Crestview
De Funiak Spring

GULF OF MEXICO

30°
Huntsville
The Woodlands
Humble
Houston
Baytown
Beaumont
Orange
Vidor
Nederland
Groves
Port Arthur
Sulphur
Lake Charles
Jennings
Crowley
Lafayette
New Iberia
Abbeville
Morgan City
Thibodaux
Houma
New Roads
Baker
Plaquemine
Hammond
Kenner
Metairie
New Orleans
Gretna
Raceland
Cut Off
Port Sulphur
Grand Isle
Gulfport
Biloxi
Pascagoula
Pensacola
Fort Walton
Santa Rosa Island
Panama
Breton Sound
Chandeleur Islands
Mississippi Sound
Mobile Bay
Mobile Point
Mississippi Delta
Lake Pontchartrain
Marsh Island
Atchafalaya Bay
Terrebonne Bay
White Lake

Sugar Land
Pasadena
Texas City
Lake Jackson
Freeport
Galveston
Galveston Island
Galveston Bay

Eufaula Lake
Lake Ouachita
Red
Arkansas
Ouachita Mountains
Ouachita
Sabine
Toledo Bend Reservoir
Sam Rayburn Reservoir
Lake Livingston
Lake Conroe
Trinity
Neches
Sabine
Red River
Mississippi
Yazoo
Pearl
Tombigbee
Cumberland
Tennesse
Kentucky Lake

METRES FEET	
5000	16404
3000	9843
2000	6562
1000	3281
500	1640
200	656
0	0
Land below sea level	
200	656
4000	13124
6000	19686

3

25°

4

B Longitude 90° west of Greenwich C

Lambert Azimuthal Equal Area Projection

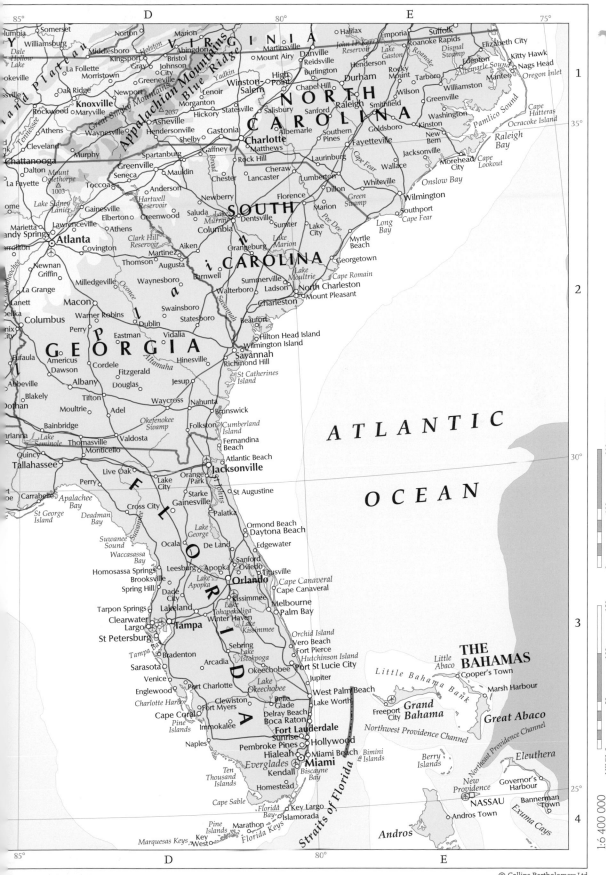

© Collins Bartholomew Ltd

A 115° 110° B 105°

NEVADA

UTAH

Caliente
Alamo
Parowan
Cedar City
Escalante
Escalante Desert
Lake Powell
Colorado
Abajo Peak 3462
Monticello
Uncompahgre Peak 4363
Silverton
COLORADO
Pueblo
Fo

St George
Hurricane
Washington
Kanab
Secret
Blanding
Bluff
Cortez
Durango
Bayfield
Pagosa Springs
Del Norte
San Juan Mountains
Monte Vista
Walsenb

Overton
Boulder City
Lake Mead
Virgin
San Juan
Page
Kayenta
Many Farms
Shiprock
Bloomfield
Farmington
Dulce
Chama
Sangre de Cristo Range
Wheeler Peak 4011
Raton
Laug

1

Dolan Springs
Black Mountains
Grand Canyon
Tuba City
Painted Desert
Polacca
Chinle
Ganado
Chaco Mesa
Los Alamos
Santa Fe
Taos
Espanola
ROCKY MOUNTAINS
Pecos
Las Vegas
Conc
Sprin

Needles
Kingman
Seligman
Williams
Bill Williams Mountain 2824
Humphreys Peak 3851
Flagstaff
Chambers
Gallup
Hosta Butte 2693
Thoreau
Grants
Rio Rancho
Santo Domingo Pueblo
Albuquerque
Clines Corners
Santa Rosa
Pecos

35°

Bullhead City
Mohave Mountains
Lake Havasu City
Chino Valley
Bagdad
Prescott
Prescott Valley
Yarnell
Sedona
Camp Verde
Winslow
Holbrook
Snowflake
Show Low
St Johns
Springerville
Alpine
Zuni Mountains
Belen
Bosque
Vaughn
Sum

Parker
Wickenburg
Verde
Mogollon Plateau
Little Colorado
Baldy Peak 3476
Magdalena
South Baldy 3287
Socorro
NEW MEXICO

Quartzsite
ARIZONA
Peoria
Glendale
Avondale
Phoenix
Buckeye
Tempe
Mesa
Chandler
Salt
Globe
Superior
Glenwood
Whitewater Baldy 3320
Truth or Consequences
Carrizozo
Ruidoso
Hondo
Roswell

Blythe
Colorado
Gila
Casa Grande
Florence
Eloy
Kearny
Clifton
Silver City
Black Range
Hatch
Bayard
Mescalero
Tularosa
Alamogordo
Sacramento Mountains
Artesia

Yuma
Wellton
San Luis Río Colorado
Gila Bend
Ajo
Marana
Tucson
Mount Graham 3265
Safford
Gila
Lordsburg
Las Cruces
Mesilla
San Andres Mountains
Carlsb

2

Desierto de Altar
Lukeville
Sonoita
Sells
Green Valley
Mount Wrightson 2881
Benson
Sierra Vista
Willcox
Chiricahua Peak 2985
Deming
Anthony
El Paso
Guadalupe Peak 2667

Puerto Peñasco
San Luisito
Caborca
Tubutama
Nogales
Sierra Cibuta 2034
Cananea
Tombstone
Bisbee
Douglas
Agua Prieta
Columbus
Ciudad Juárez
Socorro
Fabens
Diablo Plateau

El Socorro
Desemboque
Pitiquito
Santa Ana
Magdalena
Nacozari de Garcia
Arizpe
Fronteras
Casa de Janos
Guzmán
El Barreal
El Porvenir
Sierra Blanca
Van Horn

Puerto Libertad
Benjamín Hill
Cumpas
Casas Grandes
Nuevo Casas Grandes
Pacheco
Villa Ahumada
Moctezuma
Rio Grande

GULF OF CALIFORNIA

Isla Ángel de la Guarda
Carbó
Opodepe
Moctezuma
Tepache
Las Varas
Buenaventura
El Sueco
Presi
Djnag

SONORA

BAJA CALIFORNIA
Rosarito
Bahía Kino
Isla Tiburón
Hermosillo
San Pedro
El Saucito
Ures
Alamos
Sonora
Madera
San José de Bavicora
El Sauz
San Lorenzo
Potrer del Llan

Santo Domingo
Pico Echegerria 1908
Mazatán
Aros
CHIHUAHUA
Conchos

Guerrero Negro
Desierto de Vizcaína
Sierra Libre 180
Tecoripa Moreno
Ciudad Guerrero
La Junta
Cuauhtémoc
Chihuahua
Aldama

3

Punta Abreojos
San Ignacio
Volcán Las Tres Vírgenes 1996
Santa Rosalía
Empalme
Guaymas
Presa Obregón
Rosario
Yécora
Yaqui
Sierra Madre Occidental
Pedernales
Creel
San Juanito
Carichic
Doctor Belisario Dominguez
Ciudad Delicia
Saucillo
MEX

Mulegé
Ciudad Obregón
Esperanza
Chinipas
Nonoava
Uruáchic
Presa de la Boquilla
Camargo
Bolsón
Map

Rosarito
Navojoa
Presa Macuzari
Creel
Conchos
Presa de la Boquilla
San Pablo Balleza
Jiménez
Ciudad Camargo
Hidalgo del Parral

BAJA CALIFORNIA SUR
Loreto
Isla Carmen
Bacobampo
Huatabampo
Alamos
Batopilas
Santa Bárbara
Villa Ocampo
Escal
Ce

San José de Comondú
Don
El Fuerte
Choix
Presa Miguel Hidalgo
Guadalupe y Calvo
3150
DURANG
Las Nieves

SINALOA
San Blas
Ahome
Los Mochis
Camino Real de Tierra A
Guanacevi
Inde

Longitude 110° west of Greenwich

METRES / FEET

5000 / 16404
3000 / 9843
2000 / 6562
1000 / 3281
500 / 1640
200 / 656
0 / 0
Land below sea level
200 / 656
4000 / 13124
6000 / 19686

Lambert Azimuthal Equal Area Projection

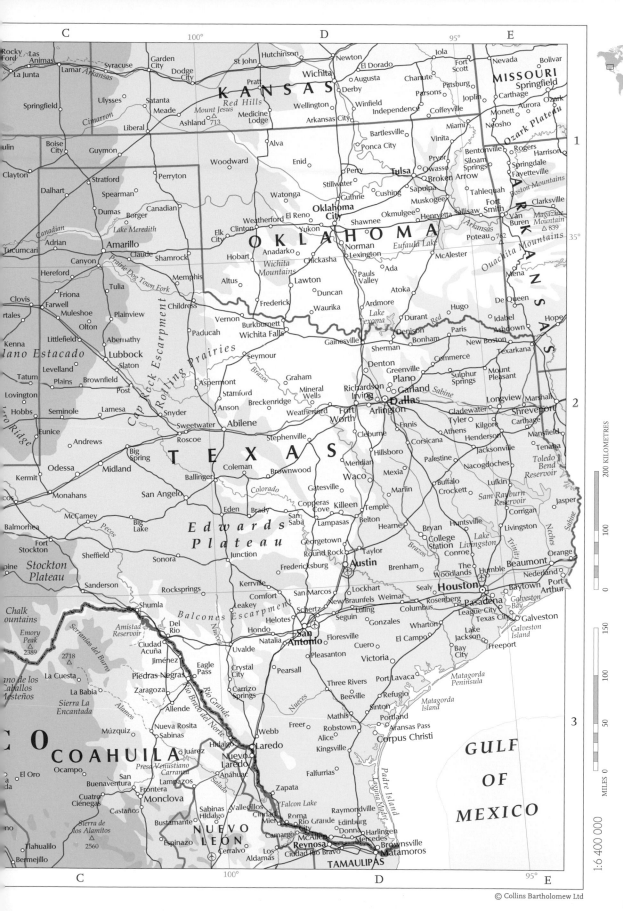

MISSOURI

KANSAS

OKLAHOMA

ARKANSAS

TEXAS

COAHUILA

NUEVO LEÓN

TAMAULIPAS

GULF OF MEXICO

1:6 400 000

200 KILOMETRES

MILES

A   110°   B

ARIZONA   NEW MEXICO   UNITE

Tecate   El Centro   Brawley   Yuma   Gila   Florence   Superior   Clifton   320   Truth or   Ruidoso   Roswell   Levelland
Tijuana   Mexicali   San Luis   Gila   Bend   Casa   Kearny   Consequences   Tularosa   Alamogordo   Brownfield   Lovington
Rosarito   Río Colorado   Grande   Ajo   Safford   Lordsburg   Silver City   Las Cruces   Carlsbad   Artesia   Hobbs   Seminole
Ensenada   Tierra de Juárez   San Felipe   Desierto de Altar   Tucson   Willcox   Chiricahua   Deming   Columbus   El Paso   Guadalupe Peak   Eunice   Andrews   Midland   Spri
San   Vicente   Puerto   Peak   Green   Benson   2985   Ciudad Juárez   Fabens   2667   Pecos
Peñasco   Valley   Sierra   Bisbee   Guzmán   Samalayuca   Van   Horn   Fort   Big La
Vicente   Picacho   El Golfo   Nogales   Vista   Douglas   El Porvenir   Stockton
Guerrero   del Diablo   de Santa Clara   3096   Nogales   Agua Prieta   Fronteras   El Barreal   Villa   Moctezuma   Marfa   Mount Livermore   Stockton
Lázaro   San   El Socorro   Tubutama   Cananea   Nacozari   Casa de Janos   Ahumada   2554   Alpine   Plateau
Cárdenas   Felipe   Caborca   Santa   de García   Nuevo Casas   Sanderson
Rosario   30°   Ana   Magdalena   Arizpe   Cumpas   Grandes   Buenaventura   Presidio
San Fernando   Benjamín Hill   Moctezuma   Tépache   Las   José   Emory   Ojinaga   2718   Serra
Puerto   Carbó   Ures   Varas   Madera   de Bavicora   Peak   2389   del Bi
Baja   Libertad   Alamos   Chihuahua   La Cue
Isla Ángel   Bahía   Mazatán   La Junta   Cuauhtémoc   Llano de los   La Babia
de la Guarda   Kino   Hermosillo   Presa Plutarco   Yécora   Meoqui   Ciudad   Caballos Mesteños
Isla   Bahía Rosarito   Pico   Sonora   Elías Calles   San   Delicias   Múzqu
Cedros   Sebastián   Echeverría   Tecoripa   Juanito   Doctor B.   Saucillo   Bolsón
Santo Domingo   Vizcaíno   1908   Empalme   Psa Obregón   Uruáchic   Carichic   Domínguez   Ciudad   de Mapimí   Ocampo
Punta   Guerrero   Guaymas   Rosario   Creel   Nonoava   Camargo   Sierra   Buenavista
Eugenia   Negro   Esperanza   Presa   Jiménez   El Oro   Cuatro Ciénegas
Bahía   Sierra   Volcán Las   Ciudad   Macuzari   San Pablo   Mojada   Monclo
Tortugas   Ignacio   Tres Vírgenes   Obregón   Navojoa   Chínipas   Balleza   Escalón   Ceballos   Casta
Punta San Hipólito   1996   Santa   Alamos   Batopilas   Hidalgo   Camino Real de   Tlahualilo
San   Rosalía   Huatabampo   Choix   del Parral   Las Nieves   Tierra Adentro   San Pedro
Mulegé   El Fuerte   Verde   Santa   Bermejillo   de las Color
San José   Ahome   Presa M.   Bárbara   Guadalupe   3150   Indé   Mapimí   Matamoros
de Comondú   Hidalgo   y Calvo   Guanaceví   Gómez Palacio   Parras
Loreto   Isla   Los Mochis   Guasave   Topia   Tepehuanes   Torreón   Viesca   Ce
Villa Insurgentes   Carmen   Guamúchil   Guanacevi   Nuevo   Victoria   Miguel   Concep
Ciudad Constitución   Dolores   Topolobampo   Perícos   Tamazula   Ideal   Nazas   Auza   Cama
2   Bahía   Puerto   Isla San José   Culiacán   Costa   Santiago   Canatlán   Guadalupe
Magdalena   Cortés   Isla Espíritu Santo   Navolato   Rica   Papasquiaro   Durango   Rio Grande
Isla Santa   La Paz   Pichilingue   El Dorado   Cosalá   Cerro   Villa   Felipe Pesc
Margarita   San Pedro   Isla Cerralvo   Huehueto   Unión   Sombrerete   3559   Sain   Villa
Tropic of Cancer   Picacho de la Laguna   La Cruz   El Salto   3150   Fresnillo   de C
Todos Santos   2163   Santiago   Villa   Alto   Salti
Cabo San Lucas   San José del Cabo   Mazatlán   Unión   Rosario   MEX   Jerez   Zacateca   Rinco
Cabo   Escuinapa   Villanueva   Rom
Falso   Teacapán   Acaponeta   Aguascalie
Nayar   Mezquitic   Colotlán
Tuxpan   Tecuala   San Martín   Calvillo   Jalpa   298
Laguna Agua Brava   Ruiz   de Bolaños   Teul de   Encarn
Santiago Ixcuintla   Tepic   González   Yahualica   Jalostotitlán   Leó
Compostela   Ortega   Tequila   Tepatitlán   Irap
Islas   Las Varas   Ixtlán   Guadalajara
Marías   Puerto Vallarta   Ameca   Zacoalco   La Piedad
Bahía de Banderas   Cocula   Sahuayo   Zamo
20°   Cabo Corrientes   Tomatlán   Laguna de Chapala   Ciudad   Zacap   Hid
Sayula   Guzmán   Pátzcu
Isla   Autlán   Nevado de Colima   4339   Colima   3859   Uru
San Benedicto   Cihuatlán   Tepalcatepec   Apatzi
Islas Revillagigedo   Isla   Manzanillo   Tecomán   Aguilill
(Mexico)   Socorro   Armería   Coalcomán
Isla Clarión   Arteaga   Infi
Lázaro Cárdenas
PACIFIC   Zihuatar
Pe

3   OCEAN

METRES   FEET

5000	16404
3000	9843
2000	6562
1000	3281
500	1640
200	656
0	0
Land below sea level	
200	656
4000	13124
6000	19686

A   Longitude 110° west of Greenwich   B

144   Lambert Azimuthal Equal Area Projection

A · 90° · B · 80°

Lake Charles · Jennings · **Baton Rouge** · Mobile · Bainbridge · Waycross · Valdosta · Brunswick
Beaumont · Orange · Lafayette · Pascagoula · Pensacola · Tallahassee · Lake City · **Jacksonville**
30° · LOUISIANA · Biloxi · Panama City · Gainesville
Morgan City · **New Orleans** · Ocala · Daytona Beach
Houma · Cross City · **Orlando** · Titusville · Cape Canaveral
Mississippi Delta · Cape San Blas · Apalachee Bay · Waccasassa Bay · **Tampa** · Lakeland · Melbourne

UNITED STATES OF AMERICA
Clearwater · FLORIDA
St Petersburg · Fort Pierce
Sarasota · Lake Okeechobee
Port Charlotte · West · Palm Beach
Fort Myers

GULF
OF
MEXICO

Grand Bahama · Little Abaco · Marsh Harbor · Great Abac
Freeport City · Eleuthe
Everglades · **Fort Lauderdale** · Berry Islands
Hollywood · Bannerman
Miami · NASSAU · Town
Florida Keys · Key Largo · Andros
Key West · George T

2

Tropic of Cancer

HAVANA (La Habana) · Archipiélago de Sabana · Matanzas
Pinar del Río · Cárdenas · Santa Clara · Esmeralda · Camagüey
Arrecife Alacrán · Guane · Sagua la Grande · Archipiélago de Camagüey
Golfo de Batabanó · Cienfuegos · Sancti Spíritus · Las Tunas · Hol
Progreso · Cabo Catoche · Ciego de Ávila
Bahía de Campeche · **Mérida** · Tizimín · CUBA · Golfo de Guacanayabo · Baya
Muna · Valladolid · Cancún · Isla de la Juventud · Manzanillo · Santi
20° · Tekax · Cozumel · Cabo Cruz · de C
Campeche · YUCATÁN · Isla de Cozumel · Grand Cayman · Little Cayman
Champotón · Montego Bay · Spanish Town · Jam
Ciudad del Carmen · MEXICO · Banco Chinchorro · Cayman Islands (U.K.) · JAMAICA
Frontera · Escárcega · KINGSTON
Laguna de Términos · Chetumal · Ambergris Caye
Villahermosa · Belize
Palenque · Tenosique · BELMOPAN · Turneffe Islands · CARIBB
Teapa · Flores · Belize · Gulf of Honduras
San Cristóbal de las Casas · La Libertad · Dangriga · Islas de la Bahía · Roatán
Punta Gorda · La Ceiba · Trujillo
Puerto Barrios · Laguna de Caratasca
GUATEMALA · San Pedro Sula · Puerto Lempira
Tapachula · Cobán · Lago de Izabal · El Progreso · Patuca
3 · Huehuetenango · Santa Rosa de Copán · HONDURAS
Quetzaltenango · GUATEMALA CITY · TEGUCIGALPA · Coco · Cayos Miskitos
Mazatenango · Santa Ana · San Vicente · Danlí · Puerto Cabezas
Puerto San José · Sonsonate · San Miguel · Somoto · Cordillera Isabelia
SAN SALVADOR · Usulután · Jinotega · Costa de Mosquitos · Isla de Providencia (Colombia)
EL SALVADOR · Matagalpa · Río Grande
Golfo de Fonseca · NICARAGUA · Isla de San Andrés (Colombia)
León · Boaco · Juigalpa · Islas del Maíz (Nicaragua)
MANAGUA · Bluefields
Jinotepe · Granada
Rivas · Lake Nicaragua
San Juan
Liberia
10° · COSTA RICA · Puerto Limón · Canal de Panamá
Puntarenas · SAN JOSÉ · Colón · Punta San Blas
Cartago · Chirripó △3819 · Changuinola · Golfo del Darién
PACIFIC · Cordillera de Talamanca · Bocas del Toro · PANAMA · PANAMA CITY · Mont
Golfo de Nicoya · Aguadulce · La Chorrera
4 · Península de Osa · La Concepción · David · Chitré · Gulf of Panama · La Palma
OCEAN · Puerto Armuelles · Golfo de Chiriquí · Santiago · Turbo
Península de Azuero · Punta Mala
Isla de Coiba

A · 90° · B · 80°

METRES FEET
5000 / 16404
3000 / 9843
2000 / 6562
1000 / 3281
500 / 1640
200 / 656
0 / 0
Land below sea level
200 / 656
4000 / 13124
6000 / 19686

Lambert Azimuthal Equal Area Projection

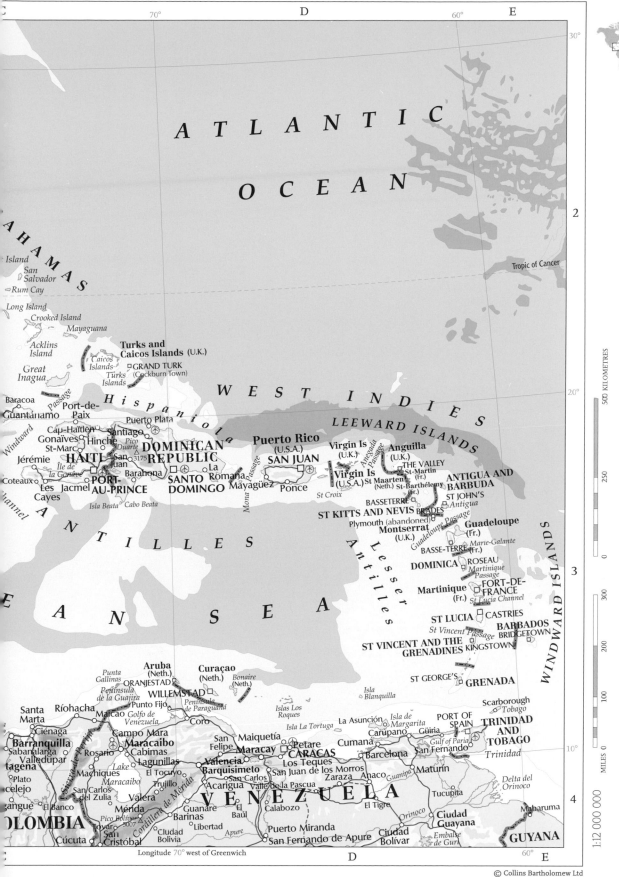

ATLANTIC

OCEAN

2

Tropic of Cancer

AHAMAS

Island

San
Salvador

Rum Cay

Long Island

Crooked Island

Mayaguana

Acklins
Island

Great
Inagua

Turks and
Caicos Islands (U.K.)

Caicos
Islands

□ **GRAND TURK**
(Cockburn Town)

Turks
Islands

W E S T   I N D I E S

20°

Baracoa

Guantánamo

Passage

Port-de-
Paix

Puerto Plata

H i s p a n i o l a

LEEWARD   ISLANDS

Cap-Haïtien

Santiago

Pico
Duarte

3175

Gonaïves

St-Marc

Hinche

DOMINICAN

Puerto Rico
(U.S.A.)

Virgin Is
(U.K.)

Anegada
Passage

Anguilla
(U.K.)

THE VALLEY

Jérémie

Île de
la Gonâve

San
Juan

REPUBLIC

SAN JUAN

Virgin Is
(U.S.A.)

St-Martin
(Fr.)

St-Barthélemy
(Fr.)

ANTIGUA AND

Coteaux

PORT-

Barahona

La
Romana

Mona
Passage

Mayagüez

Ponce

St Maarten
(Neth.)

BARBUDA

Les
Cayes

Jacmel

AU-PRINCE

SANTO
DOMINGO

St Croix

BASSETERRE □

ST JOHN'S
□ *Antigua*

Isla Beata *Cabo Beata*

ST KITTS AND NEVIS BRADES

Plymouth (abandoned)

Guadeloupe Passage

Guadeloupe
(Fr.)

A N T I L L E S

Montserrat
(U.K.)

Marie-Galante

BASSE-TERRE □

Channel

DOMINICA

□ **ROSEAU**

Martinique
Passage

Martinique
(Fr.)

FORT-DE-
FRANCE

3

E A N   S E A

St Lucia Channel

ST LUCIA □ **CASTRIES**

BARBADOS

St Vincent Passage

BRIDGETOWN

ST VINCENT AND THE
GRENADINES KINGSTOWN

W I N D W A R D   I S L A N D S

L e s s e r

A n t i l l e s

Aruba
(Neth.)

Curaçao
(Neth.)

Bonaire
(Neth.)

ST GEORGE'S
□

GRENADA

Punta
Gallinas

ORANJESTAD

WILLEMSTAD □

Isla
Blanquilla

Península
de la Guajira

Punto Fijo

Península
de Paraguaná

Scarborough

Tobago

Santa
Marta

Ríohacha

Maicao

Golfo de
Venezuela

Coro

Islas Los
Roques

Isla La Tortuga

La Asunción

Isla de
Margarita

PORT OF
SPAIN

TRINIDAD

Ciénaga

Campo Mara

San
Felipe

Maiquetía

Cumaná

Carúpano

Güiria

AND

Barranquilla

Rosario

Cabimas

Maracaibo

Maracay

Petare

Barcelona

Gulf of Paria

San Fernando

TOBAGO

Sabanalarga

Lagunillas

Valencia

CARACAS

San Juan de los Morros

Anaco

Maturín

Trinidad

10°

Valledupar

Lake

El Tocuyo

Barquisimeto

San
Carlos

Guanipa

tagena

Machiques

Maracaibo

Trujillo

Acarigua

Valle de la Pascua

Zaraza

El Tigre

Delta del
Orinoco

Tucupita

elejo

San Carlos
del Zulia

Valera

Cordillera de Mérida

Guanare

El
Baúl

Calabozo

Mabaruma

angue

El Banco

Mérida

Pico Bolívar
5007

Barinas

Libertad

Orinoco

Ciudad
Guayana

4

OLOMBIA

Tovar

San
Cristóbal

VENEZUELA

Puerto Miranda

Ciudad
Bolívar

Embalse
de Guri

GUYANA

Cúcuta

Ciudad
Bolivia

Apure

San Fernando de Apure

500 KILOMETRES

250

0

300

200

100

MILES 0

1:12 000 000

© Collins Bartholomew Ltd

ATLANTIC

OCEAN

CARIBBEAN SEA

NICARAGUA

COSTA RICA

PANAMA

Lake Nicaragua

Isla de Malpelo (Colombia)

Equator

Barranquilla

Maracaibo

Barquisimeto

Aruba (Neth.)
Curaçao (Neth.)
Bonaire (Neth.)

Caracas

VENEZUELA

Ciudad Bolívar

Orinoco

ST VINCENT AND THE GRENADINES
ST LUCIA
BARBADOS
GRENADA
TRINIDAD AND TOBAGO

Georgetown

Paramaribo

Cayenne

GUYANA

SURINAME

French Guiana

Mouths of the Amazon

Ilha de Marajó

Belém

São Luís

Fortaleza

Natal

João Pessoa

Recife

Maceió

Aracaju

Salvador

Ilha da Trinidade (Brazil)

Teresina

Parnaíba

São Francisco

Barragem de Sobradinho

Brasília

Goiânia

Belo Horizonte

Vitória

Rio de Janeiro

Campinas

São Paulo

Medellín

Bogotá

COLOMBIA

Cali

Quito

ECUADOR

Guayaquil

Trujillo

Callao

Lima

Arequipa

Ayacucho

Arica

PERU

Pucallpa

Iquitos

Rio Branco

Porto Velho

BOLIVIA

La Paz

Sucre

Santa Cruz

Lake Titicaca

ANDES

SELVAS

A Desert

Boa Vista

Manaus

Santarém

Marabá

Cuiabá

Campo Grande

PARAGU

BRAZIL

Branco

Negro

Amazon

Japurá

Purus

Juruá

Madeira

Tapajós

Xingu

Tocantins

Araguaia

Iriri

Amazon

Marañón

Paraguay

Paraná

Macapá

Bi-Polar Oblique Projection

Equator

10°

80°

70°

60°

50°

40°

20°

10°

0°

10°

20°

1

2

3

4

A   B   C   D   E   F   G

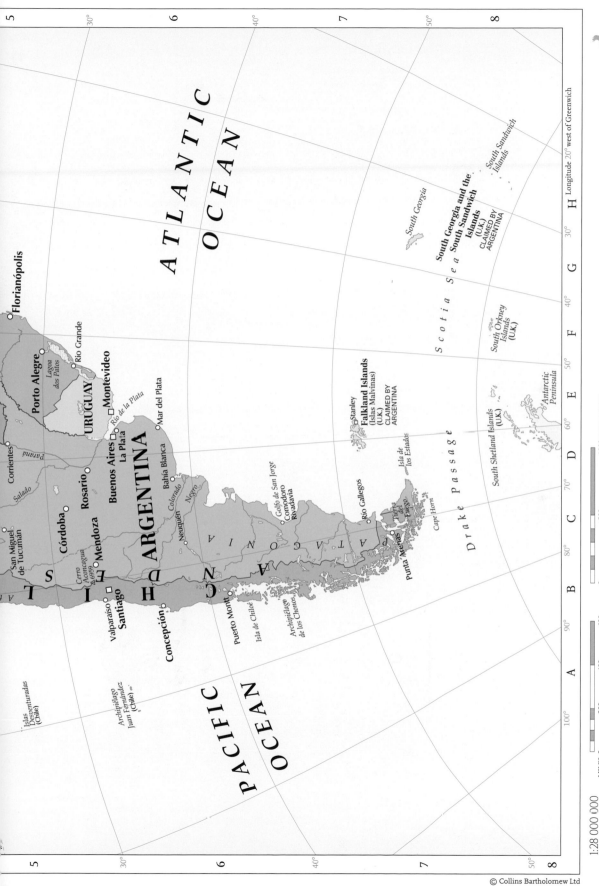

5 · 30° · 6 · 40° · 7 · 50° · 8

Longitude 20° west of Greenwich

H · G · F · E · D · C · B · A

A T L A N T I C

O C E A N

PACIFIC

OCEAN

Florianópolis

Porto Alegre

Rio Grande

Lagoa dos Patos

URUGUAY

Montevideo

Rio de la Plata

La Plata

Mar del Plata

Corrientes

Paraná

Salado

Buenos Aires

Rosario

Córdoba

San Miguel de Tucumán

Mendoza

Valparaíso

Santiago

Concepción

Puerto Montt

Isla de Chiloé

Archipiélago de los Chonos

ARGENTINA

Bahía Blanca

Colorado

Negro

Neuquén

P A T A G O N I A

Golfo de San Jorge

Comodoro Rivadavia

Río Gallegos

Tierra del Fuego

Cape Horn

Punta Arenas

Isla de los Estados

D r a k e   P a s s a g e

S c o t i a   S e a

South Georgia

South Georgia and the South Sandwich Islands (U.K.) CLAIMED BY ARGENTINA

South Sandwich Islands

South Orkney Islands (U.K.)

South Shetland Islands (U.K.)

Antarctic Peninsula

Stanley

Falkland Islands (Islas Malvinas) (U.K.) CLAIMED BY ARGENTINA

A N D E S

Cerro Aconcagua 6959

Islas Desventuradas (Chile)

Archipiélago Juan Fernández (Chile)

30°

40°

50°

60°

70°

80°

90°

100°

30°

40°

50°

© Collins Bartholomew Ltd

1:28 000 000

1000 KILOMETRES

600 400 200 0

500 0

MILES 0

149

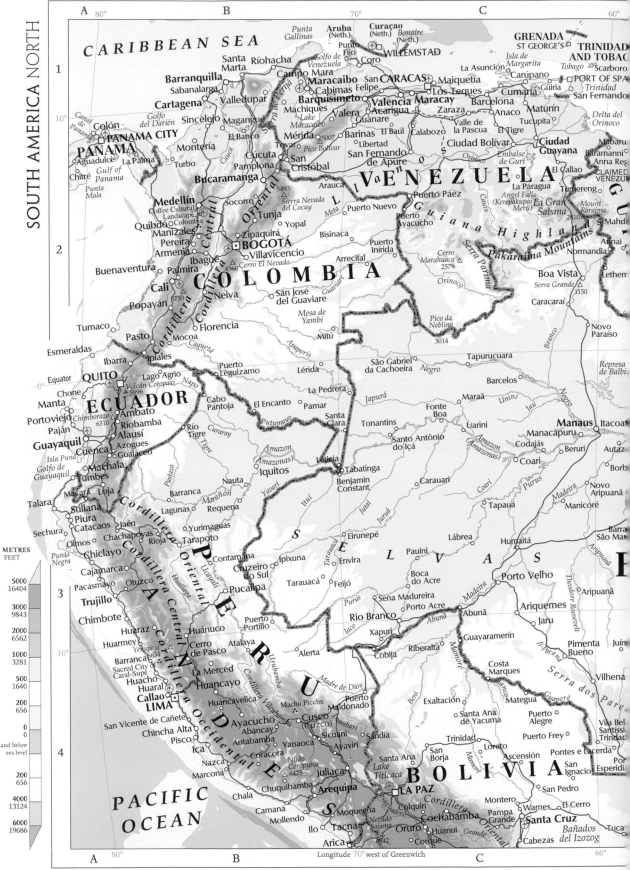

CARIBBEAN SEA

GRENADA
ST GEORGE'S
TRINIDAD AND TOBAG
Tobago Scarboro
Isla de Margarita **PORT OF SPA**
La Asunción *Trinidad*
Carúpano Güiria San Fernando
Cumana
Barcelona Maturín
Anaco Tucupita
El Tigre **Ciudad**
Guayana Mabaru
Embalse de Guri Baramanni
Anna Reg
El Callao
La Páragua CLAIMED
VENEZU
Tumereng
Angel Falls El Gran
(Kerepakupai Sabana
Merú) Mount
Roraima Arinai
2810 Normandia
Serra Grande Lethen
1150 Boa Vista

Punta Gallinas
Aruba (Neth.) Curaçao (Neth.) Bonaire (Neth.)
Punto Fijo **WILLEMSTAD**
Golfo de Venezuela Coro
Santa Marta Ríohacha
Barranquilla Campo Mara
Sabanalarga Maracaibo San CARACAS Maiquetía
Cartagena Valledupar Cabimas Felipe Los Teques
Machiques **Valencia Maracay**
Sincelejo Magangue Lake Maracaibo Acarigua Zaraza
Colón Golfo del Darién El Banco Mérida Valera Guanare
PANAMA CITY Montería 5007 Barinas El Baúl Calabozo
PANAMA Pico Bolívar Libertad **Ciudad Bolívar**
Aguadulce La Palma Turbo Cúcuta San Fernando de Apure o
Chitré Gulf of Panama Pamplona **San Cristóbal** Arauca
Punta Mala **Bucaramanga** Puerto Páez
Medellín Socorro 5493 Puerto Nuevo
Coffee Cultural Sierra Nevada del Cocuy Puerto
Landscape Tunja Ayacucho
Quibdó Colombia Zipaquirá Meta Puerto Inírida
Manizales **BOGOTÁ**
Pereira Villavicencio Arrecifal
Armenia Cerro El Nevado 4560 Bisinaca
Ibagué Cerro
Buenaventura Palmira 6750 **COLOMBIA** Marahuaca
Cali Neiva San José 2579 Pico da
Popayán del Guaviare Neblina
Cordillera Central Cordillera Oriental Mesa de Yambi 3014
Tumaco Florencia Guaviare
Pasto Mocoa Mitú

o
VENEZUELA
L
Meta
Guiana Highland
Orinoco
Serra Parima
Pakaraima Mountains
Caracaraí
Novo Paraíso

Esmeraldas Cordillera Caquetá Apaporis
Ibarra Ipiales Puerto São Gabriel Tapurucuara
Equator Leguizamo Lérida da Cachoeira Barcelos Represa de Balbi
QUITO Agrio Negro
Chone Lago Napo La Pedrera Maraã Urini Iau
Volcán Cotopaxi Cabo Marañón Fonte
Manta 5896 **ECUADOR** Pantoja El Encanto Pamar Santa Boa Manacapuru **MANAUS** Itacoa
Chimborazo Ambato Putumayo Clara Santo Antônio Codajás Beruri Autaz
Portoviejo 6310 Riobamba Río Japurá do Içá Coari
Paján Alausí Tigre Iquitos Tonantins Amazon (Amazonas) Borb
Guayaquil Azogues Curaray Nauta Léticia Coari
Isla Puná Cuenca Gualaceo Amazon Tabatinga Carauari Purus Madeira
Golfo de Guayaquil Machala Amazonas Benjamin Novo
Tumbes Barranca Constant Tapauá Aripuanã
Macará Loja Lagunas Yavari Juruá Manicoré
Talara Sullana Requena Eirunepé Itui Juruá Barra
Piura Cordillera Oriental Yurimaguas Jutaí Lábrea São Ma
Catacaos Jaén Tarapoto Pauini Humaitá
Punta Negra Chachapoyas Rioja Ucayali Contamana Ipixuna Envira Boca Porto Velho Aripuanã
Chiclayo Cruzeiro Tarauacá do Acre Madeira Theodore Roosevelt
Cajamarca Otuzco do Sul Feijó Sena Madureira Pimenta
PERU Pucallpa Rio Branco Abunã Ariquemes Bueno Juín
Pacasmayo Iaco Xapuri Porto Acre Jaru
Trujillo Cerro Puerto Alerta Cobija Riberalta Abunã Guayaramerín Serra dos Pare Vilhena
Chimbote Huaraz de Pasco Atalaya Portillo Mamoré
Huarmey Huánuco Urubamba Costa Por
Barranca Huancayo Madre de Dios Marques Esperidi
Sacred City of Huancavelica Machu Picchu Mategua Vila Bel
Caral-Supé La Merced Ucayali Puerto Santíssi
Huacho Cordillera Vilcabamba Maldonado Exaltación Trinad
Huaral Cerro de Pasco Cordillera Beni Santa Ana
Callao Huancavelica Ayacucho Cusco Tambari de Yacuma Puerto Frey
LIMA San Vicente de Cañete (CUZCO) Sicuani Sandia Alegre Pontes e Lacerda
Chincha Alta Abancay Ayaviri Santa Ana
Pisco Antabamba Yanaoca Trinidad Loreto Vila Bel
Ica Coracora Sicuani San Ascensión
Nazca Marcona Nudo Coropuna Juliaca Borja **BOLIVIA** San Ignacio Por
Chuquibamba 6425 Lake San Pedro Esperidi
Arequipa Titicaca Montero El Cerro
Chala **LA PAZ** Warnes **Santa Cruz**
Camaná Colquiri Cordillera Oriental Pampa
Mollendo Moquegua Oruro Cochabamba Grande
Ilo Tacna Nevado Sajama Huanui Bañados del Izozog
Arica 6542 Grande Corque Tuca

PACIFIC OCEAN

METRES
FEET

5000	16404
3000	9843
2000	6562
1000	3281
500	1640
200	656
0	0

Land below sea level

200	656
4000	13124
6000	19686

A 80° B C 60°
1
2
3
4

Longitude 70° west of Greenwich

Lambert Azimuthal Equal Area Projection

ATLANTIC

OCEAN

METRES
FEET

5000
16404

3000
9843

2000
6562

1000
3281

500
1640

200
656

0
0

Land below
sea level

200
656

4000
13124

6000
19686

Lambert Azimuthal Equal Area Projection

ATLANTIC

OCEAN

URUGUAY

MONTEVIDEO

BUENOS AIRES

ARGENTINA

Mar del Plata

SANTIAGO

CHILE

Concepción

Puerto Montt

Valdivia

PATAGONIA

Comodoro Rivadavia

Golfo San Matías

Bahía Blanca

Neuquén

Río Gallegos

Punta Arenas

Tierra del Fuego

Cape Horn

Falkland Islands
(Islas Malvinas)
(U.K.)
CLAIMED BY
ARGENTINA

West
Falkland

East
Falkland

STANLEY

Port
Stephens

Darwin

**South Georgia
and the South
Sandwich Islands**
(U.K.)
CLAIMED BY
ARGENTINA

Cape
Alexandra

Mount Paget
2934

Grytviken

Cape
Disappointment

Longitude 50° west of Greenwich

1:16 000 000

MILES

KILOMETRES

© Collins Bartholomew Ltd

153

55° · B · 50° · C

Rio das Mortes · Serra do Taquaral

Ceres · Rialma · Goianésia · Brazlândia · DISTRITO
Coronel Ponce · Presidente Murtinho · Araguaiana · Itapuranga · Rianápolis · **BRASÍLIA** · Planaltina · FEDERAL · Formos
Cabeceira Rio Manso · Poxoréo · Batovi · Barra do Garças · Jussara · Goiás · Serra da Canastra · Jaraguá · Corumbá de Goiás · Gama · Cabecei
Jaciara · Tesouro · Torixoréu · Aragarças · Bom Jardim de Goiás · Goiás · Itaberaí · Pirenópolis · Nerópolis · **Anápolis** · Luziânia · Rio
São Lourenço · Rondonópolis · Guiratinga · Diamantino · Iporá · Anicuns · Trindade · Silvânia · Vianópolis · Una

M A T O · Anhumas · Ponte de Pedra · Alto Garças · Caiapônia · Paraúna · **Goiânia** · Hidrolândia · Orizona · Cristalina
G R O S S O · Alto Araguaia · Santa Rita do Araguaia · Serra do Caiapó · Piracanjuba · Pires do Rio · Guarda-Mor · Parac
Corrientes · Itiquira · Mineiros · Jataí · Montividiu · Santa Helena de Goiás · Edéia · Pontalina · Morrinhos · Caldas Novas · Ipameri · Vazant
Pedro Gomes · Serranópolis · **G O I Á S** · Goiatuba · Goiandira · Catalão · Araguari · Coromandel · Parac

1

Coxim · Serra do Taquari · Baús · Serra da Mombuca · **Rio Verde** · Buriti Alegre · Itumbiara · Tupaciguara · Represa de Emborcação · Monte Carmelo
Jauru · Costa Rica · Aporé · Quirinópolis · Santa Vitória · **Barragem Itumbiara** · Uberlândia · Patrocínio
Rio Verde de Mato Grosso · Paraíso · Itarumã · Caçu · Cachoeira Alta · São Simão · Monte Alegre de Minas · Ituiutaba · Nova Ponte
Camapuã · Cassilândia · Alto Sucuriú · **B R A Z** · Gurinhatã · Prata · Campina Verde · Campo Florido · Uberaba · Araxá · Perc
Rochedo · Corguinho · Ponte do Rio Verde · Paranaíba · Iturama · Itapajipe · Planura · Igarapava · Sacramen

M A T O · G R O S S O · Inocência · Santa Fé do Sul · Frutal · Colômbia · Pedregulho
Jaraguari · Aparecida do Tabuado · Jales · Cardoso · Votuporanga · Nova Granada · Olímpia · São Joaquim da Barra · Represa Peixoto
Terenos · **Represa Ilha Solteira** · Pereira Barreto · Fernandópolis · General Salgado · Barretos · Franca Cas · São Sebasti
Palmeiras · **Campo Grande** · Ribas do Rio Pardo · Água Clara · Garcias · Ferreiros · Represa Três Irmãos · Andradina · **São José do Rio Preto** · Bebedouro · Morro Agudo · Batatais · do Para
Sidrolândia · Três Lagoas · Mirandópolis · Valparaíso · Araçatuba · Penápolis · Catanduva · Jaboticabal · Sertãozinho · **Ribeirão Preto** · Mococ

2

Prudêncio Thomaz · Porto Alegre · Bataguassu · Presidente Epitácio · Santo Anastácio · Panorama · Birigui · Promissão · Represa Promissão · Novo Horizonte · Tabatinga · Taquaritinga · Cravinhos · Casa Bra
Maracaju · Rio Brilhante · Ivinhema · Represa Jupiá · Dracena · Lucélia · Lins · Cafelândia · Araraquara · São Carlos · Pirassunung
Dourados · Teodoro Sampaio · Represa Porto Primavera · **Presidente Prudente** · Tupã · Pirajuí · Garça · **S Ã O** · Lem

Ponta Porã · Bocajá · Caarapó · Porto São José · Nova Londrina · Iepê · Rancharia · Marília · Vera Cruz · Bauru · São Manuel · **P A U L O** · Rio Claro · Arara
Amambaí · Juti · Loanda · Paranavaí · Porecatu · Assis · Palmital · Ourinhos · Agudos · Jaú · Piracicaba · America
Capitán Bado · Querência do Norte · Nova Esperança · Rolândia · Cornélio Procópio · Piraju · Avaré · Botucatu · Conchas · Tietê · Salto · **Campinas**
Coronel Sapucaia · Iguatemi · Rondon · Maringá · Araponas · Santo Antônio da Platina · Itaí · Santo Antônio · Boituva · Tatuí · Itu · Jund
Ypé-Jhú · Porto Camargo · Cianorte · Apucarana · **Londrina** · Tomazina · Itaporanga · Capão Bonito · Itapetininga · **Sorocaba** · Pa
Ygatimí · Iguatemi · Umuarama · Campo Mourão · Telêmaco Borba · Ibaiti · Jaguariaíva · Itararé · Buri · Piedade · Itanh

Salto del Guairá · Goioerê · Cândido de Abreu · Tibagi · Apiaí · Eldorado · Juquiá · Peruíb
Porto Mendes · Guaíra · Campos Eré · Reserva · Castro · Cerro Azul · Registro · Dedo de Deus
Iguatemi · Toledo · Pitanga · Ipiranga · **P A R A N Á** · Campo Largo · Rio Branco do Sul · Jacupiranga · Iguape
Represa de Itaipu · Cascavel · Catanduvas · Prudentópolis · Guarapuava · Ponta Grossa · Antonina · Cananéia

25°

Represa de Acaray · Hernandarias · Laranjeiras do Sul · Serra das Araras · Guarapuava · Palmeira · **Curitiba** · Guaraqueçaba
Ciudad del Este · Foz do Iguaçu · Iguaçu · Rio Azul · Lapa · São José dos Pinhais · Ilha das Peças · Paranaguá
Iguaçu Falls · Represa de Salto Santiago · Chopinzinho · São Mateus do Sul · Guaratuba · Ilha de São Francisco

3

Dionísio Cerqueira · Represa de Foz de Areia · União da Vitória · Canoinhas · Rio Negro · Mafra · **Joinville** · São Francisco do Sul
Wanda · Rato Branco · Clevelândia · Palmas · Porto União · Itaiópolis · Araquari · Jaraguá do Sul
Montecarlo · Eldorado · Campos de Palmas · Serra da Fartura · Caçador · Serra do Espigão · Serra Jaraguá · Itajaí
Puerto Rico · **ARGENTINA** · Xanxerê · **S A N T A · C A T A R I N A** · Blumenau

P A R A G U A Y

55° · A · B · Longitude 50° west of Greenwich · C

Lambert Azimuthal Equal Area Projection

METRES / FEET
5000 / 16404
3000 / 9843
2000 / 6562
1000 / 3281
500 / 1640
200 / 656
0 / 0
Land below sea level
200 / 656
4000 / 13124
6000 / 19686

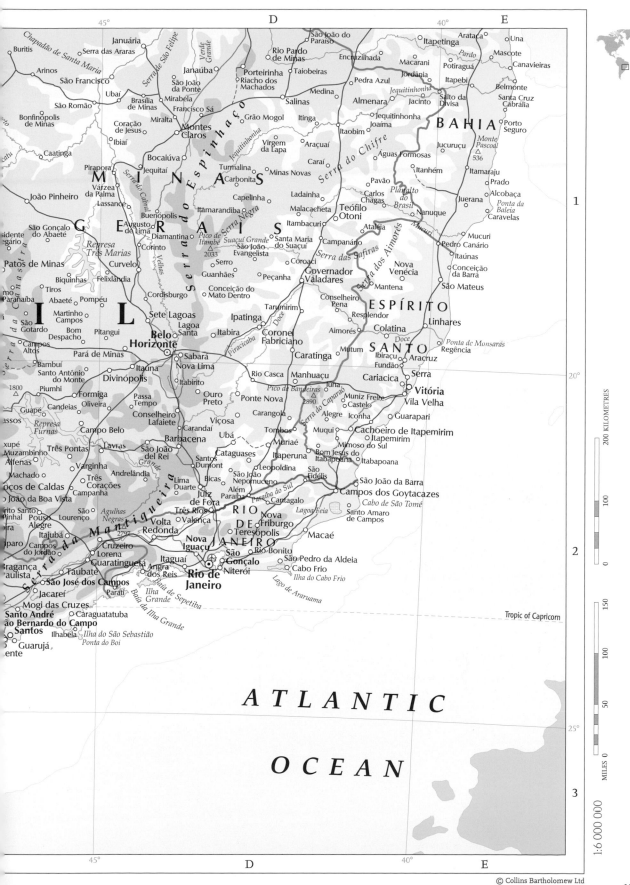

45° 40°

Buritis
Chapadão de Santa Maria
Januária
Serra das Araras
São João do Paraíso
Itapetinga
Arataca
Una
São Francisco
Arinos
Janaúba
Porteirinha
Riacho dos Machados
Rio Pardo de Minas
Taioeiras
Macarani
Potiraguá
Itapebi
Mascote
Canavieiras
Ubaí
Brasília de Minas
São João da Ponte
Mirabéla
Salinas
Medina
Pedra Azul
Jordânia
Itapebi
Belmonte
Bonfinópolis de Minas
São Romão
Francisco Sá
Miralta
Grão Mogol
Almenara
Jacinto
Salto da Divisa
Santa Cruz Cabrália
Coração de Jesus
Montes Claros
Jequitinhonha
Itinga
Joaíma
Jequitinhonha
BAHIA
Porto Seguro
Caatinga
Ibiaí
Bocaiúva
Virgem da Lapa
Araçuaí
Caraí
Águas Formosas
Itanhém
Itamaraju
Prado
Piripora
Jequitaí
Turmalina
Carbonita
Minas Novas
Pavão
Monte Pascoal △ 536
João Pinheiro
Várzea da Palma
Lassance
Capelinha
Ladainha
Malacacheta
Carlos Chagas
Planalto do Brasil
Alcobaça
Caravelas
M I N A S
Buenópolis
Augusto de Lima
Itamarandiba
Teófilo Otoni
Nanuque
Ponta da Baleia
G E R A I S
Diamantina
Serra Negra
Itambacuri
Mucuri
Pedro Canário
Patos de Minas
Represa Três Marias
Corinto
Pico de Itambé △ 2033
Suaçuí Grande
São João Evangelista
Santa Maria do Suaçuí
Campanário
Ataléia
Itaúnas
Conceição da Barra
São Gonçalo do Abaeté
Biquinhas
Felixlândia
Curvelo
Serro
Guanhães
Coroaci
Governador Valadares
Nova Venécia
São Mateus
Martinho Campos
Cordisburgo
Conceição do Mato Dentro
Peçanha
Mantena
Conselheiro Pena
ESPÍRITO
Abaeté
Pompéu
Sete Lagoas
Lagoa Santa
Ipatinga
Tarumirim
Resplendor
Colatina
Linhares
Bambuí
Pitangui
Belo Horizonte
Itabira
Coronel Fabriciano
Aimorés
SANTO
Regência
Pará de Minas
Sabará
Caratinga
Mutum
Ibiraçu
Serra
Piumhi
Itaúna
Nova Lima
Rio Casca
Manhuaçu
Cariacica
Vitória
Divinópolis
Itabirito
Pico da Bandeiras △ 2890
Iúna
Muniz Freire
Vila Velha
Formiga
Oliveira
Ouro Preto
Ponte Nova
Carangola
Castelo
Iconha
Guarapari
Conselheiro Lafaiete
Viçosa
Alegre
Muqui
Cachoeiro de Itapemirim
Campo Belo
Barbacena
Carandaí
Ubá
Tombos
Muriaé
Itaperuna
Mimoso do Sul
Itabapoana
Lavras
São João del Rei
Santos Dumont
Cataguases
Bom Jesus do Itabapoana
São João da Barra
Andrelândia
Bicas
Leopoldina
São Fidélis
Campos dos Goytacazes
Juiz de Fora
Nepomuceno
Cabo de São Tomé
RIO
Nova Friburgo
Santo Amaro de Campos
DE
Teresópolis
Macaé
Volta Redonda
Valença
Rio Bonito
São Pedro da Aldeia
JANEIRO
Nova Iguaçu
São Gonçalo
Cabo Frio
Rio de Janeiro
Niterói
Ilha do Cabo Frio
Lago de Araruama

A T L A N T I C

O C E A N

1:6 000 000

200 KILOMETRES
100
0

150
100
50
0
MILES

© Collins Bartholomew Ltd

A 90° B 120° C 150° D 180°

3 45° 2 60° Arctic Circle Chukchi Sea Bering Str

Heilong Jiang Sea of Okhotsk Bering Sea Nunivak Island

30° A S I A Sakhalin Ostrov Beringa Aleutian Basin Aleutian Islands

Vladivostok Kuril Basin Kuril Islands (Kuril'skiye Ostrova) 7822 Attu Island Aleutian Trench

4 Yellow River .3510 Hokkaido 9550 Kuril Trench 6671 Emperor Seamount Chain 1240 Emperor Trough

Tropic of Cancer Sea of Japan Honshu Northwest Pacific Basin Shatsky Rise .7900

Ganges Yellow Sea Yellow Sea Shikoku Tōkyō 8412

Kolkata Shanghai East China Sea Kyūshū Izu-Ogasawara Trench 9780

Yangtze Ryukyu Islands (Nansei-shotō) 7460 Volcano Islands (Kazan-rettō) .6345 18. Kure Atoll Hawai'i

15° Bay of Bengal Taiwan 7181 South Honshu Ridge Mapmaker Seamounts Midway Islands Hawai'i

Rangoon Hainan Dao Luzon Strait Ryukyu Trench West Mariana Basin Mid - Pacific Mountains Necker Island Hawai

Andaman Islands South China Sea Luzon Philippine Basin Kyushu - Palau Ridge Mariana Ridge Mariana Trench Saipan 6530

Andaman Basin 5560 10057 Challenger Deep Guam .1564 Central Pacific Basin

5 Sri Lanka Palawan Philippines 10920 Mariana M I C R O N E S I A

Nicobar Islands Sulu Sea Mindanao Palau Islands 8967 Kwajalein Marshall Islands P O L Y N

8054 Chuuk Caroline Islands Kosrae 6530

Singapore Celebes Sea .5484 West Caroline Basin East Caroline Basin Melanesian Gilbert Islands Phoenix Islan

Kepulauan Mentawai Halmahera 7208 Basin

0° Equator Borneo Celebes Seram Admiralty Islands Gilbert Ridge

2302 Cocos Basin Bangka New Britain M E L A N E S I A

Sumatra Laut Jawa Laut Banda 7288 New Guinea Solomon Sea 8940 Solomon Islands Funafuti Fakaofo Savai'i .13 Samoa Basin

Jakarta Java Laut Flores Arafura Sea 8322 Niue

7125 Timor Sumba Torres Strait Cape York Coral Sea Basin Vanua Levu

6 Java Trench (Sunda Trench) Timor Sea Great Barrier Reef Coral Sea Espíritu Santo Viti Levu Tonga Trench

Investigator Ridge North Australian Basin 7633 New Hebrides Trench Horizon Deep 10800

I N D I A N .6360 Grande Terre New Caledonia Trough South Fiji Basin

O C E A N West Australian Basin Exmouth Plateau Norfolk Island Kermadec Islands 10047 Kermadec Trench Sou

North West Cape Lord Howe Rise Pacif

15° 1924. A U S T R A L I A Auckland North Island

Tropic of Capricorn Sydney Tasman Sea New Zealand

Perth Basin Melbourne 5176. Wellington Chatham Rise Chatham Islands

549. Perth Great Australian Bight Tasman Basin South Island

7 Broken Plateau Cape Leeuwin South Australian Basin .5670 Tasmania South Tasman Rise 60. Campbell Plateau Antipodes Islands

7102. Diamantina Deep 6602 Auckland Islands Macquarie Ridge

30° Southeast Indian Ridge Indian - Antarctic Ridge 1646. 956. S O U T H

Île Amsterdam Île St-Paul 1840. Australian - Antarctic Basin 4650 Balleny Islands

8 Cape Adare R

9 4181. Antarctic Circle

90° 60° 120° 150° 18

A N T A

METRES / FEET scale:
0 / 0
200 / 656
2000 / 6562
3000 / 9843
4000 / 13124
5000 / 16404
6000 / 19686
7000 / 22967
9000 / 29529

Lambert Azimuthal Equal Area Projection

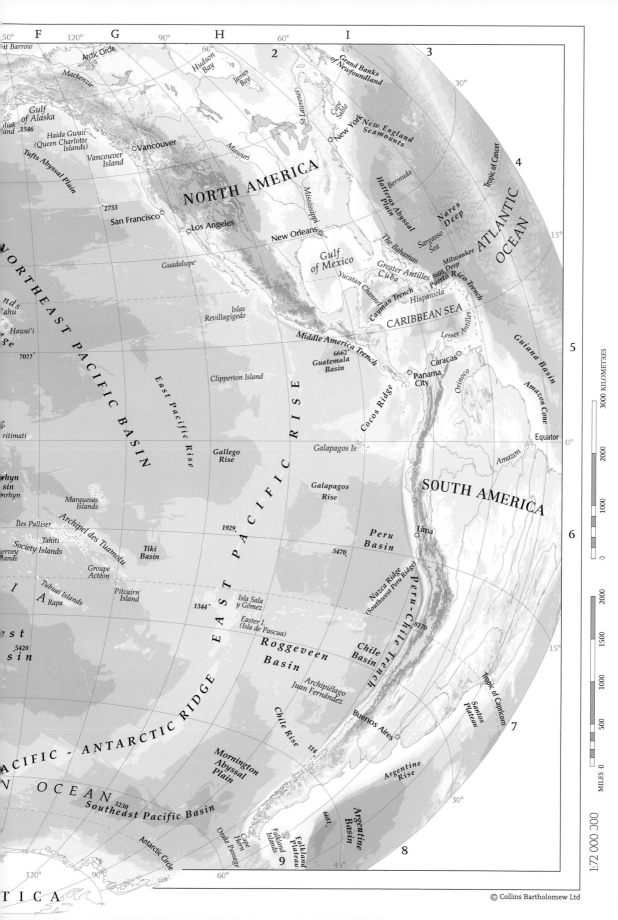

F   G   H   I

50°   120°   90°   60°   45°   3

Point Barrow   Arctic Circle   60°   2   Grand Banks
of Newfoundland   30°

Mackenzie   Hudson Bay   James Bay   Cape Sable

Gulf of Alaska   Tropic of Cancer   4

Kodiak Island   .1546   Haida Gwaii
(Queen Charlotte Islands)   Vancouver   New York   New England Seamounts   Bermuda

Tufts Abyssal Plain   Vancouver Island   Missouri   NORTH AMERICA   Hatteras Abyssal Plain   ATLANTIC   15°

.2733   San Francisco   Mississippi   Nares Deep   OCEAN

Los Angeles   New Orleans   Sargasso Sea   Milwaukee 8605 Deep

NORTHEAST   Guadalupe   Gulf of Mexico   The Bahamas   Puerto Rico Trench

slands   Oahu   Islas Revillagigedo   Greater Antilles
Cuba   Yucatan Channel   Cayman Trench   Hispaniola   CARIBBEAN SEA   Lesser Antilles   Guiana Basin   5

Hawai'i   PACIFIC   Middle America Trench   Caracas   Amazon Cone

7022   Guatemala Basin   Panama City   Orinoco   Equator   0°

kiritimati   East Pacific Rise   Clipperton Island   Cocos Ridge   Galapagos Is   Amazon

BASIN   Gallego Rise   Galapagos Rise   SOUTH AMERICA

nyhn   sin   rhyn   Marquesas Islands   Galapagos Rise

Îles Palliser   Archipel des Tuamotu   1929.   Peru Basin   Lima   6

Society Islands   Tahiti   Tiki Basin   5470.

rvey   nds   Groupe Actéon   Nazca Ridge
(Southwest Peru Ridge)

A   Tubuai Islands   Rapa   Pitcairn Island   Isla Sala y Gómez   1344.   8170

est   5420   Easter I.
(Isla de Pascua)   Chile Basin   Tropic of Capricorn   15°

sin   Roggeveen Basin   Santos Plateau

EAST   PACIFIC   RISE   Archipiélago Juan Fernández   Buenos Aires

PACIFIC   –   ANTARCTIC   RIDGE   Chile Rise   114   Argentine Rise   7

OCEAN   5230   Mornington Abyssal Plain   Argentine Basin   30°

Southeast Pacific Basin   6687.

Cape Horn   Falkland Islands   Falkland Plateau   8

Drake Passage

Antarctic Circle   9

TICA   120°   90°   60°   45°

© Collins Bartholomew Ltd

1:72 000 000

3000 KILOMETRES
2000
1000
0

2000
1500
1000
500
0
MILES

157

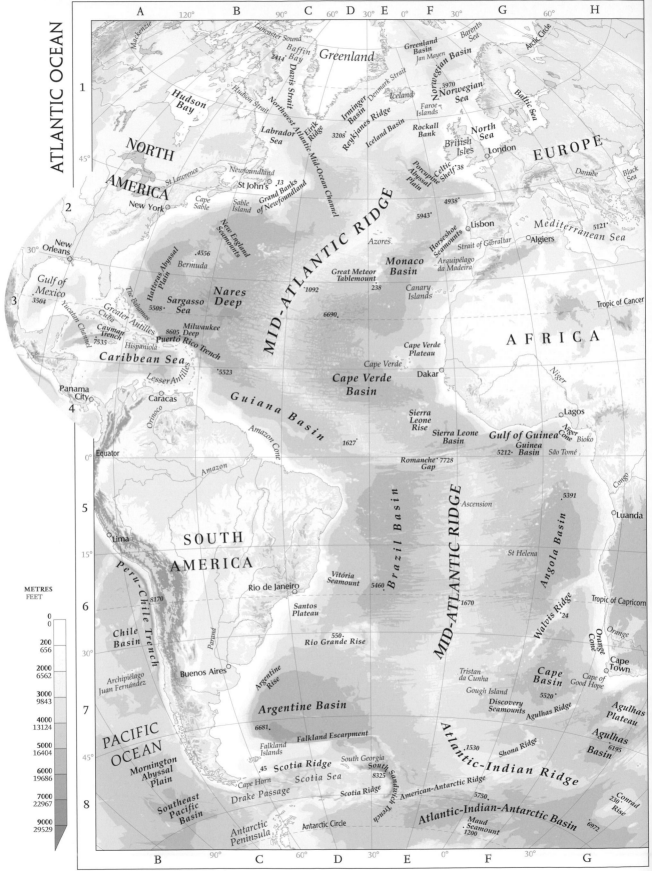

ATLANTIC OCEAN

Place names and features:

Mackenzie, Lancaster Sound, Baffin Bay, 2414, Davis Strait, Greenland, Greenland Basin, Jan Mayen, Barents Sea, Arctic Circle

Hudson Bay, Hudson Strait, Northwest Atlantic Mid-Ocean Channel, Eirik Ridge, Imminger Basin, Denmark Strait, Iceland, Reykjanes Ridge, Norwegian Basin, 3970, Norwegian Sea, Faroe Islands, Baltic Sea

NORTH AMERICA, St Lawrence, Newfoundland, St John's, 3208, Iceland Basin, Rockall Bank, British Isles, North Sea, London, EUROPE, Danube, Black Sea

New York, Cape Sable, Sable Island, Grand Banks of Newfoundland, .13, Porcupine Abyssal Plain, Celtic Shelf .38, 4938.

New Orleans, New England Seamounts, 5943., Horseshoe Seamounts, Lisbon, Mediterranean Sea, 5121., Algiers, Strait of Gibraltar

.4556, Bermuda, Hatteras Abyssal Plain, Monaco Basin, Azores, Arquipélago da Madeira

Gulf of Mexico, 3504, The Bahamas, Greater Antilles, Cuba, Nares Deep, Sargasso Sea, 5508, Great Meteor Tablemount, .1092, 238, Canary Islands, Tropic of Cancer

Cayman Trench, 7535, Hispaniola, Milwaukee Deep, 8605, Puerto Rico Trench, 6690., AFRICA

Caribbean Sea, Lesser Antilles, Cape Verde Plateau, Cape Verde, Dakar

Panama City, Caracas, .5523, Cape Verde Basin, Sierra Leone Rise

Orinoco, Guiana Basin, Amazon Cone, 1627., Sierra Leone Basin, Gulf of Guinea, Niger, Lagos, Niger Cone, Bioko, Guinea Basin, 5212., São Tomé

Equator, Amazon, Romanche Gap 7728.

SOUTH AMERICA, Brazil Basin, Ascension, 5391., Luanda

Lima, St Helena, Angola Basin

Peru–Chile Trench, 8170, Rio de Janeiro, Vitória Seamount, 5460., MID-ATLANTIC RIDGE, Tropic of Capricorn, Walvis Ridge, .24, Orange Cone, Orange

Santos Plateau, 1670., Cape Town

Chile Basin, Paraná, 550., Rio Grande Rise, Tristan da Cunha, Cape Basin, Cape of Good Hope, Agulhas Plateau

Archipiélago Juan Fernández, Buenos Aires, Argentine Rise, Gough Island, Discovery Seamounts, 5520., Agulhas Ridge, Agulhas Basin, 6195.

PACIFIC OCEAN, Argentine Basin, 6681., Falkland Escarpment, Atlantic–Indian Ridge, .1530, Shona Ridge

Mornington Abyssal Plain, Falkland Islands, .45, Scotia Ridge, South Georgia, 8325, South Sandwich Trench, Conrad Rise, 2316.

Cape Horn, Scotia Sea, Scotia Ridge, American–Antarctic Ridge, 5750., Atlantic–Indian–Antarctic Basin, 6972.

Southeast Pacific Basin, Drake Passage, Antarctic Circle, Maud Seamount, 1200

Antarctic Peninsula

Depth scale:

METRES / FEET

METRES	FEET
0	0
200	656
2000	6562
3000	9843
4000	13124
5000	16404
6000	19686
7000	22967
9000	29529

Lambert Azimuthal Equal Area Projection

158

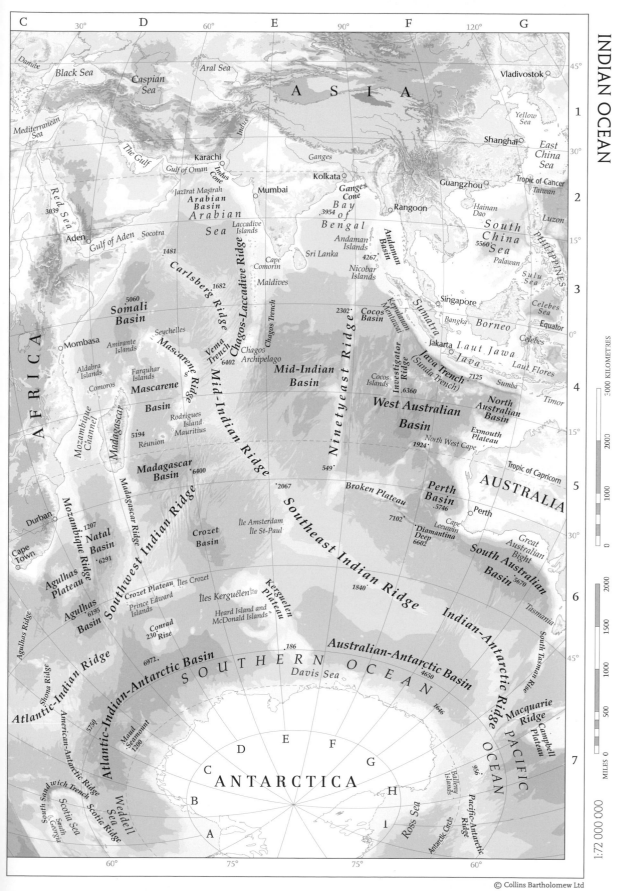

C   D   E   F   G

30°   60°   90°   120°

Danube
Black Sea
Caspian
Sea
Aral Sea
ASIA
Vladivostok
45°
Mediterranean
Sea
The Gulf
Karachi
Indus
Ganges
Shanghai
Yellow
Sea
East
China
Sea
30°
1
3039
Red Sea
Gulf of Oman
Jazīrat Maşīrah
Indus
Cone
Mumbai
Kolkata
Ganges
Cone
Bay
of
Bengal
.3954
Rangoon
Guangzhou
Hainan
Dao
Tropic of Cancer
Taiwan
South
China
Sea
Luzon
2
Aden
Gulf of Aden
Socotra
Arabian
Basin
Arabian
Sea
Laccadive
Islands
Andaman
Islands
Andaman
Basin
5560
PHILIPPINES
15°
1481
Cape
Comorin
Sri Lanka
4267
Palawan
Carlsberg Ridge
1682
Maldives
Chagos Trench
Nicobar
Islands
Sumatra
Singapore
Sulu
Sea
Celebes
Sea
3
Somali
Basin
5060
Mombasa
Amirante
Islands
Seychelles
Mascarene Ridge
Vema Trench
6402
Chagos-Laccadive Ridge
Chagos
Archipelago
2302
Cocos
Basin
Kepulauan
Mentawai
Investigator
Ridge
Jakarta
Bangka
Borneo
Celebes
Equator
0°
AFRICA
Aldabra
Islands
Comoros
Farquhar
Islands
Mascarene
Basin
8'
Mid-Indian
Basin
Cocos
Islands
6360
Laut Jawa
Java
Laut Flores
4
Madagascar
5194
Rodrigues
Island
Mauritius
Mid-Indian Ridge
Ninetyeast Ridge
West Australian
Basin
Java Trench
(Sunda Trench)
7125
North
Australian
Basin
Sumba
Timor
Réunion
Mozambique Channel
549
North West Cape
Exmouth
Plateau
15°
Madagascar
Basin
6400
1924
Tropic of Capricorn
AUSTRALIA
5
Durban
1207
2067
Broken Plateau
Perth
Basin
5746
7102
Great
Australian
Bight
Perth
Mozambique Ridge
Madagascar Ridge
Crozet
Basin
Île Amsterdam
Île St-Paul
Cape
Leeuwin
Diamantina
Deep
6602
South Australian
Basin
30°
Natal
Basin
6291
Southwest Indian Ridge
Southeast Indian Ridge
5670
Cape
Town
Agulhas
Plateau
Agulhas
Basin
6195
Crozet Plateau
Îles Crozet
Prince Edward
Islands
Îles Kerguélen
Kerguelen
Plateau
Heard Island and
McDonald Islands
1840
Indian-Antarctic Ridge
Tasmania
6
45°
Agulhas Ridge
Atlantic-Indian Ridge
Conrad
Rise
230
6972
SOUTHERN   OCEAN
186
Australian-Antarctic Basin
4650
1646
South Tasman Rise
Macquarie
Ridge
Campbell
Plateau
Shona Ridge
American-Antarctic Ridge
Maud
Seamount
1200
5790
Atlantic-Indian-Antarctic Basin
Davis Sea
E
F
G
PACIFIC
956
Ballery
Islands
Pacific-Antarctic
Ridge
7
South Sandwich Trench
Scotia
Sea
Scotia Ridge
Weddell
Sea
D
C
B
A
ANTARCTICA
G
H
1
Ross Sea
Antarctic Circle
South
Georgia

60°   75°   75°   60°

3000 KILOMETRES
2000
1000
0
2000
1500
1000
500
0 MILES
1:72 000 000

© Collins Bartholomew Ltd

159

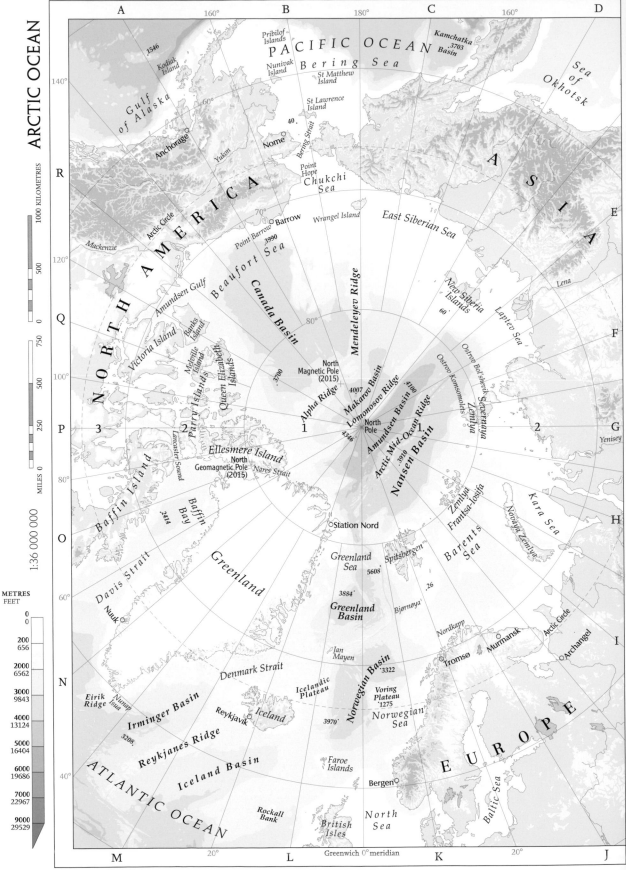

ARCTIC OCEAN

1:36 000 000

1000 KILOMETRES

500

0

750

500

250

0

MILES

METRES
FEET

0	0
200	656
2000	6562
3000	9843
4000	13124
5000	16404
6000	19686
7000	22967
9000	29529

160

Polar Stereographic Projection

PACIFIC OCEAN

Bering Sea

Sea of Okhotsk

Pribilof Islands

Kamchatka Basin
.3703

Nunivak Island

St Matthew Island

Gulf of Alaska

Kodiak Island

.1546

St Lawrence Island

40.

Bering Strait

Anchorage

Nome

Yukon

A S I A

Point Hope

Chukchi Sea

Arctic Circle

70°

Barrow

Wrangel Island

East Siberian Sea

Point Barrow

3990

Mackenzie

N O R T H   A M E R I C A

Amundsen Gulf

Beaufort Sea

Canada Basin

80°

Mendeleyev Ridge

New Siberia Islands

60

Lena

Laptev Sea

Victoria Island

Banks Island

Melville Island

Parry Islands

Queen Elizabeth Islands

.3700

North Magnetic Pole (2015)

Alpha Ridge

4007

Makarov Basin

Lomonosov Ridge

Amundsen Basin

4000

4346

Ostrov Bol'shevik

Severnaya Zemlya

Ostrov Komsomolets

Yenisey

Lancaster Sound

Baffin Island

Ellesmere Island

North Geomagnetic Pole (2015)

Nares Strait

North Pole

Arctic Mid-Ocean Ridge

Nansen Basin

.3910

Zemlya Frantsa-Iosifa

Kara Sea

Novaya Zemlya

Baffin Bay

.2414

Station Nord

Spitsbergen

Barents Sea

Davis Strait

Greenland

Greenland Sea

5608

Greenland Basin

Bjørnøya

.26

Nordkapp

Nuuk

3884.

Jan Mayen

Murmansk

Tromsø

Arctic Circle

Archangel

Denmark Strait

Norwegian Basin

.3322

Eirik Ridge

Numap Isua

Irminger Basin

Icelandic Plateau

Voring Plateau 1275

Norwegian Sea

Reykjavik

Iceland

3970.

E U R O P E

3208.

Reykjanes Ridge

Faroe Islands

Baltic Sea

Iceland Basin

Rockall Bank

Bergen

North Sea

British Isles

ATLANTIC OCEAN

Greenwich 0° meridian

3   2   1   1   2

ARCTIC OCEAN

WORLD FACTS AND FIGURES

	Total Population	2050 Projected Population	Gross National Income (GNI) Per Capita (US$)	Literacy Rate (%)	International Dialling Code	Time Zone	Official Website *Tourism Website*
WORLD	7 162 119 000	9 550 945 000	10 564	89.4	...	...	
AFGHANISTAN	30 552 000	56 551 000	700	47.0	93	+4.5	www.president.gov.af ...
ALBANIA	3 173 000	3 094 000	4 700	98.8	355	+1	www.km.gov.al *www.akt.gov.al*
ALGERIA	39 208 000	54 522 000	5 290	...	213	+1	www.el-mouradia.dz *www.mta.gov.dz*
ANDORRA	79 000	95 000	...	...	376	+1	www.govern.ad *www.visitandorra.com*
ANGOLA	21 472 000	54 324 000	5 010	73.0	244	+1	www.governo.gov.ao *www.angola.org/index.php?page=tourism*
ANTIGUA AND BARBUDA	90 000	115 000	12 910	...	1 268	-4	www.ab.gov.ag *www.visitantiguabarbuda.com*
ARGENTINA	41 446 000	51 024 000	...	99.2	54	-3	www.argentina.gov.ar *www.turismo.gov.ar*
ARMENIA	2 977 000	2 782 000	3 790	99.7	374	+4	www.gov.am *www.mfa.am*
AUSTRALIA	23 343 000	33 735 000	65 520	...	61	+8 to +10.5	www.australia.gov.au *www.australia.com*
AUSTRIA	8 495 000	9 354 000	48 590	...	43	+1	www.bundeskanzleramt.at *www.austria.info*
AZERBAIJAN	9 413 000	10 492 000	7 350	99.9	994	+4	www.president.az *http://azerbaijan.tourism.az*
THE BAHAMAS	377 000	494 000	20 600	...	1 242	-5	www.bahamas.gov.bs *www.bahamas.com*
BAHRAIN	1 332 000	1 835 000	19 560	98.2	973	+3	www.bahrain.bh *www.bahraintourism.com*
BANGLADESH	156 595 000	201 948 000	900	79.9	880	+6	www.bangladesh.gov.bd *www.visitbangladesh.gov.bd*
BARBADOS	285 000	314 000	15 080	...	1 246	-4	www.barbados.gov.bb *www.visitbarbados.org*
BELARUS	9 357 000	7 359 000	6 720	99.8	375	+2	www.belarus.by *eng.belarustourism.by*
BELGIUM	11 104 000	12 055 000	45 210	...	32	+1	www.belgium.be www.visitflanders.com *www.opt.be www.eastbelgium.com*
BELIZE	332 000	590 000	4 660	...	501	-6	www.belize.gov.bz *www.travelbelize.org*
BENIN	10 323 000	22 137 000	790	...	229	+1	www.gouv.bj ...
BHUTAN	754 000	980 000	2 460	...	975	+6	www.bhutan.gov.bt *www.tourism.gov.bt*
BOLIVIA	10 671 000	16 621 000	2 550	99.0	591	-4	www.bolivia.gob.bo *www.bolivia.travel*
BOSNIA AND HERZEGOVINA	3 829 000	3 332 000	4 740	99.7	387	+1	www.fbihvlada.gov.ba *www.bhtourism.ba*
BOTSWANA	2 021 000	2 780 000	7 730	96.0	267	+2	www.gov.bw *www.botswanatourism.co.bw*
BRAZIL	200 362 000	231 120 000	11 690	98.6	55	-2 to -5	www.brazil.gov.br *www.visitbrasil.com*
BRUNEI	418 000	546 000	...	99.8	673	+8	www.pmo.gov.bn *www.bruneitourism.travel*
BULGARIA	7 223 000	5 077 000	7 030	97.9	359	+2	www.government.bg *http://bulgariatravel.org*
BURKINA FASO	16 935 000	40 932 000	670	39.3	226	UTC	www.gouvernement.gov.bf *www.culture.gov.bf*
BURUNDI	10 163 000	26 691 000	280	88.9	257	+2	www.burundi-gov.bi *www.burundi-tourism.com*
CAMBODIA	15 135 000	22 569 000	950	87.1	855	+7	www.cambodia.gov.kh *www.tourismcambodia.org*
CAMEROON	22 254 000	48 599 000	1 270	80.6	237	+1	www.spm.gov.cm *www.cameroon-tourism.org*
CANADA	35 182 000	45 228 000	52 200	...	1	-3.5 to -8	www.canada.ca *www.canada.travel*
CAPE VERDE (CABO VERDE)	499 000	636 000	3 630	98.1	238	-1	www.governo.cv *www.caboverde.com/ilhas/maio/guide-e.htm*
CENTRAL AFRICAN REPUBLIC	4 616 000	8 491 000	320	36.4	236	+1	www.centrafricaine.info *www.centrafricaine.info*
CHAD	12 825 000	33 516 000	1 020	48.9	235	+1	www.presidencetchad.org ...
CHILE	17 620 000	20 839 000	15 230	98.9	56	-4	www.gob.cl *www.chile.travel*

	Total Population	2050 Projected Population	Gross National Income (GNI) Per Capita (US$)	Literacy Rate (%)	International Dialling Code	Time Zone	Official Website *Tourism Website*
CHINA	1 369 993 000	1 384 977 000	6 560	99.6	86	+8	www.gov.cn *www.cnto.org*
COLOMBIA	48 321 000	62 942 000	7 560	98.2	57	-5	www.gobiernoenlinea.gov.co *www.colombia.travel*
COMOROS	735 000	1 508 000	880	86.4	269	+3	www.beit-salam.km *www.tourisme.gouv.km*
CONGO	4 448 000	10 577 000	2 660	80.9	242	+1	www.presidence.cg *www.ambacongo-us.org/VisitCongo/Beforeyougo.aspx*
CONGO, DEMOCRATIC REPUBLIC OF THE	67 514 000	155 291 000	400	65.8	243	+1 to +2	www.presidentrdc.cd *...*
COSTA RICA	4 872 000	6 189 000	9 550	99.1	506	-6	www.presidencia.go.cr *www.visitcostarica.com*
CÔTE D'IVOIRE (IVORY COAST)	20 316 000	42 339 000	1 380	48.3	225	UTC	www.gouv.ci *www.tourismeci.org*
CROATIA	4 290 000	3 606 000	13 330	99.7	385	+1	www.vlada.hr *www.croatia.hr*
CUBA	11 266 000	9 392 000	5 890	100	53	-5	www.cubagob.gov.cu *www.cubatravel.tur.cu*
CYPRUS	1 141 000	1 356 000	26 390	99.8	357	+2	www.cyprus.gov.cy *www.visitcyprus.com*
CZECH REPUBLIC	10 702 000	11 218 000	18 060	...	420	+1	www.czech.cz *www.czechtourism.com*
DENMARK	5 619 000	6 361 000	61 110	...	45	+1	www.denmark.dk *www.visitdenmark.com*
DJIBOUTI	873 000	1 244 000	...	...	253	+3	www.presidence.dj *www.visitdjibouti.dj*
DOMINICA	72 000	76 000	6 760	...	1 767	-4	www.dominica.gov.dm *www.dominica.dm*
DOMINICAN REPUBLIC	10 404 000	13 320 000	5 620	97.5	1 809	-4	www.cig.gob.do *www.godominicanrepublic.com*
EAST TIMOR (TIMOR-LESTE)	1 133 000	2 087 000	3 580	79.5	670	+9	http://timor-leste.gov.tl *...*
ECUADOR	15 738 000	23 061 000	5 510	98.6	593	-5	www.presidencia.gob.ec *http://ecuador.travel*
EGYPT	82 056 000	121 798 000	3 160	89.3	20	+2	www.egypt.gov.eg *www.egypt.travel*
EL SALVADOR	6 340 000	6 912 000	3 720	96.5	503	-6	www.presidencia.gob.sv *www.elsalvador.travel*
EQUATORIAL GUINEA	757 000	1 623 000	14 320	98.1	240	+1	www.guineaecuatorialpress.com *www.embassyofequatorialguinea.co.uk*
ERITREA	6 333 000	14 314 000	490	91.0	291	+3	www.shabait.com *...*
ESTONIA	1 287 000	1 121 000	17 370	99.9	372	+2	www.valitsus.ee *www.visitestonia.com*
ETHIOPIA	94 101 000	187 573 000	470	55.0	251	+3	www.ethiopia.gov.et *www.tourismethiopia.org*
FIJI	881 000	918 000	4 430	...	679	+12	www.fiji.gov.fj *www.fijime.com*
FINLAND	5 426 000	5 693 000	47 110	...	358	+2	www.valtioneuvosto.fi *www.visitfinland.com*
FRANCE	64 291 000	73 212 000	42 250	...	33	+1	www.premier-ministre.gouv.fr *http://int.rendezvousenfrance.com*
GABON	1 672 000	3 302 000	10 650	88.5	241	+1	www.legabon.org *www.legabon.org*
THE GAMBIA	1 849 000	4 866 000	510	69.4	220	UTC	www.assembly.gov.gm *www.visitthegambia.gm*
GEORGIA	4 341 000	3 563 000	3 570	99.8	995	+4	www.parliament.ge *www.exploregeorgia.org*
GERMANY	82 727 000	72 566 000	46 100	...	49	+1	www.bundesregierung.de *www.germany.travel*
GHANA	25 905 000	45 670 000	1 760	85.7	233	UTC	www.ghana.gov.gh *www.touringghana.com*
GREECE	11 128 000	10 668 000	22 530	99.4	30	+2	www.primeminister.gr *www.visitgreece.gr*
GRENADA	106 000	95 000	7 460	...	1 473	-4	www.gov.gd *www.grenadagrenadines.com*
GUATEMALA	15 468 000	31 426 000	3 340	93.7	502	-6	www.guatemala.gob.gt *www.visitguatemala.com*
GUINEA	11 745 000	24 466 000	460	31.4	224	UTC	www.assemblee.gov.gn *...*
GUINEA-BISSAU	1 704 000	3 504 000	520	74.3	245	UTC	www.guinebissaurepublic.com *http://bissautourism.com*

	Total Population	2050 Projected Population	Gross National Income (GNI) Per Capita (US$)	Literacy Rate (%)	International Dialling Code	Time Zone	Official Website *Tourism Website*
GUYANA	800 000	815 000	3 750	93.1	592	-4	www.gina.gov.gy *www.guyana-tourism.com*
HAITI	10 317 000	14 353 000	810	...	509	-5	http://primature.gouv.ht *www.haititourisme.gouv.ht*
HONDURAS	8 098 000	13 484 000	2 180	95.0	504	-6	http://congresonacional.hn/ *www.letsgohonduras.com*
HUNGARY	9 955 000	8 954 000	12 410	99.4	36	+1	www.magyarorszag.hu *http://gotohungary.com*
ICELAND	330 000	415 000	43 930	...	354	UTC	www.iceland.is *www.visiticeland.com*
INDIA	1 252 140 000	1 620 051 000	1 570	...	91	+5.5	www.india.gov.in *www.incredibleindia.org*
INDONESIA	249 866 000	321 377 000	3 580	98.8	62	+7 to +9	www.indonesia.go.id *www.indonesia.travel*
IRAN	77 447 000	100 598 000	5 780	98.0	98	3.5	www.president.ir *www.tourismiran.ir*
IRAQ	33 765 000	71 336 000	6 710	82.2	964	+3	www.cabinet.iq *...*
IRELAND	4 627 000	5 994 000	39 110	...	353	UTC	www.gov.ie *www.discoverireland.ie*
ISRAEL	7 733 000	11 843 000	34 120	99.5	972	+2	www.knesset.gov.il *www.goisrael.com*
ITALY	60 990 000	60 015 000	34 400	99.9	39	+1	www.governo.it *www.italia.it*
JAMAICA	2 784 000	2 808 000	5 220	95.9	1 876	-5	http://jis.gov.jm *www.visitjamaica.com*
JAPAN	127 144 000	108 329 000	46 140	...	81	+9	www.japan.go.jp *www.jnto.go.jp*
JORDAN	7 274 000	11 510 000	4 950	99.1	962	+2	www.jordan.gov.jo *www.visitjordan.com*
KAZAKHSTAN	16 441 000	20 186 000	11 380	99.8	7	+5 to +6	www.government.kz *http://visitkazakhstan.kz*
KENYA	44 354 000	97 173 000	930	82.4	254	+3	www.president.go.ke *www.magicalkenya.com*
KIRIBATI	102 000	156 000	2 620	...	686	+12 to +14	www.parliament.gov.ki *www.kiribatitourism.gov.ki*
KOSOVO	1 815 606	...	3 890	...	381	+1	www.rks-gov.net *...*
KUWAIT	3 369 000	6 342 000	44 940	98.8	965	+3	www.e.gov.kw *www.e.gov.kw*
KYRGYZSTAN	5 548 000	7 976 000	1 200	99.8	996	+6	www.gov.kg *http://mfa.kg/index_en.html*
LAOS	6 770 000	10 579 000	1 460	...	856	+7	www.na.gov.la *www.tourismlaos.org*
LATVIA	2 050 000	1 674 000	14 060	99.8	371	+2	www.saeima.lv *www.latvia.travel*
LEBANON	4 822 000	5 316 000	9 870	98.7	961	+2	www.presidency.gov.lb *www.mot.gov.lb*
LESOTHO	2 074 000	2 818 000	1 550	83.2	266	+2	www.gov.ls *http://visitlesotho.travel*
LIBERIA	4 294 000	9 392 000	410	49.1	231	UTC	www.emansion.gov.lr *...*
LIBYA	6 202 000	8 350 000	...	99.9	218	+1	www.libyanmission-un.org *...*
LIECHTENSTEIN	37 000	44 000	...	...	423	+1	www.liechtenstein.li *www.tourismus.li*
LITHUANIA	3 017 000	2 557 000	13 820	99.9	370	+2	www.lrv.lt *www.lithuania.travel*
LUXEMBOURG	530 000	706 000	71 810	...	352	+1	www.gouvernement.lu *www.visitluxembourg.com*
MACEDONIA (F.Y.R.O.M.)	2 107 000	1 881 000	4 800	98.6	389	+1	www.vlada.mk *www.exploringmacedonia.com*
MADAGASCAR	22 925 000	55 498 000	440	64.9	261	+3	www.madagascar.gov.mg *www.madagascar-tourisme.com*
MALAWI	16 363 000	41 203 000	270	72.1	265	+2	www.malawi.gov.mw *www.guide2malawi.com*
MALAYSIA	29 717 000	42 113 000	10 400	98.4	60	+8	www.malaysia.gov.my *www.tourism.gov.my*
MALDIVES	345 000	504 000	5 600	...	960	+5	www.presidencymaldives.gov.mv *www.visitmaldives.com*
MALI	15 302 000	45 168 000	670	47.1	223	UTC	www.primature.gov.ml *www.maliembassy.us*

	Total Population	2050 Projected Population	Gross National Income (GNI) Per Capita (US$)	Literacy Rate (%)	International Dialling Code	Time Zone	Official Website Tourism Website
MALTA	429 000	417 000	19 730	…	356	+1	www.gov.mt www.visitmalta.com
MARSHALL ISLANDS	53 000	67 000	4 200	…	692	+12	www.rmi-op.net www.visitmarshallislands.com
MAURITANIA	3 890 000	7 921 000	1 060	56.1	222	UTC	www.mauritania.mr www.mauritania.mr
MAURITIUS	1 244 000	1 231 000	9 300	98.1	230	+4	www.gov.mu www.tourism-mauritius.mu
MEXICO	122 332 000	156 102 000	9 940	98.9	52	-6 to -8	www.presidencia.gob.mx www.visitmexico.com
MICRONESIA, FEDERATED STATES OF	104 000	130 000	3 430	…	691	+10 to +11	www.fsmgov.org www.visit-micronesia.fm
MOLDOVA	3 487 000	2 484 000	2 460	100	373	+2	www.moldova.md www.travel.md
MONACO	38 000	53 000	…	…	377	+1	www.monaco.gouv.mc www.visitmonaco.com
MONGOLIA	2 839 000	3 753 000	3 770	98.5	976	+8	www.pmis.gov.mn www.mongoliatourism.gov.mn
MONTENEGRO	621 000	557 000	7 260	99.2	382	+1	www.gov.me www.visit-montenegro.com
MOROCCO	33 008 000	42 884 000	3 030	81.5	212	UTC	www.maroc.ma www.visitmorocco.com
MOZAMBIQUE	25 834 000	59 929 000	590	67.1	258	+2	www.portaldogoverno.gov.mz www.visitmozambique.net
MYANMAR (BURMA)	53 259 000	58 645 000	…	96.0	95	+6.5	www.president-office.gov.mm www.myanmartourism.org
NAMIBIA	2 303 000	3 744 000	5 840	87.1	264	+1	www.gov.na www.namibiatourism.com.na
NAURU	10 000	11 000	…	…	674	+12	www.naurugov.nr www.naurugov.nr
NEPAL	27 797 000	36 479 000	730	82.4	977	+5.75	www.nepalgov.gov.np http://welcomenepal.com
NETHERLANDS	16 759 000	16 919 000	47 440	…	31	+1	www.overheid.nl www.holland.com
NEW ZEALAND	4 506 000	5 778 000	35 520	…	64	+12 to +12.75	www.govt.nz www.newzealand.com
NICARAGUA	6 080 000	8 355 000	1 780	…	505	-6	www.presidencia.gob.ni www.visit-nicaragua.com
NIGER	17 831 000	69 410 000	410	23.5	227	+1	www.presidence.ne …
NIGERIA	173 615 000	440 355 000	2 760	66.4	234	+1	www.nigeria.gov.ng www.tourism.gov.ng
NORTH KOREA	24 895 000	27 076 000	…	100	850	+9	www.korea-dpr.com www.korea-dpr.com/kfa_travel.html
NORWAY	5 043 000	6 556 000	102 610	…	47	+1	www.norge.no www.visitnorway.com
OMAN	3 632 000	5 065 000	25 250	97.7	968	+4	www.oman.om www.omantourism.gov.om
PAKISTAN	182 143 000	271 082 000	1 380	70.8	92	+5	www.pakistan.gov.pk www.tourism.gov.pk
PALAU	21 000	28 000	10 970	99.8	680	+9	http://palaugov.org www.visit-palau.com
PANAMA	3 864 000	5 774 000	10 700	97.6	507	-5	www.presidencia.gob.pa www.visitpanama.com
PAPUA NEW GUINEA	7 321 000	13 092 000	2 010	71.2	675	+10	www.pm.gov.pg www.papuanewguinea.travel
PARAGUAY	6 802 000	10 445 000	4 040	98.6	595	-4	www.presidencia.gov.py www.senatur.gov.py
PERU	30 376 000	41 084 000	6 390	98.7	51	-5	www.peru.gob.pe www.peru.travel
PHILIPPINES	98 394 000	157 118 000	3 270	97.8	63	+8	http://president.gov.ph/ www.wowphilippines.com.ph
POLAND	38 217 000	34 079 000	12 960	100	48	+1	www.poland.gov.pl www.poland.travel
PORTUGAL	10 608 000	9 843 000	20 670	99.4	351	UTC	www.portugal.gov.pt www.visitportugal.com
QATAR	2 169 000	2 985 000	85 550	99.1	974	+3	www.gov.qa www.qatartourism.gov.qa
ROMANIA	21 699 000	17 809 000	9 060	99.0	40	+2	www.guv.ro www.romaniatourism.com
RUSSIA	142 834 000	120 896 000	13 860	99.7	7	+2 to +12	www.gov.ru www.russiatourism.ru

	Total Population	2050 Projected Population	Gross National Income (GNI) Per Capita (US$)	Literacy Rate (%)	International Dialling Code	Time Zone	Official Website Tourism Website
RWANDA	11 777 000	25 378 000	620	77.3	250	+2	www.gov.rw www.rwandatourism.com
ST KITTS AND NEVIS	54 000	67 000	13 460	...	1 869	-4	www.gov.kn www.stkittstourism.kn
ST LUCIA	182 000	207 000	7 090	...	1 758	-4	www.stlucia.gov.lc http://stlucianow.co.uk
ST VINCENT AND THE GRENADINES	109 000	111 000	6 580	...	1 784	-4	www.gov.vc http://discoversvg.com
SAMOA	190 000	242 000	3 430	99.5	685	+13	www.samoagovt.ws samoa.travel
SAN MARINO	31 000	33 000	...	...	378	+1	www.consigliograndeegenerale.sm www.visitsanmarino.com
SÃO TOMÉ AND PRÍNCIPE	193 000	388 000	1 470	80.2	239	UTC	www.gov.st ...
SAUDI ARABIA	28 829 000	40 388 000	26 200	99.2	966	+3	www.saudi.gov.sa http://sauditourism.sa
SENEGAL	14 133 000	32 933 000	1 070	66.0	221	UTC	www.gouv.sn www.senegal-tourism.com
SERBIA	7 181 505	7 074 000	5 730	99.3	381	+1	www.srbija.gov.rs www.serbia.travel
SEYCHELLES	93 000	100 000	12 530	99.1	248	+4	www.virtualseychelles.sc www.seychelles.travel
SIERRA LEONE	6 092 000	10 296 000	680	62.7	232	UTC	www.statehouse.gov.sl www.visitsierraleone.org
SINGAPORE	5 412 000	7 065 000	54 040	99.8	65	+8	www.gov.sg www.yoursingapore.com
SLOVAKIA	5 450 000	4 990 000	17 200	...	421	+1	www.government.gov.sk www.slovakia.travel
SLOVENIA	2 072 000	2 023 000	22 830	99.9	386	+1	www.gov.si www.slovenia.info
SOLOMON ISLANDS	561 000	1 010 000	1 610	...	677	+11	www.pmc.gov.sb www.visitsolomons.com.sb
SOMALIA	10 496 000	27 076 000	...	...	252	+3	www.somaligov.net ...
SOUTH AFRICA	52 776 000	63 405 000	7 190	98.9	27	+2	www.gov.za www.southafrica.net
SOUTH KOREA	49 263 000	51 034 000	25 920	...	82	+9	www.korea.net english.visitkorea.or.kr
SOUTH SUDAN	11 296 000	24 760 000	1 120	...	211	+3	www.goss.org ...
SPAIN	46 927 000	48 224 000	29 180	99.7	34	+1	www.la-moncloa.es www.spain.info
SRI LANKA	21 273 000	23 834 000	3 170	98.2	94	+5.5	www.priu.gov.lk http://srilanka.travel
SUDAN	37 964 000	77 138 000	1 130	...	249	+3	www.presidency.gov.sd ...
SURINAME	539 000	621 000	9 260	98.4	597	-3	www.president.gov.sr www.surinametourism.sr
SWAZILAND	1 250 000	1 815 000	3 080	93.5	268	+2	www.gov.sz www.thekingdomofswaziland.com
SWEDEN	9 571 000	11 934 000	59 130	...	46	+1	www.sweden.se www.visitsweden.com
SWITZERLAND	8 078 000	10 977 000	80 950	...	41	+1	www.swissworld.org www.myswitzerland.com
SYRIA	21 898 000	36 706 000	...	95.6	963	+2	http://parliament.sy ...
TAIWAN	23 344 000	...	...	...	886	+8	www.gov.tw www.go2taiwan.net
TAJIKISTAN	8 208 000	15 093 000	990	99.9	992	+5	www.prezident.tj http://tdc.tj
TANZANIA	49 253 000	129 417 000	630	74.6	255	+3	www.tanzania.go.tz www.tanzaniatouristboard.com
THAILAND	67 011 000	61 740 000	5 370	96.6	66	+7	www.thaigov.go.th www.tourismthailand.org
TOGO	6 817 000	14 521 000	530	79.9	228	UTC	www.republicoftogo.com ...
TONGA	105 000	140 000	4 490	99.4	676	+13	www.tongaportal.gov.to www.thekingdomoftonga.com
TRINIDAD AND TOBAGO	1 341 000	1 155 000	15 760	99.6	1 868	-4	www.ttconnect.gov.tt www.gotrinidadandtobago.com
TUNISIA	10 997 000	13 192 000	4 360	97.3	216	+1	www.ministeres.tn www.tourismtunisia.com

	Total Population	2050 Projected Population	Gross National Income (GNI) Per Capita (US$)	Literacy Rate (%)	International Dialling Code	Time Zone	Official Website / Tourism Website
TURKEY	74 933 000	94 606 000	10 950	99.0	90	+2	www.tccb.gov.tr / *www.goturkey.com*
TURKMENISTAN	5 240 000	6 570 000	6 880	99.8	993	+5	www.turkmenistan.gov.tm / *www.turkmenistanembassy.org*
TUVALU	10 000	12 000	6 630	...	688	+12	... / *www.timelesstuvalu.com*
UGANDA	37 579 000	104 078 000	510	87.4	256	+3	www.statehouse.go.ug / *www.visituganda.com*
UKRAINE	45 239 000	33 658 000	3 960	99.8	380	+2, +3 (Crimea)	www.kmu.gov.ua / *www.traveltoukraine.org*
UNITED ARAB EMIRATES	9 346 000	15 479 000	38 620	...	971	+4	www.government.ae / *www.uaetourism.ae*
UNITED KINGDOM	63 136 000	73 131 000	39 110	...	44	UTC	www.gov.uk / *www.visitbritain.com*
UNITED STATES OF AMERICA	320 051 000	400 853 000	53 670	...	1	-5 to -10	www.usa.gov / *www.discoveramerica.com*
URUGUAY	3 407 000	3 641 000	15 180	99.0	598	-3	www.presidencia.gub.uy / *www.turismo.gub.uy*
UZBEKISTAN	28 934 000	36 330 000	1 900	99.9	998	+5	www.gov.uz / *www.uzbektourism.uz*
VANUATU	253 000	473 000	3 130	94.9	678	+11	www.governmentofvanuatu.gov.vu / *http://vanuatu.travel*
VATICAN CITY	800	1 000	...	...	39 or 379	+1	www.vaticanstate.va / *www.vaticanstate.va*
VENEZUELA	30 405 000	42 376 000	12 550	98.5	58	-4.5	www.presidencia.gob.ve / *http://venezuela-us.org/tourism-10*
VIETNAM	91 680 000	103 697 000	1 730	97.1	84	+7	www.na.gov.vn / *www.vietnamtourism.com*
YEMEN	24 407 000	42 497 000	1 330	87.4	967	+3	www.yemen-nic.info / *www.yementourism.com*
ZAMBIA	14 539 000	44 206 000	1 480	64.0	260	+2	www.parliament.gov.zm / *www.zambiatourism.com*
ZIMBABWE	14 150 000	26 254 000	820	90.9	263	+2	www.zim.gov.zw / *www.zimbabwetourism.net*

INDICATOR	DEFINITION
Total population	Interpolated mid-year population, 2013.
2050 projected population	Projected total population for the year 2050.
GNI per capita	Gross National Income per person in U.S. dollars using the World Bank Atlas method, from latest available data.
Literacy rate	Percentage of population aged 15–24 with at least a basic ability to read and write, from latest available data.
International dialling code	The country code prefix to be used when dialling from another country.
Time zone	Time difference in hours between local standard time and Coordinated Universal Time (UTC).
Official website	The official country website where available.
Tourism website	The country website for tourists where available.

MAIN STATISTICAL SOURCES

World Bank World Development Indicators online

International Telecommunications Union (ITU)

United Nations Department of Economic and Social Affairs (UDESA) World Population Prospects: The 2012 Revision

UNESCO Education Data Centre

WEB LINKS

http://data.worldbank.org/indicator

http://www.itu.int/en/Pages/default.aspx

http://esa.un.org/unpd/wpp/index.htm

www.uis.unesco.org/datacentre/pages/default.aspx

The system of timekeeping throughout the world is based on twenty-four time zones, each stretching over fifteen degrees of longitude – the distance equivalent to a time difference of one hour. The Prime, or Greenwich Meridian (0 degrees west), is the basis for Coordinated Universal Time (UTC) or Greenwich Mean Time (GMT), by which other times are measured. This universal reference point was agreed at an international conference in 1884.

Times are the local Standard Times observed compared with 00:00 (midnight) Coordinated Universal Time

	Web Address	Theme
Greenwich Royal Observatory	www.rmg.co.uk/royal-observatory	The home of time
Greenwich Mean Time	wwp.greenwichmeantime.com	World time since 1884
World time zones	www.worldtimezones.com	Detailed time zones information
The Official US time	www.time.gov/	The home of US time
International Date Line	aa.usno.navy.mil/faq/docs/international_date.php	Understanding the international date line

| 14₊₂ | 15₊₃ | 16₊₄ | 17₊₅ | 18₊₆ | 19₊₇ | 20₊₈ | 21₊₉ | 22₊₁₀ | 23₊₁₁ | MIDNIGHT PM\AM | 1 ₋₁₁ | 2 ₋₁₀ | 3 ₋₉ | 4 ₋₈ |

14₊₂ **15**₊₃ **16**₊₄ **17**₊₅ **18**₊₆ **19**₊₇ **20**₊₈ **21**₊₉ **22**₊₁₀ **23**₊₁₁ MIDNIGHT **1** ₋₁₁ **2** ₋₁₀ **3** ₋₉ **4** ₋₈

22.00 23.00

24.00

19.00   21.00

Yakutsk   Magadan
15.00   22.00
Moscow   Yekaterinburg
            17.00
                Krasnoyarsk
        Novosibirsk   23.00
14.00   18.00   Irkutsk

16.00   Dushanbe   Ulaanbaatar
Ankara   20.00   Beijing 20.30
   Tehrān 16.30   Tōkyō
Cairo 15.30   Chengdu   Shanghai
          17.00 Delhi   21.00
Riyadh   17.45
       17.30   18.30 Hong Kong
amena   18.00
   Addis Ababa   Bangkok   Manila
            17.30

                Singapore
         Equator
 shasa
Dar es Salaam   Jakarta   2.00 1.00
Harare   18.30   Tues Tues
                                1.00
etoria   CENTRAL   Tues
              STANDARD
    WESTERN   TIME
    STANDARD 21.30   EASTERN
pe Town TIME   STANDARD 23.30
         Perth   TIME
              Sydney 22.30 Auckland
   17.00
                      0.45
                      Tues

Monday   INTERNATIONAL DATE LINE

30° 45° 60° 75° 90° 105° 120° 135° 150° 165° 180° 165° 150°

UTC). Daylight Saving Time, normally one hour ahead of local Standard Time, which
s observed by certain countries for part of the year, is not shown on the map.

ime zone boundaries can be altered to suit international or internal boundaries.
China uses only one time zone although it should theoretically have five, while Russia
tretches over eleven zones. The four mainland USA time zones do not always follow
tate boundaries.

he International Date Line is an imaginary line at approximately 180° west (or east)
f Greenwich, across which the date changes by one day. The line has no international
egal status and countries near to the line can choose which date they will observe.

aylight Saving Time allows nations to adjust their clocks to extend daylight during
he working day. It was first introduced to the UK during the First World War to
educe the demand for artificial heating and lighting.

TIME DIFFERENCES FOR
MAJOR CITIES FROM UTC

	hours
Los Angeles	-8
New York	-5
Buenos Aires	-3
Berlin	+1
Cape Town	+2
Mumbai	+5.5
Singapore	+8
Beijing	+8
Tōkyō	+9
Sydney	+10

© Collins Bartholomew Ltd

HIGHEST MOUNTAINS	Height metres	feet	Location
Mt Everest	8 848	29 028	China/Nepal
K2	8 611	28 251	China/Pakistan
Kangchenjunga	8 586	28 169	India/Nepal
Lhotse	8 516	27 939	China/Nepal
Makalu	8 463	27 765	China/Nepal
Cho Oyu	8 201	26 906	China/Nepal
Dhaulagiri I	8 167	26 794	Nepal
Manaslu	8 163	26 781	Nepal
Nanga Parbat	8 126	26 660	Pakistan
Annapurna I	8 091	26 545	Nepal
Gasherbrum I	8 068	26 469	China/Pakistan
Broad Peak	8 047	26 401	China/Pakistan
Gasherbrum II	8 035	26 361	China/Pakistan
Xixabangma Feng	8 027	26 335	China
Annapurna II	7 937	26 040	Nepal

LONGEST RIVERS	Length km	miles	Continent
Nile	6 695	4 160	Africa
Amazon	6 516	4 049	South America
Yangtze	6 380	3 965	Asia
Mississippi-Missouri	5 969	3 709	North America
Ob'-Irtysh	5 568	3 460	Asia
Yenisey-Angara-Selenga	5 550	3 449	Asia
Yellow River	5 464	3 395	Asia
Congo	4 667	2 900	Africa
Río de la Plata-Paraná	4 500	2 796	South America
Irtysh	4 440	2 759	Asia
Mekong	4 425	2 750	Asia
Heilong Jiang-Argun'	4 416	2 744	Asia
Lena-Kirenga	4 400	2 734	Asia
MacKenzie-Peace-Finlay	4 241	2 635	North America
Niger	4 184	2 600	Africa

LARGEST LAKES	Area sq km	sq miles	Continent
Caspian Sea	371 000	143 243	Asia/Europe
Lake Superior	82 100	31 699	North America
Lake Victoria	68 870	26 591	Africa
Lake Huron	59 600	23 012	North America
Lake Michigan	57 800	22 317	North America
Lake Tanganyika	32 600	12 587	Africa
Great Bear Lake	31 328	12 096	North America
Lake Baikal	30 500	11 776	Asia
Lake Nyasa	29 500	11 390	Africa
Great Slave Lake	28 568	11 030	North America
Lake Erie	25 700	9 923	North America
Lake Winnipeg	24 387	9 416	North America
Lake Ontario	18 960	7 320	North America
Lake Ladoga	18 390	7 100	Europe
Lake Balkhash	17 400	6 718	Asia

LARGEST DRAINAGE BASINS	Area sq km	sq miles	Continent
Amazon	7 050 000	2 722 000	South America
Congo	3 700 000	1 429 000	Africa
Nile	3 349 000	1 293 000	Africa
Mississippi-Missouri	3 250 000	1 255 000	North America
Río de la Plata-Paraná	3 100 000	1 197 000	South America
Ob'-Irtysh	2 990 000	1 154 000	Asia
Yenisey-Angara-Selenga	2 580 000	996 000	Asia
Lena-Kirenga	2 490 000	961 000	Asia
Yangtze	1 959 000	756 000	Asia
Niger	1 890 000	730 000	Africa
Heilong Jiang-Argun'	1 855 000	716 000	Asia
Mackenzie-Peace-Finlay	1 805 000	697 000	North America
Ganges-Brahmaputra	1 621 000	626 000	Asia
St Lawrence-St Louis	1 463 000	565 000	North America
Volga	1 380 000	533 000	Europe

ATLANTIC OCEAN	Area sq km	sq miles	Deepest Point metres		feet
Total extent	86 557 000	33 420 000	8 605	Milwaukee Deep	28 231
Arctic Ocean	9 485 000	3 662 000	5 450		17 880
Caribbean Sea	2 512 000	970 000	7 680		25 197
Mediterranean Sea	2 510 000	969 000	5 121		16 801
Gulf of Mexico	1 544 000	596 000	3 504		11 496
Hudson Bay	1 233 000	476 000	259		850
North Sea	575 000	222 000	661		2 169
Black Sea	508 000	196 000	2 245		7 365
Baltic Sea	382 000	147 000	460		1 509

INDIAN OCEAN	Area sq km	sq miles	Deepest Point metres		feet
Total extent	73 427 000	28 350 000	7 125	Java Trench	23 376
Bay of Bengal	2 172 000	839 000	4 500		14 764
Red Sea	453 000	175 000	3 040		9 974
The Gulf	238 000	92 000	73		239

PACIFIC OCEAN	Area sq km	sq miles	Deepest Point metres		feet
Total extent	166 241 000	64 186 000	10 920	Challenger Deep	35 826
South China Sea	2 590 000	1 000 000	5 514		18 090
Bering Sea	2 261 000	873 000	4 150		13 615
Sea of Okhotsk	1 392 000	537 000	3 363		11 033
East China Sea and Yellow Sea	1 202 000	464 000	2 717		8 914
Sea of Japan (East Sea)	1 013 000	391 000	3 743		12 280

LARGEST ISLANDS	Area		Continent
	sq km	sq miles	
Greenland	2 175 600	839 999	North America
New Guinea	808 510	312 166	Oceania
Borneo	745 561	287 861	Asia
Madagascar	587 040	266 656	Africa
Baffin Island	507 451	195 927	North America
Sumatra	473 606	182 859	Asia
Honshū	227 414	87 805	Asia
Great Britain	218 476	84 354	Europe
Victoria Island	217 291	83 896	North America
Ellesmere Island	196 236	75 767	North America
Celebes	189 216	73 056	Asia
South Island, New Zealand	151 215	58 384	Oceania
Java	132 188	51 038	Asia
North Island, New Zealand	115 777	44 701	Oceania
Cuba	110 860	42 803	North America

DEEPEST LAKES	Depth		Continent
	metres	feet	
Lake Baikal	1 642	5 387	Asia
Lake Tanganyika	1 471	4 826	Africa
Caspian Sea	1 025	3 363	Europe/Asia
Lake Nyasa	706	2 316	Africa
Ysyk-Köl	702	2 303	Asia

LOWEST POINTS ON LAND	Depth below sea level		Location
	metres	feet	
Dead Sea	-428	-1 404	Asia
Lake Assal	-156	-512	Djibouti
Turpan Pendi	-154	-505	China
Qattara Depression	-133	-436	Egypt
Poluostrov Mangyshlak	-132	-433	Kazakhstan

HIGHEST WATERFALLS	Height		Location
	metres	feet	
Angel Falls (Kerepakupai Merú)	979	3 212	Venezuela
Tugela	948	3 110	South Africa
Utigård	800	2 625	Norway
Mongfossen	774	2 539	Norway
Mtarazi	762	2 500	Zimbabwe

EARTH'S DIMENSIONS	
Mass	5.974 x 10^{21} tonnes
Total area	509 450 000 sq km /196 698 645 sq miles
Land area	149 450 000 sq km / 57 702 645 sq miles
Water area	360 000 000 sq km /138 996 000 sq miles
Volume	1 083 207 x 10^6 cubic km / 259 911 x 10^6 cubic miles
Equatorial diameter	12 756 km / 7 927 miles
Polar diameter	12 714 km / 7 900 miles
Equatorial circumference	40 075 km / 24 903 miles
Meridional circumference	40 008 km / 24 861 miles

LARGEST COUNTRIES BY POPULATION	Population
China	1 369 993 000
India	1 252 140 000
United States of America	320 051 000
Indonesia	249 866 000
Brazil	200 362 000
Pakistan	182 143 000
Nigeria	173 615 000
Bangladesh	156 595 000
Russia	142 834 000
Japan	127 144 000

LARGEST COUNTRIES BY AREA	Area	
	sq km	sq miles
Russia	17 075 400	6 592 849
Canada	9 984 670	3 855 103
United States of America	9 826 635	3 794 085
China	9 606 802	3 709 186
Brazil	8 514 879	3 287 613
Australia	7 692 024	2 969 907
India	3 166 620	1 222 632
Argentina	2 766 889	1 068 302
Kazakhstan	2 717 300	1 049 155
Algeria	2 381 741	919 595

LARGEST CITIES	Population	Location
Tōkyō	38 197 000	Japan
Delhi	25 629 000	India
Shanghai	22 963 000	China
Mexico City	21 706 000	Mexico
New York	21 326 000	USA
Mumbai	21 214 000	India
São Paulo	21 028 000	Brazil
Beijing	18 079 000	China
Dhaka	17 382 000	Bangladesh
Karachi	15 500 000	Pakistan
Kolkata	15 076 000	India
Buenos Aires	14 151 000	Argentina
Los Angeles	14 081 000	USA
Lagos	13 121 000	Nigeria
Manila	12 856 000	Philippines

BUSIEST AIRPORTS (2013)	Location	Passengers
Atlanta (ATL)	USA	94 431 224
Beijing (PEK)	China	83 712 355
London (LHR)	UK	72 368 061
Tōkyō (HND)	Japan	68 906 509
Chicago (ORD)	USA	66 777 161
Los Angeles (LAX)	USA	66 667 619
Dubai (DXB)	UAE	66 431 533
Paris (CDG)	France	62 052 917
Dallas/Fort Worth (DFW)	USA	60 470 507
Jakarta (CGK)	Indonesia	60 137 347
Hong Kong (HKG)	China	59 594 290
Frankfurt (FRA)	Germany	58 036 948
Singapore (SIN)	Singapore	53 726 087
Amsterdam (AMS)	Netherlands	52 569 200
Denver (DEN)	USA	52 556 359

Climate is defined by the long-term weather conditions prevalent in any part of the world. The classification of climate types is based on the relationship between temperature and humidity and also on how these are affected by latitude, altitude, ocean currents and wind. Weather is how climatic conditions affect local areas. Weather stations collect data on temperature and rainfall, which can be plotted on graphs as shown here. These are based on average monthly figures over a minimum period of thirty years and can help to monitor climate change.

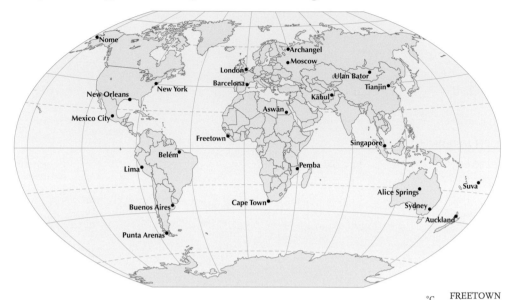

Temperature conversion							
°C	-20	-10	0	10	20	30	40
°F	-4	14	32	50	68	86	104

Rainfall conversion							
mm	25.4	127	254	381	508	635	762
ins	1	5	10	15	20	25	30

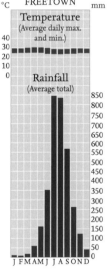

AFRICA

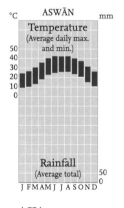

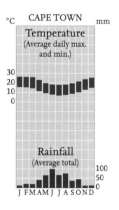

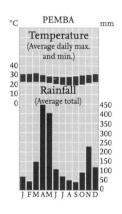

ASIA

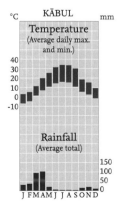

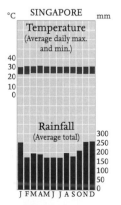

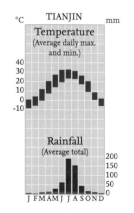

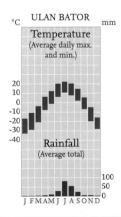

EUROPE

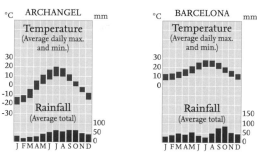

ARCHANGEL °C / mm
Temperature (Average daily max. and min.)
Rainfall (Average total)
J F M A M J J A S O N D

BARCELONA °C / mm
Temperature (Average daily max. and min.)
Rainfall (Average total)
J F M A M J J A S O N D

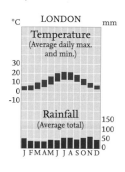

LONDON °C / mm
Temperature (Average daily max. and min.)
Rainfall (Average total)
J F M A M J J A S O N D

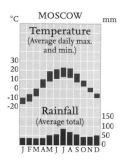

MOSCOW °C / mm
Temperature (Average daily max. and min.)
Rainfall (Average total)
J F M A M J J A S O N D

NORTH AMERICA

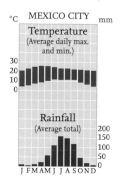

MEXICO CITY °C / mm
Temperature (Average daily max. and min.)
Rainfall (Average total)
J F M A M J J A S O N D

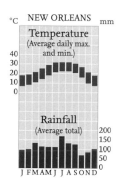

NEW ORLEANS °C / mm
Temperature (Average daily max. and min.)
Rainfall (Average total)
J F M A M J J A S O N D

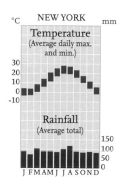

NEW YORK °C / mm
Temperature (Average daily max. and min.)
Rainfall (Average total)
J F M A M J J A S O N D

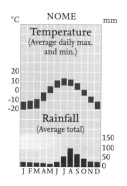

NOME °C / mm
Temperature (Average daily max. and min.)
Rainfall (Average total)
J F M A M J J A S O N D

SOUTH AMERICA

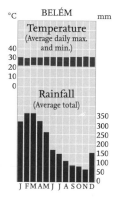

BELÉM °C / mm
Temperature (Average daily max. and min.)
Rainfall (Average total)
J F M A M J J A S O N D

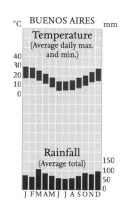

BUENOS AIRES °C / mm
Temperature (Average daily max. and min.)
Rainfall (Average total)
J F M A M J J A S O N D

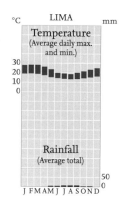

LIMA °C / mm
Temperature (Average daily max. and min.)
Rainfall (Average total)
J F M A M J J A S O N D

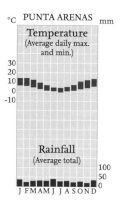

PUNTA ARENAS °C / mm
Temperature (Average daily max. and min.)
Rainfall (Average total)
J F M A M J J A S O N D

OCEANIA

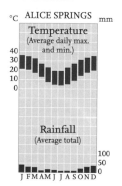

ALICE SPRINGS °C / mm
Temperature (Average daily max. and min.)
Rainfall (Average total)
J F M A M J J A S O N D

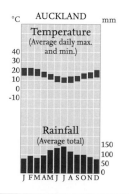

AUCKLAND °C / mm
Temperature (Average daily max. and min.)
Rainfall (Average total)
J F M A M J J A S O N D

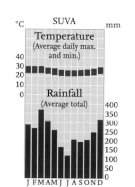

SUVA °C / mm
Temperature (Average daily max. and min.)
Rainfall (Average total)
J F M A M J J A S O N D

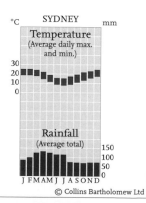

SYDNEY °C / mm
Temperature (Average daily max. and min.)
Rainfall (Average total)
J F M A M J J A S O N D

© Collins Bartholomew Ltd

ENVIRONMENT

The earth has a rich environment with a wide range of habitats. Forest and woodland form the predominant natural land cover and tropical rain forests are believed to be home to the majority of the world's bird, animal and plant species. These forests are part of a delicate land-atmosphere relationship disturbed by changes in land use. Grassland, shrubland and deserts cover most of the unwooded areas of the earth with low-growing tundra in the far northern latitudes. Grassland and shrubland regions in particular have been altered greatly by man through agriculture, livestock grazing and settlements.

Organization	Web address	Theme
NASA Earth Observatory	earthobservatory.nasa.gov	Observing the earth
USGS National Earthquake Information Center	earthquakes.usgs.gov/regional/neic	Monitoring earthquakes
Scripps Institution of Oceanography	scripps.ucsd.edu	Exploration of the oceans
NASA Visible Earth	visibleearth.nasa.gov	Satellite images of the earth
USGS Volcano Hazards Program	volcanoes.usgs.gov	Volcanic activity
UNESCO World Heritage Convention	whc.unesco.org	World Heritage Sites
British Geological Survey	bgs.ac.uk	Geology
International Union for Conservation of Nature	iucn.org	Global environmental conservation
Rainforest Action Network	www.ran.org	Rainforest information and resources
United Nations Environment Programme	unep.org	Voice for the environment within the UN
World Conservation Monitoring Centre	unep-wcmc.org	Conservation and the environment
World Resources Institute	www.wri.org	Monitoring the environment and resources
IUCN Red List	www.iucnredlist.org	Threatened species

OCEANS

Between them, the world's oceans cover approximately 70 per cent of the earth's surface. They contain 96 per cent of the earth's water and a vast range of flora and fauna. They are a major influence on the world's climate, particularly through ocean currents – the circulation of water within and between the oceans. Our understanding of the oceans has increased enormously over the last twenty years through the development of new technologies, including that of satellite images, which can generate vast amounts of data relating to the sea floor, ocean currents and sea surface temperatures.

Organization	Web address	Theme
International Maritime Organization	www.imo.org/Pages/home.aspx	Shipping and the environment
General Bathymetric Chart of the Oceans	www.gebco.net	Mapping the oceans
National Oceanography Centre, Southampton	www.noc.soton.ac.uk	Researching the oceans
Scott Polar Research Institute	www.spri.cam.ac.uk	Polar research

CLIMATE

The Earth's climate system is highly complex. It is recognized and accepted that man's activities are affecting this system, and monitoring climate change, including human influences upon it, is now a major issue. Future climate change depends critically on how quickly and to what extent the concentration of greenhouse gases in the atmosphere increase. Change will not be uniform across the globe and the information from sophisticated mathematical climate models is invaluable in helping governments and industry to assess the impacts climate change will have.

Organization	Web address	Theme
BBC Weather	news.bbc.co.uk/weather	Worldwide weather forecasts
Climatic Research Unit	www.cru.uea.ac.uk	Climatic research
Met Office	www.metoffice.gov.uk	Weather information and climatic research
National Climatic Data Center	www.ncdc.noaa.gov	Global climate data
National Hurricane Center	www.nhc.noaa.gov	Tracking hurricanes
National Oceanic and Atmospheric Administration	www.noaa.gov	Monitoring climate and the oceans
World Meteorological Organization	www.wmo.int/pages/index_en.html	The world's atmosphere and climate
NOAA El Niño	www.elnino.noaa.gov	El Niño research and observations

POPULATION

The world's population reached 6 billion in 1999. Rates of population growth vary between continents, but overall, the rate of growth has been increasing and it is predicted that by 2050 another 3 billion people will inhabit the planet. The process of urbanization, in particular migration from countryside to city, has led to the rapid growth of many cities. In mid 2009, for the first time, urban dwellers outnumbered those living in traditionally rural areas. It is estimated that by 2015 there will be 489 cities with over 1 million inhabitants and twenty-two with over 10 million.

Organization	Web address	Theme
Office for National Statistics	www.ons.gov.uk/census/index.html	UK census information
City Population	www.citypopulation.de	Statistics and maps about population
US Census Bureau	www.census.gov	US and world population
UN World Urbanization Prospects	esa.un.org/unpd/wup	Population estimates and projections
UN Population Information Network	www.un.org/popin	World population statistics
UN Population Division	www.un.org/en/development/desa/population	Monitoring world population

COUNTRIES

The present picture of the political world is the result of a long history of exploration, colonialism, conflict and negotiation. In 1950 there were eighty-two independent countries. Since then there has been a significant trend away from colonial influences and although many dependent territories still exist, there are now 196 independent countries. The newest country is South Sudan which declared independence from Sudan in July 2011. The shapes of countries reflect a combination of natural features, such as mountain ranges, and political agreements. There are still areas of the world where boundaries are disputed or only temporarily settled as ceasefire lines.

Organization	Web address	Theme
European Union	europa.eu	Gateway to the European Union
Permanent Committee on Geographical Names	www.pcgn.org.uk	Place names guidance for UK and abroad
The World Factbook	www.cia.gov/library/publications/the-world-factbook	Country profiles
US Board on Geographic Names	geonames.usgs.gov	Place names standardization in US and abroad
United Nations	www.un.org	The United Nations
International Boundaries Research Unit	www.dur.ac.uk/ibru	International boundaries resources and research
Organisation for Economic Co-operation and Development	www.oecd.org	Economic statistics
The World Bank	data.worldbank.org	World development data and statistics

TRAVEL

Travelling as a tourist or on business to some countries, or travelling within certain areas can be dangerous because of wars and political unrest. The UK Foreign Office provides the latest travel advice and security warnings. Some areas of the world, particularly tropical regions in the developing world, also carry many risks of disease. Advice should be sought on precautions to take and medications required.

Organization	Web address	Theme
UK Foreign and Commonwealth Office	www.fco.gov.uk	Travel, trade and country information
US Department of State	www.state.gov	Travel, trade and country information
World Health Organization	www.who.int/en	Health advice and world health issues
Centers for Disease Control and Prevention	www.cdc.gov	Advice for travellers
Airports Council International	www.aci.aero	The voice of the world's airports
Travel Daily News	www.traveldailynews.com	Travel and tourism newsletter

ORGANIZATIONS

Throughout the world there are many international, national and local organizations representing the interests of individual countries, groups of countries, regions and specialist groups. These can provide enormous amounts of information on economic, social, cultural, environmental and general geographical issues facing the world. The following is a selection of such sites.

Organization	Web address	Theme
United Nations	www.un.org	The United Nations
United Nations Educational, Scientific and Cultural Organization	en.unesco.org	International collaboration
United Nations Children's Fund	www.unicef.org	Children's health, education, equality and protection
United Nations High Commissioner for Refugees	www.unhcr.org/uk	The UN refugee agency
Food and Agriculture Organization of the United Nations	www.fao.org	Agriculture and food security
United Nations Development Programme	www.undp.org	The UN global development network
North Atlantic Treaty Organization	www.nato.int	North Atlantic freedom and security
European Environment Agency	www.eea.europa.eu	Europe's environment
European Centre for Nature Conservation	www.ecnc.org	Nature conservation in Europe
Europa – Gateway to the European Union	europa.eu	European Union facts and statistics
World Health Organization	www.who.int/en	Health issues and advice
Association of Southeast Asian Nations	www.asean.org	Economic, social and cultural development
Joint United Nations Programme on HIV/AIDS	www.unaids.org/en	The AIDS crisis
African Union	au.int	African international relations
World Lakes Network	worldlakes.org	Lakes around the world
Secretariat of the Pacific Community	www.spc.int	The Pacific community
The Maori world	www.maori.org.nz	Maori culture
US National Park Service	www.nps.gov/index.htm	National Parks of the USA
Parks Canada	www.pc.gc.ca	Natural and cultural heritage of Canada
Panama Canal Authority	www.pancanal.com	Explore the Panama Canal
Caribbean Community Secretariat	caricom.org	Caribbean Community
Organization of American States	www.oas.org/en/default.asp	Inter-American cooperation
The Latin American Network Information Center	lanic.utexas.edu	Latin America information
World Wildlife Fund	wwf.panda.org	Global environmental conservation
The Amazon Conservation Team	amazonteam.org	Preserving South American rainforests

DISTANCES

This table shows air distances in both kilometres and *miles* for twenty-seven cities around the world. These are the shortest distances between cities and are known as Great Circle routes.

Abu Dhabi
8075 9793 5905 6918 4303 11764 5247 11040 3422 3735 10647 14374 11689 5632 13481 5478 2043 2987 2317 7498 2367 13534 4637 5972 4795 14244
5018 6085 3669 4299 2674 7310 3260 6860 2126 2321 6616 8932 7263 3500 8377 3404 1270 1856 1440 4659 1471 8410 2881 3711 3091 8851

Auckland
8811 2161 8411 9596 18400 12288 18540 14187 13966 16194 14379 10947 2629 19592 10479 18330 16287 17042 12482 11796 16573 10372 17743 10388 9566
5475 1343 5227 5963 11521 7636 11521 8816 8678 10063 8935 6802 1634 12174 6512 11390 10121 10590 7756 7330 10298 6445 11025 6455 5944

Bangkok
4610 7523 1427 3720 8842 16081 9457 13949 7218 7070 13417 15760 7359 10196 13319 9544 6895 7477 2917 10144 7279 16885 8613 3291
2865 4675 887 2312 5494 9993 5877 8668 4485 4393 8337 9739 4573 6336 8276 5931 4285 4646 1813 6303 4523 10492 5352 2045

Beijing
2104 8923 4465 958 8144 17325 8236 11012 9216 5809 10490 12478 9093 9243 10082 8160 7135 7072 3788 12947 7557 19265 7375
1307 5545 2775 595 5061 10766 5118 6843 5727 3610 6518 7754 5650 5744 6265 5071 4434 4394 2354 8045 4696 11971 4583

Berlin
8942 16090 9927 8150 1182 9989 880 6403 6353 1612 6018 9746 15970 1871 9332 934 2903 1739 5791 9588 2891 11890
5556 9998 6169 5064 735 6207 547 3979 3948 1002 3740 6056 9924 1163 5799 580 1804 1081 3599 5958 1796 7388

Buenos Aires
18365 11821 15889 19429 11135 1968 11029 8490 10416 13461 9001 7366 11629 10024 9828 11105 12236 12235 15800 6891 11811
11412 7345 9873 12073 6919 1223 6853 5276 6472 8365 5593 4577 7226 6229 6107 6901 7603 7603 9818 4282 7339

Cairo
9587 14415 8270 6804 2135 9882 3215 9042 3518 2899 8733 12392 13966 3355 12223 3513 426 1234 4436 7208
5957 8957 5139 5284 1327 6141 1998 5618 2186 1801 5427 7700 8678 2085 7595 2183 265 767 2757 4479

Cape Town
14737 11034 9671 13710 8417 6075 9307 12551 4090 10101 12744 13703 10338 8536 16054 9635 7481 8367 9284
9157 6856 6009 8519 5230 3775 5783 7799 2542 6277 7919 8515 6424 5304 9976 5987 4649 5199 5769

Delhi
5857 10415 4142 4699 5929 14080 6601 11779 5428 4349 11286 14679 10192 7288 12882 6724 4032 4560
3640 6472 2574 2920 3684 8749 4102 7319 3373 2702 7013 9121 6333 4529 8005 4178 2505 2834

İstanbul
8970 14944 8652 7975 1379 10268 2261 8089 4751 1755 7730 11448 14628 2744 11043 2504 1170
5574 9286 5376 4956 857 6380 1405 5026 2952 1091 4803 7114 9090 1705 6862 1556 727

Jerusalem
9171 14126 7924 8083 2310 10308 3339 9190 3662 2671 8854 12552 13713 3602 12210 3615
5699 8778 4924 5023 1435 6405 2075 5711 2276 1660 5502 7800 8521 2238 7587 2246

London
9585 16990 10860 8882 1434 9254 341 5586 6805 5206 5240 8947 16902 1264 8778
5956 10557 6748 5519 891 5750 212 3471 4229 1557 3256 5560 10503 785 5455

Los Angeles
8828 12065 14136 9605 10212 10129 9106 3945 15553 9793 3973 2492 12762 9387
5486 7497 8784 5968 6346 6294 5658 2451 9664 6085 2469 1549 7930 5833

Madrid
10789 17687 10021 11396 1365 8118 1054 5785 6177 3446 5551 9083 17315
6704 10990 7081 6227 848 5044 655 3595 3838 2141 3449 5644 10759

Melbourne
8159 711 6050 8551 15987 13227 16793 16671 11513 14418 16730 13557
5070 442 3759 5314 9934 8219 10435 10359 7154 8959 10396 8424

Mexico City
11319 12972 16623 12071 10260 7669 9213 3362 14834 10740 3728
7034 8061 10329 7501 6375 4765 5725 2089 9218 6674 2317

Montréal
10409 16026 14816 10577 6601 8175 5522 533 11692 7077
6468 9958 9207 6572 4102 5080 3431 331 7265 4398

Moscow
7502 14487 8426 6626 2378 11529 7530 6323
4662 9002 5236 4117 1478 7164 1549 3929

Nairobi
11266 12162 7467 10115 5374 8941 6471 11849
7001 7557 4640 6285 3339 5556 4021 7363

New York
10870 15990 15349 11078 6907 7729 5851
6755 9936 9538 6884 4292 4803 3636

Paris
9738 16959 10743 8990 1108 9146
6051 10538 6676 5586 689 5683

Rio de Janiero
18557 13539 15740 18135 9181
10288 8413 9781 11269 5705

Rome
9881 16322 10030 8991
6140 10142 6232 5587

Seoul
1160 8298 4666
721 5156 2899

Singapore
5317 6293
3304 3910

Sydney
7794
4843

Tōkyō

CONVERSION CHARTS

To convert	into	multiply by
LENGTH AND AREA		
millimetres	inches	0.0394
centimetres	inches	0.3937
metres	feet	3.2808
metres	yards	1.0936
kilometres	miles	0.6214
inches	millimetres	25.4
inches	centimetres	2.54
feet	metres	0.3048
yards	metres	0.9144
miles	kilometres	1.6093
acres	hectares	0.4047
hectares	acres	2.4711
square miles	square kilometres	2.5900
square kilometres	square miles	0.3861
TEMPERATURE		
°C	°F	multiply by 1.8 and add 32
°F	°C	subtract 32 and divide by 1.8

To convert	into	multiply by
WEIGHT		
grams	ounces	0.0353
kilograms	pounds	2.2046
metric tonnes (1000 kg)	tons (2 240lbs)	0.9842
ounces	grams	28.3495
pounds	kilograms	0.4536
tons (2 240lbs)	metric tonnes (1000 kg)	1.0161
VOLUME		
pints (20fl oz)	litres	0.5683
imperial gallons	litres	4.5461
litres	pints (20fl oz)	1.7598
litres	imperial gallons	0.2200
SPEED		
km/h	mph	0.6214
mph	km/h	1.6093

© Collins Bartholomew Ltd

INTRODUCTION TO THE INDEX

The index includes all names shown on the maps in the Atlas of the World. Names are referenced by page number and by a grid reference. The grid reference correlates to the alphanumeric values which appear within each map frame. Each entry also includes the country or geographical area in which the feature is located. Entries relating to names appearing on insets are indicated by a small box symbol: □, followed by a grid reference if the inset has its own alphanumeric values.

Name forms are as they appear on the maps, with additional alternative names or name forms included as cross-references which refer the user to the entry for the map form of the name. Names beginning with Mc or Mac are alphabetized exactly as they appear. The terms Saint, Sainte, Sankt, etc., are abbreviated to St, Ste, St, etc., but alphabetized as if in the full form.

Names of physical features beginning with generic geographical terms are permuted – the descriptive term is placed after the main part of the name. For example, Lake Superior is indexed as Superior, Lake; Mount Everest as Everest, Mount. This policy is applied to all languages.

Entries, other than those for towns and cities, include a descriptor indicating the type of geographical feature. Descriptors are not included where the type of feature is implicit in the name itself.

Administrative divisions are included to differentiate entries of the same name and feature type within the one country. In such cases, duplicate names are alphabetized in order of administrative division.

INDEX ABBREVIATIONS

admin. div.	administrative division	Ger.	Germany	pt	point
Afgh.	Afghanistan	Guat.	Guatemala	r.	river
Alg.	Algeria	h.	hill	r. mouth	river mouth
Arg.	Argentina	hd	headland	reg.	region
Austr.	Australia	Hond.	Honduras	resr	reservoir
aut. comm.	autonomous community	i.	island	S.	South
aut. reg.	autonomous region	imp. l.	impermanent lake	salt l.	salt lake
Azer.	Azerbaijan	Indon.	Indonesia	sea chan.	sea channel
b.	bay	is	islands	str.	strait
B.I.O.T.	British Indian Ocean Territory	isth.	isthmus	Switz.	Switzerland
		Kazakh.	Kazakhstan	Tajik.	Tajikistan
Bangl.	Bangladesh	Kyrg.	Kyrgyzstan	Tanz.	Tanzania
Bol.	Bolivia	l.	lake	terr.	territory
Bos. & Herz.	Bosnia and Herzegovina	lag.	lagoon	Thai.	Thailand
Bulg.	Bulgaria	Lith.	Lithuania	Trin. and Tob.	Trinidad and Tobago
c.	cape	Lux.	Luxembourg	Turkm.	Turkmenistan
Can.	Canada	Madag.	Madagascar	U.A.E.	United Arab Emirates
C.A.R.	Central African Republic	Maur.	Mauritania	U.K.	United Kingdom
chan.	channel	Mex.	Mexico	Ukr.	Ukraine
Col.	Colombia	Moz.	Mozambique	Uru.	Uruguay
Czech Rep.	Czech Republic	mt.	mountain	U.S.A.	United States of America
Dem. Rep. Congo	Democratic Republic of the Congo	mts	mountains	Uzbek.	Uzbekistan
		mun.	municipality	val.	valley
depr.	depression	N.	North	Venez.	Venezuela
des.	desert	Neth.	Netherlands	vol.	volcano
disp. terr.	disputed territory	Nic.	Nicaragua	vol. crater	volcanic crater
Dom. Rep.	Dominican Republic	N.Z.	New Zealand		
Equat. Guinea	Equatorial Guinea	Pak.	Pakistan		
esc.	escarpment	Para.	Paraguay		
est.	estuary	pen.	peninsula		
Eth.	Ethiopia	Phil.	Philippines		
Fin.	Finland	plat.	plateau		
for.	forest	P.N.G.	Papua New Guinea		
Fr. Guiana	French Guiana	Pol.	Poland		
Fr. Polynesia	French Polynesia	Port.	Portugal		
g.	gulf	prov.	province		

1

128 B2 100 Mile House Can.

A

93 E4 Aabenraa Denmark
100 C2 Aachen Ger.
93 E4 Aalborg Denmark
102 C2 Aalen Ger.
62 A1 Aalo India
100 B2 Aalst Belgium
93 I3 Äänekoski Fin.
105 D2 Aarau Switz.
93 F4 Aarhus Denmark
100 B2 Aarschot Belgium
70 A2 Aba China
119 D2 Aba Dem. Rep. Congo
115 C4 Aba Nigeria
81 C2 Ābādān Iran
81 D2 Ābādeh Iran
81 D3 Ābādeh Ṭashk Iran
114 B1 Abadla Alg.
155 C1 Abaeté Brazil
        Abagnar Qi China see Xilinhot
135 E3 Abajo Peak U.S.A.
115 C4 Abakaliki Nigeria
83 H3 Abakan Russia
150 B4 Abancay Peru
81 D2 Abarkūh Iran
66 D2 Abashiri Japan
66 D2 Abashiri-wan b. Japan
59 D3 Abau P.N.G.
        Abaya, Lake Eth. see Lake Abaya
        Ābay Wenz r. Eth. see Blue Nile
83 H3 Abaza Russia
108 A2 Abbasanta Italy
104 C1 Abbeville France
141 C2 Abbeville AL U.S.A.
140 B3 Abbeville LA U.S.A.
97 B2 Abbeyfeale Ireland
55 M2 Abbot Ice Shelf Antarctica
74 B1 Abbottabad Pak.
115 E3 Abéché Chad
114 B4 Abengourou Côte d'Ivoire
114 C4 Abeokuta Nigeria
99 A3 Aberaeron U.K.
96 C2 Aberchirder U.K.
        Abercorn Zambia see Mbala
99 B4 Aberdare U.K.
99 A3 Aberdaron U.K.
53 D2 Aberdeen Austr.
122 B3 Aberdeen S. Africa
96 C2 Aberdeen U.K.
139 D3 Aberdeen MD U.S.A.
137 D1 Aberdeen SD U.S.A.
134 B1 Aberdeen WA U.S.A.
129 E1 Aberdeen Lake Can.
96 C2 Aberfeldy U.K.
96 B2 Aberfoyle U.K.
99 B4 Abergavenny U.K.
        Abergwaun U.K. see Fishguard
        Aberhonddu U.K. see Brecon
143 C2 Abernathy U.S.A.
134 B2 Abert, Lake U.S.A.
        Abertawe U.K. see Swansea
        Aberteifi U.K. see Cardigan
99 B4 Abertillery U.K.
99 A3 Aberystwyth U.K.
86 F2 Abez' Russia
78 B3 Abhā Saudi Arabia
81 C2 Abhar Iran
        Abiad, Bahr el r. Africa see
        White Nile
114 B4 Abidjan Côte d'Ivoire
137 D3 Abilene KS U.S.A.
143 D2 Abilene TX U.S.A.
99 C4 Abingdon U.K.
138 C3 Abingdon U.S.A.
91 D3 Abinsk Russia
130 B3 Abitibi, Lake Can.
81 C1 Abkhazia disp. terr. Georgia
        Åbo Fin. see Turku
74 B1 Abohar India
114 B4 Aboisso Côte d'Ivoire
114 C4 Abomey Benin
60 A1 Abongabong, Gunung mt. Indon.
118 B2 Abong Mbang Cameroon
64 A3 Aborlan Phil.
115 D3 Abou Déia Chad
106 B2 Abrantes Port.
152 B2 Abra Pampa Arg.
142 A3 Abreojos, Punta pt Mex.
116 B2 'Abri Sudan
136 A2 Absaroka Range mts U.S.A.
81 C1 Abşeron Yarımadası pen. Azer.
78 B3 Abū 'Arīsh Saudi Arabia
116 A2 Abū Ballāş h. Egypt
79 C2 Abu Dhabi U.A.E.
116 B3 Abu Hamed Sudan
116 B3 Abu Haraz Sudan
115 C4 Abuja Nigeria
81 C2 Abū Kamāl Syria
118 C2 Abumombazi Dem. Rep. Congo
152 B1 Abunã r. Bol./Brazil
150 C3 Abunã Brazil
74 B2 Abu Road India
78 B2 Abū Şādi, Jabal h. Saudi Arabia
116 B2 Abū Sunbul Egypt
116 A3 Abu Zabad Sudan
        Abū Ẓaby U.A.E. see Abu Dhabi

117 A4 Abyei Sudan
145 B2 Acambaro Mex.
120 B2 Acampamento de Caça do Mucusso Angola
106 B1 A Cañiza Spain
144 B2 Acaponeta Mex.
145 C3 Acapulco Mex.
151 E3 Acará Brazil
154 A3 Acaray, Represa de resr Para.
150 C2 Acarigua Venez.
110 B1 Acâş Romania
145 C3 Acatlán Mex.
145 C3 Acayucán Mex.
114 B4 Accra Ghana
98 B2 Accrington U.K.
74 B2 Achalpur India
97 A2 Achill Island Ireland
101 D1 Achim Ger.
96 B2 Achnasheen U.K.
91 D2 Achuyevo Russia
111 C3 Acıpayam Turkey
109 C3 Acireale Italy
147 C2 Acklins Island Bahamas
153 A3 Aconcagua, Cerro mt. Arg.
106 B1 A Coruña Spain
108 A2 Acqui Terme Italy
103 D2 Ács Hungary
49 N6 Actéon, Groupe is Fr. Polynesia
145 C2 Actopán Mex.
143 D2 Ada U.S.A.
        Adabazar Turkey see Adapazarı
79 C2 Adam Oman
111 B3 Adamas Greece
135 B3 Adams Peak U.S.A.
        'Adan Yemen see Aden
80 B2 Adana Turkey
111 C2 Adapazarı Turkey
97 B2 Adare Ireland
55 M2 Adare, Cape Antarctica
108 A1 Adda r. Italy
78 B2 Ad Dafinah Saudi Arabia
78 B2 Ad Dahnā' des. Saudi Arabia
79 B2 Ad Dahnā' des. Saudi Arabia
        Ad Dammām Saudi Arabia see Dammam
78 A2 Ad Dār al Ḥamrā' Saudi Arabia
78 B3 Ad Darb Saudi Arabia
78 B2 Ad Dawādimī Saudi Arabia
        Ad Dawḥah Qatar see Doha
        Aḑ Ḑiffah Egypt/Libya see Libyan Plateau
78 B2 Ad Dilam Saudi Arabia
78 B2 Ad Dir'īyah Saudi Arabia
117 B4 Addis Ababa Eth.
81 C2 Ad Dīwānīyah Iraq
141 D2 Adel U.S.A.
52 A2 Adelaide Austr.
55 A3 Adelaide Antarctica
50 C1 Adelaide River Austr.
101 D2 Adelebsen Ger.
55 K2 Adélie Land reg. Antarctica
78 B3 Aden Yemen
116 C3 Aden, Gulf of Somalia/Yemen
100 C2 Adenau Ger.
79 C2 Adh Dhayd U.A.E.
59 C3 Adi i. Indon.
116 B3 Āḍī Ārk'ay Eth.
108 B1 Adige r. Italy
116 B3 Ādīgrat Eth.
78 A3 Adi Keyih Eritrea
74 B3 Adilabad India
115 D2 Adīrī Libya
139 E2 Adirondack Mountains U.S.A.
        Ādīs Ābeba Eth. see Addis Ababa
117 B4 Ādīs Alem Eth.
80 B2 Adıyaman Turkey
110 C1 Adjud Romania
50 B1 Admiralty Gulf Austr.
128 A2 Admiralty Island U.S.A.
59 D3 Admiralty Islands P.N.G.
73 B3 Adoni India
104 B3 Adour r. France
106 C2 Adra Spain
108 B3 Adrano Italy
114 C2 Adrar Alg.
138 C2 Adrian MI U.S.A.
143 C1 Adrian TX U.S.A.
108 B2 Adriatic Sea Europe
        Adua Eth. see Adwa
116 B3 Ādwa Eth.
83 K2 Adycha r. Russia
91 D2 Adygeysk Russia
114 B4 Adzopé Côte d'Ivoire
111 B3 Aegean Sea Greece/Turkey
101 D1 Aerzen Ger.
106 B1 A Estrada Spain
116 B3 Afabet Eritrea
        Affreville Alg. see Khemis Miliana
76 C3 Afghanistan country Asia
78 B2 'Afīf Saudi Arabia
136 A2 Afton U.S.A.
80 B2 Afyon Turkey
115 C3 Agadez Niger
114 B1 Agadir Morocco
113 I7 Agalega Islands Mauritius
        Agana Guam see Hagåtña
74 B2 Agar India
119 D2 Āgaro Eth.
75 D2 Agartala India
81 C2 Ağdam (abandoned) Azer.
105 C3 Agde France
104 C3 Agen France
122 A2 Aggeneys S. Africa

111 C3 Agia Varvara Greece
111 B3 Agios Dimitrios Greece
111 C3 Agios Efstratios i. Greece
111 C3 Agios Kirykos Greece
111 C3 Agios Nikolaos Greece
78 A3 Agirwat Hills Sudan
123 C2 Agisanang S. Africa
110 B1 Agnita Romania
74 B2 Agra India
81 C2 Ağrı Turkey
        Ağrı Dağı mt. Turkey see Ararat, Mount
108 B3 Agrigento Italy
111 B3 Agrinio Greece
109 B2 Agropoli Italy
87 E3 Agryz Russia
144 B2 Agua Brava, Laguna lag. Mex.
154 A3 Agua Clara Brazil
145 C3 Aguada Mex.
146 B4 Aguadulce Panama
144 B2 Aguanaval r. Mex.
144 B1 Agua Prieta Mex.
144 B2 Aguascalientes Mex.
155 D1 Aguas Formosas Brazil
154 C2 Agudos Brazil
106 B1 Agueda Port.
114 C3 Aguelhok Mali
106 C1 Aguilar de Campoo Spain
107 C2 Aguilas Spain
144 B3 Aguililla Mex.
122 B3 Agulhas, Cape S. Africa
158 F7 Agulhas Basin Southern Ocean
155 D2 Agulhas Negras mt. Brazil
158 F7 Agulhas Plateau Southern Ocean
158 F7 Agulhas Ridge S. Atlantic Ocean
111 C2 Ağva Turkey
115 C2 Ahaggar plat. Alg.
115 C2 Ahaggar, Tassili oua-n- plat. Alg.
81 C2 Ahar Iran
100 C1 Ahaus Ger.
81 C2 Ahlat Turkey
100 C2 Ahlen Ger.
74 B2 Ahmadabad India
74 B3 Ahmadnagar India
74 B2 Ahmadpur East Pak.
74 B1 Ahmadpur Sial Pak.
117 C4 Ahmar mts Eth.
        Ahmedabad India see Ahmadabad
        Ahmednagar India see Ahmadnagar
144 B2 Ahome Mex.
81 D3 Ahram Iran
101 E1 Ahrensburg Ger.
104 C2 Ahun France
93 F4 Åhus Sweden
81 C2 Ahvāz Iran
        Ahvenanmaa is Fin. see Åland Islands
122 A2 Ai-Ais Namibia
74 A1 Aībak Afgh.
80 B2 Aigialousa Cyprus
111 B3 Aigio Greece
        Aihui China see Heihe
        Aijal India see Aizawl
141 D2 Aiken U.S.A.
97 B1 Ailt an Chorráin Ireland
155 D1 Aimorés Brazil
155 D1 Aimorés, Serra dos hills Brazil
105 D2 Ain r. France
107 E2 Aïn Azel Alg.
115 C1 Aïn Beïda Alg.
114 B2 'Aïn Ben Tili Maur.
107 D2 Aïn Defla Alg.
114 B1 Aïn Sefra Alg.
136 D2 Ainsworth U.S.A.
        Aintab Turkey see Gaziantep
107 D2 Aïn Taya Alg.
107 D2 Aïn Tédélès Alg.
107 C2 Aïn Temouchent Alg.
115 C3 Aïr, Massif de l' mts Niger
60 A1 Airbangis Indon.
128 C2 Airdrie Can.
96 C3 Airdrie U.K.
104 B3 Aire-sur-l'Adour France
101 E3 Aisch r. Ger.
128 A1 Aishihik Lake Can.
100 A3 Aisne r. France
59 D3 Aitape P.N.G.
137 E1 Aitkin U.S.A.
110 B1 Aiud Romania
105 D3 Aix-en-Provence France
        Aix-la-Chapelle Ger. see Aachen
105 D2 Aix-les-Bains France
75 D2 Aizawl India
88 C2 Aizkraukle Latvia
88 B2 Aizpute Latvia
67 C3 Aizuwakamatsu Japan
105 D3 Ajaccio France
        Ajayameru India see Ajmer
115 E1 Ajdābiyā Libya
115 C2 Ajjer, Tassili n' plat. Alg.
79 C2 'Ajman U.A.E.
74 B2 Ajmer India
        Ajmer-Merwara India see Ajmer
142 A2 Ajo U.S.A.
77 D2 Akadyr Kazakh.
119 D2 Āk'ak'i Beseka Eth.
        Akamagaseki Japan see Shimonoseki
54 B2 Akaroa N.Z.
87 E3 Akbulak Russia
80 B2 Akçakale Turkey
114 A3 Akchâr reg. Maur.
111 C3 Akdağ mt. Turkey
80 B2 Akdağmadeni Turkey
88 C2 Åkersberga Sweden
118 C2 Aketi Dem. Rep. Congo

81 C1 Akhalkalaki Georgia
81 C1 Akhaltsikhe Georgia
79 C2 Akhḍar, Jabal mts Oman
111 C3 Akhisar Turkey
87 D4 Akhtubinsk Russia
118 B3 Akiéni Gabon
130 B2 Akimiski Island Can.
66 D3 Akita Japan
114 A3 Akjoujt Maur.
        Akkerman Ukr. see Bilhorod-Dnistrovs'kyy
77 D1 Akkol' Kazakh.
        Ak-Mechet Kazakh. see Kyzylorda
88 B2 Akmenrags pt Latvia
        Akmola Kazakh. see Astana
        Akmolinsk Kazakh. see Astana
67 B4 Akō Japan
117 B4 Akobo S. Sudan
74 B2 Akola India
118 B2 Akonolinga Cameroon
78 A3 Akordat Eritrea
131 D1 Akpatok Island Can.
77 D2 Aqki China
92 □A3 Akranes Iceland
136 C3 Akron CO U.S.A.
138 C2 Akron OH U.S.A.
75 B1 Aksai Chin disp. terr. Asia
80 B2 Aksaray Turkey
86 F2 Aksarka Russia
76 B1 Aksay Kazakh.
91 D2 Aksay Russia
80 B2 Akşehir Turkey
76 C2 Akshiganak Kazakh.
77 E2 Aksu China
116 B3 Āksum Eth.
76 B2 Aktau Kazakh.
76 B1 Aktobe Kazakh.
77 D2 Aktogay Kazakh.
88 C3 Aktsyabrski Belarus
        Aktyubinsk Kazakh. see Aktobe
67 B4 Akune Japan
115 C4 Akure Nigeria
92 □B2 Akureyri Iceland
        Akyab Myanmar see Sittwe
111 C3 Akyazı Turkey
77 C2 Akzhaykyn, Ozero salt l. Kazakh.
140 C2 Alabama r. U.S.A.
140 C2 Alabama state U.S.A.
140 C2 Alabaster U.S.A.
111 C3 Alaçatı Turkey
145 D2 Alacrán, Arrecife reef Mex.
81 C1 Alagir Russia
151 F4 Alagoinhas Brazil
107 C1 Alagón Spain
78 B2 Al Aḥmadi Kuwait
77 E2 Alakol', Ozero salt l. Kazakh.
92 J2 Alakurtti Russia
78 B3 Al 'Alayyah Saudi Arabia
81 C2 Al 'Amārah Iraq
80 A2 Al 'Āmirīyah Egypt
143 C2 Alamitos, Sierra de los mt. Mex.
135 C3 Alamo U.S.A.
142 B2 Alamogordo U.S.A.
144 A2 Alamos Sonora Mex.
144 B2 Alamos Sonora Mex.
144 B2 Alamos r. Mex.
136 B3 Alamosa U.S.A.
        Åland Fin. see Åland Islands
93 G3 Åland Islands is Fin.
80 B2 Alanya Turkey
73 B4 Alappuzha India
80 B3 Al 'Aqabah Jordan
78 B2 Al 'Aqiq Saudi Arabia
107 C2 Alarcón, Embalse de resr Spain
80 B2 Al 'Arīsh Egypt
78 B2 Al Arṭāwīyah Saudi Arabia
61 C2 Alas Indon.
111 C3 Alaşehir Turkey
128 A2 Alaska state U.S.A.
124 D4 Alaska, Gulf of U.S.A.
126 B3 Alaska Peninsula U.S.A.
126 C2 Alaska Range mts U.S.A.
81 C2 Älät Azer.
87 D3 Alatyr' Russia
150 B3 Alausí Ecuador
93 H3 Alavus Fin.
52 B2 Alawoona Austr.
79 C2 Al 'Ayn U.A.E.
108 B2 Alba Italy
107 C2 Albacete Spain
78 A2 Al Badā'i' Saudi Arabia
78 B2 Al Badī' Saudi Arabia
110 B1 Alba Iulia Romania
109 C2 Albania country Europe
50 A3 Albany Austr.
130 B2 Albany r. Can.
141 D2 Albany GA U.S.A.
139 F2 Albany NY U.S.A.
134 B2 Albany OR U.S.A.
        Al Başrah Iraq see Basra
51 D1 Albatross Bay Austr.
116 A2 Al Bawīṭī Egypt
115 E1 Al Baydā' Libya
78 B3 Al Bayḑā' Yemen
141 D1 Albemarle U.S.A.
141 E1 Albemarle Sound sea chan. U.S.A.
108 A2 Albenga Italy
51 C2 Alberga watercourse Austr.
119 D2 Albert, Lake Dem. Rep. Congo/Uganda
128 C2 Alberta prov. Can.
100 B2 Albert Kanaal canal Belgium
137 E2 Albert Lea U.S.A.
117 B4 Albert Nile r. S. Sudan/Uganda

Bad Lauterberg im Harz

117 C3 **Arta** Djibouti
111 B3 **Arta** Greece
144 B3 **Arteaga** Mex.
66 B2 **Artem** Russia
91 D2 **Artemivs'k** Ukr.
104 C2 **Artenay** France
142 C2 **Artesia** U.S.A.
51 E2 **Arthur Point** Austr.
54 B2 **Arthur's Pass** N.Z.
152 C3 **Artigas** Uru.
129 D1 **Artillery Lake** Can.
123 C1 **Artisia** Botswana
104 C1 **Artois** *reg.* France
90 B2 **Artsyz** Ukr.
Artur de Paiva Angola *see* **Kuvango**
77 D3 **Artux** China
81 C1 **Artvin** Turkey
59 C3 **Aru, Kepulauan** *is* Indon.
119 D2 **Arua** Uganda
147 D3 **Aruba** *terr.* West Indies
75 C2 **Arun** *r.* Nepal
62 A1 **Arunachal Pradesh** *terr.* India
119 D3 **Arusha** Tanz.
136 B3 **Arvada** U.S.A.
68 C1 **Arvayheer** Mongolia
129 E1 **Arviat** Can.
92 G2 **Arvidsjaur** Sweden
93 F4 **Arvika** Sweden
108 A2 **Arzachena** Italy
87 D3 **Arzamas** Russia
107 C2 **Arzew** Alg.
100 C2 **Arzfeld** Ger.
Arzila Morocco *see* **Asilah**
101 F2 **Aš** Czech Rep.
115 C4 **Asaba** Nigeria
74 B1 **Asadābād** Afgh.
66 D2 **Asahi-dake** *vol.* Japan
66 D2 **Asahikawa** Japan
78 B3 **Āsalē** *l.* Eth.
75 C2 **Asansol** India
117 C3 **Āsayita** Eth.
130 C3 **Asbestos** Can.
122 B2 **Asbestos Mountains** S. Africa
119 E2 **Āsbe Teferi** Eth.
109 C2 **Ascea** Italy
152 B1 **Ascensión** Bol.
113 B6 **Ascension** *i.* S. Atlantic Ocean
145 D3 **Ascensión, Bahía de la** *b.* Mex.
101 D3 **Aschaffenburg** Ger.
100 C2 **Ascheberg** Ger.
101 E2 **Aschersleben** Ger.
108 B2 **Ascoli Piceno** Italy
119 D2 **Āsela** Eth.
92 G3 **Åsele** Sweden
111 B2 **Asenovgrad** Bulg.
76 B3 **Aşgabat** Turkm.
78 B2 **Asharat** Saudi Arabia
50 A2 **Ashburton** *watercourse* Austr.
54 B2 **Ashburton** N.Z.
140 B2 **Ashdown** U.S.A.
141 D1 **Asheville** U.S.A.
53 D1 **Ashford** Austr.
97 C2 **Ashford** Ireland
99 D4 **Ashford** U.K.
66 D2 **Ashibetsu** Japan
98 C2 **Ashington** U.K.
67 B4 **Ashizuri-misaki** *pt* Japan
Ashkhabad Turkm. *see* **Aşgabat**
136 D3 **Ashland** *KS* U.S.A.
138 C3 **Ashland** *KY* U.S.A.
138 C2 **Ashland** *OH* U.S.A.
134 B2 **Ashland** *OR* U.S.A.
138 A1 **Ashland** *WI* U.S.A.
53 C1 **Ashley** Austr.
88 C3 **Ashmyany** Belarus
66 D2 **Ashoro** Japan
81 C2 **Ash Shabakah** Iraq
78 B3 **Ash Sharawrah** Saudi Arabia
Ash Shāriqah U.A.E. *see* **Sharjah**
81 C2 **Ash Sharqāţ** Iraq
81 C2 **Ash Shaţrah** Iraq
78 B3 **Ash Shaykh 'Uthman** Yemen
79 B3 **Ash Shiḩr** Yemen
79 C2 **Ash Shināş** Oman
78 B2 **Ash Shu'aybah** Saudi Arabia
78 B2 **Ash Shu'bah** Saudi Arabia
78 B2 **Ash Shubaykīyah** Saudi Arabia
78 B2 **Ash Shumlūl** Saudi Arabia
78 B3 **Ash Shuqayq** Saudi Arabia
115 D2 **Ash Shuwayrif** Libya
138 C2 **Ashtabula** U.S.A.
131 D2 **Ashuanipi Lake** Can.
75 B3 **Asifabad** India
106 B2 **Asilah** Morocco
108 A2 **Asinara, Golfo dell'** *b.* Italy
108 A2 **Asinara, Isola** *i.* Italy
82 G3 **Asino** Russia
88 C3 **Asipovichy** Belarus
78 B2 **'Asīr** *reg.* Saudi Arabia
93 F4 **Asker** Norway
93 F4 **Askim** Norway
68 C1 **Askiz** Russia
116 B3 **Asmara** Eritrea
93 F4 **Åsnen** *l.* Sweden
116 B2 **Asoteriba, Jebel** *mt.* Sudan
103 D2 **Aspang-Markt** Austria
136 B3 **Aspen** U.S.A.
143 C2 **Aspermont** U.S.A.
54 A2 **Aspiring, Mount** N.Z.
116 C3 **Assab** Eritrea
79 B3 **Aş Şadārah** Yemen
Aş Şaḩrā' al Gharbīyah *des.* Egypt *see* **Western Desert**
Aş Şaḩrā' ash Sharqīyah *des.* Egypt *see* **Eastern Desert**
78 B2 **As Salamiyah** Saudi Arabia
78 B3 **Aş Şalīf** Yemen
75 D2 **Assam** *state* India
81 C2 **As Samāwah** Iraq
79 C2 **Aş Şanām** *reg.* Saudi Arabia
115 E2 **As Sarīr** *reg.* Libya
108 A3 **Assemini** Italy
100 C1 **Assen** Neth.
100 B2 **Assesse** Belgium
115 D1 **As Sidrah** Libya
129 D3 **Assiniboia** Can.
128 C2 **Assiniboine, Mount** Can.
154 B2 **Assis** Brazil
78 B2 **Aş Şubayḩīyah** Kuwait
81 C2 **As Sulaymānīyah/Slêmānî** Iraq
78 B2 **As Sulaymī** Saudi Arabia
78 B2 **As Sulayyil** Saudi Arabia
78 B2 **As Sūq** Saudi Arabia
80 B2 **As Suwaydā'** Syria
79 C2 **As Suwayq** Oman
As Suways Egypt *see* **Suez**
111 B3 **Astakos** Greece
77 D1 **Astana** Kazakh.
81 C2 **Āstārā** Iran
100 B2 **Asten** Neth.
Asterabad Iran *see* **Gorgān**
108 A2 **Asti** Italy
74 B1 **Astor** Pak.
106 B1 **Astorga** Spain
134 B1 **Astoria** U.S.A.
Astrabad Iran *see* **Gorgān**
87 D4 **Astrakhan'** Russia
Astrakhan' Bazar Azer. *see* **Cälilabad**
88 C3 **Astravyets** Belarus
Astrida Rwanda *see* **Butare**
106 B1 **Asturias** *aut. comm.* Spain
111 C3 **Astypalaia** *i.* Greece
152 C2 **Asunción** Para.
116 B2 **Aswān** Egypt
116 B2 **Asyūţ** Egypt
Atacama, Desierto de *des.* Chile *see* **Atacama Desert**
152 B2 **Atacama, Puna de** *plat.* Arg.
152 B2 **Atacama, Salar de** *salt flat* Chile
152 B3 **Atacama Desert** *des.* Chile
114 C4 **Atakpamé** Togo
111 B3 **Atalanti** Greece
150 B4 **Atalaya** Peru
155 D1 **Ataléia** Brazil
77 C3 **Atamyrat** Turkm.
78 B3 **'Ataq** Yemen
114 A2 **Atâr** Maur.
135 B3 **Atascadero** U.S.A.
77 D2 **Atasu** Kazakh.
111 C3 **Atavyros** *mt.* Greece
116 B3 **Atbara** Sudan
116 B3 **Atbara** *r.* Sudan
77 C1 **Atbasar** Kazakh.
140 B3 **Atchafalaya Bay** U.S.A.
137 D3 **Atchison** U.S.A.
Ateransk Kazakh. *see* **Atyrau**
108 B2 **Aterno** *r.* Italy
108 B2 **Atessa** Italy
100 A2 **Ath** Belgium
128 C2 **Athabasca** Can.
129 C2 **Athabasca** *r.* Can.
129 D2 **Athabasca, Lake** Can.
97 C2 **Athboy** Ireland
97 B2 **Athenry** Ireland
111 B3 **Athens** Greece
140 C2 **Athens** *AL* U.S.A.
141 D2 **Athens** *GA* U.S.A.
138 C3 **Athens** *OH* U.S.A.
141 D1 **Athens** *TN* U.S.A.
143 D2 **Athens** *TX* U.S.A.
51 D1 **Atherton** Austr.
Athina Greece *see* **Athens**
97 C2 **Athlone** Ireland
111 B2 **Athos** *mt.* Greece
97 C2 **Athy** Ireland
115 D3 **Ati** Chad
130 A3 **Atikokan** Can.
87 D3 **Atkarsk** Russia
141 D2 **Atlanta** U.S.A.
137 D2 **Atlantic** U.S.A.
141 D2 **Atlantic Beach** U.S.A.
139 E3 **Atlantic City** U.S.A.
55 D3 **Atlantic-Indian-Antarctic Basin** S. Atlantic Ocean
158 E8 **Atlantic-Indian Ridge** Southern Ocean
158 **Atlantic Ocean**
122 A3 **Atlantis** S. Africa
114 B1 **Atlas Mountains** Africa
114 C1 **Atlas Saharien** *mts* Alg.
128 A2 **Atlin** Can.
128 A2 **Atlin Lake** Can.
140 C2 **Atmore** U.S.A.
152 B2 **Atocha** Bol.
143 D2 **Atoka** U.S.A.
75 C2 **Atrai** *r.* India
80 B2 **Aţ Ţafīlah** Jordan
78 B2 **Aţ Ţā'if** Saudi Arabia
63 B2 **Attapu** Laos
130 B2 **Attawapiskat** Can.
130 B2 **Attawapiskat** *r.* Can.
130 B2 **Attawapiskat Lake** Can.
100 C2 **Attendorn** Ger.
100 A3 **Attichy** France
116 B2 **Aţ Ţūr** Egypt
78 B3 **At Turbah** Yemen
135 B3 **Atwater** U.S.A.
76 B2 **Atyrau** Kazakh.
105 D3 **Aubagne** France
105 C3 **Aubenas** France
126 D2 **Aubry Lake** Can.
140 C2 **Auburn** *AL* U.S.A.
135 B3 **Auburn** *CA* U.S.A.
138 B2 **Auburn** *IN* U.S.A.
139 E2 **Auburn** *ME* U.S.A.
137 D2 **Auburn** *NE* U.S.A.
139 D2 **Auburn** *NY* U.S.A.
104 C2 **Aubusson** France
104 C3 **Auch** France
54 B1 **Auckland** N.Z.
48 H9 **Auckland Islands** N.Z.
117 C4 **Audo** *mts* Eth.
101 F2 **Aue** Ger.
101 F2 **Auerbach** Ger.
102 C2 **Augsburg** Ger.
50 A3 **Augusta** Austr.
109 C3 **Augusta** Italy
141 D2 **Augusta** *GA* U.S.A.
137 D3 **Augusta** *KS* U.S.A.
139 F2 **Augusta** *ME* U.S.A.
155 D1 **Augusto de Lima** Brazil
103 E1 **Augustów** Pol.
50 A2 **Augustus, Mount** Austr.
100 A2 **Aulnoye-Aymeries** France
Aumale Alg. *see* **Sour el Ghozlane**
62 A2 **Aunglan** Myanmar
122 B2 **Auob** *watercourse* Namibia/S. Africa
131 D2 **Aupaluk** Can.
74 B3 **Aurangabad** India
104 B2 **Auray** France
100 C1 **Aurich** Ger.
154 B1 **Aurilândia** Brazil
104 C3 **Aurillac** France
136 C3 **Aurora** *CO* U.S.A.
138 B2 **Aurora** *IL* U.S.A.
137 E3 **Aurora** *MO* U.S.A.
137 D2 **Aurora** *NE* U.S.A.
122 A2 **Aus** Namibia
138 C2 **Au Sable** *r.* U.S.A.
137 E2 **Austin** *MN* U.S.A.
135 C3 **Austin** *NV* U.S.A.
143 D2 **Austin** *TX* U.S.A.
Australes, Îles *is* Fr. Polynesia *see* **Tubuai Islands**
50 B2 **Australia** *country* Oceania
55 K4 **Australian-Antarctic Basin** *sea feature* Southern Ocean
53 C3 **Australian Capital Territory** *admin. div.* Austr.
102 C2 **Austria** *country* Europe
150 D3 **Autazes** Brazil
144 B3 **Autlán** Mex.
105 C2 **Autun** France
104 C3 **Auvergne** *reg.* France
105 C2 **Auvergne, Monts d'** *mts* France
105 C2 **Auxerre** France
105 D2 **Auxonne** France
105 C2 **Avallon** France
131 E3 **Avalon Peninsula** Can.
154 C2 **Avaré** Brazil
91 D2 **Avdiyivka** Ukr.
106 B1 **Aveiro** Port.
109 B2 **Avellino** Italy
108 B2 **Aversa** Italy
100 A2 **Avesnes-sur-Helpe** France
93 G3 **Avesta** Sweden
104 C3 **Aveyron** *r.* France
108 B2 **Avezzano** Italy
96 C2 **Aviemore** U.K.
109 C2 **Avigliano** Italy
105 C3 **Avignon** France
106 C1 **Ávila** Spain
106 B1 **Avilés** Spain
52 B3 **Avoca** Austr.
139 D2 **Avoca** U.S.A.
109 C3 **Avola** Italy
99 B3 **Avon** *r. England* U.K.
99 B4 **Avon** *r. England* U.K.
99 C4 **Avon** *r. England* U.K.
142 A2 **Avondale** U.S.A.
104 B2 **Avranches** France
54 B1 **Awanui** N.Z.
78 B3 **Awārik, 'Urūq al** *des.* Saudi Arabia
119 D2 **Āwasa** Eth.
117 C4 **Āwash** Eth.
117 C3 **Āwash** *r.* Eth.
115 D2 **Awbārī** Libya
115 D2 **Awbārī, Idhān** *des.* Libya
117 C4 **Aw Dheegle** Somalia
96 B2 **Awe, Loch** *l.* U.K.
117 A4 **Aweil** S. Sudan
115 C4 **Awka** Nigeria
126 F1 **Axel Heiberg Island** Can.
114 B4 **Axim** Ghana
150 B4 **Ayacucho** Peru
77 E2 **Ayagoz** Kazakh.
Ayaguz Kazakh. *see* **Ayagoz**
68 B2 **Ayakkum Hu** *salt l.* China
106 B2 **Ayamonte** Spain
83 K3 **Ayan** Russia
150 B4 **Ayaviri** Peru
76 A2 **Aybas** Kazakh.
91 D2 **Aydar** *r.* Ukr.
77 C2 **Aydarko'l ko'li** *l.* Uzbek.
111 C3 **Aydın** Turkey
Ayers Rock *h.* Austr. *see* **Uluru**
83 I2 **Aykhal** Russia
99 C4 **Aylesbury** U.K.
106 C1 **Ayllón** Spain
129 D1 **Aylmer Lake** Can.
117 B4 **Ayod** S. Sudan
83 M2 **Ayon, Ostrov** *i.* Russia
114 B3 **'Ayoûn el 'Atroûs** Maur.
51 D1 **Ayr** Austr.
96 B3 **Ayr** U.K.
98 A2 **Ayre, Point of** Isle of Man
76 C2 **Ayteke Bi** Kazakh.
110 C2 **Aytos** Bulg.
145 C3 **Ayutla** Mex.
63 B2 **Ayutthaya** Thai.
111 C3 **Ayvacık** Turkey
111 C3 **Ayvalık** Turkey
91 C3 **Ayya, Mys** *pt* Ukr.
107 C2 **Azahar, Costa del** *coastal area* Spain
Azania *reg.* Somalia *see* **Jubaland**
114 B3 **Azaouâd** *reg.* Mali
114 C3 **Azaouagh, Vallée de** *watercourse* Mali/Niger
115 D3 **Azare** Nigeria
77 C2 **Azat, Gory** *h.* Kazakh.
114 B3 **Azawad** *reg.* Mali
Azbine *mts* Niger *see* **Aïr, Massif de l'**
81 C1 **Azerbaijan** *country* Asia
Azerbaydzhanskaya S.S.R. *country* Asia *see* **Azerbaijan**
77 C1 **Azhibeksor, Ozero** *salt l.* Kazakh.
150 B3 **Azogues** Ecuador
86 D2 **Azopol'ye** Russia
112 A2 **Azores** *aut. reg.* Port.
91 D2 **Azov** Russia
91 D2 **Azov, Sea of** Russia/Ukr.
Azraq, Bahr el *r.* Sudan *see* **Blue Nile**
106 B2 **Azuaga** Spain
146 B4 **Azuero, Península de** *pen.* Panama
153 C3 **Azul** Arg.
105 D3 **Azur, Côte d'** *coastal area* France
108 A3 **Azzaba** Alg.
Az Zahrān Saudi Arabia *see* **Dhahran**
80 B2 **Az Zaqāzīq** Egypt
80 B2 **Az Zarqā'** Jordan
115 D1 **Az Zāwiyah** Libya
78 B3 **Az Zaydīyah** Yemen
114 C2 **Azzel Matti, Sebkha** *salt pan* Alg.
78 B2 **Az Zilfī** Saudi Arabia
78 B3 **Az Zuqur** *i.* Yemen

B

63 B2 **Ba, Sông** *r.* Vietnam
117 C4 **Baardheere** Somalia
77 C3 **Bābā, Kōh-e** *mts* Afgh.
111 C3 **Baba Burnu** *pt* Turkey
110 C2 **Babadag** Romania
111 C2 **Babaeski** Turkey
116 C3 **Bāb al Mandab** *str.* Africa/Asia
61 C2 **Babana** Indon.
119 C1 **Babanusa** Sudan
59 C3 **Babar** *i.* Indon.
119 D3 **Babati** Tanz.
89 E2 **Babayevo** Russia
59 C2 **Babeldaob** *i.* Palau
128 B2 **Babine** *r.* Can.
128 B2 **Babine Lake** Can.
59 C3 **Babo** Indon.
81 D2 **Bābol** Iran
122 A3 **Baboon Point** S. Africa
88 C3 **Babruysk** Belarus
Babu China *see* **Hezhou**
64 B2 **Babuyan** *i.* Phil.
64 B2 **Babuyan Channel** Phil.
64 B2 **Babuyan Islands** Phil.
151 E3 **Bacabal** Brazil
145 D3 **Bacalar** Mex.
59 C3 **Bacan** *i.* Indon.
110 C1 **Bacău** Romania
52 B3 **Bacchus Marsh** Austr.
62 B1 **Băc Giang** Vietnam
77 D3 **Bachu** China
129 E1 **Back** *r.* Can.
109 C1 **Bačka Palanka** Serbia
109 C1 **Bačka Topola** Serbia
52 A3 **Backstairs Passage** Austr.
63 B3 **Bac Liêu** Vietnam
142 B3 **Bacobampo** Mex.
64 B2 **Bacolod** Phil.
130 C2 **Bacqueville, Lac** *l.* Can.
Bada China *see* **Xilin**
70 A1 **Badain Jaran Shamo** *des.* China
106 B2 **Badajoz** Spain
Badaojiang China *see* **Baishan**
75 D2 **Badarpur** India
101 F1 **Bad Belzig** Ger.
101 E2 **Bad Berka** Ger.
101 D2 **Bad Berleburg** Ger.
101 E1 **Bad Bevensen** Ger.
101 D3 **Bad Dürkheim** Ger.
100 C2 **Bad Ems** Ger.
103 D2 **Baden** Austria
102 B2 **Baden-Baden** Ger.
101 D1 **Bad Fallingbostel** Ger.
101 E2 **Bad Harzburg** Ger.
101 D2 **Bad Hersfeld** Ger.
102 C2 **Bad Hofgastein** Austria
101 D2 **Bad Homburg vor der Höhe** Ger.
101 D1 **Bad Iburg** Ger.
74 A2 **Badin** Pak.
Bādiyat ash Shām *des.* Asia *see* **Syrian Desert**
101 E2 **Bad Kissingen** Ger.
100 C3 **Bad Kreuznach** Ger.
136 C1 **Badlands** *reg. ND* U.S.A.
136 C2 **Badlands** *reg. SD* U.S.A.
101 E2 **Bad Lauterberg im Harz** Ger.

52 B2 Barmera Austr.
99 A3 Barmouth U.K.
101 D1 Barmstedt Ger.
98 C2 Barnard Castle U.K.
53 B2 Barnato Austr.
82 G3 Barnaul Russia
127 H2 Barnes Icecap Can.
100 B1 Barneveld Neth.
98 C3 Barnsley U.K.
99 A4 Barnstaple U.K.
99 A4 Barnstaple Bay U.K.
141 D2 Barnwell U.S.A.
Baroda India see Vadodara
150 C1 Barquisimeto Venez.
96 A2 Barra i. U.K.
53 D2 Barraba Austr.
151 D4 Barra do Bugres Brazil
151 E3 Barra do Corda Brazil
154 B1 Barra do Garças Brazil
150 D3 Barra do São Manuel Brazil
Barraigh i. U.K. see Barra
150 B4 Barranca Lima Peru
150 B3 Barranca Loreto Peru
152 C2 Barranqueras Arg.
150 B1 Barranquilla Col.
105 D3 Barre des Écrins mt. France
151 E4 Barreiras Brazil
63 A2 Barren Island India
154 C2 Barretos Brazil
128 C2 Barrhead Can.
130 C3 Barrie Can.
128 B2 Barrière Can.
52 B2 Barrier Range hills Austr.
53 D2 Barrington, Mount Austr.
129 C2 Barrington Lake Can.
53 C1 Barringun Austr.
97 C2 Barrow r. Ireland
126 B2 Barrow U.S.A.
126 B2 Barrow, Point U.S.A.
51 C2 Barrow Creek Austr.
98 B2 Barrow-in-Furness U.K.
50 A2 Barrow Island Austr.
126 F2 Barrow Strait Can.
99 B4 Barry U.K.
122 B3 Barrydale S. Africa
130 C3 Barrys Bay Can.
74 B2 Barsalpur India
101 D1 Barsinghausen Ger.
135 C4 Barstow U.S.A.
105 C2 Bar-sur-Aube France
102 C1 Barth Ger.
80 B1 Bartın Turkey
51 D1 Bartle Frere, Mount Austr.
143 D1 Bartlesville U.S.A.
137 D2 Bartlett NE U.S.A.
140 C1 Bartlett TN U.S.A.
98 C3 Barton-upon-Humber U.K.
103 E1 Bartoszyce Pol.
61 C2 Barung i. Indon.
69 D1 Baruun-Urt Mongolia
91 D2 Barvinkove Ukr.
53 C2 Barwon r. Austr.
88 C3 Barysaw Belarus
118 B2 Basankusu Dem. Rep. Congo
110 C2 Basarabi Romania
64 A1 Basco Phil.
105 D2 Basel Switz.
71 C3 Bashi Channel Phil./Taiwan
91 C2 Bashtanka Ukr.
64 B3 Basilan i. Phil.
64 B3 Basilan i. Phil.
99 D4 Basildon U.K.
99 C4 Basingstoke U.K.
81 C2 Başkale Turkey
130 C2 Baskatong, Réservoir resr Can.
Basle Switz. see Basel
118 C2 Basoko Dem. Rep. Congo
81 C2 Basra Iraq
128 C2 Bassano Can.
114 C4 Bassar Togo
51 D3 Bassein Myanmar
147 D3 Basse-Terre Guadeloupe
147 D3 Basseterre St Kitts and Nevis
114 B3 Bassikounou Maur.
114 C4 Bassila Benin
51 D3 Bass Strait Austr.
79 C2 Bastak Iran
101 E2 Bastheim Ger.
75 C2 Basti India
105 D3 Bastia France
100 B2 Bastogne Belgium
140 B2 Bastrop U.S.A.
Basuo China see Dongfang
Basutoland country Africa see Lesotho
118 A2 Bata Equat. Guinea
146 B2 Batabanó, Golfo de b. Cuba
83 J2 Batagay Russia
154 B2 Bataguassu Brazil
74 B1 Batala India
106 B2 Batalha Port.
60 B1 Batam Indon.
64 B1 Batan i. Phil.
118 B2 Batangafo C.A.R.
64 B2 Batangas Phil.
64 B2 Batanghari r. Indon.
64 B1 Batan Islands Phil.
141 C2 Batataís Brazil
139 D2 Batavia U.S.A.
91 D2 Bataysk Russia
130 B3 Batchawana Mountain h. Can.
50 C1 Batchelor Austr.
63 B2 Bătdâmbâng Cambodia
118 B3 Batéké, Plateaux Congo
53 D3 Batemans Bay Austr.
140 B1 Batesville AR U.S.A.

140 C2 Batesville MS U.S.A.
89 D2 Batetskiy Russia
99 B4 Bath U.K.
96 C3 Bathgate U.K.
74 B1 Bathinda India
53 C2 Bathurst Austr.
131 D3 Bathurst Can.
Bathurst Gambia see Banjul
126 E2 Bathurst Inlet inlet Can.
50 C1 Bathurst Island Austr.
126 F1 Bathurst Island Can.
78 B1 Bāţin, Wādī al watercourse Asia
53 C3 Batlow Austr.
81 C2 Batman Turkey
115 C1 Batna Alg.
140 B2 Baton Rouge U.S.A.
144 B2 Batopilas Mex.
118 B2 Batouri Cameroon
154 B1 Batovi Brazil
92 I1 Båtsfjord Norway
73 C4 Batticaloa Sri Lanka
109 B2 Battipaglia Italy
128 D2 Battle r. Can.
138 B2 Battle Creek U.S.A.
135 C2 Battle Mountain U.S.A.
74 B1 Battura Glacier Pak.
117 B4 Batu mt. Eth.
60 A2 Batu, Pulau-pulau is Indon.
61 D2 Batuata i. Indon.
61 D2 Batudaka i. Indon.
64 B3 Batulaki Phil.
Batum Georgia see Bat'umi
81 C1 Bat'umi Georgia
60 B1 Batu Pahat Malaysia
61 D2 Baubau Indon.
115 C3 Bauchi Nigeria
137 E1 Baudette U.S.A.
Baudouinville Dem. Rep. Congo see Moba
104 B2 Baugé-en-Anjou France
105 D2 Baume-les-Dames France
154 C2 Bauru Brazil
154 B1 Baús Brazil
88 B2 Bauska Latvia
102 C1 Bautzen Ger.
144 B2 Bavispe r. Mex.
87 E3 Bavly Russia
62 A1 Bawdwin Myanmar
61 C2 Bawean i. Indon.
114 B3 Bawku Ghana
146 C2 Bayamo Cuba
Bayan Gol China see Dengkou
68 C1 Bayanhongor Mongolia
70 A2 Bayan Hot China
70 A1 Bayannur China
70 B1 Bayan Obo China
69 D2 Bayan Shutu China
69 D1 Bayan-Uul Mongolia
136 C2 Bayard NE U.S.A.
142 B2 Bayard NM U.S.A.
64 B3 Bayawan Phil.
81 C1 Bayburt Turkey
138 C2 Bay City MI U.S.A.
143 D3 Bay City TX U.S.A.
86 F2 Baydaratskaya Guba Russia
117 C4 Baydhabo Somalia
102 C2 Bayern reg. Ger.
104 B2 Dayeux France
136 B3 Bayfield U.S.A.
78 B3 Bayhān al Qişāb Yemen
Bay Islands is Hond. see Bahía, Islas de la
81 C2 Bayjī Iraq
Baykal, Ozero l. Russia see Baikal, Lake
Baykal Range mts Russia see Baykal'skiy Khrebet
83 I3 Baykal'skiy Khrebet mts Russia
76 C2 Baykonyr Kazakh.
87 E3 Baymak Russia
64 B2 Bayombong Phil.
104 B3 Bayonne France
76 C3 Baýramaly Turkm.
111 C3 Bayramiç Turkey
101 E3 Bayreuth Ger.
78 B3 Bayt al Faqīh Yemen
143 D3 Baytown U.S.A.
106 C2 Baza Spain
106 C2 Baza, Sierra de mts Spain
74 B1 Bāzā'ī Gunbad Afgh.
74 A1 Bāzārak Afgh.
76 A2 Bazardyuzyu, Gora mt. Azer./Russia
104 B3 Bazas France
74 A2 Bazdar Pak.
70 A2 Bazhong China
79 D2 Bazmān Iran
79 D2 Bazmān, Kūh-e mt. Iran
Bé, Nossi i. Madag. see Bé, Nosy
121 □D2 Bé, Nosy i. Madag.
136 C1 Beach U.S.A.
52 B3 Beachport Austr.
99 D4 Beachy Head hd U.K.
123 C3 Beacon Bay S. Africa
50 B1 Beagle Gulf Austr.
121 □D2 Bealanana Madag.
97 B1 Béal an Mhuirthead Ireland
121 □D3 Beampingaratra mts Madag.
134 D2 Bear r. U.S.A.
130 B3 Beardmore Can.
126 C2 Bear Island i. Arctic Ocean see Bjørnøya
134 E1 Bear Paw Mountain U.S.A.

147 C3 Beata, Cabo c. Dom. Rep.
147 C3 Beata, Isla i. Dom. Rep.
137 D2 Beatrice U.S.A.
135 C3 Beatty U.S.A.
53 D1 Beaudesert Austr.
52 B3 Beaufort Austr.
61 C1 Beaufort Malaysia
141 D2 Beaufort U.S.A.
126 D2 Beaufort Sea Can./U.S.A.
122 B3 Beaufort West S. Africa
96 B2 Beauly U.K.
96 B2 Beauly r. U.K.
100 B2 Beaumont Belgium
54 A3 Beaumont N.Z.
143 E2 Beaumont U.S.A.
105 C2 Beaune France
100 B2 Beauraing Belgium
129 E2 Beauséjour Can.
104 C2 Beauvais France
129 D2 Beauval Can.
129 D2 Beaver r. Can.
135 D3 Beaver U.S.A.
128 A1 Beaver Creek Can.
138 B2 Beaver Dam U.S.A.
129 E2 Beaver Hill Lake Can.
138 B1 Beaver Island U.S.A.
128 C2 Beaverlodge Can.
74 B2 Beawar India
154 C2 Bebedouro Brazil
101 D2 Bebra Ger.
99 D3 Beccles U.K.
109 D1 Bečej Serbia
106 B1 Becerreá Spain
114 B1 Béchar Alg.
138 C3 Beckley U.S.A.
117 B4 Bedelë Eth.
99 C3 Bedford U.K.
138 B3 Bedford U.S.A.
98 C2 Bedlington U.K.
100 C1 Bedum Neth.
53 C3 Beechworth Austr.
53 D2 Beecroft Peninsula Austr.
101 F1 Beelitz Ger.
53 D1 Beenleigh Austr.
80 B2 Beersheba Israel
Be'ér Sheva' Israel see Beersheba
143 D3 Beeville U.S.A.
121 □D2 Befandriana Avaratra Madag.
53 C3 Bega Austr.
107 D1 Begur, Cap de c. Spain
81 D2 Behbahān Iran
128 C1 Behchokò Can.
81 D2 Behshahr Iran
69 E1 Bei'an China
71 A3 Beihai China
70 B2 Beijing China
100 C1 Beilen Neth.
118 B2 Béinamar Chad
96 B3 Beinn an Oir h. U.K.
96 A2 Beinn Mhòr h. U.K.
Beinn na Faoghla i. U.K. see Benbecula
121 C2 Beira Moz.
80 B2 Beirut Lebanon
123 C1 Beitbridge Zimbabwe
106 B2 Beja Port.
115 C1 Bejaïa Alg.
106 B1 Béjar Spain
74 A2 Beji r. Pak.
103 E2 Békés Hungary
103 E2 Békéscsaba Hungary
121 □D3 Bekily Madag.
114 B4 Bekwai Ghana
75 C2 Bela India
74 A2 Bela Pak.
123 C1 Bela-Bela S. Africa
118 B2 Bélabo Cameroon
109 D2 Bela Crkva Serbia
61 C1 Belaga Malaysia
Belagavi India see Belgaum
88 C3 Belarus country Europe
121 C3 Bela Vista Moz.
60 A1 Belawan Indon.
83 M2 Belaya r. Russia
Belchatow Pol. see Bełchatów
103 D1 Bełchatów Pol.
130 C2 Belcher Islands Can.
87 E3 Belebey Russia
117 C4 Beledweyne Somalia
118 B2 Bélèl Cameroon
151 E3 Belém Brazil
142 B2 Belen U.S.A.
110 C2 Belene Bulg.
89 E3 Belev Russia
97 C1 Belfast U.K.
139 F2 Belfast U.S.A.
136 C1 Belfield U.S.A.
105 D2 Belfort France
73 B3 Belgaum India
Belgian Congo country Africa see Congo, Democratic Republic of the
100 B2 Belgium country Europe
91 D1 Belgorod Russia
109 D2 Belgrade Serbia
134 D1 Belgrade U.S.A.
109 C1 Beli Manastir Croatia
60 B2 Belinyu Indon.
60 B2 Belitung i. Indon.
118 B3 Belize Angola
146 B3 Belize Belize
146 B3 Belize country Central America
83 K1 Bel'kovskiy, Ostrov i. Russia

128 B2 Bella Bella Can.
104 C2 Bellac France
128 B2 Bella Coola Can.
73 B3 Bellary India
53 C1 Bellata Austr.
138 C2 Bellefontaine U.S.A.
136 C2 Belle Fourche U.S.A.
136 C2 Belle Fourche r. U.S.A.
141 D3 Belle Glade U.S.A.
104 B2 Belle-Île i. France
131 E2 Belle Isle i. Can.
131 E2 Belle Isle, Strait of Can.
130 C3 Belleville Can.
138 B3 Belleville IL U.S.A.
137 D3 Belleville KS U.S.A.
134 D2 Bellevue ID U.S.A.
134 B1 Bellevue WA U.S.A.
Bellin Can. see Kangirsuk
53 C2 Bellingen Austr.
134 B1 Bellingham U.S.A.
55 R2 Bellingshausen Sea Antarctica
105 D2 Bellinzona Switz.
96 C2 Bell Rock i. U.K.
108 B1 Belluno Italy
122 A3 Bellville S. Africa
53 D2 Belmont Austr.
155 E1 Belmonte Brazil
146 B3 Belmopan Belize
Belmullet Ireland see Béal an Mhuirthead
69 E1 Belogorsk Russia
121 □D3 Beloha Madag.
155 D1 Belo Horizonte Brazil
138 B2 Beloit U.S.A.
86 C2 Belomorsk Russia
89 E3 Beloomut Russia
91 D3 Belorechensk Russia
Belorechenskaya Russia see Belorechensk
87 E3 Beloretsk Russia
Belorussia country Europe see Belarus
Belorusskaya S.S.R. country Europe see Belarus
Belostok Pol. see Białystok
121 □D2 Belo Tsiribihina Madag.
86 F2 Beloyarskiy Russia
89 E1 Beloye, Ozero l. Russia
Beloye More sea Russia see White Sea
89 E1 Belozersk Russia
52 A2 Beltana Austr.
143 D2 Belton U.S.A.
Bel'ts' Moldova see Bălţi
Bel'tsy Moldova see Bălţi
97 C1 Belturbet Ireland
77 E2 Belukha, Gora mt. Kazakh./Russia
86 D2 Belush'ye Russia
138 B2 Belvidere U.S.A.
51 D2 Belyando r. Austr.
89 D2 Belyy Russia
82 F2 Belyy, Ostrov i. Russia
137 E1 Bemidji U.S.A.
118 C3 Bena Dibele Dem. Rep. Congo
53 C3 Benalla Austr.
Benares India see Varanasi
115 D1 Ben Arous Tunisia
118 C3 Bena-Sungu Dem. Rep. Congo
106 B1 Benavente Spain
96 A2 Benbecula i. U.K.
134 B2 Bend U.S.A.
123 C3 Bendearg mt. S. Africa
90 B2 Bender Moldova
Bendery Moldova see Bender
52 B3 Bendigo Austr.
121 C2 Bene Moz.
102 C2 Benešov Czech Rep.
109 B2 Benevento Italy
73 C3 Bengal, Bay of sea Indian Ocean
Bengaluru India see Bangalore
70 B2 Bengbu China
115 E1 Benghazi Libya
60 B1 Bengkalis Indon.
60 B1 Bengkayang Indon.
60 B2 Bengkulu Indon.
120 A2 Benguela Angola
Benha Egypt see Banhā
96 B2 Ben Hope h. U.K.
152 B1 Beni r. Bol.
119 C2 Beni Dem. Rep. Congo
114 B1 Beni Abbès Alg.
107 C2 Benidorm Spain
114 B1 Beni Mellal Morocco
114 C3 Benin country Africa
114 C4 Benin, Bight of g. Africa
115 C4 Benin City Nigeria
107 C2 Beni Saf Alg.
Beni Suef Egypt see Banī Suwayf
153 C3 Benito Juárez Arg.
150 C3 Benjamin Constant Brazil
144 A1 Benjamín Hill Mex.
59 C3 Benjina Indon.
136 C2 Benkelman U.S.A.
96 B2 Ben Lawers mt. U.K.
96 B2 Ben Lomond h. U.K.
96 C2 Ben Macdui mt. U.K.
96 A2 Ben More mt. U.K.
96 B2 Ben More mt. U.K.
54 B2 Benmore, Lake N.Z.
96 B1 Ben More Assynt h. U.K.
128 A2 Bennett (abandoned) Can.
83 K1 Bennetta, Ostrov i. Russia
Bennett Island Russia see Bennetta, Ostrov
96 B2 Ben Nevis mt. U.K.

Breytovo

151 E3 Boa Esperança, Açude resr Brazil
134 C1 Boardman U.S.A.
123 C1 Boatlaname Botswana
151 F3 Boa Viagem Brazil
150 C2 Boa Vista Brazil
53 C2 Bobadah Austr.
71 B3 Bobai China
121 □D2 Bobaomby, Tanjona c. Madag.
114 B3 Bobo-Dioulasso Burkina Faso
121 B3 Bobonong Botswana
Bobriki Russia see Novomoskovsk
89 F3 Bobrov Russia
91 C1 Bobrovytsya Ukr.
91 C2 Bobrynets' Ukr.
121 □D3 Boby mt. Madag.
150 C3 Boca do Acre Brazil
155 D1 Bocaiúva Brazil
154 A2 Bocajá Brazil
118 B2 Bocaranga C.A.R.
141 D3 Boca Raton U.S.A.
146 B4 Bocas del Toro Panama
103 E2 Bochnia Pol.
100 B2 Bocholt Belgium
100 C2 Bocholt Ger.
100 C2 Bochum Ger.
101 E1 Bockenem Ger.
110 B1 Bocşa Romania
118 B2 Boda C.A.R.
83 I3 Bodaybo Russia
96 D2 Boddam U.K.
115 D3 Bodélé reg. Chad
92 H2 Boden Sweden
Bodensee l. Ger./Switz. see Constance, Lake
99 A4 Bodmin U.K.
99 A4 Bodmin Moor moorland U.K.
92 F2 Bodø Norway
111 C3 Bodrum Turkey
118 C3 Boende Dem. Rep. Congo
63 A2 Bogale Myanmar
140 C2 Bogalusa U.S.A.
114 B3 Bogandé Burkina Faso
118 B2 Bogangolo C.A.R.
80 B2 Boğazlıyan Turkey
68 B2 Bogda Shan mts China
53 D1 Boggabilla Austr.
53 D2 Boggabri Austr.
97 B2 Boggeragh Mountains hills Ireland
Boghari Alg. see Ksar el Boukhari
59 D3 Bogia P.N.G.
100 B3 Bogny-sur-Meuse France
97 C2 Bog of Allen reg. Ireland
53 C3 Bogong, Mount Austr.
60 B2 Bogor Indon.
89 E3 Bogoroditsk Russia
150 B2 Bogotá Col.
83 G3 Bogotol Russia
Bogoyavlenskoye Russia see Pervomayskiy
83 H3 Boguchany Russia
91 E2 Boguchar Russia
114 A3 Bogué Maur.
70 B2 Bo Hai g. China
100 A3 Bohain-en-Vermandois France
70 B2 Bohai Wan b. China
Bohemian Forest mts Ger. see Böhmer Wald
123 C2 Bohlokong S. Africa
101 F3 Böhmer Wald mts Ger.
91 D1 Bohodukhiv Ukr.
64 B3 Bohol i. Phil.
64 B3 Bohol Sea Phil.
77 E2 Bohu China
155 C2 Boi, Ponta do pt Brazil
123 C2 Boikhutso S. Africa
154 B3 Boi Preto, Serra de hills Brazil
154 B1 Bois r. Brazil
126 D2 Bois, Lac des l. Can.
134 C2 Boise U.S.A.
143 C1 Boise City U.S.A.
129 D3 Boissevain Can.
123 C2 Boitumelong S. Africa
154 C2 Boituva Brazil
101 E1 Boizenburg/Elbe Ger.
76 B3 Bojnūrd Iran
75 C2 Bokaro India
118 B3 Bokatola Dem. Rep. Congo
114 A3 Boké Guinea
118 C3 Bokele Dem. Rep. Congo
93 E4 Boknafjorden sea chan. Norway
115 D3 Bokoro Chad
63 A2 Bokpyin Myanmar
89 D2 Boksitogorsk Russia
122 B2 Bokspits Botswana
118 C3 Bokungu Dem. Rep. Congo
115 D3 Bol Chad
114 A3 Bolama Guinea-Bissau
63 B2 Bolavén, Phouphiang plat. Laos
104 C2 Bolbec France
77 E2 Bole China
118 B3 Boleko Dem. Rep. Congo
114 B3 Bolgatanga Ghana
90 B2 Bolhrad Ukr.
66 B1 Boli China
118 B3 Bolia Dem. Rep. Congo
92 H3 Boliden Sweden
62 B2 Bolikhamxai Laos
110 C2 Bolintin-Vale Romania
137 E3 Bolivar MO U.S.A.
140 C1 Bolivar TN U.S.A.
150 B2 Bolívar, Pico mt. Venez.
152 B1 Bolivia country S. America
89 E3 Bolkhov Russia
105 C3 Bollène France
93 G3 Bollnäs Sweden
53 C1 Bollon Austr.
101 E2 Bollstedt Ger.
93 F4 Bolmen l. Sweden
118 B3 Bolobo Dem. Rep. Congo
108 B2 Bologna Italy
89 D2 Bologovo Russia
89 D2 Bologoye Russia
123 C2 Bolokanang S. Africa
118 B2 Bolomba Dem. Rep. Congo
108 B2 Bolsena, Lago di l. Italy
83 H1 Bol'shevik, Ostrov i. Russia
86 E2 Bol'shezemel'skaya Tundra lowland Russia
66 B2 Bol'shoy Kamen' Russia
Bol'shoy Kavkaz mts Asia/Europe see Caucasus
83 K2 Bol'shoy Lyakhovskiy, Ostrov i. Russia
Bol'shoy Tokmak Kyrg. see Tokmok
Bol'shoy Tokmak Ukr. see Tokmak
100 B1 Bolsward Neth.
98 B3 Bolton U.K.
80 B1 Bolu Turkey
59 E3 Bolubolu P.N.G.
92 □A2 Bolungarvík Iceland
108 B1 Bolzano Italy
118 B3 Boma Dem. Rep. Congo
53 D2 Bomaderry Austr.
53 C3 Bombala Austr.
Bombay India see Mumbai
155 C1 Bom Despacho Brazil
75 D2 Bomdila India
154 B1 Bom Jardim de Goiás Brazil
151 E4 Bom Jesus da Lapa Brazil
155 D2 Bom Jesus do Itabapoana Brazil
115 D1 Bon, Cap c. Tunisia
147 D3 Bonaire mun. West Indies
134 C1 Bonaparte, Mount U.S.A.
50 B1 Bonaparte Archipelago is Austr.
131 E3 Bonavista Can.
131 E3 Bonavista Bay Can.
118 C2 Bondo Dem. Rep. Congo
114 B4 Bondoukou Côte d'Ivoire
Bône Alg. see Annaba
61 D2 Bonerate, Kepulauan is Indon.
155 C1 Bonfinópolis de Minas Brazil
117 B4 Bonga Eth.
75 D2 Bongaigaon India
118 C2 Bongandanga Dem. Rep. Congo
122 B2 Bongani S. Africa
118 C2 Bongo, Massif des mts C.A.R.
121 □D2 Bongolava mts Madag.
115 D3 Bongor Chad
114 B4 Bongouanou Côte d'Ivoire
63 B2 Bông Sơn Vietnam
143 D2 Bonham U.S.A.
105 D3 Bonifacio France
108 A2 Bonifacio, Strait of France/Italy
69 F3 Bonin Islands is Japan
100 C2 Bonn Ger.
134 C1 Bonners Ferry U.S.A.
105 D2 Bonneville France
50 A3 Bonnie Rock Austr.
129 C2 Bonnyville Can.
108 A2 Bonorva Italy
53 D1 Bonshaw Austr.
61 C1 Bontang Indon.
114 A4 Bonthe Sierra Leone
64 B2 Bontoc Phil.
61 C2 Bontosunggu Indon.
123 C3 Bontrug S. Africa
53 C1 Boolba Austr.
52 B2 Booligal Austr.
53 C1 Boomi Austr.
53 D1 Boonah Austr.
137 E2 Boone U.S.A.
140 C2 Booneville U.S.A.
137 E3 Boonville U.S.A.
52 B2 Booroorban Austr.
53 C2 Boorowa Austr.
117 C3 Boosaaso Somalia
126 G2 Boothia, Gulf of Can.
126 F2 Boothia Peninsula Can.
118 B3 Booué Gabon
100 C2 Boppard Ger.
144 B2 Boquilla, Presa de la resr Mex.
109 D2 Bor Serbia
117 B4 Bor S. Sudan
80 B2 Bor Turkey
119 E2 Bor, Lagh watercourse Kenya/Somalia
121 □E2 Boraha, Nosy i. Madag.
76 B2 Borankul Kazakh.
93 F4 Borås Sweden
81 D3 Borāzjān Iran
150 D3 Borba Brazil
104 B3 Bordeaux France
126 E1 Borden Island Can.
127 G2 Borden Peninsula Can.
52 B3 Bordertown Austr.
107 D2 Bordj Bou Arréridj Alg.
107 D2 Bordj Bounaama Alg.
114 B2 Bordj Flye Ste-Marie Alg.
115 C1 Bordj Messaouda Alg.
114 C2 Bordj Mokhtar Alg.
Bordj Omar Driss Alg. see Bordj Omer Driss
115 C2 Bordj Omer Driss Alg.
94 B1 Borðoy i. Faroe Is
Borgå Fin. see Porvoo
92 □A3 Borgarnes Iceland
143 C1 Borger U.S.A.
93 G4 Borgholm Sweden
100 B2 Borgloon Belgium
108 A1 Borgosesia Italy
87 D3 Borisoglebsk Russia
89 E2 Borisoglebskiy Russia
91 D1 Borisovka Russia
119 C2 Bo River S. Sudan
100 C2 Borken Ger.
92 G2 Borkenes Norway
100 C1 Borkum Ger.
100 C1 Borkum i. Ger.
93 G3 Borlänge Sweden
101 F2 Borna Ger.
100 C1 Borne Neth.
61 C1 Borneo i. Asia
93 F4 Bornholm i. Denmark
111 C3 Bornova Turkey
90 B1 Borodyanka Ukr.
77 E2 Borohoro Shan mts China
114 B3 Boron Mali
89 D2 Borovichi Russia
89 E2 Borovsk Russia
76 C1 Borovskoy Kazakh.
107 C2 Borriana Spain
51 C1 Borroloola Austr.
110 B1 Borşa Romania
76 B2 Borsakelmas sho'rxogi salt marsh Uzbek.
90 B2 Borshchiv Ukr.
69 D1 Borshchovochnyy Khrebet mts Russia
101 E1 Börßum Ger.
Bortala China see Bole
81 C2 Borūjerd Iran
65 B2 Boryeong S. Korea
90 A2 Boryslav Ukr.
90 C1 Boryspil' Ukr.
91 C1 Borzna Ukr.
69 D1 Borzya Russia
109 C1 Bosanska Dubica Bos. & Herz.
109 C1 Bosanska Gradiška Bos. & Herz.
109 C2 Bosanska Krupa Bos. & Herz.
109 C1 Bosanski Novi Bos. & Herz.
109 C2 Bosansko Grahovo Bos. & Herz.
123 C2 Boshof S. Africa
109 C2 Bosnia and Herzegovina country Europe
118 B2 Bosobolo Dem. Rep. Congo
111 C2 Bosporus str. Turkey
142 B2 Bosque U.S.A.
118 B2 Bossangoa C.A.R.
118 B2 Bossembélé C.A.R.
140 B2 Bossier City U.S.A.
122 A2 Bossiesvlei Namibia
68 B2 Bosten Hu l. China
99 C3 Boston U.K.
139 E2 Boston U.S.A.
140 B1 Boston Mountains U.S.A.
53 D2 Botany Bay Austr.
120 B3 Boteti r. Botswana
110 B2 Botev mt. Bulg.
80 A1 Botevgrad Bulg.
92 G3 Bothnia, Gulf of Fin./Sweden
110 C1 Botoşani Romania
70 B2 Botou China
123 C2 Botshabelo S. Africa
120 B3 Botswana country Africa
109 C3 Botte Donato, Monte mt. Italy
136 C1 Bottineau U.S.A.
100 C2 Bottrop Ger.
154 C2 Botucatu Brazil
114 B4 Bouaké Côte d'Ivoire
118 B2 Bouar C.A.R.
114 B1 Bou Arfa Morocco
131 D3 Bouctouche Can.
107 E2 Bougaa Alg.
48 G4 Bougainville Island P.N.G.
Bougie Alg. see Bejaïa
114 B3 Bougouni Mali
100 B3 Bouillon Belgium
107 D2 Bouira Alg.
114 A2 Boujdour Western Sahara
50 B3 Boulder Austr.
136 B2 Boulder CO U.S.A.
134 D1 Boulder MT U.S.A.
135 D3 Boulder City U.S.A.
Boulhaut Morocco see Ben Slimane
51 C2 Boulia Austr.
104 C2 Boulogne-Billancourt France
104 C1 Boulogne-sur-Mer France
118 C2 Boulouba C.A.R.
118 B3 Boumango Gabon
118 B2 Boumba r. Cameroon
107 D2 Boumerdès Alg.
114 B4 Bouna Côte d'Ivoire
114 B4 Boundiali Côte d'Ivoire
134 D2 Bountiful U.S.A.
49 I8 Bounty Islands N.Z.
114 B3 Bourem Mali
104 C2 Bourganeuf France
105 D2 Bourg-en-Bresse France
104 C2 Bourges France
Bourgogne reg. France see Burgundy
105 D2 Bourgoin-Jallieu France
53 C2 Bourke Austr.
99 C3 Bourne U.K.
99 C4 Bournemouth U.K.
118 C1 Bourtoutou Chad
115 C1 Bou Saâda Alg.
115 D3 Bousso Chad
100 A2 Boussu Belgium
114 A3 Boutilimit Maur.
128 C3 Bow r. Can.
Bowa China see Muli
51 D2 Bowen Austr.
53 C3 Bowen, Mount Austr.
129 C3 Bow Island Can.
138 B3 Bowling Green KY U.S.A.
137 E3 Bowling Green MO U.S.A.
138 C2 Bowling Green OH U.S.A.
136 C1 Bowman U.S.A.
53 D2 Bowral Austr.
101 D3 Boxberg Ger.
100 B2 Boxtel Neth.
80 B1 Boyabat Turkey
Boyang China see Poyang
97 B2 Boyle Ireland
97 C2 Boyne r. Ireland
136 B2 Boysen Reservoir U.S.A.
152 B2 Boyuibe Bol.
111 C3 Bozburun Turkey
111 C3 Bozcaada i. Turkey
111 C3 Bozdağ mt. Turkey
111 C3 Boz Dağları mts Turkey
111 C3 Bozdoğan Turkey
134 D1 Bozeman U.S.A.
118 B2 Bozoum C.A.R.
111 D3 Bozüyük Turkey
109 C2 Brač i. Croatia
130 C3 Bracebridge Can.
93 G3 Bräcke Sweden
99 C4 Bracknell U.K.
109 C2 Bradano r. Italy
141 D3 Bradenton U.S.A.
147 B3 Brades Montserrat
98 C3 Bradford U.K.
139 D2 Bradford U.S.A.
143 D2 Brady U.S.A.
96 C2 Braemar U.K.
106 B1 Braga Port.
151 E3 Bragança Brazil
106 B1 Bragança Port.
155 C2 Bragança Paulista Brazil
89 D3 Brahin Belarus
75 D2 Brahmanbaria Bangl.
75 C3 Brahmapur India
62 A1 Brahmaputra r. China/India
53 C3 Braidwood Austr.
110 C1 Brăila Romania
137 E1 Brainerd U.S.A.
99 D4 Braintree U.K.
100 B2 Braives Belgium
101 D1 Brake (Unterweser) Ger.
122 A1 Brakwater Namibia
98 B2 Brampton U.K.
101 D1 Bramsche Ger.
150 C3 Branco r. Brazil
101 F1 Brandenburg an der Havel Ger.
129 E3 Brandon Can.
140 C2 Brandon U.S.A.
97 A2 Brandon Mountain h. Ireland
122 B3 Brandvlei S. Africa
103 D1 Braniewo Pol.
130 B3 Brantford Can.
53 D2 Branxton Austr.
131 D3 Bras d'Or Lake Can.
155 D1 Brasil, Planalto do plat. Brazil
154 C1 Brasília Brazil
155 D1 Brasília de Minas Brazil
88 C2 Braslaw Belarus
110 C1 Braşov Romania
103 D2 Bratislava Slovakia
83 H3 Bratsk Russia
102 C2 Braunau am Inn Austria
101 E1 Braunschweig Ger.
92 □A2 Brautarholt Iceland
Bravo del Norte, Río r. Mex./U.S.A. see Rio Grande
135 C4 Brawley U.S.A.
97 C2 Bray Ireland
150 D2 Brazil country S. America
158 E6 Brazil Basin S. Atlantic Ocean
154 C1 Brazlândia Brazil
143 D3 Brazos r. U.S.A.
118 B3 Brazzaville Congo
109 C2 Brčko Bos. & Herz.
96 C2 Brechin U.K.
100 B2 Brecht Belgium
143 D2 Breckenridge U.S.A.
103 D2 Břeclav Czech Rep.
99 B4 Brecon U.K.
99 B4 Brecon Beacons reg. U.K.
100 B2 Breda Neth.
122 B3 Bredasdorp S. Africa
102 B2 Bregenz Austria
92 H1 Breivikbotn Norway
92 E3 Brekstad Norway
101 D1 Bremen Ger.
101 D1 Bremerhaven Ger.
Bremersdorp Swaziland see Manzini
134 B1 Bremerton U.S.A.
101 D1 Bremervörde Ger.
143 D2 Brenham U.S.A.
105 E2 Brennero Italy
102 C2 Brenner Pass Austria/Italy
99 D4 Brentwood U.K.
108 B1 Brescia Italy
100 A2 Breskens Neth.
105 E2 Bressanone Italy
96 □ Bressay i. U.K.
104 B2 Bressuire France
88 B3 Brest Belarus
104 B2 Brest France
Brest-Litovsk Belarus see Brest
Bretagne reg. France see Brittany
140 C3 Breton Sound b. U.S.A.
151 D3 Breves Brazil
53 C1 Brewarrina Austr.
134 C1 Brewster U.S.A.
89 E2 Breytovo Russia

120 A2 Cubal Angola
120 B2 Cubango r. Angola/Namibia
150 B2 Cúcuta Col.
73 B3 Cuddalore India
Cuddapah India see Kadapa
50 A2 Cue Austr.
106 C1 Cuéllar Spain
120 A2 Cuemba Angola
150 B3 Cuenca Ecuador
107 C1 Cuenca Spain
107 C1 Cuenca, Serranía de mts Spain
145 C3 Cuernavaca Mex.
143 D3 Cuero U.S.A.
151 D4 Cuiabá Brazil
151 D4 Cuiabá r. Brazil
96 A2 Cuillin Sound sea chan. U.K.
120 A1 Cuilo Angola
120 B2 Cuito r. Angola
120 A2 Cuito Cuanavale Angola
60 B1 Cukai Malaysia
64 B2 Culasi Phil.
53 C3 Culcairn Austr.
100 B2 Culemborg Neth.
53 C1 Culgoa r. Austr.
144 B2 Culiacán Mex.
64 A2 Culion i. Phil.
107 C2 Cullera Spain
140 C2 Cullman U.S.A.
97 C1 Cullybackey U.K.
139 D3 Culpeper U.S.A.
151 D4 Culuene r. Brazil
54 B2 Culverden N.Z.
150 C1 Cumaná Venez.
139 D3 Cumberland U.S.A.
138 B3 Cumberland r. U.S.A.
141 D3 Cumberland Island U.S.A.
129 D2 Cumberland Lake Can.
127 H2 Cumberland Peninsula Can.
140 C1 Cumberland Plateau U.S.A.
127 H2 Cumberland Sound sea chan. Can.
96 C3 Cumbernauld U.K.
135 B3 Cummings U.S.A.
96 B3 Cumnock U.K.
144 B1 Cumpas Mex.
145 C3 Cunduacán Mex.
108 A2 Cuneo Italy
53 C1 Cunnamulla Austr.
108 A1 Cuorgnè Italy
96 C2 Cupar U.K.
110 B2 Ćuprija Serbia
147 D3 Curaçao terr. West Indies
150 B3 Curaray r. Ecuador
153 A3 Curicó Chile
154 C3 Curitiba Brazil
52 A2 Curnamona Austr.
135 C3 Currant U.S.A.
51 D3 Currie Austr.
135 D2 Currie U.S.A.
51 E2 Curtis Island Austr.
151 D3 Curuá r. Brazil
61 B2 Curup Indon.
151 E3 Cururupu Brazil
155 D1 Curvelo Brazil
150 B4 Cusco Peru
97 C1 Cushendun U.K.
143 D1 Cushing U.S.A.
136 C2 Custer U.S.A.
134 D1 Cut Bank U.S.A.
140 B3 Cut Off U.S.A.
75 C2 Cuttack India
120 A2 Cuvelai Angola
101 D1 Cuxhaven Ger.
64 B2 Cuyo Islands Phil.
Cuzco Peru see Cusco
99 B4 Cwmbrân U.K.
119 C3 Cyangugu Rwanda
111 B3 Cyclades is Greece
129 C3 Cypress Hills Can.
80 B2 Cyprus country Asia
80 B2 Cyprus i. Asia
102 C2 Czech Republic country Europe
103 D1 Czersk Pol.
103 D1 Częstochowa Pol.

D

Đa, Sông r. Vietnam see Black River
69 D2 Daban China
103 D2 Dabas Hungary
114 A3 Dabola Guinea
103 D1 Dąbrowa Górnicza Pol.
110 B2 Dăbuleni Romania
Dacca Bangl. see Dhaka
102 C2 Dachau Ger.
Dachuan China see Dazhou
119 E2 Dadaab Kenya
141 D3 Dade City U.S.A.
Dadong China see Donggang
Dadra India see Achalpur
74 B2 Dadra and Nagar Haveli union terr. India
74 A2 Dadu Pak.
65 B2 Daegu S. Korea
65 B2 Daejeon S. Korea
65 B3 Daejeong S. Korea
64 B2 Daet Phil.
114 A3 Dagana Senegal
119 D3 Daga Post S. Sudan
88 C2 Dagda Latvia
64 B2 Dagupan Phil.

Dahalach, Isole is Eritrea see Dahlak Archipelago
69 D2 Da Hinggan Ling mts China
116 C3 Dahlak Archipelago is Eritrea
100 C2 Dahlem Ger.
78 B3 Dahm, Ramlat des. Saudi Arabia/Yemen
74 B2 Dahod India
Dahomey country Africa see Benin
Dahra Senegal see Dara
81 C2 Dahūk/Dihok Iraq
60 B1 Daik Indon.
106 C2 Daimiel Spain
97 A2 Daingean Uí Chúis Ireland
Dairen China see Dalian
51 C2 Dajarra Austr.
70 A2 Dajing China
114 A3 Dakar Senegal
117 C4 Daketa Shet' watercourse Eth.
114 A2 Dakhla Western Sahara
Dakhla Oasis Egypt see Wāḩāt ad Dākhilah
63 A3 Dakoank India
88 C3 Dakol'ka r. Belarus
109 C1 Đakovo Croatia
120 A2 Dala Angola
70 A1 Dalain Hob China
93 G3 Dalälven r. Sweden
111 C3 Dalaman Turkey
111 C3 Dalaman r. Turkey
68 C2 Dalandzadgad Mongolia
64 B2 Dalanganem Islands Phil.
63 B2 Đa Lat Vietnam
Dalatando Angola see N'dalatando
74 A2 Dalbandin Pak.
96 C3 Dalbeattie U.K.
53 D1 Dalby Austr.
93 E3 Dale Norway
141 C1 Dale Hollow Lake U.S.A.
53 C3 Dalgety Austr.
143 C1 Dalhart U.S.A.
131 D3 Dalhousie Can.
62 B1 Dali China
70 C2 Dalian China
96 C3 Dalkeith U.K.
143 D2 Dallas U.S.A.
128 A2 Dall Island U.S.A.
Dalmacija reg. Bos. & Herz./Croatia see Dalmatia
96 B3 Dalmally U.K.
109 C2 Dalmatia reg. Bos. & Herz./Croatia
96 B3 Dalmellington U.K.
66 C2 Dal'negorsk Russia
66 B1 Dal'nerechensk Russia
Dalny China see Dalian
114 B4 Daloa Côte d'Ivoire
71 A3 Dalou Shan mts China
51 D2 Dalrymple, Mount Austr.
92 □A3 Dalsmynni Iceland
Daltenganj India see Daltonganj
141 D2 Dalton U.S.A.
75 C2 Daltonganj India
60 B1 Daludalu Indon.
71 B3 Daluo Shan mt. China
92 □B2 Dalvík Iceland
96 B2 Dalwhinnie U.K.
50 C1 Daly r. Austr.
51 C1 Daly Waters Austr.
74 B2 Daman India
74 B2 Daman and Diu union terr. India
116 B1 Damanhūr Egypt
59 C3 Damar i. Indon.
118 B2 Damara C.A.R.
80 B2 Damascus Syria
115 D3 Damaturu Nigeria
76 B3 Dāmāvand, Qolleh-ye mt. Iran
120 A1 Damba Angola
118 B1 Damboa Nigeria
81 D2 Dāmghān Iran
Damietta Egypt see Dumyāţ
79 B3 Dammam Saudi Arabia
101 D1 Damme Ger.
75 B2 Damoh India
114 B4 Damongo Ghana
50 A2 Dampier Austr.
59 C3 Dampir, Selat sea chan. Indon.
75 C2 Damqoq Zangbo r. China
Damxung China see Gongtang
117 C3 Danakil reg. Africa
114 B4 Danané Côte d'Ivoire
63 B2 Đa Nẵng Vietnam
139 E2 Danbury U.S.A.
70 C1 Dandong China
117 B3 Dangila Eth.
146 B3 Dangriga Belize
70 B2 Dangshan China
89 F2 Danilov Russia
89 E2 Danilovskaya Vozvyshennost' hills Russia
70 B2 Danjiangkou China
79 C2 Dank Oman
89 E3 Dankov Russia
146 B3 Danlí Hond.
101 E1 Dannenberg (Elbe) Ger.
54 C2 Dannevirke N.Z.
62 B2 Dan Sai Thai.
Dantu China see Zhenjiang
110 A1 Danube r. Europe
110 C2 Danube Delta Romania/Ukr.
138 B3 Danville IL U.S.A.
138 C3 Danville KY U.S.A.
139 D3 Danville VA U.S.A.

Danxian China see Danzhou
71 A4 Danzhou China
Danzig, Gulf of Pol./Russia see Gdańsk, Gulf of
Daojiang China see Daoxian
115 D2 Dao Timmi Niger
Daoud Alg. see Aïn Beïda
114 B4 Daoukro Côte d'Ivoire
71 B3 Daoxian China
64 B3 Dapa China
114 C3 Dapaong Togo
64 B3 Dapitan Phil.
68 C2 Da Qaidam China
69 E1 Daqing China
114 A3 Dara Senegal
80 B2 Dar'ā Syria
81 D3 Dārāb Iran
81 D2 Dārān Iran
Đaravica mt. Kosovo see Gjeravicë
75 C2 Darbhanga India
Dardo China see Kangding
119 D3 Dar es Salaam Tanz.
117 A3 Darfur reg. Sudan
74 B1 Dargai Pak.
54 B1 Dargaville N.Z.
53 C3 Dargo Austr.
68 D1 Darhan Mongolia
150 B1 Darién, Golfo del g. Col.
Darjeeling India see Darjiling
75 C2 Darjiling India
52 B2 Darling r. Austr.
50 A3 Darling Range hills Austr.
98 C2 Darlington U.K.
53 C2 Darlington Point Austr.
103 D1 Darłowo Pol.
101 D3 Darmstadt Ger.
115 E1 Darnah Libya
52 B2 Darnick Austr.
55 H3 Darnley, Cape Antarctica
107 C1 Daroca Spain
99 D4 Dartford U.K.
99 A4 Dartmoor hills U.K.
131 D3 Dartmouth Can.
99 B4 Dartmouth U.K.
59 D3 Daru P.N.G.
59 C2 Daruba Indon.
50 C1 Darwin Austr.
153 C5 Darwin Falkland Is
79 C2 Dārzīn Iran
65 A1 Dashiqiao China
Dashkhovuz Turkm. see Daşoguz
74 A2 Dasht r. Pak.
76 B2 Daşoguz Turkm.
61 C1 Datadian Indon.
111 C3 Datça Turkey
66 D2 Date Japan
71 B3 Datian China
70 B1 Datong China
64 B3 Datu Piang Phil.
74 B1 Daud Khel Pak.
88 B2 Daugava r. Latvia
88 C2 Daugavpils Latvia
100 C2 Daun Ger.
129 D2 Dauphin Can.
129 E2 Dauphin Lake Can.
73 B3 Davangere India
64 B3 Davao Phil.
64 B3 Davao Gulf Phil.
137 C2 Davenport U.S.A.
99 C3 Daventry U.K.
123 C2 Daveyton S. Africa
146 B4 David Panama
129 C2 Davidson Can.
126 F3 Davidson Lake Can.
135 B3 Davis U.S.A.
131 D2 Davis Inlet (abandoned) Can.
55 I3 Davis Sea Antarctica
160 P3 Davis Strait str. Can./Greenland
105 D2 Davos Switz.
88 C3 Davyd-Haradok Belarus
Dawei Myanmar see Tavoy
78 A2 Dawmat al Jandal Saudi Arabia
79 C3 Dawqah Oman
78 B3 Dawqah Saudi Arabia
126 C2 Dawson Can.
141 D2 Dawson U.S.A.
128 B2 Dawson Creek Can.
128 B2 Dawsons Landing Can.
68 C2 Dawu China
71 C3 Dawu China
Dawukou China see Shizuishan
79 C2 Dawwah Oman
104 B3 Dax France
Daxian China see Dazhou
68 C2 Da Xueshan mts China
52 B3 Daylesford Austr.
Dayong China see Zhangjiajie
81 C2 Dayr az Zawr Syria
138 C3 Dayton U.S.A.
141 D3 Daytona Beach U.S.A.
71 B3 Dayu China
Da Yunhe canal China see Jinghang Yunhe
70 A2 Dazhou China
122 B3 De Aar S. Africa
141 D3 Deadman Bay U.S.A.
80 B2 Dead Sea salt l. Asia
99 D4 Deal U.K.
71 B3 De'an China
152 B3 Deán Funes Arg.
128 B2 Dease Lake Can.
126 E2 Dease Strait Can.
135 C3 Death Valley depr. U.S.A.

104 C2 Deauville France
61 C1 Debak Malaysia
111 B2 Debar Macedonia
103 E1 Dębica Pol.
103 E1 Deblin Pol.
114 B3 Débo, Lac l. Mali
103 E2 Debrecen Hungary
117 B3 Debre Markos Eth.
119 D2 Debre Sina Eth.
117 B3 Debre Tabor Eth.
117 B4 Debre Zeyit Eth.
140 C2 Decatur AL U.S.A.
138 B3 Decatur IL U.S.A.
73 B3 Deccan plat. India
53 D1 Deception Bay Austr.
71 A3 Dechang China
102 C1 Děčín Czech Rep.
137 E2 Decorah U.S.A.
154 C2 Dedo de Deus mt. Brazil
88 C2 Dedovichi Russia
121 C2 Dedza Malawi
99 B3 Dee r. England/Wales U.K.
96 C2 Dee r. Scotland U.K.
130 C3 Deep River Can.
53 D1 Deepwater Austr.
131 E3 Deer Lake Can.
134 D1 Deer Lodge U.S.A.
138 C3 Defiance U.S.A.
140 C2 De Funiak Springs U.S.A.
68 C2 Dêgê China
117 C4 Degeh Bur Eth.
139 F1 Dégelis Can.
102 C2 Deggendorf Ger.
91 E2 Degtevo Russia
81 C2 Dehlorān Iran
74 B1 Dehra Dun India
75 C2 Dehri India
69 E2 Dehui China
100 A2 Deinze Belgium
110 B1 Dej Romania
138 B2 De Kalb U.S.A.
116 B3 Dekemhare Eritrea
118 C2 Dekese Dem. Rep. Congo
118 B2 Dékoa C.A.R.
141 D3 De Land U.S.A.
135 C3 Delano U.S.A.
135 D3 Delano Peak U.S.A.
48 I3 Delap-Uliga-Djarrit Marshall Is
123 C2 Delareyville S. Africa
129 D2 Delaronde Lake Can.
138 C2 Delaware U.S.A.
139 D3 Delaware r. U.S.A.
139 D3 Delaware state U.S.A.
139 D3 Delaware Bay U.S.A.
53 C3 Delegate Austr.
118 C2 Délembé C.A.R.
105 D2 Delémont Switz.
100 B1 Delft Neth.
100 C1 Delfzijl Neth.
121 D2 Delgado, Cabo c. Moz.
68 C1 Delgerhaan Mongolia
Delhi China see Delingha
74 B2 Delhi India
60 B2 Deli i. Indon.
128 B1 Déline Can.
68 C2 Delingha China
101 F2 Delitzsch Ger.
107 D2 Dellys Alg.
135 C4 Del Mar U.S.A.
101 D1 Delmenhorst Ger.
109 B1 Delnice Croatia
136 B3 Del Norte U.S.A.
83 L1 De-Longa, Ostrova is Russia
De Long Islands Russia see De-Longa, Ostrova
De Long Strait Russia see Longa, Proliv
129 C2 Deloraine Can.
111 B3 Delphi tourist site Greece
141 D3 Delray Beach U.S.A.
143 C3 Del Rio U.S.A.
136 B3 Delta CO U.S.A.
135 D3 Delta UT U.S.A.
126 C2 Delta Junction U.S.A.
109 D3 Delvinë Albania
106 C1 Demanda, Sierra de la mts Spain
Demavend mt. Iran see Damāvand, Qolleh-ye
118 C3 Demba Dem. Rep. Congo
119 D1 Dembech'a Eth.
117 B4 Dembī Dolo Eth.
Demerara Guyana see Georgetown
91 C3 Demerdzhi mt. Ukr.
89 D2 Demidov Russia
142 B2 Deming U.S.A.
111 C3 Demirci Turkey
111 C2 Demirköy Turkey
102 C1 Demmin Ger.
140 C2 Demopolis U.S.A.
60 B2 Dempo, Gunung vol. Indon.
89 D2 Demyansk Russia
122 B3 De Naawte S. Africa
98 B3 Denbigh U.K.
100 B1 Den Burg Neth.
62 B2 Den Chai Thai.
60 B2 Dendang Indon.
100 B2 Dendermonde Belgium
100 B2 Dendre r. Belgium
Dengjiabu China see Yujiang
70 A1 Dengkou China
70 B2 Dengzhou Henan China
Dengzhou Shandong China see Penglai

Den Haag

Den Haag Neth. *see* **The Hague**
50 A2 **Denham** Austr.
100 B1 **Den Helder** Neth.
107 D2 **Dénia** Spain
52 B3 **Deniliquin** Austr.
134 C2 **Denio** U.S.A.
137 D2 **Denison** *IA* U.S.A.
143 D2 **Denison** *TX* U.S.A.
111 C3 **Denizli** Turkey
53 D2 **Denman** Austr.
50 A3 **Denmark** Austr.
93 E4 **Denmark** *country* Europe
84 B2 **Denmark Strait** Greenland/Iceland
77 C3 **Denov** Uzbek.
61 C2 **Denpasar** Indon.
143 D2 **Denton** U.S.A.
50 A3 **D'Entrecasteaux, Point** Austr.
59 E3 **D'Entrecasteaux Islands** P.N.G.
141 D2 **Dentsville** U.S.A.
136 B3 **Denver** U.S.A.
75 C2 **Deogarh** *Odisha* India
74 B2 **Deogarh** *Rajasthan* India
75 C2 **Deoghar** India
138 B2 **De Pere** U.S.A.
83 K2 **Deputatskiy** Russia
62 A1 **Dêqên** China
140 B2 **De Queen** U.S.A.
74 A2 **Dera Bugti** Pak.
74 B1 **Dera Ghazi Khan** Pak.
74 B1 **Dera Ismail Khan** Pak.
87 D4 **Derbent** Russia
50 B1 **Derby** Austr.
99 C3 **Derby** U.K.
137 D3 **Derby** U.S.A.
99 D3 **Dereham** U.K.
97 B2 **Derg, Lough** *l.* Ireland
91 D1 **Derhachi** Ukr.
140 B2 **DeRidder** U.S.A.
91 D2 **Derkul** *r.* Russia/Ukr.
Derry U.K. *see* **Londonderry**
75 B1 **Dêrub** China
116 B3 **Derudeb** Sudan
122 B3 **De Rust** S. Africa
109 C2 **Derventa** Bos. & Herz.
98 C3 **Derwent** *r. England* U.K.
98 C3 **Derwent** *r. England* U.K.
98 B2 **Derwent Water** *l.* U.K.
77 C1 **Derzhavinsk** Kazakh.
Derzhavinskiy Kazakh. *see* **Derzhavinsk**
152 B1 **Desaguadero** *r.* Bol.
49 M5 **Désappointement, Îles du** *is* Fr. Polynesia
129 D2 **Deschambault Lake** Can.
134 B1 **Deschutes** *r.* U.S.A.
117 B3 **Desē** Eth.
153 B4 **Deseado** Arg.
153 B4 **Deseado** *r.* Arg.
142 A2 **Desemboque** Mex.
137 E2 **Des Moines** U.S.A.
137 E2 **Des Moines** *r.* U.S.A.
91 C1 **Desna** *r.* Russia/Ukr.
89 D3 **Desnogorsk** Russia
101 F2 **Dessau-Roßlau** Ger.
Dessye Eth. *see* **Desē**
128 A1 **Destruction Bay** Can.
149 C4 **Desventuradas, Islas** *is* S. Pacific Ocean
128 C1 **Detah** Can.
120 B2 **Dete** Zimbabwe
101 D2 **Detmold** Ger.
138 C2 **Detroit** U.S.A.
137 D1 **Detroit Lakes** U.S.A.
Dett Zimbabwe *see* **Dete**
100 B2 **Deurne** Neth.
110 B1 **Deva** Romania
100 C1 **Deventer** Neth.
96 C2 **Deveron** *r.* U.K.
103 D2 **Devét skal** *h.* Czech Rep.
137 D1 **Devils Lake** U.S.A.
128 A2 **Devils Paw** *mt.* U.S.A.
99 C4 **Devizes** U.K.
74 B2 **Devli** India
110 C2 **Devnya** Bulg.
128 C2 **Devon** Can.
126 F1 **Devon Island** Can.
51 D4 **Devonport** Austr.
74 B2 **Dewas** India
137 F3 **Dexter** U.S.A.
70 A2 **Deyang** China
59 D3 **Deyong, Tanjung** *pt* Indon.
81 C2 **Dezfūl** Iran
70 B2 **Dezhou** China
79 C2 **Dhahran** Saudi Arabia
75 D2 **Dhaka** Bangl.
78 B3 **Dhamār** Yemen
75 C2 **Dhamtari** India
75 C2 **Dhanbad** India
74 B2 **Dhandhuka** India
75 C2 **Dhankuta** Nepal
74 B2 **Dhar** India
75 D2 **Dharmanagar** India
73 B3 **Dharmapuri** India
75 C2 **Dharmjaygarh** India
73 B3 **Dharwad** India
Dharwar India *see* **Dharwad**
74 B2 **Dhasa** India
75 C2 **Dhaulagiri I** *mt.* Nepal
78 B3 **Dhubāb** Yemen
74 B2 **Dhule** India
Dhulia India *see* **Dhule**
117 C4 **Dhuusa Marreeb** Somalia
144 A1 **Diablo, Picacho del** *mt.* Mex.
142 B2 **Diablo Plateau** U.S.A.
121 C2 **Diaca** Moz.
51 C2 **Diamantina** *watercourse* Austr.
155 D1 **Diamantina** Brazil
151 E4 **Diamantina, Chapada** *plat.* Brazil
159 F6 **Diamantina Deep** *sea feature* Indian Ocean
151 D4 **Diamantino** *Mato Grosso* Brazil
154 B1 **Diamantino** *Mato Grosso* Brazil
71 B3 **Dianbai** China
151 E4 **Dianópolis** Brazil
114 B4 **Dianra** Côte d'Ivoire
114 C3 **Diapaga** Burkina Faso
79 C2 **Dibā al Ḥiṣn** U.A.E.
79 C2 **Ḍibab** Oman
118 C3 **Dibaya** Dem. Rep. Congo
122 B2 **Dibeng** S. Africa
72 D2 **Dibrugarh** India
136 C1 **Dickinson** U.S.A.
140 C1 **Dickson** U.S.A.
Dicle *r.* Turkey *see* **Tigris**
105 D3 **Die** France
Diedenhofen France *see* **Thionville**
129 D2 **Diefenbaker, Lake** Can.
Diégo Suarez Madag. *see* **Antsirañana**
114 B3 **Diéma** Mali
101 D2 **Diemel** *r.* Ger.
62 B1 **Điên Biên Phu** Vietnam
62 B2 **Diên Châu** Vietnam
101 D1 **Diepholz** Ger.
104 C2 **Dieppe** France
100 B2 **Diest** Belgium
115 D3 **Diffa** Niger
131 D3 **Digby** Can.
105 D3 **Digne-les-Bains** France
105 C2 **Digoin** France
64 B3 **Digos** Phil.
59 D3 **Digul** *r.* Indon.
105 D2 **Dijon** France
115 D4 **Dik** Chad
117 C3 **Dikhil** Djibouti
111 C3 **Dikili** Turkey
100 A2 **Diksmuide** Belgium
82 G2 **Dikson** Russia
115 D3 **Dikwa** Nigeria
117 B4 **Dīla** Eth.
74 A1 **Dilārām** Afgh.
59 C3 **Dili** East Timor
101 D2 **Dillenburg** Ger.
117 A3 **Dilling** Sudan
126 B3 **Dillingham** U.S.A.
134 D1 **Dillon** *MT* U.S.A.
141 E2 **Dillon** *SC* U.S.A.
118 C4 **Dilolo** Dem. Rep. Congo
72 D2 **Dimapur** India
Dimashq Syria *see* **Damascus**
52 B3 **Dimboola** Austr.
110 C2 **Dimitrovgrad** Bulg.
87 D3 **Dimitrovgrad** Russia
Dimitrovo Bulg. *see* **Pernik**
64 B2 **Dinagat** *i.* Phil.
75 C2 **Dinajpur** Bangl.
104 B2 **Dinan** France
100 B2 **Dinant** Belgium
111 D3 **Dinar** Turkey
81 D2 **Dīnār, Kūh-e** *mt.* Iran
104 B2 **Dinard** France
Dinbych U.K. *see* **Denbigh**
73 B3 **Dindigul** India
118 B1 **Dindima** Nigeria
123 D1 **Dindiza** Moz.
101 E2 **Dingelstädt** Ger.
75 C2 **Dinggyê** China
Dingle Ireland *see* **Daingean Uí Chúis**
97 A2 **Dingle Bay** Ireland
71 B3 **Dingnan** China
102 C2 **Dingolfing** Ger.
114 A3 **Dinguiraye** Guinea
96 B2 **Dingwall** U.K.
70 A2 **Dingxi** China
123 C1 **Dinokwe** Botswana
91 D2 **Dinskaya** Russia
100 C2 **Dinslaken** Ger.
135 C3 **Dinuba** U.S.A.
114 B3 **Dioïla** Mali
154 B3 **Dionísio Cerqueira** Brazil
114 A3 **Diourbel** Senegal
75 D2 **Diphu** India
74 B1 **Dir** Pak.
51 D1 **Direction, Cape** Austr.
117 C4 **Dirē Dawa** Eth.
120 B2 **Dirico** Angola
115 D1 **Dirj** Libya
50 A2 **Dirk Hartog Island** Austr.
53 C1 **Dirranbandi** Austr.
78 B3 **Ḍirs** Saudi Arabia
153 E5 **Disappointment, Cape** S. Georgia
134 B1 **Disappointment, Cape** U.S.A.
50 B2 **Disappointment, Lake** *imp. l.* Austr.
52 B3 **Discovery Bay** Austr.
Disko *i.* Greenland *see* **Qeqertarsuaq**
141 E1 **Dismal Swamp** U.S.A.
99 D3 **Diss** U.K.
154 C1 **Distrito Federal** *admin. div.* Brazil
108 B3 **Dittaino** *r.* Italy
74 B2 **Diu** India
155 D2 **Divinópolis** Brazil
87 D4 **Divnoye** Russia
114 B4 **Divo** Côte d'Ivoire
80 B2 **Divriği** Turkey
74 A2 **Diwana** Pak.
138 B2 **Dixon** U.S.A.
128 A2 **Dixon Entrance** *sea chan.* Can./U.S.A.
81 C2 **Diyarbakır** Turkey
74 A2 **Diz** Pak.
115 D2 **Djado** Niger
115 D2 **Djado, Plateau du** Niger
Djakarta Indon. *see* **Jakarta**
118 B3 **Djambala** Congo
115 C2 **Djanet** Alg.
115 D3 **Djédaa** Chad
115 C1 **Djelfa** Alg.
119 C2 **Djéma** C.A.R.
114 B3 **Djenné** Mali
118 B2 **Djibloho** Equat. Guinea
114 B3 **Djibo** Burkina Faso
117 C3 **Djibouti** *country* Africa
117 C3 **Djibouti** Djibouti
Djidjelli Alg. *see* **Jijel**
118 C2 **Djolu** Dem. Rep. Congo
114 C4 **Djougou** Benin
118 B2 **Djoum** Cameroon
115 D3 **Djourab, Erg du** *des.* Chad
92 □C3 **Djúpivogur** Iceland
89 F3 **Dmitriyevka** Russia
89 E3 **Dmitriyev-L'govskiy** Russia
Dmitriyevsk Ukr. *see* **Makiyivka**
89 E2 **Dmitrov** Russia
Dmytriyevs'k Ukr. *see* **Makiyivka**
Dnepr *r.* Russia *see* **Dnieper**
89 D3 **Dnieper** *r.* Russia
91 C2 **Dnieper** *r.* Ukr.
90 B2 **Dniester** *r.* Ukr.
Dnipro *r.* Ukr. *see* **Dnieper**
91 C2 **Dniprodzerzhyns'k** Ukr.
91 D2 **Dnipropetrovs'k** Ukr.
91 C2 **Dniprorudne** Ukr.
Dnister *r.* Ukr. *see* **Dniester**
90 C2 **Dnistrovs'kyy Lyman** *lag.* Ukr.
88 C2 **Dno** Russia
121 C2 **Doa** Moz.
115 D4 **Doba** Chad
88 B2 **Dobele** Latvia
101 F2 **Döbeln** Ger.
59 C3 **Doberai, Jazirah** *pen.* Indon.
Doberai Peninsula Indon. *see* **Doberai, Jazirah**
59 C3 **Dobo** Indon.
109 C2 **Doboj** Bos. & Herz.
103 E1 **Dobre Miasto** Pol.
110 C2 **Dobrich** Bulg.
89 F3 **Dobrinka** Russia
89 E3 **Dobroye** Russia
89 D3 **Dobrush** Belarus
86 E3 **Dobryanka** Russia
155 E1 **Doce** *r.* Brazil
145 B2 **Doctor Arroyo** Mex.
144 B2 **Doctor Belisario Domínguez** Mex.
Doctor Petru Groza Romania *see* **Ştei**
111 C3 **Dodecanese** *is* Greece
Dodekanisos *is* Greece *see* **Dodecanese**
136 C3 **Dodge City** U.S.A.
119 D3 **Dodoma** Tanz.
100 C1 **Doesburg** Neth.
100 C2 **Doetinchem** Neth.
59 C3 **Dofa** Indon.
75 C1 **Dogai Coring** *salt l.* China
128 B2 **Dog Creek** Can.
67 B3 **Dōgo** *i.* Japan
115 C3 **Dogondoutchi** Niger
81 C2 **Doğubeyazıt** Turkey
79 C2 **Doha** Qatar
62 A2 **Doi Saket** Thai.
81 D2 **Dokali** Iran
100 B1 **Dokkum** Neth.
88 C3 **Dokshytsy** Belarus
91 D2 **Dokuchayevs'k** Ukr.
142 A1 **Dolan Springs** U.S.A.
130 C3 **Dolbeau-Mistassini** Can.
104 B2 **Dol-de-Bretagne** France
105 D2 **Dole** France
91 D2 **Dolgaya Kosa** *spit* Russia
99 B3 **Dolgellau** U.K.
89 E3 **Dolgorukovo** Russia
89 E3 **Dolgoye** Russia
69 F1 **Dolinsk** Russia
103 D2 **Dolný Kubín** Slovakia
59 D3 **Dolok, Pulau** *i.* Indon.
108 B1 **Dolomites** *mts* Italy
Dolomiti *mts* Italy *see* **Dolomites**
Dolonnur China *see* **Duolun**
117 C4 **Dolo Odo** Eth.
144 A2 **Dolores** Mex.
126 E2 **Dolphin and Union Strait** Can.
90 A2 **Dolyna** Ukr.
102 C2 **Domažlice** Czech Rep.
93 E3 **Dombås** Norway
103 D2 **Dombóvár** Hungary
Dombrovitsa Ukr. *see* **Dubrovytsya**
Dombrowa Pol. *see* **Dąbrowa Górnicza**
128 B2 **Dome Creek** Can.
147 D3 **Dominica** *country* West Indies
147 C3 **Dominican Republic** *country* West Indies
118 C3 **Domiongo** Dem. Rep. Congo
117 C4 **Domo** Eth.
89 E2 **Domodedovo** Russia
111 B3 **Domokos** Greece
61 C2 **Dompu** Indon.
153 A3 **Domuyo, Volcán** *vol.* Arg.
142 B3 **Don** Mex.
89 E3 **Don** *r.* Russia
96 C2 **Don** *r.* U.K.
97 D1 **Donaghadee** U.K.
52 B3 **Donald** Austr.
Donau *r.* Austria/Ger. *see* **Danube**
102 C2 **Donauwörth** Ger.
106 B2 **Don Benito** Spain
98 C3 **Doncaster** U.K.
120 A1 **Dondo** Angola
121 C2 **Dondo** Moz.
73 C4 **Dondra Head** *hd* Sri Lanka
97 B1 **Donegal** Ireland
97 B1 **Donegal Bay** Ireland
91 D2 **Donets'k** Ukr.
91 D2 **Donets'kyy Kryazh** *hills* Russia/Ukr.
118 B2 **Donga** Nigeria
50 A2 **Dongara** Austr.
71 A3 **Dongchuan** China
65 B2 **Dongducheon** S. Korea
71 A4 **Dongfang** China
66 B1 **Dongfanghong** China
61 C2 **Donggala** Indon.
65 A2 **Donggang** China
Donggou China *see* **Donggang**
71 B3 **Dongguan** *Guangdong* China
71 B3 **Dongguan** *Guangdong* China
62 B2 **Đông Ha** Vietnam
65 B2 **Donghae** S. Korea
Dong Hai *sea* N. Pacific Ocean *see* **East China Sea**
62 B2 **Đông Hơi** Vietnam
116 B3 **Dongola** Sudan
118 B2 **Dongou** Congo
Dong Phaya Yen Range *mts* Thai. *see* **San Khao Phang Hoei**
63 B2 **Dong Phraya Yen** *esc.* Thai.
Dongping China *see* **Anhua**
71 B3 **Dongshan** China
70 B2 **Dongsheng** China
70 C2 **Dongtai** China
71 B3 **Dongting Hu** *l.* China
Dong Ujimqin Qi China *see* **Uliastai**
71 C3 **Dongyang** China
70 B2 **Dongying** China
63 B2 **Don Kêv** Cambodia
143 D3 **Donna** U.S.A.
54 B1 **Donnellys Crossing** N.Z.
100 C3 **Donnersberg** *h.* Ger.
Donostia Spain *see* **San Sebastián**
81 D1 **Donyztau, Sor** *dry lake* Kazakh.
51 C1 **Doomadgee** Austr.
138 B2 **Door Peninsula** U.S.A.
50 B2 **Dora, Lake** *imp. l.* Austr.
99 B4 **Dorchester** U.K.
122 A1 **Dordabis** Namibia
104 B2 **Dordogne** *r.* France
100 B2 **Dordrecht** Neth.
123 C3 **Dordrecht** S. Africa
122 A1 **Doreenville** Namibia
129 D2 **Doré Lake** Can.
101 D1 **Dorfmark** Ger.
68 C1 **Dörgön Nuur** *salt l.* Mongolia
114 B3 **Dori** Burkina Faso
122 A3 **Doring** *r.* S. Africa
100 C2 **Dormagen** Ger.
96 B2 **Dornoch** U.K.
96 B2 **Dornoch Firth** *est.* U.K.
114 B3 **Doro** Mali
89 D3 **Dorogobuzh** Russia
110 C1 **Dorohoi** Romania
92 G3 **Dorotea** Sweden
50 A2 **Dorre Island** Austr.
53 D2 **Dorrigo** Austr.
100 C2 **Dorsten** Ger.
100 C2 **Dortmund** Ger.
100 C2 **Dortmund-Ems-Kanal** *canal* Ger.
81 C2 **Dorūd** Iran
153 B4 **Dos Bahías, Cabo** *c.* Arg.
77 C3 **Dōshī** Afgh.
101 F1 **Dosse** *r.* Ger.
114 C3 **Dosso** Niger
141 C2 **Dothan** U.S.A.
101 D1 **Dötlingen** Ger.
105 C1 **Douai** France
118 A2 **Douala** Cameroon
104 B2 **Douarnenez** France
105 D2 **Doubs** *r.* France/Switz.
54 A3 **Doubtful Sound** N.Z.
114 B3 **Douentza** Mali
98 A2 **Douglas** Isle of Man
122 B2 **Douglas** S. Africa
128 A2 **Douglas** *AK* U.S.A.
142 B2 **Douglas** *AZ* U.S.A.
141 D2 **Douglas** *GA* U.S.A.
136 B2 **Douglas** *WY* U.S.A.
71 C3 **Douliu** Taiwan
104 C1 **Doullens** France
154 B1 **Dourada, Serra** *hills* Brazil
154 B2 **Dourados** Brazil
154 B2 **Dourados** *r.* Brazil
154 B2 **Dourados, Serra dos** *hills* Brazil
106 B1 **Douro** *r.* Port.
99 D4 **Dover** U.K.
139 D3 **Dover** U.S.A.
95 D3 **Dover, Strait of** France/U.K.
139 F1 **Dover-Foxcroft** U.S.A.
99 B3 **Dovey** *r.* U.K.
121 C2 **Dowa** Malawi
79 C2 **Dowlatābād** *Fārs* Iran
79 C2 **Dowlatābād** *Kermān* Iran
97 D1 **Downpatrick** U.K.
67 B3 **Dōzen** *is* Japan
130 C3 **Dozois, Réservoir** *resr* Can.
114 B2 **Drâa, Hamada du** *plat.* Alg.
154 B2 **Dracena** Brazil
100 C1 **Drachten** Neth.
110 B2 **Drăgănești-Olt** Romania

Elgin

110 B2 **Drăgășani** Romania
105 D3 **Draguignan** France
88 C3 **Drahichyn** Belarus
53 D1 **Drake** Austr.
123 C2 **Drakensberg** *mts* Lesotho/S. Africa
123 C2 **Drakensberg** *mts* S. Africa
149 C8 **Drake Passage** S. Atlantic Ocean
111 B2 **Drama** Greece
93 F4 **Drammen** Norway
109 C1 **Drava** *r.* Europe
128 C2 **Drayton Valley** Can.
101 D2 **Dreieich** Ger.
102 C1 **Dresden** Ger.
104 C2 **Dreux** France
100 B1 **Driemond** Neth.
98 C2 **Driffield** U.K.
137 D1 **Drift Prairie** *reg.* U.S.A.
109 C2 **Drina** *r.* Bos. & Herz./Serbia
140 B2 **Driskill Mountain** *h.* U.S.A.
Drissa Belarus *see* **Vyerkhnyadzvinsk**
109 C2 **Drniš** Croatia
110 B2 **Drobeta-Turnu Severin** Romania
90 B2 **Drochia** Moldova
101 D1 **Drochtersen** Ger.
97 C2 **Drogheda** Ireland
90 A2 **Drohobych** Ukr.
99 B3 **Droitwich Spa** U.K.
Drokiya Moldova *see* **Drochia**
97 B1 **Dromahair** Ireland
97 C1 **Dromore** U.K.
100 B1 **Dronten** Neth.
74 B1 **Drosh** Pak.
53 C3 **Drouin** Austr.
128 C2 **Drumheller** Can.
138 C1 **Drummond Island** U.S.A.
131 C3 **Drummondville** Can.
96 B3 **Drummore** U.K.
96 B2 **Drumnadrochit** U.K.
Druskieniki Lith. *see* **Druskininkai**
88 B3 **Druskininkai** Lith.
88 C2 **Druya** Belarus
91 D2 **Druzhkivka** Ukr.
88 D2 **Druzhnaya Gorka** Russia
89 D3 **Drybin** Belarus
130 A3 **Dryden** Can.
50 B1 **Drysdale** *r.* Austr.
147 C3 **Duarte, Pico** *mt.* Dom. Rep.
78 A2 **Ḑubā** Saudi Arabia
79 C2 **Dubai** U.A.E.
90 B2 **Dubăsari** Moldova
129 D1 **Dubawnt Lake** Can.
Dubayy U.A.E. *see* **Dubai**
78 A2 **Dubbagh, Jabal ad** *mt.* Saudi Arabia
53 C2 **Dubbo** Austr.
Dubesar' Moldova *see* **Dubăsari**
97 C2 **Dublin** Ireland
141 D2 **Dublin** U.S.A.
89 E2 **Dubna** Russia
90 B1 **Dubno** Ukr.
139 D2 **Du Bois** U.S.A.
Dubossary Moldova *see* **Dubăsari**
114 A4 **Dubréka** Guinea
109 C2 **Dubrovnik** Croatia
90 B1 **Dubrovytsya** Ukr.
89 D3 **Dubrowna** Belarus
137 E2 **Dubuque** U.S.A.
63 B2 **Đưc Bôn** Vietnam
135 D2 **Duchesne** U.S.A.
129 D2 **Duck Bay** Can.
101 E2 **Duderstadt** Ger.
82 G2 **Dudinka** Russia
99 B3 **Dudley** U.K.
106 B1 **Duero** *r.* Spain
48 H4 **Duff Islands** Solomon Is
131 C2 **Duffreboy, Lac** *l.* Can.
96 C2 **Dufftown** U.K.
109 C2 **Dugi Otok** *i.* Croatia
109 C2 **Dugi Rat** Croatia
100 C2 **Duisburg** Ger.
123 C3 **Dukathole** S. Africa
117 B4 **Duk Fadiat** S. Sudan
79 C2 **Dukhān** Qatar
89 D2 **Dukhovshchina** Russia
Dukou China *see* **Panzhihua**
88 C2 **Dūkštas** Lith.
68 C2 **Dulan** China
152 B3 **Dulce** *r.* Arg.
142 B1 **Dulce** U.S.A.
97 C2 **Duleek** Ireland
100 C2 **Dülmen** Ger.
110 C2 **Dulovo** Bulg.
137 E1 **Duluth** U.S.A.
64 B3 **Dumaguete** Phil.
60 B1 **Dumai** Indon.
64 B2 **Dumaran** *i.* Phil.
142 C2 **Dumas** *AR* U.S.A.
143 C1 **Dumas** *TX* U.S.A.
96 B3 **Dumbarton** U.K.
103 D2 **Ďumbier** *mt.* Slovakia
96 C3 **Dumfries** U.K.
89 E3 **Duminichi** Russia
75 C2 **Dumka** India
55 L3 **Dumont d'Urville Sea** Antarctica
116 B1 **Dumyāṭ** Egypt
Duna *r.* Hungary *see* **Danube**
Dünaburg Latvia *see* **Daugavpils**
Dunaj *r.* Slovakia *see* **Danube**
103 D2 **Dunakeszi** Hungary
97 C2 **Dunany Point** Ireland
Dunărea *r.* Romania *see* **Danube**
Dunării, Delta Romania/Ukr. *see* **Danube Delta**
103 D2 **Dunaújváros** Hungary
Dunav *r.* Bulg./Croatia/Serbia *see* **Danube**
90 B2 **Dunayivtsi** Ukr.
96 C2 **Dunbar** U.K.
96 C1 **Dunbeath** U.K.
128 B3 **Duncan** Can.
143 D2 **Duncan** U.S.A.
96 C1 **Duncansby Head** *hd* U.K.
88 B2 **Dundaga** Latvia
97 C1 **Dundalk** Ireland
139 D3 **Dundalk** U.S.A.
97 C2 **Dundalk Bay** Ireland
Dún Dealgan Ireland *see* **Dundalk**
123 D2 **Dundee** S. Africa
96 C2 **Dundee** U.K.
97 D1 **Dundrum Bay** U.K.
54 B3 **Dunedin** N.Z.
53 C2 **Dunedoo** Austr.
96 C2 **Dunfermline** U.K.
97 C1 **Dungannon** U.K.
74 B2 **Dungarpur** India
97 C2 **Dungarvan** Ireland
99 D4 **Dungeness** *hd* U.K.
97 C1 **Dungiven** U.K.
53 D2 **Dungog** Austr.
119 C2 **Dungu** Dem. Rep. Congo
60 B1 **Dungun** Malaysia
116 B2 **Dungunab** Sudan
69 E2 **Dunhua** China
68 C2 **Dunhuang** China
96 C2 **Dunkeld** U.K.
Dunkerque France *see* **Dunkirk**
104 C1 **Dunkirk** France
139 D2 **Dunkirk** U.S.A.
97 C2 **Dún Laoghaire** Ireland
97 B3 **Dunmanway** Ireland
97 D1 **Dunmurry** U.K.
96 C1 **Dunnet Head** *hd* U.K.
96 C3 **Duns** U.K.
134 B2 **Dunsmuir** U.S.A.
99 C4 **Dunstable** U.K.
100 B3 **Dun-sur-Meuse** France
96 A2 **Dunvegan** U.K.
70 B1 **Duolun** China
Duperré Alg. *see* **Aïn Defla**
110 B2 **Dupnitsa** Bulg.
136 C1 **Dupree** U.S.A.
Duque de Bragança Angola *see* **Calandula**
138 B3 **Du Quoin** U.S.A.
50 B1 **Durack** *r.* Austr.
105 C3 **Durance** *r.* France
144 B2 **Durango** Mex.
106 C1 **Durango** Spain
136 B3 **Durango** U.S.A.
143 D2 **Durant** U.S.A.
153 C3 **Durazno** Uru.
Durazzo Albania *see* **Durrës**
123 D2 **Durban** S. Africa
105 C3 **Durban-Corbières** France
122 A3 **Durbanville** S. Africa
100 B2 **Durbuy** Belgium
100 C2 **Düren** Ger.
75 C2 **Durg** India
98 C2 **Durham** U.K.
141 E1 **Durham** U.S.A.
60 B1 **Duri** Indon.
109 C2 **Durmitor** *mt.* Montenegro
96 B1 **Durness** U.K.
109 C2 **Durrës** Albania
97 A3 **Dursey Island** Ireland
111 C3 **Dursunbey** Turkey
117 C4 **Durukhsi** Somalia
59 D3 **D'Urville, Tanjung** *pt* Indon.
54 B2 **D'Urville Island** N.Z.
71 A3 **Dushan** China
77 C3 **Dushanbe** Tajik.
100 C2 **Düsseldorf** Ger.
Dutch East Indies *country* Asia *see* **Indonesia**
Dutch Guiana *country* S. America *see* **Suriname**
123 C3 **Dutywa** S. Africa
117 C4 **Duudka, Taagga** *reg.* Somalia
Duvno Bos. & Herz. *see* **Tomislavgrad**
71 A3 **Duyun** China
80 B1 **Düzce** Turkey
91 D2 **Dvorichna** Ukr.
74 A2 **Dwarka** India
123 C1 **Dwarsberg** S. Africa
134 C1 **Dworshak Reservoir** U.S.A.
89 D3 **Dyat'kovo** Russia
96 C2 **Dyce** U.K.
127 H2 **Dyer, Cape** Can.
140 C1 **Dyersburg** U.S.A.
Dyfrdwy *r.* U.K. *see* **Dee**
103 D2 **Dyje** *r.* Austria/Czech Rep.
103 D1 **Dylewska Góra** *h.* Pol.
91 D2 **Dymytrov** Ukr.
123 C3 **Dyoki** S. Africa
51 D2 **Dysart** Austr.
122 B3 **Dysselsdorp** S. Africa
87 E3 **Dyurtyuli** Russia
69 D2 **Dzamïn Üüd** Mongolia
121 D2 **Dzaoudzi** Mayotte
91 D2 **Dzerzhyns'k** Ukr.
Dzhaltyr Kazakh. *see* **Zhaltyr**
Dzhambul Kazakh. *see* **Taraz**
91 C2 **Dzhankoy** Ukr.
Dzharkent Kazakh. *see* **Zharkent**
Dzhetygara Kazakh. *see* **Zhitikara**
Dzhezkazgan Kazakh. *see* **Zhezkazgan**
Dzhizak Uzbek. *see* **Jizzax**
91 D3 **Dzhubga** Russia
83 K3 **Dzhugdzhur, Khrebet** *mts* Russia
103 E1 **Działdowo** Pol.
145 D2 **Dzilam de Bravo** Mex.
69 D1 **Dzuunmod** Mongolia
88 C3 **Dzyarzhynsk** Belarus
88 C3 **Dzyatlavichy** Belarus

E

131 E2 **Eagle** *r.* Can.
134 C1 **Eagle Cap** *mt.* U.S.A.
130 A3 **Eagle Lake** Can.
134 B2 **Eagle Lake** U.S.A.
143 C3 **Eagle Pass** U.S.A.
126 C2 **Eagle Plain** Can.
Eap *i.* Micronesia *see* **Yap**
130 A2 **Ear Falls** Can.
135 C3 **Earlimart** U.S.A.
96 C2 **Earn** *r.* U.K.
55 J2 **East Antarctica** *reg.* Antarctica
East Bengal *country* Asia *see* **Bangladesh**
99 D4 **Eastbourne** U.K.
59 D2 **East Caroline Basin** N. Pacific Ocean
69 E2 **East China Sea** N. Pacific Ocean
54 B1 **East Coast Bays** N.Z.
East Dereham U.K. *see* **Dereham**
129 D3 **Eastend** Can.
157 G7 **Easter Island** S. Pacific Ocean
123 C3 **Eastern Cape** *prov.* S. Africa
116 B2 **Eastern Desert** Egypt
73 B3 **Eastern Ghats** *mts* India
Eastern Samoa *terr.* S. Pacific Ocean *see* **American Samoa**
Eastern Sayan Mountains Russia *see* **Vostochnyy Sayan**
Eastern Transvaal *prov.* S. Africa *see* **Mpumalanga**
129 E2 **Easterville** Can.
153 C5 **East Falkland** *i.* Falkland Is
100 C1 **East Frisian Islands** Ger.
135 C3 **Eastgate** U.S.A.
137 D1 **East Grand Forks** U.S.A.
96 B3 **East Kilbride** U.K.
138 C2 **East Lansing** U.S.A.
99 C4 **Eastleigh** U.K.
96 C3 **East Linton** U.K.
138 C2 **East Liverpool** U.S.A.
123 C3 **East London** S. Africa
130 C2 **Eastmain** Can.
130 C2 **Eastmain** *r.* Can.
141 D2 **Eastman** U.S.A.
157 G8 **East Pacific Rise** N. Pacific Ocean
East Pakistan *country* Asia *see* **Bangladesh**
138 A3 **East St Louis** U.S.A.
East Sea N. Pacific Ocean *see* **Japan, Sea of**
83 K2 **East Siberian Sea** Russia
59 C3 **East Timor** *country* Asia
53 C2 **East Toorale** Austr.
139 D2 **East York** Can.
138 A2 **Eau Claire** U.S.A.
130 C2 **Eau Claire, Lac à l'** *l.* Can.
59 D2 **Eauripik** *atoll* Micronesia
145 C2 **Ebano** Mex.
99 B4 **Ebbw Vale** U.K.
118 B2 **Ebebiyin** Equat. Guinea
102 C1 **Eberswalde** Ger.
66 D2 **Ebetsu** Japan
77 E2 **Ebinur Hu** *salt l.* China
118 C2 **Ebola** *r.* Dem. Rep. Congo
109 C2 **Eboli** Italy
118 B2 **Ebolowa** Cameroon
107 D1 **Ebro** *r.* Spain
Echeng China *see* **Ezhou**
144 A2 **Echeverria, Pico** *mt.* Mex.
67 C3 **Echizen** Japan
129 E2 **Echoing** *r.* Can.
100 C3 **Echternach** Lux.
52 B3 **Echuca** Austr.
106 B2 **Écija** Spain
102 B1 **Eckernförde** Ger.
127 G2 **Eclipse Sound** *sea chan.* Can.
150 B3 **Ecuador** *country* S. America
116 C3 **Ed** Eritrea
96 C1 **Eday** *i.* U.K.
117 A3 **Ed Da'ein** Sudan
117 B3 **Ed Damazin** Sudan
116 B3 **Ed Damer** Sudan
116 B3 **Ed Debba** Sudan
116 B3 **Ed Dueim** Sudan
51 D4 **Eddystone Point** Austr.
100 B1 **Ede** Neth.
118 B2 **Edéa** Cameroon
154 C1 **Edéia** Brazil
53 C3 **Eden** Austr.
98 B2 **Eden** *r.* U.K.
143 D2 **Eden** U.S.A.
123 C2 **Edenburg** S. Africa
97 C2 **Edenderry** Ireland
52 B3 **Edenhope** Austr.
141 E1 **Edenton** U.S.A.
111 B2 **Edessa** Greece
139 E2 **Edgartown** U.S.A.
Edge Island Svalbard *see* **Edgeøya**
82 C1 **Edgeøya** *i.* Svalbard
141 D3 **Edgewater** U.S.A.
138 C2 **Edinboro** U.S.A.
143 D3 **Edinburg** U.S.A.
96 C3 **Edinburgh** U.K.
90 B2 **Edineț** Moldova
111 C2 **Edirne** Turkey
Edith Ronne Land Antarctica *see* **Ronne Ice Shelf**
128 C2 **Edmonton** Can.
131 D3 **Edmundston** Can.
111 C3 **Edremit** Turkey
111 C3 **Edremit Körfezi** *b.* Turkey
128 C2 **Edson** Can.
119 C3 **Edward, Lake** Dem. Rep. Congo/Uganda
Edwardesabad Pak. *see* **Bannu**
143 D2 **Edwards Plateau** U.S.A.
55 O2 **Edward VII Peninsula** Antarctica
96 C2 **Edzell** U.K.
100 A2 **Eeklo** Belgium
135 B2 **Eel** *r.* U.S.A.
100 C1 **Eemskanaal** *canal* Neth.
100 C1 **Eenrum** Neth.
138 B3 **Effingham** U.S.A.
135 C3 **Egan Range** *mts* U.S.A.
103 E2 **Eger** Hungary
93 E4 **Egersund** Norway
100 B2 **Eghezée** Belgium
92 □C2 **Egilsstaðir** Iceland
80 B2 **Eğirdir** Turkey
80 B2 **Eğirdir Gölü** *l.* Turkey
104 C2 **Égletons** France
Egmont, Mount *vol.* N.Z. *see* **Taranaki, Mount**
83 M2 **Egvekinot** Russia
116 A2 **Egypt** *country* Africa
70 A2 **Ehen Hudag** China
106 C1 **Eibar** Spain
100 C1 **Eibergen** Neth.
100 C2 **Eifel** *hills* Ger.
96 A2 **Eigg** *i.* U.K.
73 B4 **Eight Degree Channel** India/Maldives
50 B1 **Eighty Mile Beach** Austr.
80 B3 **Eilat** Israel
101 F2 **Eilenburg** Ger.
101 D2 **Einbeck** Ger.
100 B2 **Eindhoven** Neth.
160 N4 **Eirik Ridge** N. Atlantic Ocean
150 C3 **Eirunepé** Brazil
120 B2 **Eiseb** *watercourse* Namibia
101 E2 **Eisenach** Ger.
101 E2 **Eisenberg** Ger.
102 C1 **Eisenhüttenstadt** Ger.
103 D2 **Eisenstadt** Austria
101 E2 **Eisleben, Lutherstadt** Ger.
Eivissa Spain *see* **Ibiza**
Eivissa *i.* Spain *see* **Ibiza**
107 C1 **Ejea de los Caballeros** Spain
121 □D3 **Ejeda** Madag.
Ejin Qi China *see* **Dalain Hob**
93 H4 **Ekenäs** Fin.
93 F4 **Eksjö** Sweden
122 A2 **Eksteenfontein** S. Africa
130 B2 **Ekwan** *r.* Can.
62 A2 **Ela** Myanmar
El Aaiún Western Sahara *see* **Laâyoune**
123 C2 **Elandsdoorn** S. Africa
El Araïche Morocco *see* **Larache**
El Asnam Alg. *see* **Chlef**
111 B3 **Elassona** Greece
Elat Israel *see* **Eilat**
80 B2 **Elazığ** Turkey
108 B2 **Elba, Isola d'** *i.* Italy
150 B2 **El Banco** Col.
142 B2 **El Barreal** *l.* Mex.
109 D2 **Elbasan** Albania
150 C2 **El Baúl** Venez.
114 C1 **El Bayadh** Alg.
101 D1 **Elbe** *r.* Ger.
136 B3 **Elbert, Mount** U.S.A.
141 D2 **Elberton** U.S.A.
104 C2 **Elbeuf** France
80 B2 **Elbistan** Turkey
103 D1 **Elbląg** Pol.
153 A4 **El Bolsón** Arg.
87 D4 **El'brus** *mt.* Russia
81 C2 **Elburz Mountains** *mts* Iran
150 C2 **El Callao** Venez.
143 D3 **El Campo** U.S.A.
135 C4 **El Centro** U.S.A.
152 B1 **El Cerro** Bol.
107 C2 **Elche-Elx** Spain
145 C3 **El Chichónal** *vol.* Mex.
107 C2 **Elda** Spain
119 D2 **Eldama Ravine** Kenya
137 E3 **Eldon** U.S.A.
154 B3 **Eldorado** Arg.
154 C2 **Eldorado** Brazil
144 B2 **El Dorado** Mex.
140 B2 **El Dorado** *AR* U.S.A.
137 D3 **El Dorado** *KS* U.S.A.
134 D1 **Electric Peak** U.S.A.
114 B2 **El Eglab** *plat.* Alg.
106 C2 **El Ejido** Spain
89 E2 **Elektrostal'** Russia
Elemi Triangle *see* **Ilemi Triangle** *disp. terr.* Africa.
150 B3 **El Encanto** Col.
146 C2 **Eleuthera** *i.* Bahamas
116 A3 **El Fasher** Sudan
144 B2 **El Fuerte** Mex.
117 A3 **El Fula** Sudan
116 A3 **El Geneina** Sudan
116 B3 **El Geteina** Sudan
96 C2 **Elgin** U.K.

F

101 F1 Falkensee Ger.
96 C3 Falkirk U.K.
158 D8 Falkland Escarpment S. Atlantic Ocean
153 C5 Falkland Islands terr. S. Atlantic Ocean
157 I9 Falkland Plateau S. Atlantic Ocean
93 F4 Falköping Sweden
135 C3 Fallon U.S.A.
139 E2 Fall River U.S.A.
137 D2 Falls City U.S.A.
99 A4 Falmouth U.K.
122 A3 False Bay S. Africa
144 B2 Falso, Cabo c. Mex.
93 F5 Falster i. Denmark
110 C1 Fălticeni Romania
93 G3 Falun Sweden
152 B2 Famailla Arg.
121 □D3 Fandriana Madag.
71 A3 Fangcheng China
Fangchenggang China see Fangcheng
71 C3 Fangshan Taiwan
66 A1 Fangzheng China
108 B2 Fano Italy
119 C2 Faradje Dem. Rep. Congo
121 □D3 Farafangana Madag.
Farafra Oasis Egypt see Wāḥāt al Farāfirah
76 C3 Farāh Afgh.
114 A3 Faranah Guinea
79 C3 Fararah Oman
78 B3 Farasān, Jazā'ir is Saudi Arabia
59 D2 Faraulep atoll Micronesia
127 J2 Farewell, Cape c. Greenland
54 B2 Farewell, Cape N.Z.
137 D1 Fargo U.S.A.
77 D2 Farg'ona Uzbek.
137 E2 Faribault U.S.A.
130 C2 Faribault, Lac l. Can.
74 B2 Faridabad India
75 C2 Faridpur Bangl.
139 E2 Farmington ME U.S.A.
142 B1 Farmington NM U.S.A.
139 D3 Farmville U.S.A.
99 C4 Farnborough U.K.
128 C2 Farnham, Mount Can.
128 A1 Faro Can.
106 B2 Faro Port.
88 A2 Fårö i. Sweden
147 C3 Faro, Punta pt Col.
106 B1 Faro, Serra do mts Spain
94 B1 Faroe Islands terr. N. Atlantic Ocean
113 I6 Farquhar Group is Seychelles
81 D3 Farrāshband Iran
Farrukhabad India see Fatehgarh
79 C3 Fartak, Ra's c. Yemen
154 B3 Fartura, Serra da mts Brazil
143 C2 Farwell U.S.A.
79 C2 Fāryāb Hormozgān Iran
79 C2 Fāryāb Kermān Iran
81 D3 Fasā Iran
109 C2 Fasano Italy
90 B1 Fastiv Ukr.
119 D2 Fataki Dem. Rep. Congo
75 B2 Fatehgarh India
75 B2 Fatehpur India
114 A3 Fatick Senegal
131 D3 Fatima Can.
123 C2 Fauresmith S. Africa
92 G2 Fauske Norway
92 □A3 Faxaflói b. Iceland
92 G3 Faxälven r. Sweden
115 D3 Faya Chad
140 B1 Fayetteville AR U.S.A.
141 E1 Fayetteville NC U.S.A.
140 C1 Fayetteville TN U.S.A.
74 B1 Fazilka India
114 A2 Fdérik Maur.
141 E2 Fear, Cape U.S.A.
54 C2 Featherston N.Z.
104 C2 Fécamp France
Federated Malay States country Asia see Malaysia
102 C1 Fehmarn i. Ger.
101 F1 Fehrbellin Ger.
155 D2 Feia, Lagoa lag. Brazil
150 B3 Feijó Brazil
54 C2 Feilding N.Z.
151 F4 Feira de Santana Brazil
107 D2 Felanitx Spain
101 F1 Feldberg Ger.
145 D3 Felipe C. Puerto Mex.
155 C1 Felixlândia Brazil
99 D4 Felixstowe U.K.
101 D2 Felsberg Ger.
108 B1 Feltre Italy
93 F3 Femunden l. Norway
Fénérive Madag. see Fenoarivo Atsinanana
Fengcheng Guangxi China see Fengshan
71 B3 Fengcheng Jiangxi China
65 A1 Fengcheng Liaoning China
62 A1 Fengqing China
71 A3 Fengshan Guangxi China see Fengqing
71 A3 Fengshan Yunnan China see Lincang
Fengyi China see Zheng'an
70 B2 Fengxian Jiangsu China
70 A2 Fengxian Shaanxi China
Fengxiang China see Lincang
71 C3 Fengyuan Taiwan
70 B1 Fengzhen China

105 D3 Feno, Capo di c. France
121 □D2 Fenoarivo Atsinanana Madag.
70 B2 Fenyang China
91 D2 Feodosiya Ukr.
108 A3 Fer, Cap de c. Alg.
137 D1 Fergus Falls U.S.A.
51 E1 Fergusson Island P.N.G.
110 D2 Ferizaj Kosovo
114 B4 Ferkessédougou Côte d'Ivoire
108 B2 Fermo Italy
131 D2 Fermont Can.
106 B1 Fermoselle Spain
97 B2 Fermoy Ireland
141 D3 Fernandina Beach U.S.A.
154 B2 Fernandópolis Brazil
Fernando Poó i. Equat. Guinea see Bioko
134 B1 Ferndale U.S.A.
99 C4 Ferndown U.K.
128 C3 Fernie Can.
135 C3 Fernley U.S.A.
97 C2 Ferns Ireland
Ferozepore India see Firozpur
108 B2 Ferrara Italy
154 B2 Ferreiros Brazil
108 A2 Ferro, Capo c. Italy
106 B1 Ferrol Spain
Ferryville Tunisia see Menzel Bourguiba
100 B1 Ferwerd Neth.
114 B1 Fès Morocco
118 B3 Feshi Dem. Rep. Congo
137 E2 Festus U.S.A.
110 C2 Feteşti Romania
97 C2 Fethard Ireland
111 C3 Fethiye Turkey
96 □ Fetlar i. U.K.
130 C2 Feuilles, Rivière aux r. Can.
Fez Morocco see Fès
121 □D3 Fianarantsoa Madag.
115 D4 Fianga Chad
117 B4 Fichë Eth.
109 C2 Fier Albania
96 C2 Fife Ness pt U.K.
104 C3 Figeac France
106 B1 Figueira da Foz Port.
107 D1 Figueres Spain
114 B1 Figuig Morocco
49 I5 Fiji country S. Pacific Ocean
152 B2 Filadelfia Para.
55 B2 Filchner Ice Shelf Antarctica
98 C2 Filey U.K.
108 B3 Filicudi, Isola i. Italy
114 C3 Filingué Niger
111 B3 Filippiada Greece
93 F4 Filipstad Sweden
92 E3 Fillan Norway
135 D3 Fillmore U.S.A.
119 E2 Fīltu Eth.
55 D2 Fimbul Ice Shelf Antarctica
96 C2 Findhorn r. U.K.
138 C2 Findlay U.S.A.
51 D4 Fingal Austr.
139 D2 Finger Lakes U.S.A.
111 D3 Finike Turkey
106 B1 Finisterre, Cape c. Spain
93 I3 Finland country Europe
93 H4 Finland, Gulf of Europe
128 B2 Finlay r. Can.
53 C3 Finley Austr.
134 C1 Finley U.S.A.
101 E2 Finne ridge Ger.
92 H2 Finnmarksvidda reg. Norway
92 G2 Finnsnes Norway
93 G4 Finspång Sweden
97 C1 Fintona U.K.
96 A2 Fionnphort U.K.
111 C3 Fira Greece
Firat r. Turkey see Euphrates
Firenze Italy see Florence
105 C2 Firminy France
74 B2 Firozabad India
74 B1 Firozpur India
81 D3 Fīrūzābād Iran
122 A2 Fish watercourse Namibia
122 B3 Fish r. S. Africa
129 F1 Fisher Strait Can.
99 A4 Fishguard U.K.
105 C2 Fismes France
Fisterra, Cabo Spain see Finisterre, Cape
139 E2 Fitchburg U.S.A.
128 C2 Fitzgerald Can.
141 D2 Fitzgerald U.S.A.
50 B1 Fitzroy Crossing Austr.
Fiume Croatia see Rijeka
54 A3 Five Rivers N.Z.
108 B2 Fivizzano Italy
119 C3 Fizi Dem. Rep. Congo
93 F3 Fjällnäs Sweden
92 G3 Fjällsjälven r. Sweden
123 C3 Flagstaff S. Africa
142 A1 Flagstaff U.S.A.
130 C2 Flaherty Island Can.
98 C2 Flamborough Head hd U.K.
101 F1 Fläming hills Ger.
136 B2 Flaming Gorge Reservoir U.S.A.
134 D1 Flathead r. U.S.A.
134 D1 Flathead Lake U.S.A.
51 D1 Flattery, Cape Austr.
134 B1 Flattery, Cape U.S.A.
98 B3 Fleetwood U.K.
93 E4 Flekkefjord Norway
102 B1 Flensburg Ger.

104 B2 Flers France
100 B2 Fleurus Belgium
104 C2 Fleury-les-Aubrais France
51 D1 Flinders r. Austr.
50 A3 Flinders Bay Austr.
51 D3 Flinders Island Austr.
52 A2 Flinders Ranges mts Austr.
129 D2 Flin Flon Can.
98 B3 Flint U.K.
138 C2 Flint U.S.A.
49 L5 Flint Island Kiribati
101 F2 Flöha r. Ger.
107 D1 Florac France
100 C3 Florange France
108 B2 Florence Italy
140 C2 Florence AL U.S.A.
142 A2 Florence AZ U.S.A.
134 B2 Florence OR U.S.A.
141 E2 Florence SC U.S.A.
150 B2 Florencia Col.
146 B3 Flores Guat.
61 D2 Flores i. Indon.
61 C2 Flores, Laut Indon.
Floreshty Moldova see Floreşti
Flores Sea Indon. see Flores, Laut
151 F3 Floresta Brazil
90 B2 Floreşti Moldova
143 D3 Floresville U.S.A.
151 E3 Floriano Brazil
152 D2 Florianópolis Brazil
153 C3 Florida Uru.
141 D2 Florida state U.S.A.
141 D4 Florida, Straits of N. Atlantic Ocean
141 D4 Florida Bay U.S.A.
141 D4 Florida Keys is U.S.A.
111 B2 Florina Greece
93 E3 Florø Norway
129 D2 Foam Lake Can.
109 C2 Foča Bos. & Herz.
110 C1 Focşani Romania
71 B3 Fogang China
109 C2 Foggia Italy
131 E3 Fogo Island Can.
104 C3 Foix France
89 D3 Fokino Russia
130 B3 Foleyet Can.
108 B2 Foligno Italy
99 D4 Folkestone U.K.
141 D2 Folkston U.S.A.
108 B2 Follonica Italy
129 D2 Fond-du-Lac Can.
129 D2 Fond du Lac r. Can.
138 B2 Fond du Lac U.S.A.
106 B1 Fondevila Spain
108 B2 Fondi Italy
146 B3 Fonseca, Golfo do b. Central America
150 C3 Fonte Boa Brazil
104 B2 Fontenay-le-Comte France
92 □C2 Fontur pt Iceland
Foochow China see Fuzhou
53 C2 Forbes Austr.
75 C2 Forbesganj India
101 E3 Forchheim Ger.
93 E3 Førde Norway
53 C1 Fords Bridge Austr.
140 B2 Fordyce U.S.A.
140 C2 Forest U.S.A.
53 C3 Forest Hill Austr.
131 D3 Forestville Can.
96 C2 Forfar U.K.
134 B1 Forks U.S.A.
108 B2 Forlì Italy
107 D2 Formentera i. Spain
107 D2 Formentor, Cap de c. Spain
155 C2 Formiga Brazil
152 C2 Formosa Arg.
Formosa country Asia see Taiwan
154 C1 Formosa Brazil
Formosa Strait China/Taiwan see Taiwan Strait
Føroyar terr. N. Atlantic Ocean see Faroe Islands
96 C2 Forres U.K.
50 B3 Forrest Austr.
140 B1 Forrest City U.S.A.
51 D1 Forsayth Austr.
93 H3 Forssa Fin.
53 D2 Forster Austr.
136 B1 Forsyth U.S.A.
74 B2 Fort Abbas Pak.
130 B2 Fort Albany Can.
151 F3 Fortaleza Brazil
Fort Archambault Chad see Sarh
128 C2 Fort Assiniboine Can.
96 B2 Fort Augustus U.K.
123 C3 Fort Beaufort S. Africa
134 D1 Fort Benton U.S.A.
Fort Brabant Can. see Tuktoyaktuk
135 B3 Fort Bragg U.S.A.
Fort Carnot Madag. see Ikongo
Fort Charlet Alg. see Djanet
Fort Chimo Can. see Kuujjuaq
129 C2 Fort Chipewyan Can.
136 B2 Fort Collins U.S.A.
Fort Crampel C.A.R. see Kaga Bandoro
Fort-Dauphin Madag. see Tôlañaro
147 D3 Fort-de-France Martinique
Fort de Polignac Alg. see Illizi
137 E2 Fort Dodge U.S.A.
139 D2 Fort Erie Can.
Fort Flatters Alg. see Bordj Omer Driss
Fort Foureau Cameroon see Kousséri

130 A3 Fort Frances Can.
Fort Franklin Can. see Déline
Fort Gardel Alg. see Zaouatallaz
Fort George Can. see Chisasibi
126 D2 Fort Good Hope Can.
Fort Gouraud Maur. see Fdérik
96 C2 Forth r. U.K.
96 C2 Forth, Firth of est. U.K.
Fort Hall Kenya see Murang'a
Fort Hertz Myanmar see Putao
152 C2 Fortín Madrejón Para.
Fort Jameson Zambia see Chipata
Fort Johnston Malawi see Mangochi
Fort Lamy Chad see Ndjamena
Fort Laperrine Alg. see Tamanrasset
141 D3 Fort Lauderdale U.S.A.
128 B1 Fort Liard Can.
128 C2 Fort Mackay Can.
137 E2 Fort Madison U.S.A.
Fort Manning Malawi see Mchinji
128 C2 Fort McMurray Can.
126 D2 Fort McPherson Can.
136 C2 Fort Morgan U.S.A.
141 D3 Fort Myers U.S.A.
128 B2 Fort Nelson Can.
128 B2 Fort Nelson r. Can.
Fort Norman Can. see Tulita
141 D3 Fort Payne U.S.A.
136 B1 Fort Peck U.S.A.
136 B1 Fort Peck Reservoir U.S.A.
141 D3 Fort Pierce U.S.A.
119 D2 Fort Portal Uganda
128 C1 Fort Providence Can.
129 D2 Fort Qu'Appelle Can.
128 C1 Fort Resolution Can.
96 B2 Fortrose U.K.
Fort Rosebery Zambia see Mansa
Fort Rousset Congo see Owando
Fort Rupert Can. see Waskaganish
128 B2 Fort St James Can.
128 B2 Fort St John Can.
Fort Sandeman Pak. see Zhob
128 C2 Fort Saskatchewan Can.
137 E3 Fort Scott U.S.A.
130 B2 Fort Severn Can.
76 B2 Fort-Shevchenko Kazakh.
128 B1 Fort Simpson Can.
128 C1 Fort Smith Can.
140 B1 Fort Smith U.S.A.
143 C2 Fort Stockton U.S.A.
142 C2 Fort Sumner U.S.A.
Fort Trinquet Maur. see Bîr Mogreïn
134 B2 Fortuna U.S.A.
131 E3 Fortune Bay Can.
128 C2 Fort Vermilion Can.
Fort Victoria Zimbabwe see Masvingo
Fort Walton U.S.A. see Fort Walton Beach
140 C2 Fort Walton Beach U.S.A.
136 B2 Fort Washakie U.S.A.
138 B2 Fort Wayne U.S.A.
96 B2 Fort William U.K.
143 D2 Fort Worth U.S.A.
126 C2 Fort Yukon U.S.A.
71 B3 Foshan China
92 F3 Fosna pen. Norway
92 E3 Fosnavåg Norway
92 □B3 Foss Iceland
92 □A2 Fússá Iceland
108 A2 Fossano Italy
137 D1 Fosston U.S.A.
53 C3 Foster Austr.
118 B3 Fougamou Gabon
104 B2 Fougères France
96 □ Foula i. U.K.
99 D4 Foulness Point U.K.
111 C3 Fournoi i. Greece
114 A3 Fouta Djallon reg. Guinea
54 A3 Foveaux Strait N.Z.
136 C3 Fowler U.S.A.
50 C3 Fowlers Bay Austr.
128 C2 Fox Creek Can.
127 G2 Foxe Basin g. Can.
127 G2 Foxe Channel Can.
127 G2 Foxe Peninsula Can.
54 B2 Fox Glacier N.Z.
128 C2 Fox Lake Can.
128 A1 Fox Mountain Can.
54 C2 Foxton N.Z.
129 D2 Fox Valley Can.
97 C1 Foyle r. Ireland/U.K.
97 C1 Foyle, Lough b. Ireland/U.K.
97 B2 Foynes Ireland
154 B1 Foz de Areia, Represa de resr Brazil
120 A2 Foz do Cunene Angola
154 B3 Foz do Iguaçu Brazil
107 C1 Fraga Spain
100 A2 Frameries Belgium
154 C2 Franca Brazil
109 C2 Francavilla Fontana Italy
104 C2 France country Europe
118 B3 Franceville Gabon
137 D2 Francis Case, Lake U.S.A.
155 D1 Francisco Sá Brazil
120 B3 Francistown Botswana
128 B2 François Lake Can.
100 B1 Franeker Neth.
101 D2 Frankenberg (Eder) Ger.
101 D3 Frankenthal (Pfalz) Ger.
101 E2 Frankenwald mts Ger.
138 C3 Frankfort KY U.S.A.
138 B2 Frankfort MI U.S.A.

Frankfurt Ger. see
Frankfurt am Main
102 C1 Frankfurt (Oder) Ger.
101 D2 Frankfurt am Main Ger.
102 C2 Fränkische Alb hills Ger.
101 E3 Fränkische Schweiz reg. Ger.
139 E2 Franklin NH U.S.A.
139 D2 Franklin PA U.S.A.
140 C1 Franklin TN U.S.A.
126 C2 Franklin Bay Can.
134 C1 Franklin D. Roosevelt Lake U.S.A.
128 C3 Franklin Mountains Can.
126 F2 Franklin Strait Can.
53 C3 Frankston Austr.
82 E1 Frantsa-Iosifa, Zemlya is Russia
54 B2 Franz Josef Glacier N.Z.
Franz Josef Land is Russia see
Frantsa-Iosifa, Zemlya
108 A3 Frasca, Capo della c. Italy
128 B3 Fraser r. B.C. Can.
131 D2 Fraser r. Nfld. and Lab. Can.
122 B3 Fraserburg S. Africa
96 C2 Fraserburgh U.K.
130 B3 Fraserdale Can.
51 E2 Fraser Island Austr.
128 B2 Fraser Lake Can.
128 B2 Fraser Plateau Can.
153 C3 Fray Bentos Uru.
93 E4 Fredericia Denmark
143 D2 Frederick MD U.S.A.
143 D2 Fredericksburg TX U.S.A.
139 D3 Fredericksburg VA U.S.A.
128 A2 Frederick Sound sea chan. U.S.A.
131 D3 Fredericton Can.
Frederikshåb Greenland see Paamiut
93 F4 Frederikshavn Denmark
Fredrikshamn Fin. see Hamina
93 F4 Fredrikstad Norway
138 B2 Freeport IL U.S.A.
143 D3 Freeport TX U.S.A.
146 C2 Freeport City Bahamas
143 D3 Freer U.S.A.
123 C2 Free State prov. S. Africa
114 A4 Freetown Sierra Leone
106 B2 Fregenal de la Sierra Spain
104 B2 Fréhel, Cap c. France
102 B2 Freiburg im Breisgau Ger.
102 C2 Freising Ger.
102 C2 Freistadt Austria
105 D3 Fréjus France
50 A3 Fremantle Austr.
135 B3 Fremont CA U.S.A.
137 D2 Fremont NE U.S.A.
138 C2 Fremont OH U.S.A.
French Congo country Africa see
Congo
151 D2 French Guiana terr. S. America
French Guinea country Africa see
Guinea
134 E1 Frenchman r. U.S.A.
49 M5 French Polynesia terr.
S. Pacific Ocean
French Somaliland country Africa see
Djibouti
French Sudan country Africa see Mali
French Territory of the Afars and
Issas country Africa see Djibouti
151 D3 Fresco r. Brazil
144 B2 Fresnillo Mex.
135 C3 Fresno U.S.A.
107 D2 Freu, Cap des c. Spain
105 D2 Freyming-Merlebach France
114 A3 Fria Guinea
152 B2 Frias Arg.
101 D2 Friedberg (Hessen) Ger.
102 B2 Friedrichshafen Ger.
101 F1 Friesack Ger.
100 C1 Friesoythe Ger.
143 C2 Friona U.S.A.
136 B3 Frisco U.S.A.
Frobisher Bay Can. see Iqaluit
127 H2 Frobisher Bay b. Can.
101 E2 Frohburg Ger.
87 D4 Frolovo Russia
103 D1 Frombork Pol.
99 B4 Frome U.K.
52 A2 Frome, Lake imp. l. Austr.
52 A2 Frome Downs Austr.
100 C2 Fröndenberg/Ruhr Ger.
143 C3 Frontera Coahuila Mex.
145 C3 Frontera Tabasco Mex.
144 B1 Fronteras Mex.
139 D3 Front Royal U.S.A.
108 B2 Frosinone Italy
92 E3 Frøya i. Norway
Frunze Kyrg. see Bishkek
154 C2 Frutal Brazil
105 D2 Frutigen Switz.
103 D2 Frýdek-Místek Czech Rep.
71 B3 Fu'an China
71 C3 Fuding China
106 C1 Fuenlabrada Spain
152 E2 Fuerte Olimpo Para.
114 A2 Fuerteventura i. Canary Is
64 B2 Fuga i. Phil.
79 C2 Fujairah U.A.E.
67 C3 Fuji Japan
71 B3 Fujian prov. China
67 C3 Fujinomiya Japan
67 C3 Fuji-san vol. Japan
Fukien prov. China see Fujian
67 C3 Fukui Japan
67 B4 Fukuoka Japan
67 D3 Fukushima Japan

101 D2 Fulda Ger.
101 D2 Fulda r. Ger.
70 A3 Fuling China
137 D3 Fulton U.S.A.
81 C2 Fūman Iran
105 C2 Fumay France
49 I4 Funafuti atoll Tuvalu
114 A1 Funchal Madeira
155 D1 Fundão Brazil
106 B1 Fundão Port.
131 D3 Fundy, Bay of g. Can.
121 C3 Funhalouro Moz.
70 B2 Funing Jiangsu China
71 A3 Funing Yunnan China
115 C3 Funtua Nigeria
79 C2 Fürgun, Küh-e mt. Iran
89 F2 Furmanov Russia
Furmanovka Kazakh. see Moyynkum
Furmanovo Kazakh. see Zhalpaktal
155 C2 Furnas, Represa resr Brazil
51 D4 Furneaux Group is Austr.
Furong China see Wan'an
100 C1 Fürstenau Ger.
101 E3 Fürth Ger.
127 G2 Fury and Hecla Strait Can.
70 C1 Fushun China
65 B1 Fusong China
79 C2 Fuwayriṭ Qatar
Fuxian Liaoning China see
Wafangdian
70 A2 Fuxian Shaanxi China
70 C1 Fuxin China
70 B2 Fuyang China
69 E1 Fuyu Heilong. China
Fuyu Jilin China see Songyuan
68 B1 Fuyun China
71 B3 Fuzhou Fujian China
71 B3 Fuzhou Jiangxi China
93 F4 Fyn i. Denmark
96 B3 Fyne, Loch inlet U.K.
F.Y.R.O.M. country Europe see
Macedonia

G

117 C4 Gaalkacyo Somalia
120 A2 Gabela Angola
Gaberones Botswana see Gaborone
115 D1 Gabès Tunisia
115 D1 Gabès, Golfe de g. Tunisia
118 B3 Gabon country Africa
123 C1 Gaborone Botswana
79 C2 Gābrīk Iran
110 C2 Gabrovo Bulg.
114 A3 Gabú Guinea-Bissau
73 B3 Gadag-Betigeri India
75 C2 Gadchiroli India
101 E1 Gadebusch Ger.
140 C2 Gadsden U.S.A.
118 B2 Gadzi C.A.R.
110 C2 Găești Romania
108 B2 Gaeta Italy
108 B2 Gaeta, Golfo di g. Italy
141 D1 Gaffney U.S.A.
115 C1 Gafsa Tunisia
89 E2 Gagarin Russia
109 C3 Gagliano del Capo Italy
114 B4 Gagnoa Côte d'Ivoire
131 D2 Gagnon Can.
Gago Coutinho Angola see
Lumbala N'guimbo
81 C1 Gagra Georgia
122 A2 Gaiab watercourse Namibia
111 C3 Gaïdouronisi i. Greece
104 C3 Gaillac France
Gaillimh Ireland see Galway
141 D3 Gainesville FL U.S.A.
141 D2 Gainesville GA U.S.A.
143 D2 Gainesville TX U.S.A.
98 C3 Gainsborough U.K.
52 A2 Gairdner, Lake imp. l. Austr.
96 B2 Gairloch U.K.
122 B2 Gakarosa mt. S. Africa
119 C3 Galana r. Kenya
103 D2 Galanta Slovakia
Galápagos, Islas is Ecuador see
Galapagos Islands
125 I10 Galapagos Islands is Ecuador
157 G2 Galapagos Rise Pacific Ocean
96 C3 Galashiels U.K.
110 C1 Galați Romania
93 E3 Galdhøpiggen mt. Norway
145 B2 Galeana Mex.
128 C2 Galena Bay Can.
138 A2 Galesburg U.S.A.
122 B2 Galeshewe S. Africa
89 F2 Galich Russia
106 B1 Galicia aut. comm. Spain
80 B2 Galilee, Sea of l. Israel
78 A3 Gallabat Sudan
140 C1 Gallatin U.S.A.
73 C4 Galle Sri Lanka
157 G3 Gallego Rise Pacific Ocean
150 B1 Gallinas, Punta pt Col.
109 C2 Gallipoli Italy
111 C2 Gallipoli Turkey
92 H2 Gällivare Sweden
142 B1 Gallup U.S.A.
117 C4 Galmudug reg. Somalia
114 A2 Galtat-Zemmour Western Sahara
97 B2 Galtymore h. Ireland
143 E3 Galveston U.S.A.

143 E3 Galveston Bay U.S.A.
143 E3 Galveston Island U.S.A.
97 B2 Galway Ireland
97 B2 Galway Bay Ireland
154 C1 Gama Brazil
123 D3 Gamalakhe S. Africa
117 B4 Gambēla Eth.
114 A3 Gambia r. Gambia
114 A3 Gambia, The country Africa
49 N6 Gambier, Îles is Fr. Polynesia
52 A3 Gambier Islands Austr.
131 E3 Gambo Can.
118 B3 Gamboma Congo
128 C1 Gamêtì Can.
92 H2 Gammelstaden Sweden
142 B1 Ganado U.S.A.
123 C2 Ga-Nala S. Africa
81 C1 Gäncä Azer.
61 C2 Gandadiwata, Bukit mt. Indon.
118 C3 Gandajika Dem. Rep. Congo
106 B1 Gándara Spain
131 E3 Gander Can.
131 E3 Gander r. Can.
101 D1 Ganderkesee Ger.
107 C1 Gandesa Spain
74 B2 Gandhidham India
74 B2 Gandhinagar India
74 B2 Gandhi Sagar resr India
107 C2 Gandia Spain
Ganga r. Bangl./India see Ganges
153 B4 Gangán Arg.
74 B2 Ganganagar India
62 A1 Gangaw Myanmar
68 C2 Gangca China
75 C1 Gangdisê Shan mts China
75 D2 Ganges r. Bangl./India
105 C3 Ganges France
75 C2 Ganges, Mouths of the Bangl./India
159 E2 Ganges Cone Indian Ocean
65 B2 Gangneung S. Korea
75 C2 Gangtok India
75 C3 Ganjam India
71 B3 Gan Jiang r. China
71 A3 Ganluo China
105 C2 Gannat France
136 B2 Gannett Peak U.S.A.
122 A3 Gansbaai S. Africa
70 A1 Gansu prov. China
115 D4 Ganye Nigeria
71 B3 Ganzhou China
114 B3 Gao Mali
Gaoleshan China see Xianfeng
97 B1 Gaoth Dobhair Ireland
114 B3 Gaoua Burkina Faso
114 A3 Gaoual Guinea
71 C3 Gaoxiong Taiwan
70 B2 Gaoyou China
70 B2 Gaoyou Hu l. China
105 D3 Gap France
64 B2 Gapan Phil.
75 C1 Gar China
97 B2 Gara, Lough l. Ireland
76 C3 Garabil Belentligi hills Turkm.
76 B2 Garabogaz Turkm.
76 B2 Garabogazköl Turkm.
76 B2 Garabogazköl Aýlagy b. Turkm.
117 C4 Garacad Somalia
53 C1 Garah Austr.
151 F3 Garanhuns Brazil
123 C2 Ga-Rankuwa S. Africa
118 C1 Garar, Plaine de plain Chad
117 C4 Garbahaarey Somalia
135 B2 Garberville U.S.A.
101 D1 Garbsen Ger.
154 C2 Garça Brazil
154 B2 Garcias Brazil
108 B1 Garda, Lake l. Italy
108 A3 Garde, Cap de c. Alg.
101 E1 Gardelegen Ger.
136 C3 Garden City U.S.A.
129 E2 Garden Hill Can.
77 C3 Gardēz Afgh.
139 F2 Gardiner U.S.A.
Gardner atoll Micronesia see
Faraulep
135 C3 Gardnerville U.S.A.
136 B3 Garfield U.S.A.
88 B2 Gargždai Lith.
123 C3 Gariep Dam dam S. Africa
122 A3 Garies S. Africa
119 D3 Garissa Kenya
88 B2 Garkalne Latvia
143 D2 Garland U.S.A.
102 C2 Garmisch-Partenkirchen Ger.
52 B2 Garnpung Lake imp. l. Austr.
104 B3 Garonne r. France
117 C4 Garoowe Somalia
74 B2 Garoth India
118 B2 Garoua Cameroon
118 B2 Garoua Boulai Cameroon
Garqêntang China see Sog
96 B2 Garry r. U.K.
126 F2 Garry Lake Can.
119 E3 Garsen Kenya
76 B2 Garşy Turkm.
122 A2 Garub Namibia
60 B2 Garut Indon.
138 B2 Gary U.S.A.
145 C2 Garza García Mex.
68 C2 Garzê China
Gascogne reg. France see Gascony
Gascogne, Golfe de g. France see
Gascony, Gulf of
104 B3 Gascony reg. France

104 B3 Gascony, Gulf of France
50 A2 Gascoyne r. Austr.
118 B2 Gashaka Nigeria
115 D3 Gashua Nigeria
59 E3 Gasmata P.N.G.
131 D2 Gaspé Can.
131 D2 Gaspésie, Péninsule de la pen. Can.
141 E1 Gaston, Lake U.S.A.
141 D1 Gastonia U.S.A.
107 C2 Gata, Cabo de c. Spain
88 C2 Gatchina Russia
98 C2 Gateshead U.K.
143 D2 Gatesville U.S.A.
130 C2 Gatineau Can.
130 C2 Gatineau r. Can.
Gatooma Zimbabwe see Kadoma
53 D1 Gatton Austr.
129 E2 Gauer Lake Can.
93 E4 Gausta mt. Norway
123 C2 Gauteng prov. S. Africa
111 B3 Gavdos i. Greece
93 G3 Gävle Sweden
93 G3 Gävlebukten b. Sweden
89 F2 Gavrilov Posad Russia
89 E2 Gavrilov-Yam Russia
122 A2 Gawachab Namibia
62 A1 Gawai Myanmar
52 A2 Gawler Austr.
52 A2 Gawler Ranges hills Austr.
75 C2 Gaya India
114 C3 Gaya Niger
114 C3 Gayéri Burkina Faso
138 C1 Gaylord U.S.A.
86 E2 Gayny Russia
116 B3 Gaza disp. terr. Asia
80 B2 Gaza Gaza
80 B2 Gaziantep Turkey
76 C2 Gazojak Turkm.
114 B4 Gbarnga Liberia
118 A2 Gboko Nigeria
103 D1 Gdańsk Pol.
88 A3 Gdańsk, Gulf of Pol./Russia
88 C2 Gdov Russia
103 D1 Gdynia Pol.
116 B3 Gedaref Sudan
101 D2 Gedern Ger.
111 C3 Gediz Turkey
111 C3 Gediz r. Turkey
93 F5 Gedser Denmark
100 B2 Geel Belgium
52 B3 Geelong Austr.
101 E1 Geesthacht Ger.
75 C1 Gê'gyai China
129 D2 Geikie r. Can.
93 E3 Geilo Norway
119 D3 Geita Tanz.
71 A3 Gejiu China
108 B3 Gela Italy
108 B3 Gela, Golfo di g. Italy
91 D3 Gelendzhik Russia
Gelibolu Turkey see Gallipoli
100 C2 Gelsenkirchen Ger.
118 C2 Gemena Dem. Rep. Congo
111 C2 Gemlik Turkey
108 B1 Gemona del Friuli Italy
117 C4 Genalē Wenz r. Eth.
153 B3 General Acha Arg.
153 B3 General Alvear Arg.
153 C3 General Belgrano Arg.
144 B2 General Cepeda Mex.
General Freire Angola see
Muxaluando
General Machado Angola see
Camacupa
153 B3 General Pico Arg.
153 B3 General Roca Arg.
154 B3 General Salgado Brazil
64 B3 General Santos Phil.
139 D2 Genesee r. U.S.A.
138 A2 Geneseo IL U.S.A.
139 D2 Geneseo NY U.S.A.
105 D2 Geneva Switz.
139 D2 Geneva U.S.A.
105 D2 Geneva, Lake l. France/Switz.
Genève Switz. see Geneva
106 B2 Genil r. Spain
100 B2 Genk Belgium
53 C3 Genoa Austr.
108 A2 Genoa Italy
108 A2 Genoa, Gulf of g. Italy
Genova Italy see Genoa
Gent Belgium see Ghent
61 C2 Genteng i. Indon.
101 F1 Genthin Ger.
50 A3 Geographe Bay Austr.
65 B2 Geongju S. Korea
131 D2 George r. Can.
122 B3 George S. Africa
52 A3 George, Lake Austr.
141 D3 George, Lake FL U.S.A.
139 E2 George, Lake NY U.S.A.
146 C2 George Town Bahamas
151 D2 Georgetown Guyana
60 B1 George Town Malaysia
138 C3 Georgetown KY U.S.A.
141 E2 Georgetown SC U.S.A.
143 D2 Georgetown TX U.S.A.
55 L2 George V Land reg. Antarctica
81 C1 Georgia country Asia
141 D2 Georgia state U.S.A.
130 B2 Georgian Bay Can.
51 C2 Georgina watercourse Austr.
Georgiu-Dezh Russia see Liski
77 E2 Georgiyevka Kazakh.

145	C3	José Cardel Mex.

145 C3 José Cardel Mex.
131 D2 Joseph, Lac l. Can.
50 B1 Joseph Bonaparte Gulf Austr.
115 C4 Jos Plateau Nigeria
93 E3 Jotunheimen mts Norway
122 B3 Joubertina S. Africa
123 C2 Jouberton S. Africa
104 C2 Joué-lès-Tours France
93 I3 Joutseno Fin.
134 B1 Juan de Fuca Strait Can./U.S.A.
Juanshui China see Tongcheng
145 B2 Juárez Mex.
144 A1 Juárez, Sierra de mts Mex.
151 E3 Juazeiro Brazil
151 F3 Juazeiro do Norte Brazil
117 B4 Juba S. Sudan
117 C4 Jubaland reg. Somalia
117 C5 Jubba r. Somalia
78 B2 Jubbah Saudi Arabia
Jubbulpore India see Jabalpur
145 C3 Juchitán Mex.
155 E1 Jucuruçu Brazil
102 C2 Judenburg Austria
155 E1 Juerana Brazil
101 D2 Jühnde Ger.
146 B3 Juigalpa Nic.
150 D4 Juína Brazil
100 C1 Juist i. Ger.
155 D2 Juiz de Fora Brazil
136 C2 Julesburg U.S.A.
150 B4 Juliaca Peru
Julianatop mt. Indon. see
Mandala, Puncak
151 D2 Juliana Top mt. Suriname
107 C2 Jumilla Spain
75 C2 Jumla Nepal
Jumna r. India see Yamuna
74 B2 Junagadh India
143 D2 Junction U.S.A.
137 D3 Junction City U.S.A.
154 C2 Jundiaí Brazil
128 A2 Juneau U.S.A.
53 C2 Junee Austr.
105 D2 Jungfrau mt. Switz.
139 D2 Juniata r. U.S.A.
153 B3 Junín Arg.
92 G3 Junsele Sweden
134 C2 Juntura U.S.A.
Junxi China see Datian
Junxian China see Danjiangkou
154 B2 Jupiá, Represa resr Brazil
141 D3 Jupiter U.S.A.
154 C2 Juquiá Brazil
151 D3 Juruá Brazil
150 D3 Juruá r. Brazil
154 C2 Juruena r. Brazil
154 C2 Jurumirim, Represa de resr Brazil
151 D3 Juruti Brazil
151 D3 Jussara Brazil
150 C3 Jutaí r. Brazil
101 F2 Jüterbog Ger.
154 B2 Juti Brazil
145 D3 Jutiapa Guat.
93 E4 Jutland pen. Denmark
146 B2 Juventud, Isla de la i. Cuba
70 B2 Juxian China
81 D3 Jūyom Iran
122 B1 Jwaneng Botswana
Jylland pen. Denmark see Jutland
93 I3 Jyväskylä Fin.

K

74 B1 K2 mt. China/Pak.
Kaakhka Turkm. see Kaka
92 I2 Kaamanen Fin.
61 D2 Kabaena i. Indon.
119 C3 Kabalo Dem. Rep. Congo
119 C3 Kabambare Dem. Rep. Congo
119 C3 Kabare Dem. Rep. Congo
119 C3 Kabemba Dem. Rep. Congo
130 B3 Kabinakagami Lake Can.
119 C3 Kabinda Dem. Rep. Congo
118 B2 Kabo C.A.R.
120 B2 Kabompo Zambia
119 C3 Kabongo Dem. Rep. Congo
77 C3 Kābul Afgh.
64 B3 Kaburuang i. Indon.
121 B2 Kabwe Zambia
109 D2 Kaçanik Kosovo
74 A2 Kachchh, Gulf of India
109 D2 Kachchh, Rann of marsh India
83 I3 Kachug Russia
81 C1 Kaçkar Dağı mt. Turkey
73 B3 Kadapa India
Kadıköy Turkey
52 A2 Kadina Austr.
114 B3 Kadiolo Mali
Kadiyevka Ukr. see Stakhanov
145 Kadmat atoll India
89 F2 Kadnikov Russia
121 B2 Kadoma Zimbabwe
63 A2 Kadonkani Myanmar
117 A3 Kadugli Sudan
115 C3 Kaduna Nigeria
89 E2 Kaduy Russia

86 E2 Kadzherom Russia
114 A3 Kaédi Maur.
118 B1 Kaélé Cameroon
65 B2 Kaesŏng N. Korea
118 C3 Kafakumba Dem. Rep. Congo
114 A3 Kaffrine Senegal
80 B2 Kafr ash Shaykh Egypt
121 B2 Kafue Zambia
120 B2 Kafue r. Zambia
67 C3 Kaga Japan
118 B2 Kaga Bandoro C.A.R.
91 E2 Kagal'nitskaya Russia
Kaganovich Pervyye Ukr. see
Polis'ke (abandoned)
60 C3 Kagologolo Indon.
67 B4 Kagoshima Japan
Kagul Moldova see Cahul
119 D3 Kahama Tanz.
90 C2 Kaharlyk Ukr.
61 C2 Kahayan r. Indon.
118 B3 Kahemba Dem. Rep. Congo
101 E2 Kahla Ger.
79 C2 Kahnūj Iran
92 H2 Kahperusvaarat mts Fin.
80 B2 Kahramanmaraş Turkey
79 C2 Kahūrak Iran
59 C3 Kai, Kepulauan is Indon.
115 C4 Kaiama Nigeria
54 B2 Kaiapoi N.Z.
59 C3 Kai Besar i. Indon.
70 B2 Kaifeng China
Kaihua China see Wenshan
122 B3 Kaiingveld reg. S. Africa
59 C3 Kai Kecil i. Indon.
54 B2 Kaikoura N.Z.
114 A4 Kailahun Sierra Leone
Kailas Range mts China see
Gangdisê Shan
71 A3 Kaili China
59 C3 Kaimana Indon.
54 C1 Kaimanawa Mountains N.Z.
72 C2 Kaimur Range hills India
88 B2 Käina Estonia
67 C4 Kainan Japan
115 C3 Kainji Reservoir Nigeria
54 B1 Kaipara Harbour N.Z.
74 B2 Kairana India
115 D1 Kairouan Tunisia
100 C3 Kaiserslautern Ger.
55 I2 Kaiser Wilhelm II Land reg.
Antarctica
54 B1 Kaitaia N.Z.
54 C1 Kaitawa N.Z.
Kaitong China see Tongyu
59 C3 Kaiwatu Indon.
65 A1 Kaiyuan Liaoning China
71 A3 Kaiyuan Yunnan China
92 I3 Kajaani Fin.
51 D2 Kajabbi Austr.
53 C1 Kajarabie, Lake Austr.
76 B3 Kaka Turkm.
122 B2 Kakamas S. Africa
119 D2 Kakamega Kenya
114 A4 Kakata Liberia
91 C2 Kakhovka Ukr.
91 C2 Kakhovs'ke Vodoskhovyshche resr
Ukr.
Kakhul Moldova see Cahul
73 C3 Kakinada India
128 C1 Kakisa Can.
67 B4 Kakogawa Japan
119 C3 Kakoswa Dem. Rep. Congo
126 C2 Kaktovik U.S.A.
Kalaallit Nunaat terr. N. America see
Greenland
59 C3 Kalabahi Indon.
120 B2 Kalabo Zambia
91 E1 Kalach Russia
119 D2 Kalacha Dida Kenya
87 D4 Kalach-na-Donu Russia
62 A1 Kaladan r. India/Myanmar
120 B3 Kalahari Desert Africa
92 H3 Kalajoki Fin.
123 C1 Kalamare Botswana
111 B3 Kalamaria Greece
111 B3 Kalamata Greece
138 B2 Kalamazoo U.S.A.
111 B3 Kalampaka Greece
88 B2 Kalana Estonia
91 C2 Kalanchak Ukr.
115 E2 Kalanshiyū ar Ramlī al Kabīr, Sarīr
des. Libya
61 C2 Kalao i. Indon.
61 D2 Kalaotoa i. Indon.
63 B2 Kalasin Thai.
79 C2 Kalāt Iran
74 A2 Kalat Pak.
50 A2 Kalbarri Austr.
111 C3 Kale Turkey
80 B1 Kalecik Turkey
119 C3 Kalema Dem. Rep. Congo
119 C3 Kalemie Dem. Rep. Congo
62 A1 Kalemyo Myanmar
86 C2 Kalevala Russia
Kalgan China see Zhangjiakou
50 B3 Kalgoorlie Austr.
109 C2 Kali Croatia
109 C2 Kaliakra, Nos pt Bulg.
60 A2 Kaliet Indon.
119 C3 Kalima Dem. Rep. Congo
61 C2 Kalimantan reg. Indon.
Kalinin Russia see Tver'
88 B3 Kaliningrad Russia

91 D2 Kalininskaya Russia
88 C3 Kalinkavichy Belarus
134 D1 Kalispell U.S.A.
103 D1 Kalisz Pol.
91 E2 Kalitva r. Russia
92 H2 Kalix Sweden
92 H2 Kalixälven r. Sweden
111 C3 Kalkan Turkey
120 A3 Kalkfeld Namibia
100 C2 Kall Ger.
92 I3 Kallavesi l. Fin.
92 F3 Kallsjön l. Sweden
93 G4 Kalmar Sweden
93 G4 Kalmarsund sea chan. Sweden
73 C4 Kalmunai Sri Lanka
119 C3 Kalole Dem. Rep. Congo
120 B2 Kalomo Zambia
128 B2 Kalone Peak Can.
74 B1 Kalpa India
73 B3 Kalpeni atoll India
75 B2 Kalpi India
126 B2 Kaltag U.S.A.
101 D1 Kaltenkirchen Ger.
118 B2 Kaltungo Nigeria
89 E3 Kaluga Russia
93 F4 Kalundborg Denmark
74 B1 Kalur Kot Pak.
90 A2 Kalush Ukr.
74 B3 Kalyan India
89 E2 Kalyazin Russia
111 C3 Kalymnos Greece
111 C3 Kalymnos i. Greece
119 C3 Kama Dem. Rep. Congo
62 A2 Kama Myanmar
86 E3 Kama r. Russia
66 D3 Kamaishi Japan
80 B2 Kaman Turkey
120 A2 Kamanjab Namibia
78 B3 Kamarān i. Yemen
Kamaran Island Yemen see Kamarān
74 A2 Kamarod Pak.
50 B3 Kambalda Austr.
119 C4 Kambove Dem. Rep. Congo
160 C4 Kamchatka Basin Bering Sea
83 L3 Kamchatka Peninsula Russia
110 C2 Kamchia r. Bulg.
108 B2 Kamenjak, Rt pt Croatia
86 D2 Kamenka Arkhangel'skaya Oblast'
Russia
87 D3 Kamenka Penzenskaya Oblast' Russia
66 C2 Kamenka Primorskiy Kray Russia
91 D1 Kamenka Voronezhskaya Oblast'
Russia
Kamenka-Strumilovskaya Ukr. see
Kam"yanka-Buz'ka
91 C3 Kamennomostskiy Russia
91 E2 Kamenolomni Russia
Kamenongue Angola see
Camanongue
83 M2 Kamenskoye Russia
Kamenskoye Ukr. see
Dniprodzerzhyns'k
91 E2 Kamensk-Shakhtinskiy Russia
86 F3 Kamensk-Ural'skiy Russia
89 F2 Kameshkovo Russia
72 C1 Kamet mt. China/India
75 B1 Kamet mt. China/India
122 A3 Kamiesberge mts S. Africa
122 A3 Kamieskroon S. Africa
129 D1 Kamilukuak Lake Can.
119 C3 Kamina Dem. Rep. Congo
129 E1 Kaminak Lake Can.
90 A1 Kamin'-Kashyrs'kyy Ukr.
119 C3 Kamituga Dem. Rep. Congo
128 B2 Kamloops Can.
54 B1 Kamo N.Z.
116 B3 Kamob Sanha Sudan
118 C3 Kamonia Dem. Rep. Congo
119 D2 Kampala Uganda
60 B1 Kampar r. Indon.
60 B1 Kampar Malaysia
100 B1 Kampen Neth.
119 C3 Kampene Dem. Rep. Congo
63 A2 Kamphaeng Phet Thai.
63 B2 Kâmpóng Cham Cambodia
63 B2 Kâmpóng Chhnăng Cambodia
Kâmpóng Saôm Cambodia see
Sihanoukville
63 B2 Kâmpóng Spœ Cambodia
63 B2 Kâmpôt Cambodia
Kampuchea country Asia see
Cambodia
129 C2 Kamsack Can.
86 E3 Kamskoye Vodokhranilishche resr
Russia
117 C4 Kamsuuma Somalia
90 B2 Kam"yanets'-Podil's'kyy Ukr.
90 A1 Kam"yanka-Buz'ka Ukr.
88 B3 Kamyanyets Belarus
91 D2 Kamyshevatskaya Russia
87 D3 Kamyshin Russia
135 D3 Kanab U.S.A.
118 C3 Kananga Dem. Rep. Congo
87 D3 Kanash Russia
138 C3 Kanawha r. U.S.A.
67 C3 Kanazawa Japan
62 A1 Kanbalu Myanmar
63 A2 Kanchanaburi Thai.
73 B3 Kanchipuram India
77 C3 Kandahār Afgh.
86 C2 Kandalaksha Russia
61 C2 Kandangan Indon.
74 A2 Kandh Kot Pak.
114 C3 Kandi Benin

74 A2 Kandiaro Pak.
74 B2 Kandla India
53 C2 Kandos Austr.
121 D2 Kandreho Madag.
73 C4 Kandy Sri Lanka
76 B2 Kandyagash Kazakh.
127 H1 Kane Bassin b. Greenland
91 D2 Kanevskaya Russia
122 B1 Kang Botswana
127 I2 Kangaatsiaq Greenland
114 B3 Kangaba Mali
80 B2 Kangal Turkey
60 B1 Kangar Malaysia
52 A3 Kangaroo Island Austr.
93 H3 Kangasala Fin.
81 C2 Kangāvar Iran
75 C2 Kangchenjunga mt. India/Nepal
70 A2 Kangding China
65 B2 Kangdong N. Korea
61 C2 Kangean, Kepulauan is Indon.
119 D2 Kangen r. S. Sudan
127 J2 Kangeq c. Greenland
127 I2 Kangerlussuaq inlet Greenland
127 J2 Kangerlussuaq inlet Greenland
127 I2 Kangersuatsiaq Greenland
65 B3 Kanggye N. Korea
131 D2 Kangiqsualujjuaq Can.
127 H2 Kangiqsujuaq Can.
131 C1 Kangirsuk Can.
75 C2 Kangmar China
65 A1 Kangping China
72 D2 Kangto mt. China/India
62 A1 Kani Myanmar
118 C3 Kaniama Dem. Rep. Congo
61 C1 Kanibongan Malaysia
86 D2 Kanin, Poluostrov pen. Russia
86 D2 Kanin Nos Russia
86 D2 Kanin Nos, Mys c. Russia
91 C2 Kaniv Ukr.
52 B3 Kaniva Austr.
93 H3 Kankaanpää Fin.
138 B2 Kankakee U.S.A.
114 B3 Kankan Guinea
75 C2 Kanker India
73 B3 Kannur India
115 C3 Kano Nigeria
122 B3 Kanonpunt pt S. Africa
67 B3 Kanoya Japan
75 C2 Kanpur India
136 C2 Kansas r. U.S.A.
137 D3 Kansas state U.S.A.
137 E3 Kansas City KS U.S.A.
137 E3 Kansas City MO U.S.A.
83 H3 Kansk Russia
Kansu prov. China see Gansu
63 B2 Kantaralak Thai.
114 B3 Kantchari Burkina Faso
91 D2 Kantemirovka Russia
49 J4 Kanton atoll Kiribati
97 B2 Kanturk Ireland
123 D2 Kanyamazane S. Africa
123 C1 Kanye Botswana
120 A2 Kaokoveld plat. Namibia
114 A3 Kaolack Senegal
120 B2 Kaoma Zambia
118 C3 Kapanga Dem. Rep. Congo
88 C3 Kapatkyevichy Belarus
100 B2 Kapellen Belgium
121 B2 Kapiri Mposhi Zambia
127 I2 Kapisillit Greenland
130 B2 Kapiskau r. Can.
61 C1 Kapit Malaysia
63 A3 Kapoe Thai.
117 B4 Kapoeta S. Sudan
103 D2 Kaposvár Hungary
102 B1 Kappeln Ger.
65 B1 Kapsan N. Korea
77 D2 Kapshagay Kazakh.
77 D2 Kapshagay, Vodokhranilishche resr
Kazakh.
Kapsukas Lith. see Marijampol
61 B2 Kapuas r. Indon.
52 A2 Kapunda Austr.
130 B3 Kapuskasing Can.
53 D2 Kaputar mt. Austr.
103 D2 Kapuvár Hungary
88 C3 Kapyl' Belarus
77 D2 Kaqung China
114 C4 Kara Togo
111 C3 Kara Ada i. Turkey
77 D2 Kara-Balta Kyrg.
76 C1 Karabalyk Kazakh.
81 D1 Karabaur, Uval hills Kazakh./Uzbek.
Kara-Bogaz-Gol Turkm. see
Garabogazköl
80 B1 Karabük Turkey
76 C2 Karabutak Kazakh.
111 C3 Karacabey Turkey
111 C2 Karacaköy Turkey
81 C1 Karachayevsk Russia
89 D3 Karachev Russia
74 A2 Karachi Pak.
77 D2 Karagandy Kazakh.
77 D2 Karagayly Kazakh.
83 L3 Karaginskiy Zaliv b. Russia
81 D2 Karaj Iran
Kara-Kala Turkm. see
Magtymguly
64 B3 Karakelong i. Indon.
Karaklis Armenia see Vanadzor
77 D2 Kara-Köl Kyrg.
77 D2 Karakol Kyrg.
74 B1 Karakoram Range mts Asia
117 B3 Kara K'orē Eth.

90 B2 Khmel'nyts'kyy Ukr.
Khmer Republic country Asia see Cambodia
Khodzheyli Uzbek. see Xo'jayli
89 E3 Khokhol'skiy Russia
74 B2 Khokhropar Pak.
89 D2 Kholm Russia
89 D2 Kholm-Zhirkovskiy Russia
122 A1 Khomas Highland hills Namibia
89 E3 Khomutovo Russia
79 C2 Khonj Iran
63 B2 Khon Kaen Thai.
62 A1 Khonsa India
83 K2 Khonuu Russia
86 E2 Khorey-Ver Russia
69 D1 Khorinsk Russia
120 A3 Khorixas Namibia
66 B2 Khorol Russia
91 C2 Khorol Ukr.
81 C2 Khorramābād Iran
81 C2 Khorramshahr Iran
77 D3 Khorugh Tajik.
86 F2 Khoshgort Russia
77 C3 Khōst Afgh.
Khotan China see Hotan
90 B2 Khotyn Ukr.
114 B1 Khouribga Morocco
88 C3 Khoyniki Belarus
62 A1 Khreum Myanmar
76 B1 Khromtau Kazakh.
Khrushchev Ukr. see Svitlovods'k
90 B2 Khrystynivka Ukr.
123 B1 Khudumelapye Botswana
77 C2 Khŭjand Tajik.
63 B2 Khu Khan Thai.
78 A2 Khulays Saudi Arabia
74 A1 Khulm Afgh.
75 C2 Khulna Bangl.
Khūnīnshahr Iran see Khorramshahr
79 B2 Khurays Saudi Arabia
74 B1 Khushab Pak.
90 A2 Khust Ukr.
123 C2 Khutsong S. Africa
74 A2 Khuzdar Pak.
81 D2 Khvānsār Iran
81 D3 Khvormüj Iran
81 C2 Khvoy Iran
89 D2 Khvoynaya Russia
74 A1 Khyber Pass Afgh./Pak.
53 D2 Kiama Austr.
64 B3 Kiamba Phil.
119 C3 Kiambi Dem. Rep. Congo
Kiangsi prov. China see Jiangxi
Kiangsu prov. China see Jiangsu
119 D3 Kibaha Tanz.
119 D3 Kibaya Tanz.
119 D3 Kibiti Tanz.
119 C3 Kibombo Dem. Rep. Congo
119 D3 Kibondo Tanz.
119 D3 Kibungo Rwanda
117 H2 Kibre Mengist Eth.
111 B2 Kičevo Macedonia
114 C3 Kidal Mali
99 B3 Kidderminster U.K.
114 A3 Kidira Senegal
74 B1 Kidmang India
54 C1 Kidnappers, Cape N.Z.
102 C1 Kiel Ger.
103 E1 Kielce Pol.
98 B2 Kielder Water resr U.K.
119 C4 Kienge Dem. Rep. Congo
90 C1 Kiev Ukr.
114 A3 Kiffa Maur.
119 D3 Kigali Rwanda
119 D3 Kigoma Tanz.
88 B2 Kihnu i. Estonia
92 I2 Kiiminki Fin.
67 B4 Kii-suidō sea chan. Japan
109 D1 Kikinda Serbia
119 C3 Kikondja Dem. Rep. Congo
59 D3 Kikori P.N.G.
59 D3 Kikori r. P.N.G.
118 B3 Kikwit Dem. Rep. Congo
65 B1 Kilchu N. Korea
97 C2 Kilcock Ireland
97 C2 Kildare Ireland
118 B3 Kilembe Dem. Rep. Congo
143 E2 Kilgore U.S.A.
119 D3 Kilifi Kenya
119 D3 Kilimanjaro vol. Tanz.
119 D3 Kilindoni Tanz.
73 C4 Kilinochchi Sri Lanka
80 B2 Kilis Turkey
90 B2 Kiliya Ukr.
97 B2 Kilkee Ireland
97 D1 Kilkeel U.K.
97 C2 Kilkenny Ireland
111 B2 Kilkis Greece
97 B1 Killala Ireland
97 B1 Killala Bay Ireland
97 B2 Killaloe Ireland
128 C2 Killam Can.
97 B2 Killarney Ireland
143 D2 Killeen U.S.A.
96 B2 Killin U.K.
131 D1 Killiniq (abandoned) Can.
97 B2 Killorglin Ireland
97 B1 Killybegs Ireland
96 B3 Kilmarnock U.K.
53 B3 Kilmore Austr.
119 D3 Kilosa Tanz.
97 C2 Kilrush Ireland
119 C3 Kilwa Dem. Rep. Congo
119 D3 Kilwa Masoko Tanz.

119 D3 Kimambi Tanz.
52 A2 Kimba Austr.
136 C2 Kimball U.S.A.
59 E3 Kimbe P.N.G.
128 C3 Kimberley Can.
122 B2 Kimberley S. Africa
50 B1 Kimberley Plateau Austr.
65 B1 Kimch'aek N. Korea
127 H2 Kimmirut Can.
89 E3 Kimovsk Russia
118 C3 Kimpanga Dem. Rep. Congo
118 B3 Kimpese Dem. Rep. Congo
89 E2 Kimry Russia
61 C1 Kinabalu, Gunung mt. Malaysia
128 C2 Kinbasket Lake Can.
96 C1 Kinbrace U.K.
130 B3 Kincardine Can.
62 A1 Kinchang Myanmar
119 C3 Kinda Dem. Rep. Congo
98 C3 Kinder Scout h. U.K.
129 D2 Kindersley Can.
114 A3 Kindia Guinea
119 C3 Kindu Dem. Rep. Congo
89 F2 Kineshma Russia
78 A2 King Abdullah Economic City Saudi Arabia
118 B3 Kingandu Dem. Rep. Congo
51 B3 Kingaroy Austr.
135 B3 King City U.S.A.
130 C2 King George Islands Can.
88 C2 Kingisepp Russia
51 D3 King Island Austr.
Kingisseppa Estonia see Kuressaare
50 B1 King Leopold Ranges hills Austr.
142 A1 Kingman U.S.A.
135 B3 Kings r. U.S.A.
52 A3 Kingscote Austr.
97 C2 Kingscourt Ireland
99 D3 King's Lynn U.K.
50 B1 King Sound b. Austr.
134 D2 Kings Peak U.S.A.
141 D1 Kingsport U.S.A.
51 D4 Kingston Austr.
130 C3 Kingston Can.
146 C3 Kingston Jamaica
139 E2 Kingston U.S.A.
52 A3 Kingston South East Austr.
98 C3 Kingston upon Hull U.K.
147 D3 Kingstown St Vincent
143 D3 Kingsville U.S.A.
99 B4 Kingswood U.K.
96 B2 Kingussie U.K.
126 F2 King William Island Can.
123 C3 King William's Town S. Africa
67 D3 Kinka-san i. Japan
96 B2 Kinlochleven U.K.
93 F4 Kinna Sweden
97 B3 Kinsale Ireland
118 B3 Kinshasa Dem. Rep. Congo
141 D1 Kinston U.S.A.
88 B2 Kintai Lith.
97 C2 Kintampo Ghana
96 C2 Kintore U.K.
96 B3 Kintyre pen. U.K.
62 A1 Kin-U Myanmar
119 D3 Kiomboi Tanz.
130 C3 Kipawa, Lac l. Can.
119 D3 Kipembawe Tanz.
119 D3 Kipengere Range mts Tanz.
129 D2 Kipling Can.
Kipling Station Can. see Kipling
119 C4 Kipushi Dem. Rep. Congo
119 C4 Kipushia Dem. Rep. Congo
101 D2 Kirchhain Ger.
101 D3 Kirchheimbolanden Ger.
83 I3 Kirenga r. Russia
83 I3 Kirensk Russia
89 E3 Kireyevsk Russia
Kirghizia country Asia see Kyrgyzstan
77 D2 Kirghiz Range mts Kazakh./Kyrg.
Kirgizskaya S.S.R. country Asia see Kyrgyzstan
49 J4 Kiribati country Pacific Ocean
80 B2 Kırıkkale Turkey
89 E2 Kirillov Russia
Kirin China see Jilin
Kirin prov. China see Jilin
Kirinyaga Kenya see Kenya, Mount
89 D2 Kirishi Russia
48 L3 Kiritimati atoll Kiribati
111 C3 Kırkağaç Turkey
98 B3 Kirkby U.K.
98 B2 Kirkby Stephen U.K.
96 C2 Kirkcaldy U.K.
96 B3 Kirkcudbright U.K.
92 J2 Kirkenes Norway
88 B1 Kirkkonummi Fin.
130 B3 Kirkland Lake Can.
111 C2 Kırklareli Turkey
137 E2 Kirksville U.S.A.
81 C2 Kirkūk Iraq
96 C1 Kirkwall U.K.
Kirov Kazakh. see Balpyk Bi
89 D3 Kirov Kaluzhskaya Oblast' Russia
86 D3 Kirov Kirovskaya Oblast' Russia
Kirovabad Азєрб. see Gäncä
Kirovakan Armenia see Vanadzor
Kirovo Ukr. see Kirovohrad
86 E3 Kirovo-Chepetsk Russia
Kirovo-Chepetskiy Russia see Kirovo-Chepetsk
91 C2 Kirovohrad Ukr.
86 C2 Kirovsk Russia
91 D2 Kirovs'ke Ukr.

Kirovskiy Kazakh. see Balpyk Bi
66 B1 Kirovskiy Russia
96 C2 Kirriemuir U.K.
86 E3 Kirs Russia
87 D3 Kirsanov Russia
80 B2 Kırşehir Turkey
74 A2 Kirthar Range mts Pak.
92 H2 Kiruna Sweden
67 C3 Kiryū Japan
89 E2 Kirzhach Russia
119 D3 Kisaki Tanz.
119 C2 Kisangani Dem. Rep. Congo
118 B3 Kisantu Dem. Rep. Congo
60 A1 Kisaran Indon.
82 G3 Kiselevsk Russia
75 C2 Kishanganj India
115 C4 Kishi Nigeria
Kishinev Moldova see Chişinău
67 C4 Kishiwada Japan
77 D1 Kishkenekol' Kazakh.
75 D2 Kishoreganj Bangl.
74 B1 Kishtwar India
119 D3 Kisii Kenya
103 D2 Kiskunfélegyháza Hungary
103 D2 Kiskunhalas Hungary
87 D4 Kislovodsk Russia
117 C5 Kismaayo Somalia
Kismayu Somalia see Kismaayo
119 C3 Kisoro Uganda
111 B3 Kissamos Greece
114 A4 Kissidougou Guinea
141 D3 Kissimmee U.S.A.
141 D3 Kissimmee, Lake U.S.A.
129 D2 Kississing Lake Can.
Kistna r. India see Krishna
119 D3 Kisumu Kenya
103 E2 Kisvárda Hungary
Kisykkamys Kazakh. see Zhanakala
114 B3 Kita Mali
67 D3 Kitaibaraki Japan
66 D3 Kitakami Japan
66 D3 Kitakami-gawa r. Japan
67 B4 Kita-Kyūshū Japan
119 D2 Kitale Kenya
66 D2 Kitami Japan
130 B3 Kitchener Can.
93 J3 Kitee Fin.
119 D2 Kitgum Uganda
128 B2 Kitimat Can.
118 B3 Kitona Dem. Rep. Congo
92 H2 Kittilä Fin.
141 E1 Kitty Hawk U.S.A.
119 D3 Kitunda Tanz.
128 B2 Kitwanga Can.
121 B2 Kitwe Zambia
102 C2 Kitzbühel Austria
101 E3 Kitzingen Ger.
59 D3 Kiunga P.N.G.
92 I3 Kiuruvesi Fin.
92 I2 Kivalo ridge Fin.
90 B1 Kivertsi Ukr.
88 C2 Kiviõli Estonia
91 D2 Kivsharivka Ukr.
119 C3 Kivu, Lac Dem. Rep. Congo/Rwanda
111 C2 Kıyıköy Turkey
86 E3 Kizel Russia
111 C3 Kızılca Dağ mt. Turkey
80 B1 Kızılırmak r. Turkey
87 D4 Kizlyar Russia
Kizyl-Arbat Turkm. see Serdar
92 I1 Kjøllefjord Norway
92 G2 Kjøpsvik Norway
102 C1 Kladno Czech Rep.
102 C2 Klagenfurt am Wörthersee Austria
88 B2 Klaipėda Lith.
94 B1 Klaksvík Faroe Is
134 B2 Klamath r. U.S.A.
134 B2 Klamath Falls U.S.A.
134 B2 Klamath Mountains U.S.A.
60 B1 Klang Malaysia
102 C2 Klatovy Czech Rep.
122 A3 Klawer S. Africa
128 A2 Klawock U.S.A.
128 B2 Kleena Kleene Can.
122 B2 Kleinbegin S. Africa
122 A2 Klein Karas Namibia
122 A2 Kleinsee S. Africa
123 C2 Klerksdorp S. Africa
90 B1 Klesiv Ukr.
89 D3 Kletnya Russia
100 C2 Kleve Ger.
88 C3 Klichaw Belarus
89 D3 Klimavichy Belarus
89 E3 Klimovo Russia
89 E2 Klimovsk Russia
89 E2 Klin Russia
101 F2 Klingenthal Ger.
101 F2 Klínovec mt. Czech Rep.
93 G4 Klintehamn Sweden
89 D3 Klintsy Russia
109 C2 Ključ Bos. & Herz.
103 D1 Kłodzko Pol.
100 C1 Kloosterhaar Neth.
103 D2 Klosterneuburg Austria
101 E1 Klötze (Altmark) Ger.
128 A1 Kluane Lake Can.
103 D1 Kluczbork Pol.
128 A2 Klukwan U.S.A.
89 F2 Klyaz'ma r. Russia
88 C3 Klyetsk Belarus
83 L3 Klyuchi Russia
98 C2 Knaresborough U.K.

93 F3 Knästen h. Sweden
129 E2 Knee Lake Can.
101 E1 Knesebeck Ger.
101 E3 Knetzgau Ger.
109 C2 Knin Croatia
103 C2 Knittelfeld Austria
109 D2 Knjaževac Serbia
Knob Lake Can. see Schefferville
97 B3 Knockboy h. Ireland
100 A2 Knokke-Heist Belgium
141 D1 Knoxville U.S.A.
127 H1 Knud Rasmussen Land reg. Greenland
122 B3 Knysna S. Africa
60 B2 Koba Indon.
76 B1 Kobda Kazakh.
67 C4 Kōbe Japan
København Denmark see Copenhagen
100 C2 Koblenz Ger.
59 C3 Kobroör i. Indon.
88 B3 Kobryn Belarus
Kocaeli Turkey see İzmit
111 B2 Kočani Macedonia
111 C2 Kocasu r. Turkey
109 B1 Kočevje Slovenia
75 C2 Koch Bihar India
89 F3 Kochetovka Russia
73 B4 Kochi India
67 B4 Kōchi Japan
87 D4 Kochubey Russia
75 C2 Kodarma India
126 B3 Kodiak U.S.A.
126 B3 Kodiak Island U.S.A.
123 C1 Kodibeleng Botswana
117 B4 Kodok S. Sudan
90 B2 Kodyma Ukr.
111 C2 Kodzhaele mt. Bulg./Greece
122 A2 Koës Namibia
122 C2 Koffiefontein S. Africa
114 B4 Koforidua Ghana
67 C3 Kōfu Japan
131 D2 Kogaluk r. Can.
117 B5 Kogelo Kenya
114 B1 Kogoni Mali
74 B1 Kohat Pak.
72 D2 Kohima India
88 C2 Kohtla-Järve Estonia
128 A1 Koidern Can.
114 A4 Koidu-Sefadu Sierra Leone
Kokand Uzbek. see Qo'qon
88 B2 Kökar Fin.
Kokchetav Kazakh. see Kokshetau
122 A2 Kokerboom Namibia
88 C3 Kokhanava Belarus
89 F2 Kokhma Russia
92 H3 Kokkola Fin.
88 C2 Koknese Latvia
138 B2 Kokomo U.S.A.
122 B1 Kokong Botswana
123 C2 Kokosi S. Africa
77 E2 Kokpekty Kazakh.
77 C1 Kokshetau Kazakh.
131 D2 Koksoak r. Can.
123 C3 Kokstad S. Africa
Koktokay China see Fuyun
61 D2 Kolaka Indon.
86 C2 Kola Peninsula Russia
92 H2 Kolari Fin.
Kolarovgrad Bulg. see Shumen
114 A3 Kolda Senegal
93 E4 Kolding Denmark
119 C2 Kole Dem. Rep. Congo
107 D2 Koléa Alg.
86 D2 Kolguyev, Ostrov i. Russia
73 B3 Kolhapur India
88 B2 Kolkasrags pt Latvia
75 C2 Kolkata India
73 B4 Kollam India
100 C1 Kollum Neth.
Köln Ger. see Cologne
103 D1 Koło Pol.
103 D1 Kołobrzeg Pol.
114 B3 Kolokani Mali
89 E2 Kolomna Russia
90 B2 Kolomyya Ukr.
114 B3 Kolondiéba Mali
61 D2 Kolonedale Indon.
122 B2 Kolonkwaneng Botswana
82 G3 Kolpashevo Russia
89 E3 Kolpny Russia
Kol'skiy Poluostrov pen. Russia see Kola Peninsula
78 B3 Koluli Eritrea
92 H3 Kolvereid Norway
119 C4 Kolwezi Dem. Rep. Congo
83 L2 Kolyma r. Russia
Kolyma Lowland Russia see Kolymskaya Nizmennost'
Kolyma Range mts Russia see Kolymskiy, Khrebet
83 M2 Kolymskaya Nizmennost' lowland Russia
83 M3 Kolymskiy, Khrebet mts Russia
122 A2 Komaggas S. Africa
67 C3 Komaki Japan
83 M3 Komandorskiye Ostrova is Russia
103 D2 Komárno Slovakia
123 D2 Komati r. S. Africa/Swaziland
123 D2 Komatipoort S. Africa
67 C3 Komatsu Japan
120 A2 Kombat Namibia
119 C3 Kombe Dem. Rep. Congo
Komintern Ukr. see Marhanets'

Kominternivs'ke

90 C2 Kominternivs'ke Ukr.
109 C2 Komiža Croatia
103 D2 Komló Hungary
Kommunarsk Ukr. see Alchevs'k
118 B3 Komono Congo
111 C2 Komotini Greece
Kompong Som Cambodia see Sihanoukville
Komrat Moldova see Comrat
122 B3 Komsberg mts S. Africa
83 H1 Komsomolets, Ostrov i. Russia
89 F2 Komsomol'sk Russia
91 C2 Komsomol's'k Ukr.
83 M2 Komsomol'skiy Chukotskiy Avtonomnyy Okrug Russia
Komsomol'skiy Khanty-Mansiyskiy Avtonomnyy Okrug-Yugra Russia see Yugorsk
87 D4 Komsomol'skiy Respublika Kalmykiya-Khalm'g-Tangch Russia
83 K3 Komsomol'sk-na-Amure Russia
89 E2 Konakovo Russia
75 C3 Kondagaon India
Kondinskoye Khanty-Mansiyskiy Avtonomnyy Okrug-Yugra Russia see Oktyabr'skoye
86 F2 Kondinskoye Khanty-Mansiyskiy Avtonomnyy Okrug-Yugra Russia
119 D3 Kondoa Tanz.
86 C2 Kondopoga Russia
89 E3 Kondrovo Russia
127 J2 Kong Christian IX Land reg. Greenland
127 K2 Kong Christian X Land reg. Greenland
127 J2 Kong Frederik VI Kyst coastal area Greenland
119 C3 Kongolo Dem. Rep. Congo
93 E4 Kongsberg Norway
93 F3 Kongsvinger Norway
77 D3 Kongur Shan mt. China
100 C2 Königswinter Ger.
103 D1 Konin Pol.
109 C2 Konjic Bos. & Herz.
122 A2 Konkiep watercourse Namibia
86 D2 Konosha Russia
91 C1 Konotop Ukr.
103 E1 Końskie Pol.
Konstantinograd Ukr. see Krasnohrad
102 B2 Konstanz Ger.
115 C3 Kontagora Nigeria
63 B2 Kon Tum Vietnam
63 B2 Kon Tum, Cao Nguyên Vietnam
80 B2 Konya Turkey
77 D2 Konyrat Kazakh.
100 C3 Konz Ger.
86 E3 Konzhakovskiy Kamen', Gora mt. Russia
134 C1 Kooskia U.S.A.
128 C3 Kootenay Lake Can.
53 D2 Kootingal Austr.
122 B3 Kootjieskolk S. Africa
92 □B2 Kópasker Iceland
108 B1 Koper Slovenia
93 G4 Köping Sweden
123 C1 Kopong Botswana
93 G4 Kopparberg Sweden
109 C1 Koprivnica Croatia
89 F3 Korablino Russia
73 C3 Koraput India
75 C2 Korba India
101 D2 Korbach Ger.
109 D2 Korçë Albania
109 C2 Korčula Croatia
109 C2 Korčula i. Croatia
70 C2 Korea Bay g. China/N. Korea
65 B3 Korea Strait Japan/S. Korea
65 B1 Korea, North country Asia
65 B2 Korea, South country Asia
89 D3 Korenevo Russia
91 D2 Korenovsk Russia
Korenovskaya Russia see Korenovsk
90 B1 Korets' Ukr.
111 C2 Körfez Turkey
114 B4 Korhogo Côte d'Ivoire
Korinthos Greece see Corinth
103 D2 Kris-hegy h. Hungary
109 D2 Koritnik mt. Albania/Kosovo
Koritsa Albania see Korçë
67 D3 Kōriyama Japan
87 F3 Korkino Russia
111 D3 Korkuteli Turkey
77 E2 Korla China
103 D2 Körmend Hungary
49 I5 Koro i. Fiji
114 B3 Koro Mali
131 D2 Koroc r. Can.
91 D1 Korocha Russia
119 D3 Korogwe Tanz.
59 C2 Koror Palau
103 E2 Körös r. Hungary
90 B1 Korosten' Ukr.
90 B1 Korostyshiv Ukr.
115 D3 Koro Toro Chad
93 H3 Korpo Fin.
66 D1 Korsakov Russia
91 C2 Korsun'-Shevchenkivs'kyy Ukr.
103 E1 Korsze Pol.
116 B3 Korti Sudan
100 A2 Kortrijk Belgium
83 L3 Koryakskaya Sopka, Vulkan vol. Russia
83 M2 Koryakskoye Nagor'ye mts Russia
86 D2 Koryazhma Russia
91 C1 Koryukivka Ukr.
111 C3 Kos Greece
111 C3 Kos i. Greece
91 D2 Kosa Biryuchyy Ostriv i. Ukr.
65 B2 Kosan N. Korea
103 D1 Kościan Pol.
Kosciusko, Mount Austr. see Kosciuszko, Mount
53 C3 Kosciuszko, Mount Austr.
77 E2 Kosh-Agach Russia
67 A4 Koshikijima-rettō is Japan
103 E2 Košice Slovakia
92 H2 Koskullskulle Sweden
65 B2 Kosŏng N. Korea
109 D2 Kosovo country Europe
Kosovska Mitrovica Kosovo see Mitrovicë
48 H3 Kosrae atoll Micronesia
114 B4 Kossou, Lac de l. Côte d'Ivoire
76 C1 Kostanay Kazakh.
110 B2 Kostenets Bulg.
123 C2 Koster S. Africa
116 B3 Kosti Sudan
92 J3 Kostomuksha Russia
90 B1 Kostopil' Ukr.
89 F2 Kostroma Russia
89 F2 Kostroma r. Russia
102 C1 Kostrzyn nad Odrą Pol.
91 D2 Kostyantynivka Ukr.
103 D1 Koszalin Pol.
103 D2 Kszeg Hungary
74 B2 Kota India
60 B2 Kotaagung Indon.
61 C2 Kotabaru Indon.
61 C1 Kota Belud Malaysia
60 B1 Kota Bharu Malaysia
60 B2 Kotabumi Indon.
61 C1 Kota Kinabalu Malaysia
75 C3 Kotaparh India
61 C1 Kota Samarahan Malaysia
86 D3 Kotel'nich Russia
87 D4 Kotel'nikovo Russia
83 K1 Kotel'nyy, Ostrov i. Russia
91 C1 Kotel'va Ukr.
101 E2 Köthen (Anhalt) Ger.
119 D2 Kotido Uganda
93 I3 Kotka Fin.
86 D2 Kotlas Russia
126 B2 Kotlik U.S.A.
109 C2 Kotor Varoš Bos. & Herz.
87 D3 Kotovo Russia
91 E1 Kotovsk Russia
90 B2 Kotovs'k Ukr.
73 C3 Kottagudem India
118 C2 Kotto r. C.A.R.
83 H2 Kotuy r. Russia
126 B2 Kotzebue U.S.A.
126 B2 Kotzebue Sound sea chan. U.S.A.
114 A3 Koubia Guinea
100 A2 Koudekerke Neth.
114 B3 Koudougou Burkina Faso
122 B3 Kougaberge mts S. Africa
118 B3 Koulamoutou Gabon
114 B3 Koulikoro Mali
118 B2 Koum Cameroon
118 B2 Koumra Chad
114 A3 Koundâra Guinea
Kounradskiy Kazakh. see Konyrat
151 D2 Kourou Fr. Guiana
114 B3 Kouroussa Guinea
115 D3 Kousséri Cameroon
114 B3 Koutiala Mali
93 I3 Kouvola Fin.
109 D1 Kovačica Serbia
92 J2 Kovdor Russia
90 A1 Kovel' Ukr.
Kovno Lith. see Kaunas
89 F2 Kovrov Russia
51 D1 Kowanyama Austr.
54 B2 Kowhitirangi N.Z.
Koyamutthoor India see Coimbatore
111 C3 Köyceğiz Turkey
86 D2 Koyda Russia
126 B2 Koyukuk r. U.S.A.
111 B2 Kozani Greece
90 C1 Kozelets' Ukr.
89 E3 Kozel'sk Russia
73 B3 Kozhikode India
90 B2 Kozyatyn Ukr.
63 A2 Kra, Isthmus of Myanmar/Thai.
63 A3 Krabi Thai.
63 A2 Kra Buri Thai.
63 B2 Krâchéh Cambodia
93 E4 Kragerø Norway
100 B1 Kraggenburg Neth.
109 D2 Kragujevac Serbia
60 B2 Krakatau i. Indon.
103 D1 Kraków Pol.
109 D2 Kraljevo Serbia
91 D2 Kramators'k Ukr.
93 G3 Kramfors Sweden
111 B3 Kranidi Greece
102 C2 Kranj Slovenia
123 D2 Kranskop S. Africa
86 E1 Krasino Russia
88 C2 Krāslava Latvia
101 F2 Kraslice Czech Rep.
89 D3 Krasnapollye Belarus
89 D3 Krasnaya Gora Russia
89 F2 Krasnaya Gorbatka Russia
Krasnoarmeysk Kazakh. see Taiynsha
87 D3 Krasnoarmeysk Russia
Krasnoarmeyskaya Russia see Poltavskaya
91 D2 Krasnoarmiys'k Ukr.
86 D2 Krasnoborsk Russia
91 D2 Krasnodar Russia
91 D2 Krasnodarskoye Vodokhranilishche resr Russia
91 D2 Krasnodon Ukr.
88 C2 Krasnogorodsk Russia
91 D2 Krasnohrad Ukr.
91 C2 Krasnohvardiys'ke Ukr.
86 E3 Krasnokamsk Russia
89 D2 Krasnomayskiy Russia
91 C2 Krasnoperekops'k Ukr.
87 D3 Krasnoslobodsk Russia
86 F3 Krasnotur'insk Russia
86 E3 Krasnoufimsk Russia
86 E2 Krasnovishersk Russia
Krasnovodsk Turkm. see Türkmenbaşy
83 H3 Krasnoyarsk Russia
89 E3 Krasnoye Russia
83 M2 Krasnoye, Ozero l. Russia
89 F2 Krasnoye-na-Volge Russia
103 E1 Krasnystaw Pol.
89 D3 Krasnyy Russia
Krasnyy Kamyshanik Russia see Komsomol'skiy
89 E2 Krasnyy Kholm Russia
91 D2 Krasnyy Luch Ukr.
91 E2 Krasnyy Sulin Russia
90 B2 Krasyliv Ukr.
Kraulshavn Greenland see Nuussuaq
100 C2 Krefeld Ger.
91 C2 Kremenchuk Ukr.
91 C2 Kremenchuts'ke Vodoskhovyshche resr Ukr.
90 B1 Kremenets' Ukr.
103 D2 Křemešník h. Czech Rep.
Kremges Ukr. see Svitlovods'k
91 D2 Kreminna Ukr.
136 B2 Kremmling U.S.A.
103 D2 Krems an der Donau Austria
89 D2 Kresttsy Russia
88 B2 Kretinga Lith.
100 C2 Kreuzau Ger.
101 C2 Kreuztal Ger.
118 A2 Kribi Cameroon
111 B3 Krikellos Greece
66 D1 Kril'on, Mys c. Russia
111 B3 Krios, Akrotirio pt Greece
73 C3 Krishna r. India
73 C3 Krishna, Mouths of the India
75 C2 Krishnanagar India
93 E4 Kristiansand Norway
93 F4 Kristianstad Sweden
92 E3 Kristiansund Norway
93 F4 Kristinehamn Sweden
Kristinopol' Ukr. see Chervonohrad
Kriti i. Greece see Crete
111 C3 Kritiko Pelagos sea Greece
110 B2 Kriva Palanka Macedonia
Krivoy Rog Ukr. see Kryvyy Rih
109 C1 Križevci Croatia
108 B1 Krk i. Croatia
92 F3 Krokom Sweden
91 C1 Krolevets' Ukr.
89 E3 Kromy Russia
101 E2 Kronach Ger.
63 B2 Krŏng Kaôh Kŏng Cambodia
127 J2 Kronprins Frederik Bjerge nunataks Greenland
123 C2 Kroonstad S. Africa
91 E2 Kropotkin Russia
103 E2 Krosno Pol.
103 D1 Krotoszyn Pol.
60 B2 Krui Indon.
122 B3 Kruisfontein S. Africa
109 C2 Krujë Albania
111 C2 Krumovgrad Bulg.
Krung Thep Thai. see Bangkok
88 C3 Krupki Belarus
109 D2 Kruševac Serbia
101 E2 Krušné hory mts Czech Rep.
128 A2 Kruzof Island U.S.A.
89 D3 Krychaw Belarus
91 D2 Krylovskaya Russia
91 D3 Krymsk Russia
Krymskaya Russia see Krymsk
Kryms'kyy Pivostriv see Crimea pen. Ukr.
Krystynopol Ukr. see Chervonohrad
91 C2 Kryvyy Rih Ukr.
90 B2 Kryzhopil' Ukr.
114 B2 Ksabi Alg.
107 D2 Ksar el Boukhari Alg.
114 B1 Ksar el Kebir Morocco
Ksar-es-Souk Morocco see Er Rachidia
89 E3 Kshenskiy Russia
78 B2 Kū', Jabal al h. Saudi Arabia
61 C1 Kuala Belait Brunei
Kuala Dungun Malaysia see Dungun
60 B1 Kuala Kangsar Malaysia
60 B1 Kuala Kerai Malaysia
60 B1 Kuala Lipis Malaysia
60 B1 Kuala Lumpur Malaysia
61 C2 Kualapembuang Indon.
60 B1 Kuala Terengganu Malaysia
60 B2 Kualatungal Indon.
61 C1 Kuamut Malaysia
65 A1 Kuandian China
60 B1 Kuantan Malaysia
91 D2 Kuban' r. Russia
89 E2 Kubenskoye, Ozero l. Russia
110 C2 Kubrat Bulg.
60 B2 Kubu Indon.
61 C1 Kubuang Indon.
60 C1 Kuching Malaysia
109 C2 Kuçovë Albania
61 C1 Kudat Malaysia
61 C2 Kudus Indon.
102 C2 Kufstein Austria
127 G2 Kugaaruk Can.
91 D2 Kugey Russia
126 E2 Kugluktuk Can.
126 D2 Kugmallit Bay Can.
92 I3 Kuhmo Fin.
79 C2 Kūhrān, Kūh-e mt. Iran
122 A1 Kuis Namibia
Kuitin China see Kuytun
120 A2 Kuito Angola
92 I2 Kuivaniemi Fin.
65 B2 Kujang N. Korea
66 D2 Kuji Japan
67 B4 Kujū-san vol. Japan
109 D2 Kukës Albania
76 B3 Kükürtli Turkm.
111 C3 Kula Turkey
75 D2 Kula Kangri mt. Bhutan/China
76 B2 Kulandy Kazakh.
88 B2 Kuldīga Latvia
Kuldja China see Yining
122 B1 Kule Botswana
101 E2 Kulmbach Ger.
77 C3 Külob Tajik.
76 B2 Kul'sary Kazakh.
111 D3 Kulübe Tepe mt. Turkey
77 D1 Kulunda Russia
77 D1 Kulundinskoye, Ozero salt l. Russia
127 J2 Kulusuk Greenland
67 C3 Kumagaya Japan
67 B4 Kumamoto Japan
67 C4 Kumano Japan
110 B2 Kumanovo Macedonia
114 B4 Kumasi Ghana
118 A2 Kumba Cameroon
Kum-Dag Turkm. see Gumdag
78 B2 Kumdah Saudi Arabia
87 E3 Kumertau Russia
119 D2 Kumi Uganda
111 C3 Kumkale Turkey
93 G4 Kumla Sweden
115 D3 Kumo Nigeria
62 A1 Kumon Range mts Myanmar
62 B2 Kumphawapi Thai.
Kumul China see Hami
120 A2 Kunene r. Angola/Namibia
77 D2 Kungei Alatau mts Kazakh./Kyrg.
93 F4 Kungsbacka Sweden
118 B2 Kungu Dem. Rep. Congo
86 E3 Kungur Russia
62 A1 Kunhing Myanmar
62 A1 Kunlong Myanmar
75 B1 Kunlun Shan mts China
71 A3 Kunming China
50 B1 Kununurra Austr.
101 D3 Künzelsau Ger.
92 I3 Kuopio Fin.
109 C1 Kupa r. Croatia/Slovenia
59 C3 Kupang Indon.
88 B2 Kupiškis Lith.
111 C2 Küplü Turkey
128 A2 Kupreanof Island U.S.A.
91 D2 Kup''yans'k Ukr.
77 E2 Kuqa China
81 C2 Kür r. Azer.
67 B4 Kurashiki Japan
75 C2 Kurasia India
67 B3 Kurayoshi Japan
89 E3 Kurchatov Russia
81 C2 Kurdistan reg. Asia
67 B4 Kure Japan
57 I7 Kure Atoll U.S.A.
88 B2 Kuressaare Estonia
87 F3 Kurgan Russia
Kuria Muria Islands Oman see Ḥalāniyāt, Juzur al
93 H3 Kurikka Fin.
156 C2 Kuril Basin Sea of Okhotsk
69 F1 Kuril Islands is Russia
69 F1 Kuril'sk Russia
Kuril'skiye Ostrova Russia see Kuril Islands
156 C3 Kuril Trench N. Pacific Ocean
89 E3 Kurkino Russia
Kurmashkino Kazakh. see Kurshim
117 B3 Kurmuk Sudan
73 B3 Kurnool India
53 D2 Kurri Kurri Austr.
88 B2 Kuršnai Lith.
78 B2 Kursh, Jabal mt. Saudi Arabia
77 E2 Kurshim Kazakh.
89 E3 Kursk Russia
109 D2 Kuršumlija Serbia
122 B2 Kuruman S. Africa
122 B2 Kuruman watercourse S. Africa
67 B4 Kurume Japan
83 I3 Kurumkan Russia
73 C4 Kurunegala Sri Lanka
81 D1 Kuryk Kazakh.
111 C3 Kuşadası Turkey
111 C3 Kuşadası, Gulf of b. Turkey
111 C2 Kuş Gölü l. Turkey
91 D2 Kushchevskaya Russia
66 D2 Kushiro Japan

Luiza

118 C3 Luiza Dem. Rep. Congo
70 B2 Lujiang China
Lukapa Angola see Lucapa
109 C2 Lukavac Bos. & Herz.
118 B3 Lukenie r. Dem. Rep. Congo
142 A2 Lukeville U.S.A.
89 E3 Lukhovitsy Russia
Lukou China see Zhuzhou
103 E1 Łuków Pol.
120 B2 Lukulu Zambia
92 H2 Luleå Sweden
92 H2 Luleälven r. Sweden
111 C2 Lüleburgaz Turkey
70 B2 Lüliang Shan mts China
143 D3 Luling U.S.A.
Luluabourg Dem. Rep. Congo see Kananga
61 C2 Lumajang Indon.
75 C1 Lumajangdong Co salt l. China
Lumbala Moxico Angola see Lumbala Kaquengue
Lumbala Moxico Angola see Lumbala N'guimbo
120 B2 Lumbala Kaquengue Angola
120 B2 Lumbala N'guimbo Angola
140 C2 Lumberton MS U.S.A.
141 E2 Lumberton NC U.S.A.
61 C1 Lumbis Indon.
106 B1 Lumbrales Spain
63 B2 Lumphăt Cambodia
129 D2 Lumsden Can.
54 A3 Lumsden N.Z.
93 F4 Lund Sweden
121 C2 Lundazi Zambia
99 A4 Lundy U.K.
101 E1 Lüneburg Ger.
101 E1 Lüneburger Heide reg. Ger.
100 C2 Lünen Ger.
105 D2 Lunéville France
120 B2 Lunga r. Zambia
114 A4 Lungi Sierra Leone
Lungleh India see Lunglei
75 D2 Lunglei India
120 B2 Lungwebungu r. Zambia
74 B2 Luni r. India
88 C3 Luninyets Belarus
104 C3 L'Union France
114 A4 Lunsar Sierra Leone
77 E2 Luntai China
71 A3 Luodian China
71 B3 Luoding China
70 B2 Luohe China
70 B2 Luoyang China
118 B3 Luozi Dem. Rep. Congo
121 B2 Lupane Zimbabwe
Lupanshui China see Liupanshui
110 B1 Lupeni Romania
121 C2 Lupilichi Moz.
101 F2 Luppa Ger.
95 B3 Lurgan U.K.
Luring China see Gêrzê
121 D2 Lúrio Moz.
121 D2 Lurio r. Moz.
92 F2 Lurøy Norway
121 B2 Lusaka Zambia
118 C3 Lusambo Dem. Rep. Congo
109 C2 Lushnjë Albania
70 C2 Lüshunkou China
123 C3 Lusikisiki S. Africa
136 C2 Lusk U.S.A.
Luso Angola see Luena
76 B3 Lūt, Kavīr-e des. Iran
99 C4 Luton U.K.
61 C1 Lutong Malaysia
129 C1 Łutselk'e Can.
90 B1 Luts'k Ukr.
55 F3 Lützow-Holm Bay Antarctica
122 B2 Lutzputs S. Africa
122 A3 Lutzville S. Africa
117 C4 Luuq Somalia
137 D2 Luverne U.S.A.
119 C3 Luvua r. Dem. Rep. Congo
120 B2 Luvuei Angola
123 D1 Luvuvhu r. S. Africa
119 D3 Luwegu r. Tanz.
119 D2 Luwero Uganda
61 D2 Luwuk Indon.
100 C3 Luxembourg country Europe
100 C3 Luxembourg Lux.
105 D2 Luxeuil-les-Bains France
Luxi China see Mangshi
123 C3 Luxolweni S. Africa
116 B2 Luxor Egypt
100 B2 Luyksgestel Neth.
86 D2 Luza Russia
Luzern Switz. see Lucerne
62 B1 Luzhai China
71 A3 Luzhou China
154 C1 Luziânia Brazil
151 E3 Luzilândia Brazil
64 B2 Luzon i. Phil.
64 B1 Luzon Strait Phil./Taiwan
109 C3 Luzzi Italy
90 A2 L'viv Ukr.
L'vov Ukr. see L'viv
Lwów Ukr. see L'viv
88 C3 Lyakhavichy Belarus
Lyallpur Pak. see Faisalabad
89 D2 Lychkovo Russia
92 G3 Lycksele Sweden
55 C2 Lyddan Ice Rise Antarctica
88 C3 Lyel'chytsy Belarus
88 C3 Lyepyel' Belarus
136 A2 Lyman U.S.A.
99 B4 Lyme Bay U.K.
99 B4 Lyme Regis U.K.
139 D3 Lynchburg U.S.A.
52 A2 Lyndhurst Austr.
129 D2 Lynn Lake Can.
134 B1 Lynnwood U.S.A.
129 D1 Lynx Lake Can.
105 C2 Lyon France
Lyons France see Lyon
89 D2 Lyozna Belarus
103 E1 Łysica h. Pol.
86 E3 Lys'va Russia
91 D2 Lysychans'k Ukr.
87 D3 Lysyye Gory Russia
98 B3 Lytham St Anne's U.K.
88 C3 Lyuban' Belarus
90 C2 Lyubashivka Ukr.
89 E2 Lyubertsy Russia
90 B1 Lyubeshiv Ukr.
89 F2 Lyubim Russia
90 A1 Lyuboml' Ukr.
91 D2 Lyubotyn Ukr.
89 D2 Lyubytino Russia
89 D3 Lyudinovo Russia

M

80 B2 Ma'ān Jordan
70 B2 Ma'anshan China
88 C2 Maardu Estonia
78 B3 Ma'āriḍ, Banī des. Saudi Arabia
80 B2 Ma'arrat an Nu'mān Syria
100 B1 Maarssen Neth.
100 B2 Maas r. Neth.
100 B2 Maaseik Belgium
64 B2 Maasin Phil.
100 B2 Maastricht Neth.
121 C3 Mabalane Moz.
78 B3 Ma'bar Yemen
150 D2 Mabaruma Guyana
98 D3 Mablethorpe U.K.
123 C2 Mabopane S. Africa
121 C3 Mabote Moz.
122 B2 Mabule Botswana
122 B1 Mabutsane Botswana
155 D2 Macaé Brazil
121 C2 Macaloge Moz.
126 F2 MacAlpine Lake Can.
71 B3 Macao aut. reg. China
151 D2 Macapá Brazil
150 B3 Macará Ecuador
155 D1 Macarani Brazil
121 C3 Macarretane Moz.
Macassar Indon. see Makassar
Macassar Strait Indon. see Makassar, Selat
121 C2 Macatanja Moz.
151 F3 Macau Brazil
98 B3 Macclesfield U.K.
50 B2 Macdonald, Lake imp. l. Austr.
50 C2 MacDonnell Ranges mts Austr.
130 A2 MacDowell Lake Can.
96 C2 Macduff U.K.
106 B1 Macedo de Cavaleiros Port.
52 B3 Macedon mt. Austr.
111 B2 Macedonia country Europe
151 F3 Maceió Brazil
108 B2 Macerata Italy
52 A2 Macfarlane, Lake imp. l. Austr.
97 B3 Macgillycuddy's Reeks mts Ireland
74 A2 Mach Pak.
155 C2 Machado Brazil
121 C3 Machaila Moz.
119 D3 Machakos Kenya
150 B3 Machala Ecuador
68 C2 Machali China
121 C3 Machanga Moz.
Machaze Moz. see Chitobe
70 B2 Macheng China
138 B2 Machesney Park U.S.A.
139 F2 Machias U.S.A.
73 C3 Machilipatnam India
121 C2 Machinga Malawi
150 B1 Machiques Venez.
150 B4 Machu Picchu tourist site Peru
99 B3 Machynlleth U.K.
123 D2 Macia Moz.
Macias Nguema i. Equat. Guinea see Bioko
110 C1 Măcin Romania
114 B3 Macina Mali
53 D1 Macintyre r. Austr.
51 D2 Mackay Austr.
50 B2 Mackay, Lake imp. l. Austr.
128 C1 MacKay Lake Can.
128 B2 Mackenzie Can.
128 A1 Mackenzie r. Can.
Mackenzie Guyana see Linden
Mackenzie atoll Micronesia see Ulithi
55 H3 Mackenzie Bay Antarctica
126 C2 Mackenzie Bay Can.
126 E1 Mackenzie King Island Can.
128 A1 Mackenzie Mountains Can.
Mackillop, Lake imp. l. Austr. see Yamma Yamma, Lake
129 D2 Macklin Can.
53 D2 Macksville Austr.
53 D1 Maclean Austr.
123 C3 Maclear S. Africa
50 A2 MacLeod, Lake dry lake Austr.
138 A2 Macomb U.S.A.
108 A2 Macomer Italy
121 D2 Macomia Moz.
105 C2 Mâcon France
141 D2 Macon GA U.S.A.
137 E3 Macon MO U.S.A.
140 C2 Macon MS U.S.A.
53 C2 Macquarie r. Austr.
48 G9 Macquarie Island S. Pacific Ocean
53 C2 Macquarie Marshes Austr.
53 C2 Macquarie Mountain Austr.
156 D9 Macquarie Ridge S. Pacific Ocean
55 H2 Mac. Robertson Land reg. Antarctica
97 B3 Macroom Ireland
52 A1 Macumba watercourse Austr.
145 C3 Macuspana Mex.
144 B2 Macuzari, Presa resr Mex.
123 D2 Madadeni S. Africa
121 □D3 Madagascar country Africa
159 D5 Madagascar Ridge Indian Ocean
115 D2 Madama Niger
111 B2 Madan Bulg.
59 D3 Madang P.N.G.
139 D1 Madawaska r. Can.
62 A1 Madaya Myanmar
150 D3 Madeira r. Brazil
114 A1 Madeira terr. N. Atlantic Ocean
131 D3 Madeleine, Îles de la is Can.
99 B3 Madeley U.K.
144 B2 Madera Mex.
135 B3 Madera U.S.A.
Madgaon India see Margao
74 B2 Madhya Pradesh state India
123 C2 Madibogo S. Africa
118 B3 Madingou Congo
121 □D2 Madirovalo Madag.
138 B3 Madison IN U.S.A.
137 D2 Madison SD U.S.A.
138 B2 Madison WI U.S.A.
138 C3 Madison WV U.S.A.
134 D1 Madison r. U.S.A.
138 B3 Madisonville U.S.A.
61 C2 Madiun Indon.
119 D2 Mado Gashi Kenya
88 C2 Madona Latvia
78 A2 Madrakah Saudi Arabia
79 C3 Madrakah, Ra's c. Oman
Madras India see Chennai
134 B2 Madras U.S.A.
145 C2 Madre, Laguna lag. Mex.
143 D3 Madre, Laguna lag. U.S.A.
150 C4 Madre de Dios r. Peru
145 B3 Madre del Sur, Sierra mts Mex.
144 B2 Madre Occidental, Sierra mts Mex.
145 B2 Madre Oriental, Sierra mts Mex.
106 C1 Madrid Spain
106 C2 Madridejos Spain
61 C2 Madura i. Indon.
61 C2 Madura, Selat sea chan. Indon.
73 B4 Madurai India
121 B2 Madzivazvido Zimbabwe
67 C3 Maebashi Japan
62 A2 Mae Hong Son Thai.
62 A1 Mae Sai Thai.
62 A2 Mae Sariang Thai.
99 B4 Maesteg U.K.
62 A2 Mae Suai Thai.
121 □D2 Maevatanana Madag.
123 C2 Mafadi mt. Lesotho/S. Africa
Mafeking S. Africa see Mahikeng
123 C2 Mafeteng Lesotho
53 C3 Maffra Austr.
119 D3 Mafia Island Tanz.
119 D3 Mafinga Tanz.
154 C3 Mafra Brazil
83 L3 Magadan Russia
Magallanes Chile see Punta Arenas
Magallanes, Estrecho de Chile see Magellan, Strait of
150 B2 Magangué Col.
140 B1 Magazine Mountain h. U.S.A.
114 A4 Magburaka Sierra Leone
69 E1 Magdagachi Russia
144 A1 Magdalena Mex.
142 B2 Magdalena U.S.A.
144 A2 Magdalena, Bahía b. Mex.
101 E1 Magdeburg Ger.
153 A5 Magellan, Strait of sea chan. Chile
Maggiore, Lago Italy see Maggiore, Lake
108 A1 Maggiore, Lake l. Italy
116 B2 Maghâghah Egypt
97 C1 Magherafelt U.K.
87 E3 Magnitogorsk Russia
140 B2 Magnolia U.S.A.
121 C2 Màgoé Moz.
130 C3 Magog Can.
131 D2 Magpie, Lac l. Can.
114 A3 Magta' Laḥjar Maur.
81 D2 Magtymguly Turkm.
119 D3 Magu Tanz.
151 E3 Maguarinho, Cabo c. Brazil
123 D2 Magude Moz.
90 B2 Măgura, Dealul h. Moldova
62 A1 Magwe Myanmar
81 C2 Mahābād Iran
74 B2 Mahajan India
121 □D2 Mahajanga Madag.
61 C2 Mahakam r. Indon.
123 C1 Mahalapye Botswana
121 □D2 Mahalevona Madag.
75 C2 Mahanadi r. India
121 □D2 Mahanoro Madag.
74 B3 Maharashtra state India
63 B2 Maha Sarakham Thai.
121 □D2 Mahavavy r. Madag.
68 B3 Mahbubnagar India
78 B2 Mahd adh Dhahab Saudi Arabia
107 D2 Mahdia Alg.
150 D2 Mahdia Guyana
113 I6 Mahé i. Seychelles
75 C3 Mahendragiri mt. India
119 D3 Mahenge Tanz.
54 B3 Maheno N.Z.
74 B2 Mahesana India
74 B2 Mahi r. India
54 C1 Mahia Peninsula N.Z.
123 C2 Mahikeng S. Africa
89 D3 Mahilyow Belarus
Mahón Spain see Maó
114 B3 Mahou Mali
Mahsana India see Mahesana
74 B2 Mahuva India
111 C2 Mahya Dağı mt. Turkey
106 B1 Maia Port.
Maiaia Moz. see Nacala
147 C3 Maicao Col.
74 A1 Maīdān Shahr Afgh.
129 D2 Maidstone Can.
99 D4 Maidstone U.K.
115 D3 Maiduguri Nigeria
75 D2 Maijdi Bangl.
75 C2 Mailani India
76 C3 Maīmanah Afgh.
101 D2 Main r. Ger.
118 B3 Mai-Ndombe, Lac l. Dem. Rep. Congo
101 E3 Main-Donau-Kanal canal Ger.
139 F1 Maine state U.S.A.
131 D3 Maine, Gulf of Can./U.S.A.
62 A1 Maingkwan Myanmar
96 C1 Mainland i. Scotland U.K.
96 □ Mainland i. Scotland U.K.
121 □D2 Maintirano Madag.
101 D2 Mainz Ger.
150 C1 Maiquetía Venez.
120 B3 Maitengwe Botswana
53 D2 Maitland N.S.W. Austr.
52 A2 Maitland S.A. Austr.
146 B3 Maíz, Islas del is Nic.
67 C3 Maizuru Japan
61 C2 Majene Indon.
119 D2 Majī Eth.
107 D2 Majorca i. Spain
Majunga Madag. see Mahajanga
123 C2 Majwemasweu S. Africa
118 B3 Makabana Congo
61 C2 Makale Indon.
119 C3 Makamba Burundi
77 E2 Makanshy Kazakh.
118 B2 Makanza Dem. Rep. Congo
90 B1 Makariv Ukr.
69 F1 Makarov Russia
160 B1 Makarov Basin Arctic Ocean
109 C2 Makarska Croatia
61 C2 Makassar Indon.
61 C2 Makassar, Selat Indon.
76 B2 Makat Kazakh.
119 D3 Makatapora Tanz.
123 D2 Makatini Flats lowland S. Africa
114 A4 Makeni Sierra Leone
120 B3 Makgadikgadi depr. Botswana
87 D4 Makhachkala Russia
123 C1 Makhado S. Africa
76 B2 Makhambet Kazakh.
119 D3 Makindu Kenya
77 D1 Makinsk Kazakh.
91 D2 Makiyivka Ukr.
Makkah Saudi Arabia see Mecca
131 E2 Makkovik Can.
103 E2 Makó Hungary
118 B2 Makokou Gabon
119 D3 Makongolosi Tanz.
122 B2 Makopong Botswana
119 C2 Makoro Dem. Rep. Congo
118 B3 Makoua Congo
111 B3 Makrakomi Greece
79 D2 Makran reg. Iran/Pak.
74 A2 Makran Coast Range mts Pak.
89 E2 Maksatikha Russia
81 C2 Mākū Iran
62 A1 Makum India
67 B4 Makurazaki Japan
115 C4 Makurdi Nigeria
92 G2 Malå Sweden
146 B4 Mala, Punta pt Panama
73 B3 Malabar Coast India
118 A2 Malabo Equat. Guinea
155 D1 Malacacheta Brazil
Malacca Malaysia see Melaka
60 A1 Malacca, Strait of Indon./Malaysia
134 D2 Malad City U.S.A.
88 C3 Maladzyechna Belarus
106 C2 Málaga Spain
Malagasy Republic country Africa see Madagascar
121 □D3 Malaimbandy Madag.
97 B1 Málainn Mhóir Ireland
48 H4 Malaita i. Solomon Is
117 B4 Malakal S. Sudan
48 H5 Malakula i. Vanuatu
61 D2 Malamala Indon.
61 C2 Malang Indon.
Malange Angola see Malanje
120 A1 Malanje Angola
93 G4 Mälaren l. Sweden
153 B3 Malargüe Arg.
130 C3 Malartic Can.
88 B3 Malaryta Belarus
80 B2 Malatya Turkey

Mölnlycke

74 B2 Merta India
99 B4 Merthyr Tydfil U.K.
119 D2 Merti Plateau Kenya
106 B2 Mértola Port.
76 B2 Mertvyy Kultuk, Sor dry lake Kazakh.
119 D3 Meru vol. Tanz.
122 B3 Merweville S. Africa
80 B1 Merzifon Turkey
100 C3 Merzig Ger.
142 A2 Mesa AZ U.S.A.
142 C2 Mesa NM U.S.A.
137 E1 Mesabi Range hills U.S.A.
109 C2 Mesagne Italy
142 B2 Mescalero U.S.A.
143 C2 Mescalero Ridge U.S.A.
101 D2 Meschede Ger.
89 E3 Meshchovsk Russia
Meshed Iran see Mashhad
91 E2 Meshkovskaya Russia
142 B2 Mesilla U.S.A.
111 B2 Mesimeri Greece
111 B3 Mesolongi Greece
115 C1 Messaad Alg.
121 D2 Messalo r. Moz.
109 C3 Messina Italy
109 C3 Messina, Strait of str. Italy
Messina, Stretta di Italy see Messina, Strait of
111 B3 Messini Greece
111 B3 Messiniakos Kolpos g. Greece
111 B2 Mesta r. Bulg.
Mesta r. Greece see Nestos
150 C2 Meta r. Col./Venez.
130 C3 Métabetchouan Can.
127 H2 Meta Incognita Peninsula Can.
140 B3 Metairie U.S.A.
152 B2 Metán Arg.
111 B3 Methoni Greece
109 C2 Metković Croatia
121 C2 Metoro Moz.
60 B2 Metro Indon.
100 C3 Mettlach Ger.
135 C3 Mettler U.S.A.
117 B4 Metu Eth.
105 D2 Metz France
100 B2 Meuse r. Belgium/France
105 C2 Meuse, Côtes de ridge France
143 D2 Mexia U.S.A.
144 A1 Mexicali Mex.
144 B2 Mexico country Central America
México Mex. see Mexico City
137 E3 Mexico U.S.A.
125 I7 Mexico, Gulf of Mex./U.S.A.
145 C3 Mexico City Mex.
81 D2 Meybod Iran
101 F1 Meyenburg Ger.
83 M2 Meynypil'gyno Russia
86 D2 Mezen' Russia
86 D2 Mezen' r. Russia
86 E1 Mezhdusharskiy, Ostrov i. Russia
103 E2 Meztúr Hungary
132 C4 Mezquital r. Mex.
144 B2 Mezquitic Mex.
88 C2 Mežvidi Latvia
121 C2 Mfuwe Zambia
89 D3 Mglin Russia
123 D2 Mhlume Swaziland
74 B2 Mhow India
145 C3 Miahuatlán Mex.
106 B2 Miajadas Spain
141 D3 Miami FL U.S.A.
143 E1 Miami OK U.S.A.
141 D3 Miami Beach U.S.A.
81 C2 Mīāndoāb Iran
121 □D2 Miandrivazo Madag.
81 C2 Mīāneh Iran
71 A3 Mianning China
74 B1 Mianwali Pak.
Mianyang Hubei China see Xiantao
70 A2 Mianyang Sichuan China
121 □D2 Miarinarivo Madag.
87 F3 Miass Russia
103 D1 Miastko Pol.
128 C2 Mica Creek Can.
103 E2 Michalovce Slovakia
138 B1 Michigan state U.S.A.
138 B2 Michigan, Lake U.S.A.
138 B2 Michigan City U.S.A.
138 B1 Michipicoten Bay Can.
130 B3 Michipicoten Island Can.
130 B3 Michipicoten River Can.
Michurin Bulg. see Tsarevo
89 F3 Michurinsk Russia
156 D5 Micronesia is Pacific Ocean
48 G3 Micronesia, Federated States of country N. Pacific Ocean
158 E6 Mid-Atlantic Ridge Atlantic Ocean
100 A2 Middelburg Neth.
123 C3 Middelburg E. Cape S. Africa
123 C2 Middelburg Mpumalanga S. Africa
100 B2 Middelharnis Neth.
134 B2 Middle Alkali Lake U.S.A.
73 D3 Middle Andaman i. India
Middle Congo country Africa see Congo
136 D2 Middle Loup r. U.S.A.
138 C3 Middlesboro U.S.A.
98 C2 Middlesbrough U.K.
139 E2 Middletown NY U.S.A.
138 C3 Middletown OH U.S.A.
78 B3 Mīdī Yemen
130 C3 Midland Can.
138 C2 Midland MI U.S.A.
143 C2 Midland TX U.S.A.
97 B3 Midleton Ireland
Midnapore India see Medinipur
94 B1 Miðvágur Faroe Is
Midway Oman see Thamarīt
57 T7 Midway Islands terr. N. Pacific Ocean
109 D2 Midzhur mt. Bulg./Serbia
103 E1 Mielec Pol.
110 C1 Miercurea-Ciuc Romania
106 B1 Mieres del Camín Spain
101 E1 Mieste Ger.
145 C3 Miguel Alemán, Presa resr Mex.
144 B2 Miguel Auza Mex.
144 B2 Miguel Hidalgo, Presa resr Mex.
63 A2 Migyaunglaung Myanmar
67 B4 Mihara Japan
89 E3 Mikhaylov Russia
Mikhaylovgrad Bulg. see Montana
66 B2 Mikhaylovka Primorskiy Kray Russia
Mikhaylovka Tul'skaya Oblast' Russia see Kimovsk
87 D3 Mikhaylovka Volgogradskaya Oblast' Russia
77 D1 Mikhaylovskoye Russia
93 I3 Mikkeli Fin.
86 E2 Mikun' Russia
67 C3 Mikuni-sanmyaku mts Japan
67 C4 Mikura-jima i. Japan
73 B4 Miladhunmadulu Maldives
108 A1 Milan Italy
121 C2 Milange Moz.
Milano Italy see Milan
111 C3 Milas Turkey
137 D1 Milbank U.S.A.
99 D3 Mildenhall U.K.
52 B2 Mildura Austr.
71 A3 Mile China
136 B1 Miles City U.S.A.
139 D3 Milford DE U.S.A.
135 D3 Milford UT U.S.A.
99 A4 Milford Haven U.K.
54 A2 Milford Sound N.Z.
Milḥ, Baḥr al l. Iraq see Razāzah, Buḥayrat ar
107 D2 Miliana Alg.
50 C1 Milikapiti Austr.
51 C1 Milingimbi Austr.
134 E1 Milk r. U.S.A.
116 B3 Milk, Wadi el watercourse Sudan
83 L3 Mil'kovo Russia
128 C3 Milk River Can.
105 C3 Millau France
141 D2 Milledgeville U.S.A.
137 E1 Mille Lacs lakes U.S.A.
130 A3 Mille Lacs, Lac des l. Can.
Millennium Island Kiribati see Caroline Island
137 D2 Miller U.S.A.
91 E2 Millerovo Russia
52 A2 Millers Creek Austr.
96 B3 Milleur Point U.K.
52 B3 Millicent Austr.
140 C1 Millington U.S.A.
139 F1 Millinocket U.S.A.
55 J3 Mill Island Antarctica
53 D1 Millmerran Austr.
98 B2 Millom U.K.
136 B2 Mills U.S.A.
128 C1 Mills Lake Can.
111 B3 Milos i. Greece
89 E3 Miloslavskoye Russia
91 E2 Milove Ukr.
52 B1 Milparinka Austr.
54 A3 Milton N.Z.
99 C3 Milton Keynes U.K.
138 B2 Milwaukee U.S.A.
158 C3 Milwaukee Deep sea feature Caribbean Sea
104 B3 Mimizan France
118 B3 Mimongo Gabon
155 D2 Mimoso do Sul Brazil
79 C2 Mīnāb Iran
61 D1 Minahasa, Semenanjung pen. Indon.
Minahassa Peninsula Indon. see Minahasa, Semenanjung
79 C2 Mīnā' Jabal 'Alī U.A.E.
Minaker Can. see Prophet River
60 B1 Minas Indon.
153 C3 Minas Uru.
79 B2 Mīnā' Sa'ūd Kuwait
155 D1 Minas Gerais state Brazil
155 D1 Minas Novas Brazil
145 C3 Minatitlán Mex.
62 A1 Minbu Myanmar
64 B3 Mindanao i. Phil.
52 B2 Mindarie Austr.
101 D1 Minden Ger.
140 B2 Minden LA U.S.A.
137 D2 Minden NE U.S.A.
64 B2 Mindoro i. Phil.
64 A2 Mindoro Strait Phil.
118 B3 Mindouli Congo
99 B4 Minehead U.K.
154 B1 Mineiros Brazil
143 D2 Mineral Wells U.S.A.
75 C1 Minfeng China
119 C4 Minga Dem. Rep. Congo
81 C1 Mingäçevir Azer.
131 D2 Mingan Can.
52 B2 Mingary Austr.
70 B2 Mingguang China
62 A1 Mingin Myanmar
107 C2 Minglanilla Spain
119 D4 Mingoyo Tanz.
69 E1 Mingshui China
96 A2 Mingulay i. U.K.
71 B3 Mingxi China
Mingzhou China see Suide
70 A2 Minhe China
73 B4 Minicoy atoll India
114 B3 Minignan Côte d'Ivoire
50 A2 Minilya Austr.
131 D2 Minipi Lake Can.
130 A2 Miniss Lake Can.
52 A2 Minlaton Austr.
115 C4 Minna Nigeria
137 E2 Minneapolis U.S.A.
129 E2 Minnedosa Can.
137 E2 Minnesota r. U.S.A.
137 E1 Minnesota state U.S.A.
106 B1 Miño r. Port./Spain
107 D1 Minorca i. Spain
136 C1 Minot U.S.A.
88 C3 Minsk Belarus
103 E1 Mińsk Mazowiecki Pol.
96 D2 Mintlaw U.K.
131 D3 Minto Can.
130 C2 Minto, Lac l. Can.
68 C1 Minusinsk Russia
62 A1 Minutang India
70 A2 Minxian China
155 D1 Mirabela Brazil
155 D1 Miralta Brazil
131 D2 Miramichi Can.
111 C3 Mirampellou, Kolpos b. Greece
152 C2 Miranda Brazil
152 C1 Miranda r. Brazil
Miranda Moz. see Macaloge
106 C1 Miranda de Ebro Spain
106 B1 Mirandela Port.
154 B2 Mirandópolis Brazil
154 B2 Mirante, Serra do hills Brazil
79 C3 Mirbāṭ Oman
61 C1 Miri Malaysia
153 C3 Mirim, Lagoa l. Brazil/Uru.
79 D2 Mīrjāveh Iran
89 D3 Mirnyy Bryanskaya Oblast' Russia
83 I2 Mirnyy Respublika Sakha (Yakutiya) Russia
101 F1 Mirow Ger.
74 A2 Mirpur Khas Pak.
Mirtoan Sea Greece see Mirtoö Pelagos
111 B3 Mirtoö Pelagos sea Greece
65 B2 Miryang S. Korea
Mirzachirla Turkm. see Murzechirla
Mirzachul Uzbek. see Guliston
75 C2 Mirzapur India
77 E3 Misalay China
66 B1 Mishan China
51 E1 Misima Island P.N.G.
146 B3 Miskitos, Cayos is Nic.
103 E2 Miskolc Hungary
59 C3 Misoöl i. Indon.
115 D1 Miṣrātah Libya
130 B2 Missinaibi r. Can.
130 B3 Missinaibi Lake Can.
128 B3 Mission Can.
130 B2 Missisa Lake Can.
140 C3 Mississippi r. U.S.A.
140 C2 Mississippi state U.S.A.
140 C3 Mississippi Delta U.S.A.
140 C2 Mississippi Sound sea chan. U.S.A.
Missolonghi Greece see Mesolongi
134 D1 Missoula U.S.A.
137 E3 Missouri r. U.S.A.
137 E3 Missouri state U.S.A.
130 C3 Mistassibi r. Can.
130 C2 Mistassini, Lac l. Can.
131 D2 Mistastin Lake Can.
103 D2 Mistelbach Austria
131 D2 Mistinibi, Lac l. Can.
130 C2 Mistissini Can.
51 D2 Mitchell Austr.
51 D1 Mitchell r. Austr.
136 C2 Mitchell NE U.S.A.
137 D2 Mitchell SD U.S.A.
97 B2 Mitchelstown Ireland
74 A2 Mithi Pak.
67 D3 Mito Japan
119 D3 Mitole Tanz.
109 D2 Mitrovicë Kosovo
53 D2 Mittagong Austr.
101 E2 Mittelhausen Ger.
101 D1 Mittellandkanal canal Ger.
101 F3 Mitterteich Ger.
Mittimatalik Can. see Pond Inlet
150 B2 Mitú Col.
119 C4 Mitumba, Chaîne des mts Dem. Rep. Congo
119 C3 Mitumba, Monts mts Dem. Rep. Congo
119 C3 Mitwaba Dem. Rep. Congo
118 B2 Mitzic Gabon
78 B2 Miyah, Wādī al watercourse Saudi Arabia
67 C4 Miyake-jima i. Japan
66 D3 Miyako Japan
67 B4 Miyakonojō Japan
76 B2 Miyaly Kazakh.
Miyang China see Mile
67 B4 Miyazaki Japan
67 C3 Miyazu Japan
115 D1 Mizdah Libya
97 B3 Mizen Head hd Ireland
90 A2 Mizhhirr"ya Ukr.
Mizo Hills state India see Mizoram
75 D2 Mizoram state India
93 G4 Mjölby Sweden
93 F3 Mjøsa l. Norway
119 D3 Mkomazi Tanz.
103 C1 Mladá Boleslav Czech Rep.
109 D2 Mladenovac Serbia
103 E1 Mława Pol.
109 C2 Mljet i. Croatia
123 C3 Mlungisi S. Africa
90 B1 Mlyniv Ukr.
123 C2 Mmabatho S. Africa
123 C2 Mmathethe Botswana
93 E3 Mo Norway
135 E3 Moab U.S.A.
123 D2 Moamba Moz.
54 B2 Moana N.Z.
118 B3 Moanda Gabon
97 C2 Moate Ireland
119 C3 Moba Dem. Rep. Congo
118 C2 Mobayi-Mbongo Dem. Rep. Congo
137 E3 Moberly U.S.A.
140 C2 Mobile U.S.A.
140 C2 Mobile Bay U.S.A.
140 C2 Mobile Point U.S.A.
136 C1 Mobridge U.S.A.
Mobutu, Lake Dem. Rep. Congo/Uganda see Albert, Lake
Mobutu Sese Seko, Lake Dem. Rep. Congo/Uganda see Albert, Lake
121 C2 Moçambicano, Planalto plat. Moz.
121 D2 Moçambique Moz.
Moçâmedes Angola see Namibe
62 B1 Môc Châu Vietnam
78 B3 Mocha Yemen
123 C1 Mochudi Botswana
121 D2 Mocimboa da Praia Moz.
101 D3 Möckmühl Ger.
150 B2 Mocoa Col.
154 C2 Mococa Brazil
144 B2 Mocorito Mex.
144 B1 Moctezuma Chihuahua Mex.
145 B2 Moctezuma San Luis Potosí Mex.
144 B2 Moctezuma Sonora Mex.
121 C2 Mocuba Moz.
105 D2 Modane France
122 B2 Modder r. S. Africa
108 B2 Modena Italy
135 B3 Modesto U.S.A.
109 B3 Modica Italy
123 C1 Modimolle S. Africa
123 D1 Modjadjiskloof S. Africa
53 C3 Moe Austr.
Moero, Lake Dem. Rep. Congo/Zambia see Mweru, Lake
100 C2 Moers Ger.
96 C3 Moffat U.K.
117 C4 Mogadishu Somalia
Mogador Morocco see Essaouira
106 B1 Mogadouro, Serra de mts Port.
123 C1 Mogalakwena r. S. Africa
62 A1 Mogaung Myanmar
155 C2 Mogi das Cruzes Brazil
Mogilev Belarus see Mahilyow
154 C2 Mogi Mirim Brazil
83 I3 Mogocha Russia
123 C1 Mogoditshane Botswana
62 A1 Mogok Myanmar
142 A2 Mogollon Plateau U.S.A.
103 D2 Mohács Hungary
123 C3 Mohale's Hoek Lesotho
74 B1 Mohali India
79 C2 Moḥammadābād Iran
107 D2 Mohammadia Alg.
142 A2 Mohave Mountains U.S.A.
139 E2 Mohawk r. U.S.A.
62 A1 Mohnyin Myanmar
119 D3 Mohoro Tanz.
90 B2 Mohyliv-Podil's'kyy Ukr.
123 C1 Moijabana Botswana
110 C1 Moinești Romania
Mointy Kazakh. see Moyynty
92 F2 Mo i Rana Norway
88 C2 Mõisaküla Estonia
104 C3 Moissac France
135 C3 Mojave U.S.A.
135 C3 Mojave Desert U.S.A.
62 B1 Mojiang China
154 C2 Moji-Guaçu r. Brazil
109 C2 Mojkovac Montenegro
54 B1 Mokau N.Z.
123 C2 Mokhotlong Lesotho
83 J2 Mokhsogollokh Russia
118 B1 Mokolo Cameroon
123 C1 Mokopane S. Africa
65 B3 Mokpo S. Korea
109 C2 Mola di Bari Italy
145 C2 Molango Mex.
Moldavia country Europe see Moldova
Moldavskaya S.S.R. country Europe see Moldova
93 E3 Molde Norway
90 B2 Moldova country Europe
110 B2 Moldova Nouă Romania
110 B1 Moldoveanu, Vârful mt. Romania
110 B1 Moldovei, Podișul plat. Romania
90 B2 Moldovei Centrale, Podișul plat. Moldova
123 C1 Molepolole Botswana
88 C2 Moltai Lith.
109 C2 Molfetta Italy
Molière Alg. see Bordj Bounaama
107 C1 Molina de Aragón Spain
107 C2 Molina de Segura Spain
119 D3 Moliro Dem. Rep. Congo
150 B4 Mollendo Peru
93 F4 Mölnlycke Sweden

Nauchas

66 B1 **Muling He** *r.* China
96 B2 **Mull** *i.* U.K.
53 C2 **Mullaley** Austr.
136 C2 **Mullen** U.S.A.
61 C1 **Muller, Pegunungan** *mts* Indon.
50 A2 **Mullewa** Austr.
97 C2 **Mullingar** Ireland
96 B3 **Mull of Galloway** *c.* U.K.
96 B3 **Mull of Kintyre** *hd* U.K.
96 A3 **Mull of Oa** *hd* U.K.
53 D1 **Mullumbimby** Austr.
120 B2 **Mulobezi** Zambia
74 B1 **Multan** Pak.
86 F2 **Mulym'ya** Russia
74 B3 **Mumbai** India
120 B2 **Mumbeji** Zambia
120 B2 **Mumbwa** Zambia
61 D2 **Muna** *i.* Indon.
145 D2 **Muna** Mex.
101 E2 **Münchberg** Ger.
München Ger. *see* **Munich**
München-Gladbach Ger. *see* **Mönchengladbach**
138 B2 **Muncie** U.S.A.
50 B3 **Mundrabilla** Austr.
138 B3 **Munfordville** U.S.A.
119 C2 **Mungbere** Dem. Rep. Congo
75 C2 **Munger** India
52 A1 **Mungeranie** Austr.
53 C1 **Mungindi** Austr.
102 C2 **Munich** Ger.
155 D2 **Muniz Freire** Brazil
101 E1 **Munster** Ger.
100 C2 **Münster** Ger.
97 B2 **Munster** *reg.* Ireland
100 C2 **Münsterland** *reg.* Ger.
62 B1 **Mường Nhe** Vietnam
92 H2 **Muonio** Fin.
92 H2 **Muonioälven** *r.* Fin./Sweden
Muqdisho Somalia *see* **Mogadishu**
155 D2 **Muqui** Brazil
103 D2 **Mur** *r.* Austria
67 C3 **Murakami** Japan
119 C3 **Muramvya** Burundi
119 D3 **Murang'a** Kenya
86 D3 **Murashi** Russia
81 B2 **Murat** *r.* Turkey
111 C2 **Muratlı** Turkey
67 D3 **Murayama** Japan
50 A2 **Murchison** *watercourse* Austr.
107 C2 **Murcia** Spain
107 C2 **Murcia** *aut. comm.* Spain
136 C2 **Murdo** U.S.A.
131 D3 **Murdochville** Can.
111 C2 **Mürefte** Turkey
110 B1 **Mureşul** *r.* Romania
104 C3 **Muret** France
140 C1 **Murfreesboro** U.S.A.
77 D3 **Murghob** Tajik.
155 D2 **Muriaé** Brazil
120 B1 **Muriege** Angola
101 F1 **Müritz** *l.* Ger.
92 J2 **Murmansk** Russia
86 C2 **Murmanskiy Bereg** *coastal area* Russia
87 D3 **Murom** Russia
66 D2 **Muroran** Japan
106 B1 **Muros** Spain
67 B4 **Muroto** Japan
67 B4 **Muroto-zaki** *pt* Japan
141 D1 **Murphy** U.S.A.
53 C1 **Murra Murra** Austr.
52 A3 **Murray** *r.* Austr.
128 B2 **Murray** *r.* Can.
138 B3 **Murray** U.S.A.
59 D3 **Murray, Lake** P.N.G.
141 D2 **Murray, Lake** U.S.A.
52 A3 **Murray Bridge** Austr.
122 B3 **Murraysburg** S. Africa
52 B3 **Murrayville** Austr.
52 B2 **Murrumbidgee** *r.* Austr.
53 C2 **Murrumburrah** Austr.
121 C2 **Murrupula** Moz.
53 D2 **Murrurundi** Austr.
109 C1 **Murska Sobota** Slovenia
54 C1 **Murupara** N.Z.
49 N6 **Mururoa** *atoll* Fr. Polynesia
Murwara India *see* **Katni**
53 D1 **Murwillumbah** Austr.
76 C3 **Murzechirla** Turkm.
115 D2 **Murzūq** Libya
115 D2 **Murzuq, Idhān** *des.* Libya
103 D2 **Mürzzuschlag** Austria
81 C2 **Muş** Turkey
110 B2 **Musala** *mt.* Bulg.
65 B1 **Musan** N. Korea
78 B3 **Musaymir** Yemen
79 C2 **Muscat** Oman
Muscat and Oman *country* Asia *see* **Oman**
137 E2 **Muscatine** U.S.A.
50 C2 **Musgrave Ranges** *mts* Austr.
118 B3 **Mushie** Dem. Rep. Congo
60 B2 **Musi** *r.* Indon.
123 D1 **Musina** S. Africa
138 B2 **Muskegon** U.S.A.
138 B2 **Muskegon** *r.* U.S.A.
138 C3 **Muskingum** *r.* U.S.A.
143 D1 **Muskogee** U.S.A.
139 D1 **Muskoka, Lake** Can.
128 B2 **Muskwa** *r.* Can.
74 A1 **Muslimbagh** Pak.
116 B3 **Musmar** Sudan
119 D3 **Musoma** Tanz.
59 D3 **Mussau Island** P.N.G.
96 C3 **Musselburgh** U.K.
117 C4 **Mustahīl** Eth.
88 B2 **Mustjala** Estonia
53 D2 **Muswellbrook** Austr.
116 A2 **Mūţ** Egypt
121 C2 **Mutare** Zimbabwe
121 C2 **Mutoko** Zimbabwe
66 D2 **Mutsu** Japan
66 D2 **Mutsu-wan** *b.* Japan
121 C2 **Mutuali** Moz.
155 D1 **Mutum** Brazil
92 I2 **Muurola** Fin.
70 A2 **Mu Us Shadi** *des.* China
120 A1 **Muxaluando** Angola
86 C2 **Muyezerskiy** Russia
119 D3 **Muyinga** Burundi
74 B1 **Muzaffargarh** Pak.
75 C2 **Muzaffarpur** India
123 D1 **Muzamane** Moz.
155 C2 **Muzambinho** Brazil
144 B2 **Múzquiz** Mex.
75 C1 **Muz Tag** *mt.* China
117 A4 **Mvolo** S. Sudan
119 C3 **Mwanza** Dem. Rep. Congo
119 D3 **Mwanza** Tanz.
118 C3 **Mweka** Dem. Rep. Congo
121 B2 **Mwenda** Zambia
118 C3 **Mwene-Ditu** Dem. Rep. Congo
121 C3 **Mwenezi** Zimbabwe
121 C3 **Mwenezi** *r.* Zimbabwe
119 C3 **Mweru, Lake** Dem. Rep. Congo/Zambia
121 B1 **Mweru Wantipa, Lake** Zambia
118 C3 **Mwimba** Dem. Rep. Congo
120 B2 **Mwinilunga** Zambia
88 C3 **Myadzyel** Belarus
62 A2 **Myanaung** Myanmar
62 A1 **Myanmar** *country* Asia
63 A2 **Myaungmya** Myanmar
63 A2 **Myeik** Myanmar
Myeik Kyunzu *is* Myanmar *see* **Mergui Archipelago**
62 A1 **Myingyan** Myanmar
62 A1 **Myitkyina** Myanmar
90 A2 **Mykolayiv** *L'vivs'ka Oblast'* Ukr.
91 C2 **Mykolayiv** *Mykolayivs'ka Oblast'* Ukr.
111 C3 **Mykonos** Greece
111 C3 **Mykonos** *i.* Greece
86 E2 **Myla** Russia
75 D2 **Mymensingh** Bangl.
67 C3 **Myōkō** Japan
65 B1 **Myŏnggan** N. Korea
88 C2 **Myory** Belarus
92 □B3 **Mýrdalsjökull** Iceland
92 G2 **Myre** Norway
91 C2 **Myrhorod** Ukr.
111 C3 **Myrina** Greece
90 C2 **Myronivka** Ukr.
141 E2 **Myrtle Beach** U.S.A.
53 C3 **Myrtleford** Austr.
134 B2 **Myrtle Point** U.S.A.
83 H1 **Mys Chelyuskin** Russia
89 E2 **Myshkin** Russia
Myshkino Russia *see* **Myshkin**
103 C1 **Myślibórz** Pol.
73 B3 **Mysore** India
83 N2 **Mys Shmidta** Russia
Mysuru India *see* **Mysore**
63 B2 **My Tho** Vietnam
111 C3 **Mytilini** Greece
89 E3 **Mytishchi** Russia
123 C3 **Mzamomhle** S. Africa
121 C2 **Mzimba** Malawi
121 C2 **Mzuzu** Malawi

N

101 F3 **Naab** *r.* Ger.
100 B1 **Naarden** Neth.
97 C2 **Naas** Ireland
122 A2 **Nababeep** S. Africa
87 E3 **Naberezhnyye Chelny** Russia
59 D3 **Nabire** Indon.
80 B2 **Nāblus** West Bank
121 D2 **Nacala** Moz.
119 D4 **Nachingwea** Tanz.
103 D1 **Náchod** Czech Rep.
73 D3 **Nachuge** India
143 E2 **Nacogdoches** U.S.A.
144 B1 **Nacozari de García** Mex.
Nada China *see* **Danzhou**
74 B2 **Nadiad** India
90 A2 **Nadvirna** Ukr.
86 C2 **Nadvoitsy** Russia
86 G2 **Nadym** Russia
93 F4 **Næstved** Denmark
111 B3 **Nafpaktos** Greece
111 B3 **Nafplio** Greece
115 D1 **Nafūsah, Jabal** *hills* Libya
78 B2 **Nafy** Saudi Arabia
64 B2 **Naga** Phil.
130 B2 **Nagagami** *r.* Can.
67 C3 **Nagano** Japan
67 C3 **Nagaoka** Japan
75 D2 **Nagaon** India
74 B1 **Nagar** India
74 B2 **Nagar Parkar** Pak.
67 A4 **Nagasaki** Japan
67 B4 **Nagato** Japan
74 B2 **Nagaur** India
73 B4 **Nagercoil** India
74 A2 **Nagha Kalat** Pak.
74 B2 **Nagina** India
81 C1 **Nagorno-Karabakh** *disp. terr.* Azer.
67 C3 **Nagoya** Japan
75 B2 **Nagpur** India
75 D1 **Nagqu** China
141 E1 **Nags Head** U.S.A.
82 E1 **Nagurskoye** Russia
103 D2 **Nagyatád** Hungary
103 D2 **Nagykanizsa** Hungary
128 B1 **Nahanni Butte** Can.
81 C2 **Nahāvand** Iran
101 E1 **Nahrendorf** Ger.
153 A4 **Nahuel Huapí, Lago** *l.* Arg.
141 D2 **Nahunta** U.S.A.
131 D2 **Nain** Can.
81 D2 **Nā'īn** Iran
121 C2 **Naiopué** Moz.
96 C2 **Nairn** U.K.
119 D3 **Nairobi** Kenya
Naissus Serbia *see* **Niš**
119 D3 **Naivasha** Kenya
81 D2 **Najafābād** Iran
78 B2 **Najd** *reg.* Saudi Arabia
106 C1 **Nájera** Spain
65 C1 **Najin** N. Korea
78 B3 **Najrān** Saudi Arabia
119 D2 **Nakasongola** Uganda
67 C3 **Nakatsugawa** Japan
78 A3 **Nakfa** Eritrea
66 B2 **Nakhodka** Russia
63 B2 **Nakhon Nayok** Thai.
63 B2 **Nakhon Pathom** Thai.
62 B2 **Nakhon Phanom** Thai.
63 B2 **Nakhon Ratchasima** Thai.
63 B2 **Nakhon Sawan** Thai.
63 A3 **Nakhon Si Thammarat** Thai.
Nakhrachi Russia *see* **Kondinskoye**
130 B2 **Nakina** Can.
126 B3 **Naknek** U.S.A.
121 C1 **Nakonde** Zambia
93 F5 **Nakskov** Denmark
119 D3 **Nakuru** Kenya
128 C2 **Nakusp** Can.
75 D2 **Nalbari** India
87 D4 **Nal'chik** Russia
115 D1 **Nālūt** Libya
123 D2 **Namaacha** Moz.
123 C2 **Namahadi** S. Africa
81 D2 **Namak, Daryācheh-ye** *imp. l.* Iran
76 B3 **Namak, Kavīr-e** *salt flat* Iran
79 C1 **Namakzar-e Shadad** *salt flat* Iran
77 D2 **Namangan** Uzbek.
119 D3 **Namanyere** Tanz.
122 A2 **Namaqualand** *reg.* S. Africa
51 E2 **Nambour** Austr.
53 D2 **Nambucca Heads** Austr.
63 B3 **Năm Căn** Vietnam
75 D1 **Nam Co** *salt l.* China
62 B1 **Nam Đinh** Vietnam
121 C2 **Namialo** Moz.
120 A3 **Namib Desert** Namibia
120 A2 **Namibe** Angola
120 A3 **Namibia** *country* Africa
72 D2 **Namjagbarwa Feng** *mt.* China
59 C3 **Namlea** Indon.
53 C2 **Namoi** *r.* Austr.
134 C2 **Nampa** U.S.A.
114 B3 **Nampala** Mali
65 B2 **Namp'o** N. Korea
121 C2 **Nampula** Moz.
72 D2 **Namrup** India
62 A1 **Namsang** Myanmar
92 F3 **Namsos** Norway
92 F3 **Namsskogan** Norway
63 A2 **Nam Tok** Thai.
83 J2 **Namtsy** Russia
62 A1 **Namtu** Myanmar
121 C2 **Namuno** Moz.
100 B2 **Namur** Belgium
120 B2 **Namwala** Zambia
65 B2 **Namwon** S. Korea
62 A1 **Namya Ra** Myanmar
62 B2 **Nan** Thai.
128 B3 **Nanaimo** Can.
71 B3 **Nan'an** China
122 A1 **Nananib Plateau** Namibia
Nan'ao China *see* **Dayu**
67 C3 **Nanao** Japan
71 B3 **Nanchang** *Jiangxi* China
71 B3 **Nanchang** *Jiangxi* China
71 B3 **Nancheng** China
70 A2 **Nanchong** China
63 A3 **Nancowry** *i.* India
105 D2 **Nancy** France
75 C1 **Nanda Devi** *mt.* India
71 A3 **Nandan** China
74 B3 **Nanded** India
Nander India *see* **Nanded**
53 D2 **Nandewar Range** *mts* Austr.
74 B2 **Nandurbar** India
73 B3 **Nandyal** India
71 B3 **Nanfeng** China
62 A1 **Nang** China
118 B2 **Nanga Eboko** Cameroon
61 C2 **Nangahpinoh** Indon.
77 D3 **Nanga Parbat** *mt.* Pak.
61 C2 **Nangatayap** Indon.
63 A2 **Nangin** Myanmar
65 B1 **Nangnim-sanmaek** *mts* N. Korea
70 B2 **Nangong** China
119 D3 **Nangulangwa** Tanz.
70 C2 **Nanhui** China
70 B2 **Nanjing** China
Nanking China *see* **Nanjing**
67 B4 **Nankoku** Japan
120 A2 **Nankova** Angola
70 B2 **Nanle** China
71 B3 **Nan Ling** *mts* China
71 A3 **Nanning** China
127 I2 **Nanortalik** Greenland
71 A3 **Nanpan Jiang** *r.* China
75 C2 **Nanpara** India
71 B3 **Nanping** China
Nanpu China *see* **Pucheng**
Nansei-shotō Japan *see* **Ryukyu Islands**
160 I1 **Nansen Basin** Arctic Ocean
126 F1 **Nansen Sound** *sea chan.* Can.
104 B2 **Nantes** France
70 C2 **Nantong** China
139 E2 **Nantucket** U.S.A.
139 F2 **Nantucket Island** U.S.A.
99 B3 **Nantwich** U.K.
49 I4 **Nanumea** *atoll* Tuvalu
155 D1 **Nanuque** Brazil
64 B3 **Nanusa, Kepulauan** *is* Indon.
71 B3 **Nanxiong** China
70 B2 **Nanyang** China
119 D3 **Nanyuki** Kenya
70 B2 **Nanzhang** China
Nanzhao China *see* **Zhao'an**
107 D2 **Nao, Cabo de la** *c.* Spain
131 C2 **Naococane, Lac** *l.* Can.
71 B3 **Naozhou Dao** *i.* China
135 B3 **Napa** U.S.A.
126 E2 **Napaktulik Lake** Can.
139 D2 **Napanee** Can.
127 I2 **Napasoq** Greenland
137 F2 **Naperville** U.S.A.
54 C1 **Napier** N.Z.
108 B2 **Naples** Italy
141 D3 **Naples** U.S.A.
150 B3 **Napo** *r.* Ecuador/Peru
Napoli Italy *see* **Naples**
Napug China *see* **Gê'gyai**
114 B3 **Nara** Mali
88 C3 **Narach** Belarus
52 B3 **Naracoorte** Austr.
53 C2 **Naradhan** Austr.
145 C2 **Naranjos** Mex.
63 B3 **Narathiwat** Thai.
74 B3 **Narayangaon** India
105 C3 **Narbonne** France
63 A2 **Narcondam Island** India
127 H1 **Nares Strait** Can./Greenland
103 E1 **Narew** *r.* Pol.
122 A1 **Narib** Namibia
87 D4 **Narimanov** Russia
67 D3 **Narita** Japan
74 B2 **Narmada** *r.* India
74 B2 **Narnaul** India
108 B2 **Narni** Italy
86 F2 **Narodnaya, Gora** *mt.* Russia
90 B1 **Narodychi** Ukr.
89 E2 **Naro-Fominsk** Russia
53 D3 **Narooma** Austr.
88 C3 **Narowlya** Belarus
93 H3 **Närpes** Fin.
53 C2 **Narrabri** Austr.
53 C2 **Narrandera** Austr.
53 C2 **Narromine** Austr.
67 B4 **Naruto** Japan
88 C2 **Narva** Estonia
88 C2 **Narva Bay** Estonia/Russia
92 G2 **Narvik** Norway
88 C2 **Narvskoye Vodokhranilishche** *resr* Estonia/Russia
86 E2 **Nar'yan-Mar** Russia
77 D2 **Naryn** Kyrg.
74 B2 **Nashik** India
139 E2 **Nashua** U.S.A.
140 C1 **Nashville** U.S.A.
137 E1 **Nashwauk** U.S.A.
109 C1 **Našice** Croatia
49 I5 **Nasinu** Fiji
117 B4 **Nasir** S. Sudan
Nasirabad Bangl. *see* **Mymensingh**
119 C4 **Nasondoye** Dem. Rep. Congo
76 B3 **Naşrābād** Iran
128 B2 **Nass** *r.* Can.
146 C2 **Nassau** Bahamas
116 B2 **Nasser, Lake** *resr* Egypt
93 F4 **Nässjö** Sweden
130 C2 **Nastapoca** *r.* Can.
130 C2 **Nastapoka Islands** Can.
67 D3 **Nasushiobara** Japan
89 D2 **Nasva** Russia
120 B3 **Nata** Botswana
151 F3 **Natal** Brazil
60 A1 **Natal** Indon.
Natal *prov.* S. Africa *see* **KwaZulu-Natal**
159 D6 **Natal Basin** Indian Ocean
143 D3 **Natalia** U.S.A.
131 D2 **Natashquan** Can.
131 D2 **Natashquan** *r.* Can.
140 B2 **Natchez** U.S.A.
140 B2 **Natchitoches** U.S.A.
53 C3 **Nathalia** Austr.
107 D1 **Nati, Punta** *pt* Spain
114 C3 **Natitingou** Benin
151 E4 **Natividade** Brazil
67 D3 **Natori** Japan
119 D3 **Natron, Lake** *salt l.* Tanz.
60 B1 **Natuna, Kepulauan** *is* Indon.
60 B1 **Natuna Besar** *i.* Indon.
122 A1 **Nauchas** Namibia

119 D3 Nkondwe Tanz.
118 A2 Nkongsamba Cameroon
123 C3 Nkululeko S. Africa
123 C3 Nkwenkwezi S. Africa
67 B4 Nobeoka Japan
138 B2 Noblesville U.S.A.
52 B1 Noccundra Austr.
144 A1 Nogales Mex.
142 A2 Nogales U.S.A.
104 C2 Nogent-le-Rotrou France
83 H2 Noginsk *Krasnoyarskiy Kray* Russia
89 E2 Noginsk *Moskovskaya Oblast'* Russia
83 K3 Nogliki Russia
74 B2 Nohar India
100 C3 Nohfelden Ger.
104 B2 Noires, Montagnes *hills* France
104 B2 Noirmoutier, Île de *i.* France
104 B2 Noirmoutier-en-l'Île France
67 C4 Nojima-zaki *c.* Japan
74 B2 Nokha India
93 H3 Nokia Fin.
74 A2 Nok Kundi Pak.
118 B2 Nola C.A.R.
86 D3 Nolinsk Russia
126 A2 Nome U.S.A.
123 C3 Nomonde S. Africa
123 D2 Nondweni S. Africa
Nonghui China *see* Guang'an
62 B2 Nong Khai Thai.
75 B2 Nongstoin India
52 A2 Nonning Austr.
144 B2 Nonoava Mex.
65 B2 Nonsan S. Korea
63 B2 Nonthaburi Thai.
122 B3 Nonzwakazi S. Africa
100 B1 Noordwijk Neth.
77 C3 Norak Tajik.
82 C1 Nordaustlandet *i.* Svalbard
128 C2 Nordegg Can.
100 C1 Norden Ger.
83 H1 Nordenshel'da, Arkhipelag *is* Russia
Nordenskjold Archipelago *is* Russia *see* Nordenshel'da, Arkhipelag
100 C1 Norderney Ger.
100 C1 Norderney *i.* Ger.
101 E2 Norderstedt Ger.
93 E3 Nordfjordeid Norway
Nordfriesische Inseln Ger. *see* North Frisian Islands
101 E2 Nordhausen Ger.
101 D2 Nordholz Ger.
100 C1 Nordhorn Ger.
Nordkapp Norway *see* North Cape
92 F3 Nordli Norway
102 C2 Nördlingen Ger.
92 G3 Nordmaling Sweden
94 B1 Norðoyar *is* Faroe Is
94 Nore *r.* Ireland
88 B3 Noreikiškis Lith.
137 D2 Norfolk *NE* U.S.A.
139 D3 Norfolk *VA* U.S.A.
48 H6 Norfolk Island *terr.* S. Pacific Ocean
93 E3 Norheimsund Norway
82 G2 Noril'sk Russia
75 C2 Norkyung China
143 D1 Norman U.S.A.
Normandes, Îles *is* English Chan. *see* Channel Islands
150 D2 Normandia Brazil
Normandie *reg.* France *see* Normandy
104 B2 Normandy *reg.* France
51 D1 Normanton Austr.
128 B1 Norman Wells Can.
93 G4 Norrköping Sweden
93 G4 Norrtälje Sweden
50 B3 Norseman Austr.
92 G3 Norsjö Sweden
55 M2 North, Cape Antarctica
98 B1 Northallerton U.K.
50 A2 Northam Austr.
99 C3 Northampton U.K.
50 A2 Northampton Austr.
99 C3 Northampton U.K.
73 D3 North Andaman *i.* India
159 F4 North Australian Basin Indian Ocean
129 D2 North Battleford Can.
130 C3 North Bay Can.
127 G2 North Belcher Islands Can.
96 C2 North Berwick U.K.
North Borneo *state* Malaysia *see* Sabah
92 I1 North Cape *c.* Norway
54 B1 North Cape N.Z.
130 A2 North Caribou Lake Can.
141 E1 North Carolina *state* U.S.A.
130 B3 North Channel *lake channel* Can.
96 A3 North Channel U.K.
141 E2 North Charleston U.S.A.
128 B3 North Cowichan Can.
136 C1 North Dakota *state* U.S.A.
98 C3 North Downs *hills* U.K.
157 E3 Northeast Pacific Basin N. Pacific Ocean
141 E3 Northeast Providence Channel Bahamas
101 D2 Northeim Ger.
122 A2 Northern Cape *prov.* S. Africa
80 B2 Northern Cyprus Asia
91 D2 Northern Donets *r.* Russia/Ukr.
Northern Dvina *r.* Russia *see* Severnaya Dvina
129 E2 Northern Indian Lake Can.
97 C1 Northern Ireland *prov.* U.K.

59 D1 Northern Mariana Islands *terr.* N. Pacific Ocean
Northern Rhodesia *country* Africa *see* Zambia
50 C1 Northern Territory *admin. div.* Austr.
Northern Transvaal *prov.* S. Africa *see* Limpopo
96 C2 North Esk *r.* U.K.
137 E2 Northfield U.S.A.
99 D4 North Foreland *c.* U.K.
102 B1 North Frisian Islands *is* Ger.
160 P2 North Geomagnetic Pole
54 B1 North Island N.Z.
129 E2 North Knife Lake Can.
65 B1 North Korea *country* Asia
72 D2 North Lakhimpur India
North Land *is* Russia *see* Severnaya Zemlya
160 North Magnetic Pole
128 B1 North Nahanni *r.* Can.
136 C2 North Platte U.S.A.
136 C2 North Platte *r.* U.S.A.
96 C1 North Ronaldsay *i.* U.K.
129 D2 North Saskatchewan *r.* Can.
94 D2 North Sea Europe
63 A2 North Sentinel Island India
130 A2 North Spirit Lake Can.
53 D1 North Stradbroke Island Austr.
131 D3 North Sydney Can.
54 B1 North Taranaki Bight *b.* N.Z.
130 C2 North Twin Island Can.
98 B2 North Tyne *r.* U.K.
96 A2 North Uist *i.* U.K.
131 D3 Northumberland Strait Can.
99 D3 North Walsham U.K.
123 C2 North West *prov.* S. Africa
158 D1 Northwest Atlantic Mid-Ocean Channel *sea chan.* N. Atlantic Ocean
50 A2 North West Cape Austr.
156 D3 Northwest Pacific Basin N. Pacific Ocean
141 E3 Northwest Providence Channel Bahamas
131 E2 North West River Can.
128 B1 Northwest Territories *admin. div.* Can.
98 C2 North York Moors *moorland* U.K.
138 C3 Norton U.S.A.
121 C2 Norton Zimbabwe
Norton de Matos Angola *see* Balombo
126 B2 Norton Sound *sea chan.* U.S.A.
55 D2 Norvegia, Cape Antarctica
138 C2 Norwalk U.S.A.
93 F3 Norway *country* Europe
129 E2 Norway House Can.
160 L2 Norwegian Basin N. Atlantic Ocean
92 Norwegian Sea N. Atlantic Ocean
99 D3 Norwich U.K.
139 E2 Norwich *CT* U.S.A.
139 D2 Norwich *NY* U.S.A.
66 D2 Noshiro Japan
91 C1 Nosivka Ukr.
122 B3 Nosop *watercourse* Africa
86 E2 Nosovaya Russia
79 C2 Noşratābād Iran
122 A1 Nossob *watercourse* Africa
103 D1 Noteć *r.* Pol.
93 E4 Notodden Norway
67 C3 Noto-hantō *pen.* Japan
131 D3 Notre-Dame, Monts *mts* Can.
131 E3 Notre Dame Bay Can.
130 C2 Nottaway *r.* Can.
99 C3 Nottingham U.K.
114 A2 Nouâdhibou Maur.
114 A3 Nouakchott Maur.
114 A3 Nouâmghâr Maur.
63 B2 Nouei Vietnam
48 H6 Nouméa New Caledonia
114 A3 Nouna Burkina Faso
122 B3 Noupoort S. Africa
Nouveau-Comptoir Can. *see* Wemindji
Nouvelle Anvers Dem. Rep. Congo *see* Makanza
Nouvelle-Calédonie *terr.* S. Pacific Ocean *see* New Caledonia
Nouvelles Hébrides *country* S. Pacific Ocean *see* Vanuatu
Nova Chaves Angola *see* Muconda
154 B2 Nova Esperança Brazil
Nova Freixa Moz. *see* Cuamba
155 D2 Nova Friburgo Brazil
109 C1 Nova Gradiška Croatia
154 C2 Nova Granada Brazil
155 D2 Nova Iguaçu Brazil
91 C2 Nova Kakhovka Ukr.
155 D1 Nova Lima Brazil
Nova Lisboa Angola *see* Huambo
154 B2 Nova Londrina Brazil
91 C2 Nova Odesa Ukr.
154 C1 Nova Ponte Brazil
108 A1 Novara Italy
131 D3 Nova Scotia *prov.* Can.
155 D1 Nova Venécia Brazil
83 K1 Novaya Sibir', Ostrov *i.* Russia
86 E1 Novaya Zemlya *is* Russia
107 D2 Novelda Spain
103 D2 Nové Zámky Slovakia
Novgorod Russia *see* Velikiy Novgorod
91 C1 Novhorod-Sivers'kyy Ukr.
110 B2 Novi Iskar Bulg.
66 D1 Novikovo Russia

108 A2 Novi Ligure Italy
109 D2 Novi Pazar Serbia
109 C1 Novi Sad Serbia
87 D3 Novoalekseyevka Kazakh. *see* Kobda
91 D2 Novoanninskiy Russia
150 A2 Novo Aripuanã Brazil
91 D2 Novoazovs'k Ukr.
91 E2 Novocherkassk Russia
89 D2 Novodugino Russia
86 D2 Novodvinsk Russia
Novoekonomicheskoye Ukr. *see* Dymytrov
152 C2 Novo Hamburgo Brazil
154 C2 Novo Horizonte Brazil
90 B1 Novohrad-Volyns'kyy Ukr.
Novokazalinsk Kazakh. *see* Ayteke Bi
91 E1 Novokhopersk Russia
68 B1 Novokuznetsk Russia
109 C1 Novo mesto Slovenia
91 D3 Novomikhaylovskiy Russia
89 E3 Novomoskovsk Russia
91 D2 Novomoskovs'k Ukr.
91 C2 Novomyrhorod Ukr.
Novonikolayevsk Russia *see* Novosibirsk
91 C2 Novooleksiyivka Ukr.
150 C2 Novo Paraíso Brazil
91 E2 Novopokrovskaya Russia
91 D2 Novopskov Ukr.
Novo Redondo Angola *see* Sumbe
91 D3 Novorossiysk Russia
88 C2 Novorzhev Russia
87 E3 Novosergiyevka Russia
89 E2 Novoshakhtinsk Russia
82 G3 Novosibirsk Russia
Novosibirskiye Ostrova *is* Russia *see* New Siberia Islands
89 E3 Novosil' Russia
89 D2 Novosokol'niki Russia
91 C2 Novotroyits'ke Ukr.
91 C2 Novoukrayinka Ukr.
90 A1 Novovolyns'k Ukr.
89 E3 Novovoronezh Russia
Novovoronezhskiy Russia *see* Novovoronezh
89 D3 Novozybkov Russia
103 D2 Nový Jičín Czech Rep.
86 E2 Nový Bor Russia
91 C2 Novyy Buh Ukr.
Novyy Donbass Ukr. *see* Dymytrov
Novyye Petushki Russia *see* Petushki
Novyy Margelan Uzbek. *see* Farg'ona
89 E2 Novyy Nekouz Russia
91 D1 Novyy Oskol Russia
86 G2 Novyy Port Russia
86 G2 Novyy Urengoy Russia
69 E1 Novyy Urgal Russia
Novyy Uzen' Kazakh. *see* Zhanaozen
103 D1 Nowogard Pol.
Noworadomsk Pol. *see* Radomsko
53 D2 Nowra Austr.
81 D2 Now Shahr Iran
74 B1 Nowshera Pak.
103 E2 Nowy Sącz Pol.
103 E2 Nowy Targ Pol.
82 G2 Noyabr'sk Russia
105 C2 Noyon France
68 C2 Noyon Mongolia
121 C2 Nsanje Malawi
121 B2 Nsombo Zambia
118 B3 Ntandembele Dem. Rep. Congo
123 C3 Ntha S. Africa
111 B3 Ntoro, Kavo *pt* Greece
118 A2 Ntoum Gabon
119 D3 Ntungamo Uganda
Nuanetsi *r.* Zimbabwe *see* Mwenezi
79 C2 Nu'aym *reg.* Oman
116 B2 Nubian Desert Sudan
143 D3 Nueces *r.* U.S.A.
129 E1 Nueltin Lake Can.
150 B2 Nueva Loja Ecuador
153 A4 Nueva Lubecka Arg.
145 B2 Nueva Rosita Mex.
144 B1 Nuevo Casas Grandes Mex.
144 B2 Nuevo Ideal Mex.
145 C2 Nuevo Laredo Mex.
117 C4 Nugaal *watercourse* Somalia
117 C4 Nugaaleed, Dooxo *val.* Somalia
105 C2 Nuits-St-Georges France
Nu Jiang *r.* China/Myanmar *see* Salween
Nukha Azer. *see* Şäki
49 J6 Nuku'alofa Tonga
49 M4 Nuku Hiva *i.* Fr. Polynesia
48 G4 Nukumanu Islands P.N.G.
76 B2 Nukus Uzbek.
50 B2 Nullagine Austr.
50 C3 Nullarbor Austr.
50 B3 Nullarbor Plain Austr.
115 D4 Numan Nigeria
67 C3 Numazu Japan
51 C1 Numbulwar Austr.
93 E3 Numedal *val.* Norway
59 C3 Numfoor *i.* Indon.
53 C3 Numurkah Austr.
Nunap Isua *c.* Greenland *see* Farewell, Cape
127 G2 Nunavik *reg.* Can.
129 E1 Nunavut *admin. div.* Can.
99 C3 Nuneaton U.K.
126 A3 Nunivak Island U.S.A.
106 B1 Nuñomoral Spain
108 A2 Nuoro Italy
78 B2 Nuqrah Saudi Arabia

77 C1 Nura *r.* Kazakh.
101 E3 Nuremberg Ger.
52 A2 Nuriootpa Austr.
92 I3 Nurmes Fin.
Nürnberg Ger. *see* Nuremberg
53 C2 Nurri, Mount *h.* Austr.
62 A1 Nu Shan *mts* China
74 A2 Nushki Pak.
127 I2 Nuuk Greenland
127 I2 Nuussuaq Greenland
127 I2 Nuussuaq *pen.* Greenland
80 B3 Nuwaybi' al Muzaywinah Egypt
122 A3 Nuwerus S. Africa
122 B3 Nuweveldberge *mts* S. Africa
86 F2 Nyagan' Russia
75 D1 Nyainqêntanglha Feng *mt.* China
75 D2 Nyainqêntanglha Shan *mts* China
Nyakh Russia *see* Nyagan'
117 A3 Nyala Sudan
119 D4 Nyamtumbo Tanz.
Nyande Zimbabwe *see* Masvingo
86 D2 Nyandoma Russia
118 B3 Nyanga Congo
118 B3 Nyanga *r.* Gabon
121 C2 Nyanga Zimbabwe
Nyang'oma Kenya *see* Kogelo
121 C1 Nyasa, Lake Africa
Nyasaland *country* Africa *see* Malawi
88 C3 Nyasvizh Belarus
62 A1 Nyaunglebin Myanmar
93 F4 Nyborg Denmark
92 I1 Nyborg Norway
93 G4 Nybro Sweden
Nyenchen Tangbla Range *mts* China *see* Nyainqêntanglha Shan
119 D3 Nyeri Kenya
68 C3 Nyingchi China
103 E2 Nyíregyháza Hungary
93 F5 Nykøbing Falster Denmark
93 G4 Nyköping Sweden
53 C2 Nymagee Austr.
93 G4 Nynäshamn Sweden
53 C2 Nyngan Austr.
88 B3 Nyoman *r.* Belarus/Lith.
105 D3 Nyons France
91 Nyrob Russia
103 D1 Nysa Pol.
134 C2 Nyssa U.S.A.
119 C3 Nyunzu Dem. Rep. Congo
83 I2 Nyurba Russia
91 C2 Nyzhni Sirohozy Ukr.
91 C2 Nyzhn'ohirs'kyy Ukr.
118 B3 Nzambi Congo
119 D3 Nzega Tanz.
114 B4 Nzérékoré Guinea
120 A1 N'zeto Angola

O

136 C2 Oahe, Lake U.S.A.
49 L1 O'ahu *i.* U.S.A.
52 B2 Oakbank Austr.
140 B2 Oakdale U.S.A.
53 D1 Oakey Austr.
138 B3 Oak Grove U.S.A.
99 C3 Oakham U.K.
134 B1 Oak Harbor U.S.A.
138 C3 Oak Hill U.S.A.
135 B3 Oakland *CA* U.S.A.
139 D3 Oakland *MD* U.S.A.
138 B2 Oak Lawn U.S.A.
136 C3 Oakley U.S.A.
50 B2 Oakover *r.* Austr.
134 B2 Oakridge U.S.A.
141 D1 Oak Ridge U.S.A.
54 B3 Oamaru N.Z.
64 B2 Oas Phil.
145 C3 Oaxaca Mex.
86 F2 Ob' *r.* Russia
Ob, Gulf of *sea chan.* Russia *see* Obskaya Guba
88 C2 Obal' Belarus
118 B2 Obala Cameroon
96 B2 Oban U.K.
106 B1 O Barco Spain
Obbia Somalia *see* Hobyo
136 C3 Oberlin U.S.A.
53 C2 Oberon Austr.
101 F3 Oberviechtach Ger.
59 C3 Obi *i.* Indon.
151 D3 Óbidos Brazil
66 D2 Obihiro Japan
69 E1 Obluch'ye Russia
89 E2 Obninsk Russia
119 C2 Obo C.A.R.
117 C3 Obock Djibouti
103 D1 Oborniki Pol.
118 B3 Obouya Congo
89 E3 Oboyan' Russia
86 D2 Obozerskiy Russia
144 B2 Obregón, Presa *resr* Mex.
109 D2 Obrenovac Serbia
134 D2 O'Brien U.S.A.
87 D3 Obshchiy Syrt *hills* Kazakh./Russia
86 G2 Obskaya Guba *sea chan.* Russia
114 B4 Obuasi Ghana
90 C1 Obukhiv Ukr.
86 D2 Ob''yachevo Russia
141 D3 Ocala U.S.A.
144 B2 Ocampo Mex.
106 C2 Ocaña Spain

93 G3 Östhammar Sweden
103 D2 Ostrava Czech Rep.
103 D1 Ostróda Pol.
89 E3 Ostrogozhsk Russia
90 B1 Ostroh Ukr.
103 E1 Ostrołęka Pol.
101 F2 Ostrov Czech Rep.
88 C2 Ostrov Russia
Ostrovets Pol. see Ostrowiec Świętokrzyski
86 C2 Ostrovnoy Russia
89 F2 Ostrovskoye Russia
103 E1 Ostrowiec Świętokrzyski Pol.
103 E1 Ostrów Mazowiecka Pol.
Ostrowo Pol. see Ostrów Wielkopolski
103 D1 Ostrów Wielkopolski Pol.
109 C2 Ostuni Italy
67 B4 Ōsumi-kaikyō sea chan. Japan
67 B4 Ōsumi-shotō is Japan
106 B2 Osuna Spain
139 D2 Oswego U.S.A.
99 B3 Oswestry U.K.
67 C3 Ōta Japan
54 B3 Otago Peninsula N.Z.
54 C2 Otaki N.Z.
77 D2 Otar Kazakh.
66 D2 Otaru Japan
120 A3 Otavi Namibia
67 D3 Ōtawara Japan
92 G2 Oteren Norway
134 C1 Othello U.S.A.
123 D2 oThongathi S. Africa
120 A3 Otjiwarongo Namibia
109 C2 Otočac Croatia
Otog Qi China see Ulan
117 B3 Otoro, Jebel mt. Sudan
Otpor Russia see Zabaykal'sk
93 E4 Otra r. Norway
109 C2 Otranto, Strait of Albania/Italy
67 C3 Ōtsu Japan
93 E3 Otta Norway
130 C3 Ottawa Can.
130 C3 Ottawa r. Can.
138 B3 Ottawa IL U.S.A.
137 D3 Ottawa KS U.S.A.
130 C3 Ottawa Islands Can.
98 B2 Otterburn U.K.
130 B2 Otter Rapids Can.
100 B2 Ottignies Belgium
137 E2 Ottumwa U.S.A.
150 B3 Otuzco Peru
52 B3 Otway, Cape Austr.
140 B2 Ouachita r. U.S.A.
140 B2 Ouachita, Lake U.S.A.
140 B2 Ouachita Mountains U.S.A.
118 C2 Ouadda C.A.R.
115 D3 Ouaddaï reg. Chad
114 B3 Ouagadougou Burkina Faso
114 B3 Ouahigouya Burkina Faso
114 B3 Oualâta Maur.
114 B2 Oualâta, Dahr hills Maur.
118 C2 Ouanda Djallé C.A.R.
114 B2 Ouarâne reg. Maur.
115 C1 Ouargla Alg.
114 B1 Ouarzazate Morocco
100 A2 Oudenaarde Belgium
122 B3 Oudtshoorn S. Africa
107 C2 Oued Tlélat Alg.
114 B1 Oued Zem Morocco
104 A2 Ouessant, Île d' i. France
118 B2 Ouesso Congo
97 B2 Oughterard Ireland
118 B2 Ouham r. C.A.R./Chad
114 B1 Oujda Morocco
92 H3 Oulainen Fin.
107 D2 Ouled Farès Alg.
92 I3 Oulu Fin.
92 I3 Oulujärvi l. Fin.
108 A1 Oulx Italy
115 E3 Oum-Chalouba Chad
115 E3 Oum-Hadjer Chad
115 E3 Ounianga Kébir Chad
100 A2 Oupeye Belgium
100 C3 Our r. Ger./Lux.
106 B1 Ourense Spain
155 D2 Ourinhos Brazil
155 D2 Ouro Preto Brazil
100 B2 Ourthe r. Belgium
98 C3 Ouse r. U.K.
Outaouais, Rivière des r. Can. see Ottawa
120 A2 Outapi Namibia
131 D3 Outardes, Rivière aux r. Can.
131 D2 Outardes Quatre, Réservoir resr Can.
96 A2 Outer Hebrides is U.K.
Outer Mongolia country Asia see Mongolia
120 A3 Outjo Namibia
129 D2 Outlook Can.
92 I3 Outokumpu Fin.
52 B3 Ouyen Austr.
108 A2 Ovace, Punta d' mt. France
152 A3 Ovalle Chile
106 B1 Ovar Port.
92 H2 Överkalix Sweden
137 E3 Overland Park U.S.A.
135 D3 Overton U.S.A.
92 H2 Övertorneå Sweden
106 B1 Oviedo Spain
141 D3 Oviedo U.S.A.
88 B2 Ovišrags hd Latvia
93 E3 Øvre Årdal Norway

93 F3 Øvre Rendal Norway
90 B1 Ovruch Ukr.
118 B3 Owando Congo
67 C4 Owase Japan
143 D1 Owasso U.S.A.
137 E2 Owatonna U.S.A.
139 D2 Owego U.S.A.
138 B3 Owensboro U.S.A.
135 C3 Owens Lake U.S.A.
130 B3 Owen Sound Can.
51 D1 Owen Stanley Range mts P.N.G.
115 C4 Owerri Nigeria
115 C4 Owo Nigeria
138 C2 Owosso U.S.A.
134 C2 Owyhee U.S.A.
134 C2 Owyhee r. U.S.A.
129 D3 Oxbow Can.
54 B2 Oxford N.Z.
99 C4 Oxford U.K.
140 C2 Oxford U.S.A.
129 E2 Oxford Lake Can.
145 D2 Oxkutzcab Mex.
52 B2 Oxley Austr.
97 B1 Ox Mountains hills Ireland
135 C4 Oxnard U.S.A.
67 C3 Oyama Japan
118 B2 Oyem Gabon
129 C2 Oyen Can.
105 D2 Oyonnax France
77 C2 Oyoqquduq Uzbek.
64 B3 Ozamis Phil.
140 C2 Ozark AL U.S.A.
137 E3 Ozark MO U.S.A.
137 E3 Ozark Plateau U.S.A.
137 E3 Ozarks, Lake of the U.S.A.
83 L3 Ozernovskiy Russia
88 B3 Ozersk Russia
89 E3 Ozery Russia
87 D3 Ozinki Russia

P

127 I2 Paamiut Greenland
122 A3 Paarl S. Africa
103 D1 Pabianice Pol.
75 C2 Pabna Bangl.
88 C3 Pabradé Lith.
74 A2 Pab Range mts Pak.
150 B3 Pacasmayo Peru
142 B2 Pacheco Mex.
109 C3 Pachino Italy
145 C2 Pachuca Mex.
135 B3 Pacifica U.S.A.
157 E9 Pacific-Antarctic Ridge S. Pacific Ocean
156 Pacific Ocean
61 C2 Pacitan Indon.
52 B2 Packsaddle Austr.
103 D1 Paczków Pol.
60 B2 Padang Indon.
60 B1 Padang Endau Malaysia
60 B2 Padangpanjang Indon.
60 A1 Padangsidimpuan Indon.
101 D2 Paderborn Ger.
Padova Italy see Padua
143 D3 Padre Island U.S.A.
99 A4 Padstow U.K.
52 B3 Padthaway Austr.
108 B1 Padua Italy
138 B3 Paducah KY U.S.A.
143 C2 Paducah TX U.S.A.
65 B1 Paegam N. Korea
Paektu-san mt. China/N. Korea see Baitou Shan
Paengnyŏng-do i. S. Korea see Baengnyeong-do
54 C1 Paeroa N.Z.
Pafos Cyprus see Paphos
109 C2 Pag Croatia
109 B2 Pag i. Croatia
64 B3 Pagadian Phil.
60 B2 Pagai Selatan i. Indon.
60 B2 Pagai Utara i. Indon.
59 D1 Pagan i. N. Mariana Is
61 C2 Pagatan Indon.
142 A1 Page U.S.A.
88 B2 Paggiai Lith.
153 E5 Paget, Mount S. Georgia
136 B3 Pagosa Springs U.S.A.
88 C2 Paide Estonia
99 B4 Paignton U.K.
93 I3 Päijänne l. Fin.
75 C2 Paikü Co l. China
60 B2 Painan Indon.
138 C2 Painesville U.S.A.
142 A1 Painted Desert U.S.A.
Paint Hills Can. see Wemindji
96 B3 Paisley U.K.
92 H2 Pajala Sweden
150 A3 Paján Ecuador
150 C2 Pakaraima Mountains mts S. America
150 D2 Pakaraima Mountains S. America
74 A2 Pakistan country Asia
62 A1 Pakokku Myanmar
88 B2 Pakruojis Lith.
103 D2 Paks Hungary
130 A2 Pakwash Lake Can.
62 B2 Pakxan Laos
63 B2 Pakxé Laos
115 D4 Pala Chad
60 B2 Palabuhanratu, Teluk b. Indon.

111 C3 Palaikastro Greece
111 B3 Palaiochora Greece
73 B3 Palakkad India
122 B1 Palamakoloi Botswana
107 D1 Palamós Spain
83 L3 Palana Russia
64 B2 Palanan Phil.
61 C2 Palangkaraya Indon.
74 B2 Palanpur India
123 C1 Palapye Botswana
83 L2 Palatka Russia
141 D3 Palatka U.S.A.
59 C2 Palau country N. Pacific Ocean
63 A2 Palaw Myanmar
64 A3 Palawan i. Phil.
64 A3 Palawan Passage str. Phil.
88 B2 Paldiski Estonia
89 F2 Palekh Russia
60 B2 Palembang Indon.
106 C1 Palencia Spain
145 C3 Palenque Mex.
108 B3 Palermo Italy
143 D2 Palestine U.S.A.
62 A1 Paletwa Myanmar
Palghat India see Palakkad
74 B2 Pali India
48 G4 Palikir Micronesia
114 C4 Palimé Togo
109 C2 Palinuro, Capo c. Italy
111 B3 Paliouri, Akrotirio pt Greece
100 B3 Paliseul Belgium
92 I3 Paljakka h. Fin.
88 C2 Palkino Russia
73 B4 Palk Strait India/Sri Lanka
Palla Bianca mt. Austria/Italy see Weißkugel
54 C2 Palliser, Cape N.Z.
157 F7 Palliser, Îles is Fr. Polynesia
106 C2 Palma del Río Spain
107 D2 Palma de Mallorca Spain
154 B3 Palmas Paraná Brazil
151 E4 Palmas Tocantins Brazil
154 B3 Palmas, Campos de hills Brazil
114 B4 Palmas, Cape Liberia
141 D3 Palm Bay U.S.A.
135 C4 Palmdale U.S.A.
154 C3 Palmeira Brazil
151 E3 Palmeirais Brazil
154 A2 Palmeiras Brazil
126 C2 Palmer U.S.A.
55 A2 Palmer Land reg. Antarctica
49 K5 Palmerston atoll Cook Is
54 C2 Palmerston North N.Z.
109 C3 Palmi Italy
145 C2 Palmillas Mex.
150 B2 Palmira Col.
154 B2 Palmital Brazil
135 C4 Palm Springs U.S.A.
Palmyra Syria see Tadmur
49 K3 Palmyra Atoll N. Pacific Ocean
135 B3 Palo Alto U.S.A.
117 B3 Paloich S. Sudan
145 C3 Palomares Mex.
61 D2 Palopo Indon.
107 C2 Palos, Cabo de c. Spain
92 I3 Paltamo Fin.
61 C2 Palu Indon.
83 M2 Palyavaam r. Russia
150 B3 Pamar Col.
121 C3 Pambarra Moz.
104 C3 Pamiers France
77 D3 Pamir mts Asia
141 E1 Pamlico Sound sea chan. U.S.A.
152 B1 Pampa Grande Bol.
153 B3 Pampas reg. Arg.
150 B2 Pamplona Col.
107 C1 Pamplona Spain
111 D2 Pamukova Turkey
60 B2 Panaitan i. Indon.
73 B3 Panaji India
146 B4 Panama country Central America
Panamá Panama see Panama City
146 C4 Panamá, Canal de canal Panama
Panamá, Golfo de Panama see Panama, Gulf of
146 C4 Panama, Gulf of g. Panama
Panama Canal canal Panama see Panamá, Canal de
146 C4 Panama City Panama
140 C2 Panama City U.S.A.
135 C3 Panamint Range mts U.S.A.
60 B1 Panarik Indon.
64 B2 Panay i. Phil.
109 D2 Pančevo Serbia
64 B2 Pandan Antique Phil.
64 B2 Pandan Catanduanes Phil.
75 C2 Pandaria India
73 B3 Pandharpur India
88 B2 Panevėžys Lith.
Panfilov Kazakh. see Zharkent
61 C2 Pangkalanbuun Indon.
60 A1 Pangkalansusu Indon.
60 B2 Pangkalpinang Indon.
61 D2 Pangkalsiang, Tanjung pt Indon.
64 A3 Panglima Sugala Phil.
127 H2 Pangnirtung Can.
86 G2 Pangody Russia
89 F3 Panino Russia
74 B2 Panipat India
74 A2 Panjgur Pak.
Panji India see Panaji
118 A2 Pankshin Nigeria
65 C1 Pan Ling mts China
75 C2 Panna India

50 A2 Pannawonica Austr.
154 B2 Panorama Brazil
65 B1 Panshi China
152 C1 Pantanal reg. Brazil
145 C2 Pánuco Mex.
145 C2 Pánuco r. Mex.
62 B1 Panzhihua China
109 C3 Paola Italy
118 B2 Paoua C.A.R.
63 B2 Paôy Pêt Cambodia
103 D2 Pápa Hungary
54 B1 Papakura N.Z.
145 C2 Papantla Mex.
96 □ Papa Stour i. U.K.
54 B1 Papatoetoe N.Z.
49 M5 Papeete Fr. Polynesia
100 C1 Papenburg Ger.
80 B2 Paphos Cyprus
137 D2 Papillion U.S.A.
59 D3 Papua, Gulf of P.N.G.
59 D3 Papua New Guinea country Oceania
89 F3 Para r. Russia
50 A2 Paraburdoo Austr.
154 C1 Paracatu Brazil
155 C1 Paracatu r. Brazil
52 A2 Parachilna Austr.
109 D2 Paraćin Serbia
155 D1 Pará de Minas Brazil
151 D2 Paradise Guyana
135 B3 Paradise U.S.A.
140 B1 Paragould U.S.A.
151 D2 Paraguai r. Brazil
147 D3 Paraguaná, Península de pen. Venez.
152 C2 Paraguay r. Arg./Para.
152 C2 Paraguay country S. America
155 D2 Paraíba do Sul r. Brazil
154 B1 Paraíso Brazil
145 C3 Paraíso Mex.
114 C4 Parakou Benin
52 A2 Parakylia Austr.
151 D2 Paramaribo Suriname
83 L3 Paramushir, Ostrov i. Russia
152 B3 Paraná Arg.
154 C1 Paraná state Brazil
154 A3 Paraná r. S. America
154 C1 Paraná, Serra do hills Brazil
154 B1 Paranaguá Brazil
154 B1 Paranaíba Brazil
154 B2 Paranaíba r. Brazil
154 B2 Paranapanema r. Brazil
154 B2 Paranapiacaba, Serra mts Brazil
154 B2 Paranavaí Brazil
90 A2 Parângul Mare, Vârful mt. Romania
54 B2 Paraparaumu N.Z.
155 D2 Parati Brazil
52 A2 Paratoo Austr.
151 D3 Parauaquara, Serra h. Brazil
154 B1 Paraúna Brazil
105 C2 Paray-le-Monial France
74 B2 Parbati r. India
74 B3 Parbhani India
101 E1 Parchim Ger.
103 E1 Parczew Pol.
155 E1 Pardo r. Bahia Brazil
154 C2 Pardo r. Mato Grosso do Sul Brazil
154 C2 Pardo r. São Paulo Brazil
103 D1 Pardubice Czech Rep.
157 C2 Parecis, Serra dos hills Brazil
130 C2 Parent Can.
130 C2 Parent, Lac l. Can.
61 C2 Parepare Indon.
89 D2 Parfino Russia
111 B3 Parga Greece
109 C3 Parghelia Italy
147 D3 Paria, Gulf of Trin. and Tob./Venez.
150 C2 Parima, Serra mts Brazil
151 D3 Parintins Brazil
104 C2 Paris France
140 C1 Paris TN U.S.A.
143 D2 Paris TX U.S.A.
93 H3 Parkano Fin.
142 A2 Parker U.S.A.
138 C3 Parkersburg U.S.A.
53 C2 Parkes Austr.
138 A1 Park Falls U.S.A.
134 B1 Parkland U.S.A.
137 D1 Park Rapids U.S.A.
106 C1 Parla Spain
108 B2 Parma Italy
134 C2 Parma U.S.A.
151 E3 Parnaíba Brazil
151 E3 Parnaíba r. Brazil
54 B2 Parnassus N.Z.
Parnassus, Mount mt. Greece see Liakoura
111 B3 Parnonas mts Greece
88 B2 Pärnu Estonia
65 B2 Paro-ho l. S. Korea
52 B2 Paroo watercourse Austr.
Paropamisus mts Afgh. see Safēd Kōh, Silsilah-ye
111 C3 Paros i. Greece
135 D3 Parowan U.S.A.
153 A3 Parral Chile
53 C2 Parramatta Austr.
144 B2 Parras Mex.
126 C2 Parry, Cape Can.
126 E1 Parry Islands Can.
130 B3 Parry Sound Can.
79 C2 Pārsīān Iran
137 D3 Parsons U.S.A.
108 B3 Partanna Italy

Port Sudan

50 A2 Pilbara reg. Austr.
75 B2 Pilibhit India
53 C2 Pilliga Austr.
154 C1 Pilões, Serra dos mts Brazil
150 C4 Pimenta Bueno Brazil
88 C3 Pina r. Belarus
153 C3 Pinamar Arg.
60 B1 Pinang i. Malaysia
80 B2 Pınarbaşı Turkey
146 B2 Pinar del Río Cuba
64 B2 Pinatubo, Mount vol. Phil.
103 E1 Pińczów Pol.
151 E3 Pindaré r. Brazil
Pindos Greece see Pindus Mountains
111 B2 Pindus Mountains mts Greece
140 B2 Pine Bluff U.S.A.
136 C2 Pine Bluffs U.S.A.
50 C1 Pine Creek Austr.
136 B2 Pinedale U.S.A.
86 D2 Pinega Russia
129 D2 Pinehouse Lake Can.
111 B3 Pineios r. Greece
141 D3 Pine Islands FL U.S.A.
141 D4 Pine Islands FL U.S.A.
128 C1 Pine Point (abandoned) Can.
136 C2 Pine Ridge U.S.A.
108 A2 Pinerolo Italy
Pines, Isle of i. Cuba see Juventud, Isla de la
123 D2 Pinetown S. Africa
140 B2 Pineville U.S.A.
70 B2 Pingdingshan China
70 B2 Pingdu China
62 B1 Pingguo China
71 B3 Pingjiang China
70 A2 Pingliang China
70 B1 Pingquan China
71 C3 P'ingtung Taiwan
Pingxi China see Yuping
71 A3 Pingxiang Guangxi China
71 B3 Pingxiang Jiangxi China
70 B2 Pingyin China
151 E3 Pinheiro Brazil
52 B3 Pinnaroo Austr.
101 D1 Pinneberg Ger.
Pinos, Isla de i. Cuba see Juventud, Isla de la
145 C3 Pinotepa Nacional Mex.
48 H6 Pins, Île des i. New Caledonia
88 C3 Pinsk Belarus
152 B2 Pinto Arg.
135 D3 Pioche U.S.A.
119 C3 Piodi Dem. Rep. Congo
108 B2 Piombino Italy
86 F2 Pionerskiy Russia
103 E1 Pionki Pol.
103 D1 Piotrków Trybunalski Pol.
137 D2 Pipestone U.S.A.
131 C3 Pipmuacan, Réservoir resr Can.
154 B2 Piquiri r. Brazil
154 C1 Piracanjuba Brazil
154 C2 Piracicaba Brazil
155 D1 Piracicaba r. Brazil
111 B3 Piraeus Greece
154 C2 Piraí do Sul Brazil
154 C2 Piraju Brazil
154 C2 Pirajuí Brazil
154 B1 Piranhas Brazil
151 F3 Piranhas r. Brazil
155 D1 Pirapora Brazil
154 C2 Pirassununga Brazil
154 C1 Pirenópolis Brazil
154 C1 Pires do Rio Brazil
Pirineos mts Europe see Pyrenees
151 E3 Piripiri Brazil
109 D2 Pirot Serbia
59 C3 Piru Indon.
108 B2 Pisa Italy
152 A1 Pisagua Chile
150 B4 Pisco Peru
102 C2 Písek Czech Rep.
79 D2 Pīshīn Iran
74 A1 Pishin Pak.
145 D2 Pisté Mex.
109 C2 Pisticci Italy
108 B2 Pistoia Italy
106 C1 Pisuerga r. Spain
134 B2 Pit r. U.S.A.
114 A3 Pita Guinea
154 B2 Pitanga Brazil
155 D1 Pitangui Brazil
49 O6 Pitcairn Island Pitcairn Is
49 O6 Pitcairn Islands terr. S. Pacific Ocean
92 H2 Piteå Sweden
92 H2 Piteälven r. Sweden
110 B2 Pitești Romania
75 C2 Pithoragarh India
142 A2 Pitiquito Mex.
86 C2 Pitkyaranta Russia
96 C2 Pitlochry U.K.
128 B2 Pitt Island Can.
137 E3 Pittsburg U.S.A.
138 D2 Pittsburgh U.S.A.
139 E2 Pittsfield U.S.A.
53 D1 Pittsworth Austr.
155 C2 Piumhi Brazil
150 A3 Piura Peru
90 C2 Pivdennyy Buh r. Ukr.
Þjórsá see Thjórsá r. Iceland
131 E3 Placentia Can.
135 B3 Placerville U.S.A.
146 C2 Placetas Cuba
143 C2 Plains U.S.A.
143 C2 Plainview U.S.A.
60 B2 Plaju Indon.
61 C2 Plampang Indon.
154 C1 Planaltina Brazil
137 D2 Plankinton U.S.A.
143 D2 Plano U.S.A.
154 C2 Planura Brazil
140 B2 Plaquemine U.S.A.
106 B1 Plasencia Spain
109 C1 Plaški Croatia
147 C4 Plato Col.
137 D2 Platte r. U.S.A.
138 A2 Platteville U.S.A.
139 E2 Plattsburgh U.S.A.
101 F1 Plau am See Ger.
101 F2 Plauen Ger.
101 F1 Plauer See l. Ger.
89 E3 Plavsk Russia
129 E2 Playgreen Lake Can.
63 B2 Plây Ku Vietnam
153 B3 Plaza Huincul Arg.
143 D3 Pleasanton U.S.A.
54 B2 Pleasant Point N.Z.
138 B3 Pleasure Ridge Park U.S.A.
104 C2 Pleaux France
130 B2 Pledger Lake Can.
54 C1 Plenty, Bay of g. N.Z.
136 C1 Plentywood U.S.A.
86 D2 Plesetsk Russia
131 C2 Plétipi, Lac l. Can.
100 C2 Plettenberg Ger.
122 B3 Plettenberg Bay S. Africa
110 B2 Pleven Bulg.
Plevna Bulg. see Pleven
109 C2 Pljevlja Montenegro
108 A2 Ploaghe Italy
109 C2 Ploče Croatia
103 D1 Płock Pol.
109 C2 Pločno mt. Bos. & Herz.
104 B2 Ploemeur France
Ploești Romania see Ploiești
110 C2 Ploiești Romania
89 F2 Ploskoye Russia
104 B2 Plouzané France
110 B2 Plovdiv Bulg.
Plozk Pol. see Płock
121 B3 Plumtree Zimbabwe
88 B2 Plung Lith.
144 B2 Plutarco Elías Calles, Presa resr Mex.
88 C3 Plyeshchanitsy Belarus
99 A4 Plymouth U.K.
138 B2 Plymouth U.S.A.
147 D3 Plymouth (abandoned) Montserrat
99 B3 Plynlimon h. U.K.
88 C2 Plyussa Russia
102 C2 Plzeň Czech Rep.
114 B3 Pô Burkina Faso
108 B1 Po r. Italy
61 C1 Po, Tanjung pt Malaysia
77 E2 Pobeda Peak China/Kyrg.
Pobedy, Pik mt. China/Kyrg. see Pobeda Peak
140 B1 Pocahontas U.S.A.
134 D2 Pocatello U.S.A.
90 B1 Pochayiv Ukr.
89 D3 Pochep Russia
89 D3 Pochinok Russia
145 C3 Pochutla Mex.
139 D3 Pocomoke City U.S.A.
155 C2 Poços de Caldas Brazil
89 D2 Poddor'ye Russia
91 D1 Podgorenskiy Russia
109 C2 Podgorica Montenegro
82 G3 Podgornoye Russia
83 H2 Podkamennaya Tunguska r. Russia
89 E2 Podol'sk Russia
109 D2 Podujevë Kosovo
Podujevo Kosovo see Podujevë
122 A2 Pofadder S. Africa
89 D3 Pogar Russia
105 E3 Poggibonsi Italy
109 D2 Pogradec Albania
66 B2 Pogranichnyy Russia
Po Hai g. China see Bo Hai
65 B2 Pohang S. Korea
48 G3 Pohnpei atoll Micronesia
90 B2 Pohrebyshche Ukr.
110 B2 Poiana Mare Romania
118 C3 Poie Dem. Rep. Congo
118 B3 Pointe-Noire Congo
126 A2 Point Hope U.S.A.
128 C1 Point Lake Can.
138 C3 Point Pleasant U.S.A.
104 C2 Poitiers France
104 C2 Poitou, Plaines et Seuil du plain France
74 B2 Pokaran India
53 C1 Pokataroo Austr.
75 C2 Pokhara Nepal
119 C2 Poko Dem. Rep. Congo
83 J2 Pokrovsk Russia
91 D2 Pokrovskoye Russia
Pola Croatia see Pula
142 A1 Polacca U.S.A.
103 D1 Poland country Europe
55 C1 Polar Plateau Antarctica
80 B2 Polatlı Turkey
88 C2 Polatsk Belarus
61 C2 Polewali Indon.
118 B2 Poli Cameroon
102 C1 Police Pol.
109 C2 Policoro Italy
105 D2 Poligny France
64 B2 Polillo Islands Phil.
80 B2 Polis Cyprus
90 B1 Polis'ke (abandoned) Ukr.
109 C3 Polistena Italy
103 D1 Polkowice Pol.
107 D2 Pollença Spain
109 C3 Pollino, Monte mt. Italy
86 F2 Polnovat Russia
91 D2 Polohy Ukr.
123 C1 Polokwane S. Africa
123 D1 Polokwane r. S. Africa
90 B1 Polonne Ukr.
134 D1 Polson U.S.A.
91 C2 Poltava Ukr.
66 B2 Poltavka Russia
91 D2 Poltavskaya Russia
88 C2 Põltsamaa Estonia
88 C2 Põlva Estonia
Polyanovgrad Bulg. see Karnobat
92 J2 Polyarnyy Russia
86 C2 Polyarnyye Zori Russia
111 B2 Polygyros Greece
111 B2 Polykastro Greece
156 E6 Polynesia is Pacific Ocean
106 B2 Pombal Port.
102 C1 Pomeranian Bay b. Ger./Pol.
108 B2 Pomezia Italy
92 I2 Pomokaira reg. Fin.
110 C2 Pomorie Bulg.
Pomorska, Zatoka b. Ger./Pol. see Pomeranian Bay
155 D1 Pompéu Brazil
143 D1 Ponca City U.S.A.
147 D3 Ponce Puerto Rico
Pondicherry India see Puducherry
127 G2 Pond Inlet Can.
Ponds Bay Can. see Pond Inlet
106 B1 Ponferrada Spain
117 A4 Pongo watercourse S. Sudan
123 D2 Pongola r. S. Africa
123 D2 Pongolapoort Dam dam S. Africa
128 C2 Ponoka Can.
154 B3 Ponta Grossa Brazil
154 C1 Pontalina Brazil
105 D2 Pont-à-Mousson France
154 A2 Ponta Porã Brazil
105 D2 Pontarlier France
105 D2 Pontcharra France
140 B2 Pontchartrain, Lake U.S.A.
154 B1 Ponte de Pedra Brazil
106 B2 Ponte de Sor Port.
154 B1 Ponte do Rio Verde Brazil
98 C3 Pontefract U.K.
129 D3 Ponteix Can.
155 D2 Ponte Nova Brazil
150 D4 Pontes e Lacerda Brazil
106 B1 Pontevedra Spain
Ponthierville Dem. Rep. Congo see Ubundu
138 B2 Pontiac IL U.S.A.
138 C2 Pontiac MI U.S.A.
60 B2 Pontianak Indon.
Pontine Islands is Italy see Ponziane, Isole
104 B2 Pontivy France
151 D2 Pontoetoe Suriname
104 C2 Pontoise France
129 E2 Ponton Can.
99 B4 Pontypool U.K.
99 B4 Pontypridd U.K.
108 B2 Ponziane, Isole is Italy
99 C4 Poole U.K.
Poona India see Pune
52 B2 Pooncarie Austr.
152 B1 Poopó, Lago de l. Bol.
150 B2 Popayán Col.
83 I2 Popigay r. Russia
52 B2 Popiltah Austr.
129 E2 Poplar r. Can.
137 E3 Poplar Bluff U.S.A.
118 B3 Popokabaka Dem. Rep. Congo
Popovichskaya Russia see Kalininskaya
110 C2 Popovo Bulg.
103 E2 Poprad Slovakia
151 E4 Porangatu Brazil
74 A2 Porbandar India
126 C2 Porcupine r. Can./U.S.A.
108 B1 Pordenone Italy
108 B1 Poreč Croatia
154 B2 Porecatu Brazil
114 C3 Porga Benin
93 H3 Pori Fin.
54 B2 Porirua N.Z.
88 C2 Porkhov Russia
104 B2 Pornic France
83 K3 Poronaysk Russia
75 D1 Porong China
111 B3 Poros Greece
92 I1 Porsangerfjorden sea chan. Norway
93 E4 Porsgrunn Norway
111 D3 Porsuk r. Turkey
97 C1 Portadown U.K.
97 D1 Portaferry U.K.
138 B2 Portage U.S.A.
129 E3 Portage la Prairie Can.
128 B3 Port Alberni Can.
106 B2 Portalegre Port.
143 C2 Portales U.S.A.
128 A2 Port Alexander U.S.A.
128 B2 Port Alice Can.
140 B2 Port Allen U.S.A.
134 B1 Port Angeles U.S.A.
97 C2 Portarlington Ireland
51 D4 Port Arthur Austr.
Port Arthur China see Lüshunkou
143 E3 Port Arthur U.S.A.
96 A3 Port Askaig U.K.
52 A2 Port Augusta Austr.
147 C3 Port-au-Prince Haiti
131 E2 Port aux Choix Can.
122 B3 Port Beaufort S. Africa
73 D3 Port Blair India
52 B3 Port Campbell Austr.
131 D2 Port-Cartier Can.
54 B3 Port Chalmers N.Z.
141 D3 Port Charlotte U.S.A.
147 C3 Port-de-Paix Haiti
128 A2 Port Edward Can.
155 D1 Porteirinha Brazil
151 D3 Portel Brazil
130 B3 Port Elgin Can.
123 C3 Port Elizabeth S. Africa
96 A3 Port Ellen U.K.
98 A2 Port Erin Isle of Man
122 A3 Porterville S. Africa
135 C3 Porterville U.S.A.
Port Étienne Maur. see Nouâdhibou
52 B3 Port Fairy Austr.
54 C1 Port Fitzroy N.Z.
Port Francqui Dem. Rep. Congo see Ilebo
118 A3 Port-Gentil Gabon
115 C4 Port Harcourt Nigeria
128 B2 Port Hardy Can.
Port Harrison Can. see Inukjuak
131 D3 Port Hawkesbury Can.
99 B4 Porthcawl U.K.
50 A2 Port Hedland Austr.
Port Herald Malawi see Nsanje
99 A3 Porthmadog U.K.
131 E2 Port Hope Simpson Can.
138 C2 Port Huron U.S.A.
106 B2 Portimão Port.
Port Keats Austr. see Wadeye
Port Láirge Ireland see Waterford
53 C2 Portland N.S.W. Austr.
52 B3 Portland Vic. Austr.
139 E2 Portland ME U.S.A.
134 B1 Portland OR U.S.A.
143 D3 Portland TX U.S.A.
99 B4 Portland, Isle of pen. U.K.
128 A2 Portland Canal inlet Can.
97 C2 Portlaoise Ireland
143 D3 Port Lavaca U.S.A.
52 A2 Port Lincoln Austr.
114 A4 Port Loko Sierra Leone
113 I8 Port Louis Mauritius
Port-Lyautrey Morocco see Kenitra
52 B3 Port MacDonnell Austr.
53 D2 Port Macquarie Austr.
Portmadoc U.K. see Porthmadog
128 B2 Port McNeill Can.
131 D3 Port-Menier Can.
59 D3 Port Moresby P.N.G.
96 A3 Portnahaven U.K.
Port Nis U.K. see Port of Ness
122 A2 Port Nolloth S. Africa
Port-Nouveau-Québec Can. see Kangiqsualujjuaq
Porto Port. see Oporto
150 C3 Porto Acre Brazil
154 B2 Porto Alegre Mato Grosso do Sul Brazil
152 C3 Porto Alegre Rio Grande do Sul Brazil
Porto Alexandre Angola see Tombua
Porto Amélia Moz. see Pemba
151 D4 Porto Artur Brazil
154 B2 Porto Camargo Brazil
151 D4 Porto dos Gaúchos Óbidos Brazil
150 D4 Porto Esperidião Brazil
108 B2 Portoferraio Italy
96 A1 Port of Ness U.K.
151 E3 Porto Franco Brazil
147 D3 Port of Spain Trin. and Tob.
108 B1 Portogruaro Italy
108 B2 Portomaggiore Italy
154 B2 Porto Mendes Brazil
152 C2 Porto Murtinho Brazil
151 E4 Porto Nacional Brazil
114 C4 Porto-Novo Benin
154 B2 Porto Primavera, Represa resr Brazil
134 B2 Port Orford U.S.A.
154 B2 Porto São José Brazil
108 A3 Portoscuso Italy
155 E1 Porto Seguro Brazil
108 B2 Porto Tolle Italy
108 A2 Porto Torres Italy
154 B3 Porto União Brazil
105 D3 Porto-Vecchio France
150 C3 Porto Velho Brazil
150 A3 Portoviejo Ecuador
96 B3 Portpatrick U.K.
52 B3 Port Phillip Bay Austr.
52 A2 Port Pirie Austr.
96 A2 Portree U.K.
128 B3 Port Renfrew Can.
97 C1 Portrush U.K.
116 B1 Port Said Egypt
141 C3 Port St Joe U.S.A.
123 C3 Port St Johns S. Africa
141 D3 Port St Lucie City U.S.A.
123 D3 Port Shepstone S. Africa
99 C4 Portsmouth U.K.
139 E2 Portsmouth NH U.S.A.
138 C3 Portsmouth OH U.S.A.
139 D3 Portsmouth VA U.S.A.
153 B5 Port Stephens Falkland Is
97 C1 Portstewart U.K.
116 B3 Port Sudan Sudan

Q

Qingjiang *Jiangsu* China see Huai'an
Qingjiang *Jiangxi* China see Zhangshu
70 B2 Qingshuihe China
70 A2 Qingtongxia China
71 B3 Qingyang China
71 B3 Qingyuan *Guangdong* China
65 A1 Qingyuan *Guangxi* China see Yizhou
Qingyuan *Liaoning* China
Qingzang Gaoyuan *plat.* China see Tibet, Plateau of
70 B2 Qingzhou China
70 B2 Qinhuangdao China
Qinjiang China see Shicheng
70 A2 Qin Ling *mts* China
Qinting China see Lianhua
70 B2 Qinyang China
71 A3 Qinzhou China
71 B4 Qionghai China
70 A2 Qionglai Shan *mts* China
71 A4 Qiongshan China
71 A4 Qiongzhong China
69 E1 Qiqihar China
81 D3 Qīr Iran
Qishan China see Qimen
79 C3 Qishn Yemen
66 B1 Qitaihe China
70 B2 Qixian *Henan* China
70 B2 Qixian *Shanxi* China
Qogir Feng *mt.* China/Pak. see K2
81 D2 Qom Iran
Qomishēh Iran see Shahrezā
Qomolangma Feng *mt.* China/Nepal see Everest, Mount
76 B2 Qo'ng'irot Uzbek.
Qoqek China see Tacheng
77 D2 Qo'qon Uzbek.
76 B2 Qoraqalpog'iston Uzbek.
80 B2 Qornet es Saouda *mt.* Lebanon
81 C2 Qorveh Iran
79 C2 Qoţbābād Iran
101 C1 Quakenbrück Ger.
53 C2 Quambone Austr.
63 B2 Quang Ngai Vietnam
63 B2 Quang Tri Vietnam
62 B1 Quan Hoa Vietnam
Quan Long Vietnam see Ca Mau
Quan Phu Quoc *i.* Vietnam see Phu Quốc, Đao
71 B3 Quanzhou *Fujian* China
71 B3 Quanzhou *Guangxi* China
108 A3 Quartu Sant'Elena Italy
142 A2 Quartzsite U.S.A.
81 C1 Quba Azer.
76 B2 Qūchān Iran
53 C3 Queanbeyan Austr.
131 C2 Québec Can.
131 C2 Québec *prov.* Can.
101 E2 Quedlinburg Ger.
Queen Adelaide Islands Chile see Reina Adelaida, Archipiélago de la
128 B2 Queen Charlotte Can.
128 B2 Queen Charlotte Sound *sea chan.* Can.
128 B2 Queen Charlotte Strait Can.
126 E1 Queen Elizabeth Islands Can.
55 A1 Queen Elizabeth Land *reg.* Antarctica
55 I2 Queen Mary Land *reg.* Antarctica
126 F2 Queen Maud Gulf Can.
55 C2 Queen Maud Land *reg.* Antarctica
55 P1 Queen Maud Mountains Antarctica
52 B3 Queenscliff Austr.
52 B1 Queensland *state* Austr.
51 D4 Queenstown Austr.
Queenstown Ireland see Cobh
54 A3 Queenstown N.Z.
123 C3 Queenstown S. Africa
121 C2 Quelimane Moz.
153 A4 Quellón Chile
Quelpart Island S. Korea see Cheju-do
142 B2 Quemado U.S.A.
Que Que Zimbabwe see Kwekwe
154 B2 Querência do Norte Brazil
145 B2 Querétaro Mex.
101 E2 Querfurt Ger.
128 B2 Quesnel Can.
128 B2 Quesnel Lake Can.
74 A1 Quetta Pak.
146 A3 Quetzaltenango Guat.
64 A3 Quezon Phil.
64 B2 Quezon City Phil.
120 A2 Quibala Angola
150 B2 Quibdó Col.
104 B2 Quiberon France
120 A2 Quilengues Angola
104 C3 Quillan France
153 C3 Quilmes Arg.
Quilon India see Kollam
51 D2 Quilpie Austr.
153 A3 Quilpué Chile
120 A1 Quimbele Angola
152 B2 Quimilí Arg.
104 B2 Quimper France
104 B2 Quimperlé France
135 B3 Quincy CA U.S.A.
141 D2 Quincy FL U.S.A.
137 E3 Quincy IL U.S.A.
139 E2 Quincy MA U.S.A.
107 C1 Quinto Spain
121 D2 Quionga Moz.
120 A2 Quirima Angola

53 D2 Quirindi Austr.
154 B1 Quirinópolis Brazil
131 D3 Quispamsis Can.
120 A2 Quitapa Angola
150 B3 Quito Ecuador
151 F3 Quixadá Brazil
71 A3 Qujing China
75 D1 Qumar He *r.* China
123 C3 Qumrha S. Africa
115 E1 Qunayyin, Sabkhat al *salt marsh* Libya
129 E1 Quoich *r.* Can.
52 A2 Quorn Austr.
79 C2 Qurayat Oman
77 C3 Qūrghonteppa Tajik.
Quxar China see Lhazê
Quyang China see Jingzhou
63 B2 Quy Nhơn Vietnam
71 B3 Quzhou China
Qyteti Stalin Albania see Kuçovë
Qyzyltü Kazakh. see Kishkenekol'

R

103 D2 Raab *r.* Austria
92 H3 Raahe Fin.
100 C1 Raalte Neth.
61 C2 Raas *i.* Indon.
61 C2 Raba Indon.
114 B1 Rabat Morocco
59 E3 Rabaul P.N.G.
50 C2 Rabbit Flat Austr.
78 A2 Rābigh Saudi Arabia
103 D2 Rabka-Zdrój Pol.
Râbniţa Moldova see Rîbniţa
Rabyānah, Ramlat *des.* Libya see Rebiana Sand Sea
131 E3 Race, Cape Can.
140 B3 Raceland U.S.A.
139 E2 Race Point U.S.A.
63 B3 Rach Gia Vietnam
103 D1 Racibórz Pol.
138 B2 Racine U.S.A.
78 B3 Radā' Yemen
110 C1 Rădăuţi Romania
138 B3 Radcliff U.S.A.
74 B2 Radhanpur India
130 C2 Radisson Can.
103 E1 Radom Pol.
103 D1 Radomsko Pol.
90 B1 Radomyshl' Ukr.
111 B2 Radoviš Macedonia
88 B2 Radviliškis Lith.
78 A2 Raḍwá, Jabal *mt.* Saudi Arabia
90 B1 Radyvyliv Ukr.
75 C2 Rae Bareli India
100 C2 Raeren Belgium
54 C1 Raetihi N.Z.
78 A2 Rāf *h.* Saudi Arabia
152 B3 Rafaela Arg.
118 C2 Rafaï C.A.R.
78 B2 Rafḥa' Saudi Arabia
79 C1 Rafsanjān Iran
119 C2 Raga S. Sudan
64 B3 Ragang, Mount *vol.* Phil.
109 B3 Ragusa Italy
61 D2 Raha Indon.
88 D3 Rahachow Belarus
78 A3 Rahad *r.* Sudan
74 B2 Rahimyar Khan Pak.
110 D2 Rahovec Kosovo
73 B3 Raichur India
75 C2 Raigarh India
128 C2 Rainbow Lake Can.
134 B1 Rainier, Mount *vol.* U.S.A.
130 A3 Rainy Lake Can./U.S.A.
129 E3 Rainy River Can.
75 C2 Raipur India
93 H3 Raisio Fin.
88 C2 Raja Estonia
73 C3 Rajahmundry India
61 C1 Rajang *r.* Malaysia
74 B2 Rajanpur Pak.
73 B4 Rajapalayam India
74 B2 Rajasthan *state* India
74 B2 Rajasthan Canal *canal* India
74 B2 Rajgarh India
60 B2 Rajik Indon.
74 B2 Rajkot India
75 C2 Rajnandgaon India
74 B2 Rajpur India
75 C2 Rajshahi Bangl.
54 B2 Rakaia *r.* N.Z.
74 B1 Rakaposhi *mt.* Pak.
90 A2 Rakhiv Ukr.
91 D1 Rakitnoye Russia
88 B2 Rakke Estonia
88 C2 Rakvere Estonia
141 E1 Raleigh U.S.A.
141 E2 Raleigh Bay U.S.A.
48 H3 Ralik Chain *is* Marshall Is
75 C2 Ramanuj Ganj India
123 C2 Ramatlabama S. Africa
104 C2 Rambouillet France
119 D2 Ramciel S. Sudan
99 A4 Rame Head *hd* U.K.
97 C1 Ramelton Ireland
89 E2 Rameshki Russia
74 B2 Ramgarh India
81 C2 Rāmhormoz Iran
110 C1 Râmnicu Sărat Romania
110 B1 Râmnicu Vâlcea Romania

89 E3 Ramon' Russia
135 C4 Ramona U.S.A.
123 C1 Ramotswa Botswana
75 B2 Rampur India
62 A2 Ramree Island Myanmar
98 A2 Ramsey Isle of Man
130 B3 Ramsey Lake Can.
99 D4 Ramsgate U.K.
81 C2 Rāmshīr Iran
62 A1 Ramsing India
75 C2 Ranaghat India
61 C1 Ranau Malaysia
153 A3 Rancagua Chile
154 B2 Rancharia Brazil
75 C2 Ranchi India
93 F4 Randers Denmark
63 B3 Rangae Thai.
75 D2 Rangapara India
54 B2 Rangiora N.Z.
49 M5 Rangiroa *atoll* Fr. Polynesia
54 C1 Rangitaiki *r.* N.Z.
60 B2 Rangkasbitung Indon.
63 A2 Rangoon Myanmar
75 C2 Rangpur Bangl.
129 E1 Rankin Inlet Can.
53 C2 Rankin's Springs Austr.
Rankovićevo Serbia see Kraljevo
96 B2 Rannoch Moor *moorland* U.K.
63 A3 Ranong Thai.
59 C3 Ransiki Indon.
61 C2 Rantaupanjang Indon.
60 A1 Rantauprapat Indon.
61 C2 Rantepang Indon.
92 I2 Ranua Fin.
93 E4 Ranum Denmark
78 B2 Ranyah, Wādī *watercourse* Saudi Arabia
49 J6 Raoul Island Kermadec Is
49 M6 Rapa *i.* Fr. Polynesia
108 A2 Rapallo Italy
136 C2 Rapid City U.S.A.
88 B2 Rapla Estonia
74 A2 Rapur India
49 L6 Rarotonga *i.* Cook Is
153 B4 Rasa, Punta *pt* Arg.
Ras al Khaimah U.A.E. see Ra's al Khaymah
79 C2 Ra's al Khaymah U.A.E.
116 B3 Ras Dejen *mt.* Eth.
88 B2 Raseiniai Lith.
116 B2 Ra's Ghārib Egypt
81 C2 Rasht Iran
74 A2 Ras Koh *mt.* Pak.
110 C1 Râşnov Romania
88 C2 Rasony Belarus
79 C2 Ra's Şirāb Oman
108 B3 Rass Jebel Tunisia
87 D3 Rasskazovo Russia
79 C2 Ras Tannūrah Saudi Arabia
101 D1 Rastede Ger.
48 I2 Ratak Chain *is* Marshall Is
93 F3 Rätan Sweden
123 C2 Ratanda S. Africa
74 B2 Ratangarh India
63 A2 Rat Buri Thai.
62 A1 Rathedaung Myanmar
101 F1 Rathenow Ger.
97 C1 Rathfriland U.K.
97 C1 Rathlin Island U.K.
Rathluirc Ireland see Charleville
100 C2 Ratingen Ger.
74 B2 Ratlam India
73 B3 Ratnagiri India
73 C4 Ratnapura Sri Lanka
90 A1 Ratne Ukr.
142 C1 Raton U.S.A.
96 D2 Rattray Head *hd* U.K.
93 G3 Rättvik Sweden
101 E1 Ratzeburg Ger.
78 B2 Raudhatain Kuwait
92 □B2 Raufarhöfn Iceland
54 C1 Raukumara Range *mts* N.Z.
93 H3 Rauma Fin.
88 C2 Rauna Latvia
61 C2 Raung, Gunung *vol.* Indon.
75 C2 Raurkela India
66 D2 Rausu Japan
90 B2 Rǎut *r.* Moldova
134 D1 Ravalli U.S.A.
81 C2 Ravānsar Iran
108 B2 Ravenna Italy
102 B2 Ravensburg Ger.
50 B3 Ravensthorpe Austr.
74 B1 Ravi *r.* Pak.
74 B1 Rawalpindi Pak.
103 D1 Rawicz Pol.
50 B3 Rawlinna Austr.
136 B2 Rawlins U.S.A.
153 B4 Rawson Arg.
75 C3 Rayagada India
69 E1 Raychikhinsk Russia
78 B3 Raydah Yemen
87 E3 Rayevskiy Russia
134 B1 Raymond U.S.A.
53 D2 Raymond Terrace Austr.
143 D3 Raymondville U.S.A.
129 D2 Raymore Can.
145 C2 Rayón Mex.
63 B2 Rayong Thai.
78 A2 Rayyis Saudi Arabia
104 B2 Raz, Pointe du *pt* France
81 C2 Razāzah, Buḩayrat ar *l.* Iraq
110 C2 Razgrad Bulg.
110 C2 Razim, Lacul *lag.* Romania

111 B2 Razlog Bulg.
104 B2 Ré, Île de *i.* France
99 C4 Reading U.K.
138 C2 Reading OH U.S.A.
139 D2 Reading PA U.S.A.
123 C2 Reagile S. Africa
115 E2 Rebiana Sand Sea *des.* Libya
66 D1 Rebun-tō *i.* Japan
50 B2 Recherche, Archipelago of the *is* Austr.
89 D3 Rechytsa Belarus
151 F3 Recife Brazil
123 C3 Recife, Cape S. Africa
100 C2 Recklinghausen Ger.
152 C2 Reconquista Arg.
137 D1 Red *r.* Can./U.S.A.
140 B2 Red *r.* U.S.A.
100 B3 Redange Lux.
141 C1 Red Bank U.S.A.
Red Basin China see Sichuan Pendi
131 E2 Red Bay Can.
135 B2 Red Bluff U.S.A.
98 C1 Redcar U.K.
129 C2 Redcliff Can.
121 B2 Redcliff Zimbabwe
52 B2 Red Cliffs Austr.
128 C2 Red Deer Can.
126 E3 Red Deer *r.* Can.
129 D2 Red Deer Lake Can.
135 B2 Redding U.S.A.
99 C3 Redditch U.K.
137 D2 Redfield U.S.A.
137 D3 Red Hills U.S.A.
130 A2 Red Lake Can.
130 A2 Red Lake *l.* Can.
137 E1 Red Lakes U.S.A.
134 E1 Red Lodge U.S.A.
134 B2 Redmond U.S.A.
137 D2 Red Oak U.S.A.
104 B2 Redon France
106 C1 Redondela Spain
106 B2 Redondo Port.
78 A2 Red Sea Africa/Asia
128 B1 Redstone *r.* Can.
100 B1 Reduzum Neth.
128 C2 Redwater Can.
137 E2 Red Wing U.S.A.
137 D2 Redwood Falls U.S.A.
97 C2 Ree, Lough *l.* Ireland
134 B2 Reedsport U.S.A.
54 B2 Reefton N.Z.
143 D3 Refugio U.S.A.
102 C2 Regen Ger.
155 E1 Regência Brazil
102 C2 Regensburg Ger.
114 C2 Reggane Alg.
109 C3 Reggio di Calabria Italy
108 B2 Reggio nell'Emilia Italy
110 B1 Reghin Romania
129 D2 Regina Can.
154 C2 Registro Brazil
122 A1 Rehoboth Namibia
80 B2 Rehovot Israel
101 F2 Reichenbach im Vogtland Ger.
141 E1 Reidsville U.S.A.
99 C4 Reigate U.K.
105 C2 Reims France
153 A5 Reina Adelaida, Archipiélago de la *is* Chile
101 E1 Reinbek Ger.
129 D2 Reindeer *r.* Can.
129 E2 Reindeer Island Can.
129 D2 Reindeer Lake Can.
92 F2 Reine Norway
106 C1 Reinosa Spain
100 C1 Reinsfeld Ger.
123 C2 Reitz S. Africa
122 B2 Reivilo S. Africa
129 D1 Reliance Can.
107 D2 Relizane Alg.
151 E3 Remanso Brazil
52 A2 Remarkable, Mount *h.* Austr.
79 C2 Remeshk Iran
105 D2 Remiremont France
100 C2 Remscheid Ger.
102 B1 Rendsburg Ger.
139 D1 Renfrew Can.
60 B2 Rengat Indon.
90 B2 Reni Ukr.
52 B2 Renmark Austr.
48 H5 Rennell *i.* Solomon Is
101 D2 Rennerod Ger.
104 B2 Rennes France
129 D1 Rennie Lake Can.
108 B2 Reno *r.* Italy
135 C3 Reno U.S.A.
70 A3 Renshou China
138 B2 Rensselaer U.S.A.
75 C2 Renukut India
54 B2 Renwick N.Z.
61 D2 Reo Indon.
136 D3 Republican *r.* U.S.A.
127 G2 Repulse Bay Can.
150 B2 Requena Peru
107 C2 Requena Spain
60 B2 Resag, Gunung *mt.* Indon.
111 B2 Resen Macedonia
154 B2 Reserva Brazil
91 C2 Reshetylivka Ukr.
152 C2 Resistencia Arg.
110 B1 Reşiţa Romania
126 F2 Resolute Can.
127 H2 Resolution Island Can.

S

Sal'sk

87 D4 Sal'sk Russia
122 B3 Salt *watercourse* S. Africa
107 D1 Salt Spain
142 A2 Salt *r.* U.S.A.
152 B2 Salta Arg.
99 A4 Saltash U.K.
96 B3 Saltcoats U.K.
145 B2 Saltillo Mex.
134 D2 Salt Lake City U.S.A.
154 C2 Salto Brazil
152 C3 Salto Uru.
155 E1 Salto da Divisa Brazil
154 B2 Salto del Guairá Para.
135 C4 Salton Sea *salt l.* U.S.A.
154 B3 Salto Osório, Represa *resr* Brazil
154 B3 Salto Santiago, Represa de *resr* Brazil
141 D2 Saluda U.S.A.
76 B3 Sālūk, Kūh-e *mt.* Iran
108 A2 Saluzzo Italy
151 F4 Salvador Brazil
79 C2 Salwah Saudi Arabia
62 A2 Salween *r.* China/Myanmar
62 A2 Salween *r.* China/Myanmar
81 C2 Salyan Azer.
138 C3 Salyersville U.S.A.
122 A1 Salzbrunn Namibia
102 C2 Salzburg Austria
101 E1 Salzgitter Ger.
101 D2 Salzkotten Ger.
101 E1 Salzwedel Ger.
144 B1 Samalayuca Mex.
66 D2 Samani Japan
64 B2 Samar *i.* Phil.
87 E3 Samara Russia
Samarahan Malaysia *see* Sri Aman
59 E3 Samarai P.N.G.
61 C2 Samarinda Indon.
77 C3 Samarqand Uzbek.
81 C2 Sāmarrā' Iraq
81 C1 Şamaxı Azer.
119 C3 Samba Dem. Rep. Congo
61 C1 Sambaliung *mts* Indon.
75 C2 Sambalpur India
60 C2 Sambar, Tanjung *pt* Indon.
60 B1 Sambas Indon.
121 □E2 Sambava Madag.
74 B2 Sambhar India
90 A2 Sambir Ukr.
61 C2 Samboja Indon.
153 C3 Samborombón, Bahía *b.* Arg.
65 B2 Samcheok S. Korea
Samch'ŏnp'o S. Korea *see* Sacheon
81 C2 Samdi Dag *mt.* Turkey
119 D3 Same Tanz.
121 B2 Samfya Zambia
78 B2 Samīrah Saudi Arabia
65 B1 Samjiyŏn N. Korea
Sam Neua Laos *see* Xam Nua
48 J5 Samoa *country* S. Pacific Ocean
156 E6 Samoa Basin S. Pacific Ocean
Samoa i Sisifo *country* S. Pacific Ocean *see* Samoa
109 C1 Samobor Croatia
110 B2 Samokov Bulg.
111 C3 Samos *i.* Greece
Samothrace *i.* Greece *see* Samothraki
111 C2 Samothraki Greece
111 C2 Samothraki *i.* Greece
61 C2 Sampit Indon.
119 C3 Sampwe Dem. Rep. Congo
63 B2 Sâmraông Cambodia
143 E2 Sam Rayburn Reservoir U.S.A.
62 B2 Sâm Sơn Vietnam
80 B1 Samsun Turkey
81 C1 Samt'redia Georgia
63 B3 Samui, Ko *i.* Thai.
63 B2 Samut Songkhram Thai.
114 B3 San Mali
78 B3 Şan'ā' Yemen
118 A2 Sanaga *r.* Cameroon
81 C2 Sanandaj Iran
146 B3 San Andrés, Isla de *i.* Caribbean Sea
106 B1 San Andrés del Rabanedo Spain
142 B2 San Andres Mountains U.S.A.
145 C3 San Andrés Tuxtla Mex.
143 C2 San Angelo U.S.A.
143 D3 San Antonio U.S.A.
135 C4 San Antonio, Mount U.S.A.
152 B2 San Antonio de los Cobres Arg.
153 B4 San Antonio Oeste Arg.
108 B2 San Benedetto del Tronto Italy
144 A3 San Benedicto, Isla *i.* Mex.
135 C4 San Bernardino U.S.A.
135 C4 San Bernardino Mountains U.S.A.
142 B3 San Blas Mex.
141 C3 San Blas, Cape U.S.A.
146 C4 San Blas, Punta *pt* Panama
152 B1 San Borja Bol.
144 B2 San Buenaventura Mex.
64 B2 San Carlos Phil.
147 D4 San Carlos Venez.
153 A4 San Carlos de Bariloche Arg.
147 C4 San Carlos del Zulia Venez.
104 C2 Sancerrois, Collines du *hills* France
135 C4 San Clemente U.S.A.
135 C4 San Clemente Island U.S.A.
105 C2 Sancoins France
48 H5 San Cristobal *i.* Solomon Is
150 B2 San Cristóbal Venez.
145 C3 San Cristóbal de las Casas Mex.
146 C2 Sancti Spíritus Cuba
61 C1 Sandakan Malaysia
93 E3 Sandane Norway
111 B2 Sandanski Bulg.
114 A3 Sandaré Mali
96 C1 Sanday *i.* U.K.
143 C2 Sanderson U.S.A.
150 C4 Sandia Peru
135 C4 San Diego U.S.A.
111 D3 Sandıklı Turkey
93 E4 Sandnes Norway
92 F2 Sandnessjøen Norway
118 C3 Sandoa Dem. Rep. Congo
103 E1 Sandomierz Pol.
89 E2 Sandovo Russia
94 B1 Sandoy *i.* Faroe Is
134 C1 Sandpoint U.S.A.
71 B3 Sandu China
94 B1 Sandur Faroe Is
138 C2 Sandusky U.S.A.
122 A3 Sandveld *mts* S. Africa
122 A2 Sandverhaar Namibia
93 F4 Sandvika Norway
93 G3 Sandviken Sweden
131 E2 Sandwich Bay Can.
135 D2 Sandy U.S.A.
129 D2 Sandy Bay Can.
51 E2 Sandy Cape Austr.
130 A2 Sandy Lake Can.
130 A2 Sandy Lake *l.* Can.
141 D2 Sandy Springs U.S.A.
144 A1 San Felipe *Baja California* Mex.
145 B2 San Felipe *Guanajuato* Mex.
150 C1 San Felipe Venez.
144 A2 San Fernando *Baja California* Mex.
145 C2 San Fernando *Tamaulipas* Mex.
64 B2 San Fernando *La Union* Phil.
64 B2 San Fernando *Pampanga* Phil.
106 B2 San Fernando Spain
147 D3 San Fernando Trin. and Tob.
150 C2 San Fernando de Apure Venez.
141 D3 Sanford *FL* U.S.A.
139 E2 Sanford *ME* U.S.A.
141 E1 Sanford *NC* U.S.A.
152 B3 San Francisco Arg.
135 B3 San Francisco U.S.A.
74 B3 Sangamner India
83 J2 Sangar Russia
108 A3 San Gavino Monreale Italy
101 E2 Sangerhausen Ger.
61 C1 Sanggau Indon.
118 B3 Sangha *r.* Congo
109 C3 San Giovanni in Fiore Italy
59 C2 Sangir *i.* Indon.
59 C2 Sangir, Kepulauan *is* Indon.
65 B2 Sangju S. Korea
63 A2 Sangkhla Buri Thai.
61 C1 Sangkulirang Indon.
73 B3 Sangli India
118 B2 Sangmélima Cameroon
121 C3 Sango Zimbabwe
San Gottardo, Passo del *pass* Switz. *see* St Gotthard Pass
136 B3 Sangre de Cristo Range *mts* U.S.A.
75 C2 Sangsang China
144 A2 San Hipólito, Punta *pt* Mex.
145 D3 San Ignacio Belize
152 B1 San Ignacio Bol.
144 A2 San Ignacio Mex.
130 C2 Sanikiluaq Can.
71 A3 Sanjiang *Guangxi* China
Sanjiang *Guizhou* China *see* Jinping
67 C3 Sanjō Japan
135 B3 San Joaquin *r.* U.S.A.
153 B4 San Jorge, Golfo de *g.* Arg.
146 B4 San José Costa Rica
64 B2 San Jose *Nueva Ecija* Phil.
64 B2 San Jose *Occidental Mindoro* Phil.
135 B3 San Jose U.S.A.
144 A2 San José, Isla *i.* Mex.
144 B2 San José de Bavicora Mex.
64 B2 San Jose de Buenavista Phil.
144 A2 San José de Comondú Mex.
144 B2 San José del Cabo Mex.
150 B2 San José del Guaviare Col.
152 B3 San Juan Arg.
146 B3 San Juan *r.* Costa Rica/Nic.
147 C3 San Juan Dom. Rep.
147 D3 San Juan Puerto Rico
135 D3 San Juan *r.* U.S.A.
152 C2 San Juan Bautista Para.
145 C3 San Juan Bautista Tuxtepec Mex.
147 D4 San Juan de los Morros Venez.
145 C2 San Juan del Río Mex.
134 B1 San Juan Islands U.S.A.
144 B2 San Juanito Mex.
136 B3 San Juan Mountains U.S.A.
153 B4 San Julián Arg.
75 C2 Sankh *r.* India
63 B2 San Khao Phang Hoei *mts* Thai.
100 C2 Sankt Augustin Ger.
105 D2 Sankt Gallen Switz.
105 D2 Sankt Moritz Switz.
Sankt-Peterburg Russia *see* St Petersburg
102 C2 Sankt Veit an der Glan Austria
100 C3 Sankt Wendel Ger.
80 B2 Şanlıurfa Turkey
142 B3 San Lorenzo Mex.
106 B2 Sanlúcar de Barrameda Spain
153 B3 San Luis Arg.
145 B2 San Luis de la Paz Mex.
142 A2 San Luisito Mex.
135 B3 San Luis Obispo U.S.A.
135 B3 San Luis Obispo Bay U.S.A.
145 B2 San Luis Potosí Mex.
144 A1 San Luis Río Colorado Mex.
143 D3 San Marcos U.S.A.
108 B2 San Marino *country* Europe
108 B2 San Marino San Marino
144 B2 San Martín de Bolaños Mex.
153 A4 San Martín de los Andes Arg.
135 B3 San Mateo U.S.A.
153 B4 San Matías, Golfo *g.* Arg.
70 B2 Sanmenxia China
146 B3 San Miguel El Salvador
152 B2 San Miguel de Tucumán Arg.
135 B4 San Miguel Island U.S.A.
145 C3 San Miguel Sola de Vega Mex.
71 B3 Sanming China
153 B3 San Nicolás de los Arroyos Arg.
135 C4 San Nicolas Island U.S.A.
110 B1 Sânnicolau Mare Romania
123 C2 Sannieshof S. Africa
114 B4 Sanniquellie Liberia
103 E2 Sanok Pol.
64 B2 San Pablo Phil.
144 B2 San Pablo Balleza Mex.
152 B2 San Pedro Arg.
152 B1 San Pedro Bol.
114 B4 San-Pédro Côte d'Ivoire
144 A2 San Pedro Mex.
142 A2 San Pedro *watercourse* U.S.A.
106 B2 San Pedro, Sierra de *mts* Spain
144 B2 San Pedro de las Colonias Mex.
152 C2 San Pedro de Ycuamandyyú Para.
142 A3 San Pedro el Saucito Mex.
146 B3 San Pedro Sula Hond.
108 A3 San Pietro, Isola di *i.* Italy
96 C3 Sanquhar U.K.
144 A1 San Quintín, Cabo *c.* Mex.
153 B3 San Rafael Arg.
108 A2 Sanremo Italy
143 D2 San Saba U.S.A.
147 C2 San Salvador *i.* Bahamas
146 B3 San Salvador El Salvador
152 B2 San Salvador de Jujuy Arg.
107 C1 San Sebastián Spain
108 B2 Sansepolcro Italy
109 C2 San Severo Italy
109 C2 Sanski Most Bos. & Herz.
152 B1 Santa Ana Bol.
146 B3 Santa Ana El Salvador
144 A1 Santa Ana Mex.
135 C4 Santa Ana U.S.A.
152 B1 Santa Ana de Yacuma Bol.
144 B2 Santa Bárbara Mex.
135 C4 Santa Barbara U.S.A.
154 B2 Santa Bárbara, Serra de *hills* Brazil
152 B2 Santa Catalina Chile
135 C4 Santa Catalina Island U.S.A.
154 B3 Santa Catarina *state* Brazil
150 C3 Santa Clara Col.
146 C2 Santa Clara Cuba
135 B3 Santa Clara U.S.A.
135 C4 Santa Clarita U.S.A.
107 D1 Santa Coloma de Gramenet Spain
Santa Comba Angola *see* Waku-Kungo
109 C3 Santa Croce, Capo *c.* Italy
153 B5 Santa Cruz *r.* Arg.
152 B1 Santa Cruz Bol.
64 B2 Santa Cruz Phil.
135 B3 Santa Cruz U.S.A.
145 C3 Santa Cruz Barillas Guat.
155 E1 Santa Cruz Cabrália Brazil
107 C2 Santa Cruz de Moya Spain
114 A2 Santa Cruz de Tenerife Canary Is
152 C2 Santa Cruz do Sul Brazil
135 C4 Santa Cruz Island U.S.A.
48 H5 Santa Cruz Islands Solomon Is
107 D2 Santa Eulària des Riu Spain
152 B3 Santa Fe Arg.
142 B1 Santa Fe U.S.A.
154 B2 Santa Fé do Sul Brazil
154 B1 Santa Helena de Goiás Brazil
153 B3 Santa Isabel Arg.
Santa Isabel Equat. Guinea *see* Malabo
48 G4 Santa Isabel *i.* Solomon Is
154 B1 Santa Luisa, Serra de *hills* Brazil
151 E3 Santa Luzia Brazil
144 A2 Santa Margarita, Isla *i.* Mex.
152 C2 Santa Maria Brazil
144 B1 Santa María *r.* Mex.
135 B4 Santa Maria U.S.A.
123 D2 Santa Maria, Cabo de *c.* Moz.
106 B2 Santa Maria, Cabo de *c.* Port.
155 C1 Santa Maria, Chapadão de *hills* Brazil
151 E3 Santa Maria das Barreiras Brazil
109 C3 Santa Maria di Leuca, Capo *c.* Italy
155 D1 Santa Maria do Suaçuí Brazil
150 B1 Santa Marta Col.
135 C4 Santa Monica U.S.A.
151 D3 Santana *Amapá* Brazil
151 E4 Santana *Bahia* Brazil
110 B1 Sântana Romania
106 C1 Santander Spain
108 A3 Sant'Antioco Italy
108 A3 Sant'Antioco, Isola di *i.* Italy
107 D2 Sant Antoni de Portmany Spain
151 D3 Santarém Brazil
106 B2 Santarém Port.
154 B1 Santa Rita do Araguaia Brazil
153 B3 Santa Rosa Arg.
152 C2 Santa Rosa Brazil
135 B3 Santa Rosa *CA* U.S.A.
142 C2 Santa Rosa *NM* U.S.A.
146 B3 Santa Rosa de Copán Hond.
135 B4 Santa Rosa Island *CA* U.S.A.
140 C2 Santa Rosa Island *FL* U.S.A.
144 A2 Santa Rosalía Mex.
134 C2 Santa Rosa Range *mts* U.S.A.
106 B1 Santa Uxía de Ribeira Spain
154 B1 Santa Vitória Brazil
107 D1 Sant Carles de la Ràpita Spain
135 C4 Santee U.S.A.
107 D2 Sant Francesc de Formentera Spain
152 C2 Santiago Brazil
153 A3 Santiago Chile
147 C3 Santiago Dom. Rep.
144 B2 Santiago Mex.
146 B4 Santiago Panama
64 B2 Santiago Phil.
106 B1 Santiago de Compostela Spain
146 C2 Santiago de Cuba Cuba
144 B2 Santiago Ixcuintla Mex.
144 B2 Santiago Papasquiaro Mex.
106 C1 Santillana Spain
107 D2 Sant Joan de Labritja Spain
107 D1 Sant Jordi, Golf de *g.* Spain
155 D2 Santo Amaro de Campos Brazil
154 B2 Santo Anastácio Brazil
155 C2 Santo André Brazil
152 C2 Santo Ângelo Brazil
154 B2 Santo Antônio da Platina Brazil
151 F4 Santo Antônio de Jesus Brazil
150 C3 Santo Antônio do Içá Brazil
155 C2 Santo Antônio do Monte Brazil
147 D3 Santo Domingo Dom. Rep.
144 A2 Santo Domingo Mex.
142 B1 Santo Domingo Pueblo U.S.A.
111 C3 Santorini *i.* Greece
155 C2 Santos Brazil
155 D2 Santos Dumont Brazil
157 I7 Santos Plateau S. Atlantic Ocean
152 C2 Santo Tomé Arg.
153 A4 San Valentín, Cerro *mt.* Chile
146 B3 San Vicente El Salvador
144 A1 San Vicente Mex.
64 B2 San Vicente Phil.
150 B4 San Vicente de Cañete Peru
108 B2 San Vincenzo Italy
108 B3 San Vito, Capo *c.* Italy
71 A4 Sanya China
155 C2 São Bernardo do Campo Brazil
152 C2 São Borja Brazil
154 C2 São Carlos Brazil
155 D1 São Felipe, Serra de *hills* Brazil
151 D4 São Félix do Araguaia Brazil
151 D3 São Félix do Xingu Brazil
155 D2 São Fidélis Brazil
155 D1 São Francisco Brazil
151 F4 São Francisco *r.* Brazil
154 C3 São Francisco, Ilha de *i.* Brazil
154 C3 São Francisco do Sul Brazil
152 C3 São Gabriel Brazil
150 C3 São Gabriel da Cachoeira Brazil
155 D2 São Gonçalo Brazil
155 C1 São Gonçalo do Abaeté Brazil
155 C1 São Gotardo Brazil
154 B1 São Jerônimo, Serra de *hills* Brazil
155 D2 São João da Barra Brazil
155 C2 São João da Boa Vista Brazil
106 B1 São João da Madeira Port.
155 D1 São João da Ponte Brazil
155 D2 São João del Rei Brazil
155 D1 São João do Paraíso Brazil
155 D1 São João Evangelista Brazil
155 D2 São João Nepomuceno Brazil
154 C2 São Joaquim da Barra Brazil
152 D2 São José Brazil
154 C2 São José do Rio Preto Brazil
155 C2 São José dos Campos Brazil
154 C3 São José dos Pinhais Brazil
154 A1 São Lourenço *Mato Grosso* Brazil
155 C2 São Lourenço *Minas Gerais* Brazil
151 E3 São Luís Brazil
154 C2 São Manuel Brazil
154 C1 São Marcos *r.* Brazil
151 E3 São Marcos, Baía de *b.* Brazil
155 E1 São Mateus Brazil
154 B3 São Mateus do Sul Brazil
105 C2 Saône *r.* France
155 C2 São Paulo Brazil
154 C1 São Paulo *state* Brazil
155 D2 São Pedro da Aldeia Brazil
151 E3 São Raimundo Nonato Brazil
155 C1 São Romão Brazil
São Salvador Angola *see* M'banza Congo
São Salvador do Congo Angola *see* M'banza Congo
155 C2 São Sebastião, Ilha do *i.* Brazil
154 C2 São Sebastião do Paraíso Brazil
154 B1 São Simão Brazil
154 B1 São Simão, Barragem de *resr* Braz
59 C2 Sao-Siu Indon.
113 D5 São Tomé São Tomé and Príncipe
113 D5 São Tomé *i.* São Tomé and Princi
155 D2 São Tomé, Cabo de *c.* Brazil
113 D5 São Tomé and Príncipe *country* Africa
155 C2 São Vicente Brazil
106 B2 São Vicente, Cabo de *c.* Port.
59 C3 Saparua Indon.
107 D2 Sa Pobla Spain
89 F3 Sapozhok Russia
66 D2 Sapporo Japan
109 C2 Sapri Italy
143 D1 Sapulpa U.S.A.
81 C2 Saqqez Iran
81 C2 Sarāb Iran
63 B2 Sara Buri Thai.
Saragossa Spain *see* Zaragoza

Severo-Kuril'sk

89 F3 Sarai Russia
109 C2 Sarajevo Bos. & Herz.
87 E3 Saraktash Russia
62 A1 Saramati mt. India/Myanmar
139 E2 Saranac Lake U.S.A.
109 D3 Sarandë Albania
64 B3 Sarangani Islands Phil.
87 D3 Saransk Russia
87 E3 Sarapul Russia
141 D3 Sarasota U.S.A.
90 B2 Sarata Ukr.
136 B2 Saratoga U.S.A.
139 E2 Saratoga Springs U.S.A.
61 C1 Saratok Malaysia
87 D3 Saratov Russia
79 D2 Sarāvān Iran
61 C1 Sarawak state Malaysia
111 C2 Saray Turkey
111 C3 Sarayköy Turkey
79 D2 Sarbāz Iran
76 B3 Sar Bīsheh Iran
74 B2 Sardarshahr India
Sardegna i. Italy see Sardinia
108 A2 Sardinia i. Italy
92 G2 Sarektjåkkå mt. Sweden
77 C3 Sar-e Pul Afgh.
158 C3 Sargasso Sea sea N. Atlantic Ocean
74 B1 Sargodha Pak.
115 D4 Sarh Chad
79 D2 Sarhad reg. Iran
81 D2 Sārī Iran
111 C3 Sarıgöl Turkey
81 C1 Sarıkamış Turkey
61 C1 Sarikei Malaysia
51 D2 Sarina Austr.
65 B2 Sariwŏn N. Korea
111 C2 Sarıyer Turkey
78 B2 Sark, Safrā' as esc. Saudi Arabia
77 D2 Sarkand Kazakh.
111 C2 Şarköy Turkey
104 C3 Sarlat-la-Canéda France
59 D3 Sarmi Indon.
153 B4 Sarmiento Arg.
138 C2 Sarnia Can.
90 B1 Sarny Ukr.
60 B2 Sarolangun Indon.
111 B3 Saronikos Kolpos g. Greece
111 C2 Saros Körfezi b. Turkey
103 E2 Sárospatak Hungary
87 D3 Sarov Russia
Sarpan i. N. Mariana Is see Rota
105 D2 Sarrebourg France
106 B1 Sarria Spain
107 C1 Sarrión Spain
105 D3 Sartène France
Sartu China see Daqing
111 C3 Saruhanlı Turkey
103 D2 Sárvár Hungary
81 D3 Sarvestān Iran
77 D1 Saryarka plain Kazakh.
76 B2 Sarykamyshskoye Ozero salt l. Turkm./Uzbek.
77 D2 Saryozek Kazakh.
77 D2 Saryshagan Kazakh.
77 C2 Sarysu watercourse Kazakh.
77 D3 Sary-Tash Kyrg.
75 C2 Sasaram India
67 A4 Sasebo Japan
129 D2 Saskatchewan prov. Can.
129 D2 Saskatchewan r. Can.
129 D2 Saskatoon Can.
83 I2 Saskylakh Russia
123 C2 Sasolburg S. Africa
87 D3 Sasovo Russia
114 B4 Sassandra Côte d'Ivoire
108 A2 Sassari Italy
102 C1 Sassnitz Ger.
114 A3 Satadougou Mali
136 C3 Satanta U.S.A.
73 B3 Satara India
123 D1 Satara S. Africa
87 E3 Satka Russia
74 B2 Satluj r. India/Pak.
75 C2 Satna India
77 C2 Satpayev Kazakh.
74 B2 Satpura Range mts India
67 B4 Satsuma-Sendai Japan
63 B2 Sattahip Thai.
110 B1 Satu Mare Romania
63 B3 Satun Thai.
144 B2 Saucillo Mex.
93 E4 Sauda Norway
92 □B2 Sauðárkrókur Iceland
78 B2 Saudi Arabia country Asia
105 C3 Saugues France
137 E1 Sauk Center U.S.A.
105 C2 Saulieu France
88 B2 Saulkrasti Latvia
130 B3 Sault Sainte Marie Can.
138 C1 Sault Sainte Marie U.S.A.
77 C1 Saumalkol' Kazakh.
59 C3 Saumlakki Indon.
104 B2 Saumur France
120 B1 Saurimo Angola
109 D2 Sava r. Europe
49 J5 Savai'i i. Samoa
91 E1 Savala r. Russia
114 C4 Savalou Benin
141 D2 Savannah GA U.S.A.
140 C1 Savannah TN U.S.A.
141 D2 Savannah r. U.S.A.
63 B2 Savannakhét Laos
130 A2 Savant Lake Can.
111 C3 Savaştepe Turkey
114 C4 Savè Benin
105 D2 Saverne France
89 F2 Savino Russia
86 D2 Savinskiy Russia
Savoie reg. France see Savoy
108 A2 Savona Italy
93 I3 Savonlinna Fin.
105 D2 Savoy reg. France
93 F4 Sävsjö Sweden
59 C3 Savu i. Indon.
92 I2 Savukoski Fin.
Savu Sea Indon. see Laut Sawu
74 B2 Sawai Madhopur India
62 A1 Sawan Myanmar
62 A2 Sawankhalok Thai.
136 B3 Sawatch Range mts U.S.A.
Sawhāj Egypt see Sūhāj
121 B2 Sawmills Zimbabwe
79 C3 Şawqirah, Dawḩat b. Oman
Şawqirah Bay Oman see Şawqirah, Dawḩat
53 D2 Sawtell Austr.
134 C2 Sawtooth Range mts U.S.A.
68 C1 Sayano-Shushenskoye Vodokhranilishche resr Russia
76 C3 Saýat Turkm.
79 C3 Sayḩūt Yemen
93 I3 Säynätsalo Fin.
69 D2 Saynshand Mongolia
139 D2 Sayre U.S.A.
144 B3 Sayula Jalisco Mex.
145 C3 Sayula Veracruz Mex.
128 B2 Sayward Can.
Sayyod Turkm. see Saýat
89 E2 Sazonovo Russia
114 B2 Sbaa Alg.
98 B2 Scafell Pike h. U.K.
109 C3 Scalea Italy
96 □ Scalloway U.K.
108 B2 Scandicci Italy
96 C1 Scapa Flow inlet U.K.
96 B2 Scarba i. U.K.
130 C3 Scarborough Can.
147 D3 Scarborough Trin. and Tob.
98 C2 Scarborough U.K.
64 A2 Scarborough Reef sea feature S. China Sea
96 A2 Scarinish U.K.
Scarpanto i. Greece see Karpathos
100 B2 Schaerbeek Belgium
105 D2 Schaffhausen Switz.
100 B1 Schagen Neth.
102 C2 Schärding Austria
100 A2 Scharendijke Neth.
101 D1 Scharhörn i. Ger.
101 D1 Scheeßel Ger.
131 D2 Schefferville Can.
135 D3 Schell Creek Range mts U.S.A.
139 E2 Schenectady U.S.A.
143 D3 Schertz U.S.A.
101 E3 Scheßlitz Ger.
100 C1 Schiermonnikoog i. Neth.
100 B2 Schilde Belgium
108 B1 Schio Italy
101 F2 Schkeuditz Ger.
101 E1 Schladen Ger.
102 C2 Schladming Austria
101 E2 Schleiz Ger.
102 B1 Schleswig Ger.
101 D2 Schloß Holte-Stukenbrock Ger.
101 D2 Schlüchtern Ger.
101 E3 Schlüsselfeld Ger.
101 E2 Schmalkalden Ger.
101 D2 Schmallenberg Ger.
Schmidt Island Russia see Shmidta, Ostrov
101 F2 Schmölln Ger.
101 D1 Schneverdingen Ger.
101 E1 Schönebeck (Elbe) Ger.
101 E1 Schöningen Ger.
100 B2 Schoonhoven Neth.
59 D3 Schouten Islands P.N.G.
97 B3 Schull Ireland
101 E3 Schwabach Ger.
102 B2 Schwäbische Alb mts Ger.
101 F3 Schwandorf Ger.
61 C2 Schwaner, Pegunungan mts Indon.
101 E1 Schwarzenbek Ger.
101 F2 Schwarzenberg/Erzgebirge Ger.
122 A2 Schwarzrand mts Namibia
Schwarzwald mts Ger. see Black Forest
102 C2 Schwaz Austria
102 C1 Schwedt/Oder Ger.
101 E2 Schweinfurt Ger.
101 E1 Schwerin Ger.
101 E1 Schweriner See l. Ger.
105 D2 Schwyz Switz.
108 B3 Sciacca Italy
95 B4 Scilly, Isles of U.K.
138 C3 Scioto r. U.S.A.
136 B1 Scobey U.S.A.
53 D2 Scone Austr.
110 B2 Scorniceşti Romania
55 C3 Scotia Ridge S. Atlantic Ocean
149 F8 Scotia Sea S. Atlantic Ocean
96 C2 Scotland admin. div. U.K.
128 B2 Scott, Cape Can.
123 D3 Scottburgh S. Africa
136 C3 Scott City U.S.A.
136 C2 Scottsbluff U.S.A.
140 C2 Scottsboro U.S.A.
96 B1 Scourie U.K.
139 D2 Scranton U.S.A.
98 C3 Scunthorpe U.K.
105 E2 Scuol Switz.
Scutari Albania see Shkodër
99 D4 Seaford U.K.
98 C2 Seaham U.K.
129 E2 Seal r. Can.
122 B3 Seal, Cape S. Africa
52 B3 Sea Lake Austr.
143 D3 Sealy U.S.A.
140 B1 Searcy U.S.A.
98 B2 Seascale U.K.
134 B1 Seattle U.S.A.
139 E2 Sebago Lake U.S.A.
144 A2 Sebastián Vizcaíno, Bahía b. Mex.
Sebastopol Ukr. see Sevastopol'
Sebenico Croatia see Šibenik
110 B1 Sebeş Romania
60 B2 Sebesi i. Indon.
88 C2 Sebezh Russia
80 B1 Şebinkarahisar Turkey
141 D3 Sebring U.S.A.
61 C2 Sebuku i. Indon.
128 B3 Sechelt Can.
150 A3 Sechura Peru
73 B3 Secunderabad India
137 E3 Sedalia U.S.A.
105 C2 Sedan France
54 B2 Seddon N.Z.
114 A3 Sédhiou Senegal
142 A2 Sedona U.S.A.
101 E2 Seeburg Ger.
101 E1 Seehausen (Altmark) Ger.
122 A2 Seeheim Namibia
104 C2 Sées France
101 E2 Seesen Ger.
101 E1 Seevetal Ger.
123 C1 Sefare Botswana
93 F3 Segalstad Norway
60 B1 Segamat Malaysia
86 C2 Segezha Russia
114 B3 Ségou Mali
106 C1 Segovia Spain
86 C2 Segozerskoye Vodokhranilishche resr Russia
115 D2 Séguédine Niger
114 B4 Séguéla Côte d'Ivoire
143 D3 Seguin U.S.A.
107 C2 Segura r. Spain
106 C2 Segura, Sierra de mts Spain
120 B3 Sehithwa Botswana
93 H3 Seinäjoki Fin.
104 C2 Seine r. France
104 B2 Seine, Baie de b. France
105 C2 Seine, Val de val. France
103 E1 Sejny Pol.
60 B2 Sekayu Indon.
114 B4 Sekondi Ghana
134 B1 Selah U.S.A.
59 C3 Selaru i. Indon.
61 C2 Selatan, Tanjung pt Indon.
126 B2 Selawik U.S.A.
61 D2 Selayar, Pulau i. Indon.
98 C3 Selby U.K.
136 C1 Selby U.S.A.
111 C3 Selçuk Turkey
120 B3 Selebi-Phikwe Botswana
Selebi-Pikwe Botswana see Selebi-Phikwe
105 D2 Sélestat France
Seletyteniz, Oz. salt l. Kazakh. see Siletyteniz, Ozero
92 □A3 Selfoss Iceland
114 A3 Sélibabi Maur.
142 A1 Seligman U.S.A.
116 A2 Selîma Oasis Sudan
111 C3 Selimiye Turkey
114 B3 Sélingué, Lac de l. Mali
89 D2 Selizharovo Russia
93 E4 Seljord Norway
129 E2 Selkirk Can.
96 C3 Selkirk U.K.
128 C2 Selkirk Mountains Can.
142 A2 Sells U.S.A.
140 C2 Selma AL U.S.A.
135 C3 Selma CA U.S.A.
105 D2 Selongey France
99 C4 Selsey Bill hd U.K.
89 D3 Sel'tso Russia
Selukwe Zimbabwe see Shurugwi
150 B3 Selvas reg. Brazil
134 C1 Selway r. U.S.A.
129 D1 Selwyn Lake Can.
128 A1 Selwyn Mountains Can.
51 C2 Selwyn Range hills Austr.
60 B2 Semangka, Teluk b. Indon.
61 C2 Semarang Indon.
60 B1 Sematan Malaysia
118 B2 Sembé Congo
81 C2 Şemdinli Turkey
91 C1 Semenivka Ukr.
87 D3 Semenov Russia
61 C2 Semeru, Gunung vol. Indon.
77 E1 Semey Kazakh.
91 E2 Semikarakorsk Russia
89 E3 Semiluki Russia
136 B2 Seminoe Reservoir U.S.A.
143 C2 Seminole U.S.A.
141 D2 Seminole, Lake U.S.A.
61 C1 Semitau Indon.
Sem Kolodezey Ukr. see Lenine
81 D2 Semnān Iran
61 C1 Semporna Malaysia
105 C2 Semur-en-Auxois France
Semyonovskoye Arkhangel'skaya Oblast' Russia see Bereznik
Semyonovskoye Kostromskaya Oblast' Russia see Ostrovskoye
150 C3 Sena Madureira Brazil
120 B2 Senanga Zambia
67 D3 Sendai Japan
141 D2 Seneca U.S.A.
114 A3 Senegal country Africa
114 A3 Sénégal r. Maur./Senegal
102 C1 Senftenberg Ger.
119 D3 Sengerema Tanz.
61 D2 Sengkang Indon.
151 E4 Senhor do Bonfim Brazil
103 D2 Senica Slovakia
108 B2 Senigallia Italy
109 B2 Senj Croatia
92 G2 Senja i. Norway
122 B2 Senlac S. Africa
105 C2 Senlis France
63 B2 Sênmônoŭrôm Cambodia
116 B3 Sennar Sudan
130 C3 Senneterre Can.
123 C3 Senqu r. Lesotho
105 C2 Sens France
109 D1 Senta Serbia
128 B2 Sentinel Peak Can.
123 C1 Senwabarwana S. Africa
65 B2 Seocheon S. Korea
65 B2 Seongnam S. Korea
75 B2 Seoni India
65 B2 Seosan S. Korea
65 B2 Seoul S. Korea
155 D2 Sepetiba, Baía de b. Brazil
59 D3 Sepik r. P.N.G.
61 C1 Sepinang Indon.
131 D2 Sept-Îles Can.
87 D4 Serafimovich Russia
100 B2 Seraing Belgium
59 C3 Seram i. Indon.
60 B2 Serang Indon.
60 B1 Serasan, Selat sea chan. Indon.
109 D2 Serbia country Europe
76 B3 Serdar Turkm.
117 C3 Serdo Eth.
89 E3 Serebryanyye Prudy Russia
60 B1 Seremban Malaysia
119 D3 Serengeti Plain Tanz.
121 C2 Serenje Zambia
90 B2 Seret r. Ukr.
87 D3 Sergach Russia
86 F2 Sergino Russia
89 E2 Sergiyev Posad Russia
Sergo Ukr. see Stakhanov
74 A1 Serhetabat Turkm.
61 C1 Seria Brunei
61 C1 Serian Malaysia
111 B3 Serifos i. Greece
80 B2 Serik Turkey
59 C3 Sermata, Kepulauan is Indon.
Sernyy Zavod Turkm. see Kükürtli
86 F3 Serov Russia
120 B3 Serowe Botswana
106 B2 Serpa Port.
Serpa Pinto Angola see Menongue
89 E3 Serpukhov Russia
155 D2 Serra Brazil
155 C1 Serra das Araras Brazil
108 A3 Serramanna Italy
154 B1 Serranópolis Brazil
100 A3 Serre r. France
111 B2 Serres Greece
151 F4 Serrinha Brazil
155 D1 Serro Brazil
154 C2 Sertãozinho Brazil
59 D3 Serui Indon.
120 B3 Serule Botswana
61 C2 Seruyan r. Indon.
68 C2 Sêrxü China
120 A2 Sesfontein Namibia
108 B2 Sessa Aurunca Italy
108 A2 Sestri Levante Italy
105 C3 Sète France
155 D1 Sete Lagoas Brazil
92 G2 Setermoen Norway
93 E4 Setesdal val. Norway
115 C1 Sétif Alg.
67 B4 Seto-naikai sea Japan
114 B1 Settat Morocco
98 B2 Settle U.K.
106 B2 Setúbal Port.
106 B2 Setúbal, Baía de b. Port.
130 A2 Seul, Lac l. Can.
81 C1 Sevan Armenia
76 A2 Sevan, Lake Armenia
Sevana Lich l. Armenia see Sevan, Lake
91 C3 Sevastopol' Ukr.
Seven Islands Can. see Sept-Îles
131 D2 Seven Islands Bay Can.
99 D4 Sevenoaks U.K.
105 C3 Sévérac-le-Château France
130 B2 Severn r. Can.
122 B2 Severn S. Africa
99 D4 Severn r. U.K.
86 D2 Severnaya Dvina r. Russia
83 H1 Severnaya Zemlya is Russia
86 D2 Severnyy Nenetskiy Avtonomnyy Okrug Russia
86 F2 Severnyy Respublika Komi Russia
82 E1 Severnyy, Ostrov i. Russia
83 I3 Severobaykal'sk Russia
86 C2 Severodvinsk Russia
83 L3 Severo-Kuril'sk Russia

South Horr

106 B2 Sines Port.
106 B2 Sines, Cabo de c. Port.
116 B3 Singa Sudan
75 C2 Singahi India
60 B1 Singapore country Asia
61 C2 Singaraja Indon.
63 B2 Sing Buri Thai.
119 D3 Singida Tanz.
62 A1 Singkaling Hkamti Myanmar
60 B1 Singkawang Indon.
60 B2 Singkep i. Indon.
60 A1 Singkil Indon.
53 D2 Singleton Austr.
Sin'gosan N. Korea see Kosan
62 A1 Singu Myanmar
Sining China see Xining
108 A2 Siniscola Italy
109 C2 Sinj Croatia
61 D2 Sinjai Indon.
116 B3 Sinkat Sudan
151 D2 Sinnamary Fr. Guiana
Sînnicolau Mare Romania see Sânnicolau Mare
Sinoia Zimbabwe see Chinhoyi
80 B1 Sinop Turkey
65 B1 Sinp'o N. Korea
61 C1 Sintang Indon.
100 B2 Sint Anthonis Neth.
100 A2 Sint-Laureins Belgium
147 D3 Sint Maarten terr. West Indies
100 B2 Sint-Niklaas Belgium
143 D3 Sinton U.S.A.
65 A1 Sinŭiju N. Korea
64 B3 Siocon Phil.
103 D2 Siófok Hungary
105 D2 Sion Switz.
137 D2 Sioux Center U.S.A.
137 D2 Sioux City U.S.A.
137 D2 Sioux Falls U.S.A.
130 A2 Sioux Lookout Can.
65 A1 Siping China
129 E2 Sipiwesk Lake Can.
55 P2 Siple, Mount Antarctica
55 P2 Siple Island Antarctica
Sipolilo Zimbabwe see Guruve
60 A2 Sipura i. Indon.
64 B3 Siquijor Phil.
93 E4 Sira r. Norway
Siracusa Italy see Syracuse
51 C1 Sir Edward Pellew Group is Austr.
110 C1 Siret Romania
110 C1 Siret r. Romania
78 A1 Sirhān, Wādī an watercourse Saudi Arabia
79 C2 Sīrīk Iran
61 C1 Sirik, Tanjung pt Malaysia
62 B2 Siri Kit, Khuan Thai.
128 B1 Sir James MacBrien, Mount Can.
79 C2 Sīrjān Iran
81 C2 Şırnak Turkey
74 B2 Sirohi India
60 A1 Sirombu Indon.
74 B2 Sirsa India
115 D1 Sirte Libya
115 D1 Sirte, Gulf of Libya
81 C2 Şirvan Azer.
88 B2 Širvintos Lith.
109 C1 Sisak Croatia
63 B2 Sisaket Thai.
145 C2 Sisal Mex.
122 B2 Sishen S. Africa
81 C2 Sisian Armenia
127 I2 Sisimiut Greenland
129 D2 Sisipuk Lake Can.
63 B2 Sisŏphŏn Cambodia
105 D3 Sisteron France
Sitang China see Sinan
75 C2 Sitapur India
111 C3 Siteia Greece
123 D2 Siteki Swaziland
128 A2 Sitka U.S.A.
100 B2 Sittard Neth.
62 A1 Sittaung Myanmar
62 A2 Sittaung r. Myanmar
62 A1 Sittwe Myanmar
61 C2 Situbondo Indon.
80 B2 Sivas Turkey
62 A1 Sivasagar India
111 C3 Sivaslı Turkey
80 B2 Siverek Turkey
88 D2 Siverskiy Russia
80 B2 Sivrihisar Turkey
116 A2 Sīwah Egypt
75 B1 Siwalik Range mts India/Nepal
Siwa Oasis Egypt see Wāḥāt Sīwah
105 D3 Six-Fours-les-Plages France
70 B2 Sixian China
123 C2 Siyabuswa S. Africa
Sjælland i. Denmark see Zealand
109 D2 Sjenica Serbia
92 G2 Sjøvegan Norway
91 C2 Skadovs'k Ukr.
93 F4 Skagen Denmark
93 E4 Skagerrak str. Denmark/Norway
134 B1 Skagit r. U.S.A.
128 A2 Skagway U.S.A.
92 G2 Skaland Norway
93 F4 Skara Sweden
74 B1 Skardu Pak.
103 E1 Skarżysko-Kamienna Pol.
103 D2 Skawina Pol.
114 A2 Skaymat Western Sahara
128 B2 Skeena r. Can.
128 B2 Skeena Mountains Can.
98 D3 Skegness U.K.
92 H3 Skellefteå Sweden
92 H3 Skellefteälven r. Sweden
97 C2 Skerries Ireland
93 F4 Ski Norway
111 B3 Skiathos i. Greece
97 B3 Skibbereen Ireland
92 □B2 Skíðadalsjökull glacier Iceland
98 B2 Skiddaw h. U.K.
93 E4 Skien Norway
103 E1 Skierniewice Pol.
115 C1 Skikda Alg.
52 B3 Skipton Austr.
98 B3 Skipton U.K.
93 E4 Skive Denmark
92 H1 Skjervøy Norway
Skobelev Uzbek. see Farg'ona
111 B3 Skopelos i. Greece
89 E3 Skopin Russia
111 B2 Skopje Macedonia
111 C3 Skoutaros Greece
93 F4 Skövde Sweden
83 J3 Skovorodino Russia
139 F2 Skowhegan U.S.A.
92 H2 Skröven Sweden
88 B2 Skrunda Latvia
128 A1 Skukum, Mount Can.
123 D1 Skukuza S. Africa
88 C3 Skuodas Lith.
90 B2 Skvyra Ukr.
96 A2 Skye i. U.K.
111 B3 Skyros Greece
111 B3 Skyros i. Greece
93 F4 Slagelse Denmark
60 B2 Slamet, Gunung vol. Indon.
97 C2 Slaney r. Ireland
88 C2 Slantsy Russia
109 C1 Slatina Croatia
110 B2 Slatina Romania
143 C2 Slaton U.S.A.
129 C1 Slave r. Can.
114 C4 Slave Coast Africa
128 C2 Slave Lake Can.
77 D1 Slavgorod Russia
88 C2 Slavkovichi Russia
Slavonska Požega Croatia see Požega
109 C1 Slavonski Brod Croatia
90 B1 Slavuta Ukr.
90 C1 Slavutych Ukr.
66 B2 Slavyanka Russia
Slavyanskaya Russia see Slavyansk-na-Kubani
91 D2 Slavyansk-na-Kubani Russia
89 D3 Slawharad Belarus
103 D1 Sławno Pol.
99 C3 Sleaford U.K.
97 A2 Slea Head hd Ireland
130 C2 Sleeper Islands Can.
97 D1 Slieve Donard h. U.K.
Slieve Gamph hills Ireland see Ox Mountains
96 A2 Sligachan U.K.
Sligeach Ireland see Sligo
97 B1 Sligo Ireland
97 B1 Sligo Bay Ireland
93 G4 Slite Sweden
110 C2 Sliven Bulg.
Sloboda Russia see Ezhva
110 C2 Slobozia Romania
128 C3 Slocan Can.
88 C3 Slonim Belarus
100 B1 Sloten Neth.
99 C4 Slough U.K.
103 D2 Slovakia country Europe
108 B1 Slovenia country Europe
91 D2 Slov"yans'k Ukr.
102 C1 Słubice Pol.
90 B1 Sluch r. Ukr.
100 A2 Sluis Neth.
103 D1 Słupsk Pol.
88 C3 Slutsk Belarus
97 A2 Slyne Head hd Ireland
68 C1 Slyudyanka Russia
131 D2 Smallwood Reservoir Can.
88 C3 Smalyavichy Belarus
88 C3 Smarhon' Belarus
129 D2 Smeaton Can.
109 D2 Smederevo Serbia
109 D2 Smederevska Palanka Serbia
91 C2 Smila Ukr.
88 C3 Smilavichy Belarus
88 C2 Smiltene Latvia
137 D3 Smith Center U.S.A.
128 B2 Smithers Can.
141 E1 Smithfield NC U.S.A.
134 D2 Smithfield UT U.S.A.
139 D3 Smith Mountain Lake U.S.A.
130 C3 Smiths Falls Can.
53 D3 Smithtown Austr.
53 D2 Smoky Cape Austr.
137 D3 Smoky Hills U.S.A.
92 E3 Smøla i. Norway
89 D3 Smolensk Russia
89 D3 Smolensko-Moskovskaya Vozvyshennost' hills Belarus/Russia
111 B2 Smolyan Bulg.
66 B2 Smolyaninovo Russia
130 B3 Smooth Rock Falls Can.
Smyrna Turkey see İzmir
91 D2 Smyrnove Ukr.
92 □B3 Snæfell mt. Iceland
98 A2 Snaefell h. Isle of Man
128 A1 Snag (abandoned) Can.
134 C1 Snake r. U.S.A.
134 D2 Snake River Plain U.S.A.
Snare Lakes Can. see Wekweètì
92 F3 Snåsvatn l. Norway
100 B1 Sneek Neth.
97 B3 Sneem Ireland
122 B3 Sneeuberge mts S. Africa
Snegurovka Ukr. see Tetiyiv
103 D1 Sněžka mt. Czech Rep.
108 B1 Snežnik mt. Slovenia
103 E1 Śniardwy, Jezioro l. Pol.
Sniečkus Lith. see Visaginas
91 C2 Snihurivka Ukr.
93 E3 Snøhetta mt. Norway
Snovsk Ukr. see Shchors
129 D1 Snowbird Lake Can.
99 A3 Snowdon mt. U.K.
Snowdrift Can. see Łutselk'e
129 C1 Snowdrift r. Can.
142 A2 Snowflake U.S.A.
129 D2 Snow Lake Can.
134 C1 Snowshoe Peak U.S.A.
52 A2 Snowtown Austr.
53 C3 Snowy r. Austr.
53 C3 Snowy Mountains Austr.
143 C2 Snyder U.S.A.
121 □D2 Soalala Madag.
121 □D2 Soanierana-Ivongo Madag.
90 B2 Sob r. Ukr.
65 B2 Sobaek-sanmaek mts S. Korea
117 B4 Sobat r. S. Sudan
89 F2 Sobinka Russia
151 E4 Sobradinho, Barragem de resr Brazil
151 E3 Sobral Brazil
91 D3 Sochi Russia
49 L5 Society Islands Fr. Polynesia
150 B2 Socorro Col.
142 B2 Socorro NM U.S.A.
142 B2 Socorro TX U.S.A.
144 A3 Socorro, Isla i. Mex.
56 B4 Socotra i. Yemen
63 B3 Soc Trăng Vietnam
106 C2 Socuéllamos Spain
92 I2 Sodankylä Fin.
134 D2 Soda Springs U.S.A.
93 G3 Söderhamn Sweden
93 G4 Södertälje Sweden
116 A3 Sodiri Sudan
117 B4 Sodo Eth.
93 G3 Södra Kvarken str. Fin./Sweden
Soerabaia Indon. see Surabaya
101 D2 Soest Ger.
53 C2 Sofala Austr.
110 B2 Sofia Bulg.
121 □D2 Sofia r. Madag.
Sofiya Bulg. see Sofia
Sofiyevka Ukr. see Vil'nyans'k
75 D1 Sog China
93 E3 Sognefjorden inlet Norway
111 D2 Söğüt Turkey
Sohâg Egypt see Sūhāj
Sohar Oman see Ṣuḥār
100 B2 Soignies Belgium
105 C2 Soissons France
90 A1 Sokal' Ukr.
65 B2 Sokcho S. Korea
111 C3 Söke Turkey
81 C1 Sokhumi Georgia
114 C4 Sokodé Togo
89 F2 Sokol Russia
101 F2 Sokolov Czech Rep.
115 C3 Sokoto Nigeria
115 C3 Sokoto r. Nigeria
90 B2 Sokyryany Ukr.
73 B3 Solapur India
135 B3 Soledad U.S.A.
89 F2 Soligalich Russia
99 C3 Solihull U.K.
86 E3 Solikamsk Russia
87 E3 Sol'-Iletsk Russia
100 C2 Solingen Ger.
122 A1 Solitaire Namibia
92 G3 Sollefteå Sweden
93 G4 Sollentuna Sweden
107 D2 Sóller Spain
101 D2 Solling hills Ger.
89 E2 Solnechnogorsk Russia
60 B2 Solok Indon.
48 H4 Solomon Islands country S. Pacific Ocean
48 G4 Solomon Sea S. Pacific Ocean
61 D2 Solor, Kepulauan is Indon.
105 D2 Solothurn Switz.
81 D2 Solṭānābād Iran
101 D1 Soltau Ger.
89 D2 Sol'tsy Russia
96 C3 Solway Firth est. U.K.
120 B2 Solwezi Zambia
111 C3 Soma Turkey
117 C4 Somalia country Africa
117 C4 Somaliland disp. terr. Somalia
61 C2 Somba Indon.
120 B1 Sombo Angola
109 C1 Sombor Serbia
144 B2 Sombrerete Mex.
138 C3 Somerset U.S.A.
123 C3 Somerset East S. Africa
126 F2 Somerset Island Can.
122 A3 Somerset West S. Africa
110 B1 Someş r. Romania
101 E2 Sömmerda Ger.
146 B3 Somoto Nic.
75 C2 Son r. India
65 C1 Sŏnbong N. Korea
93 E5 Sønderborg Denmark
101 E2 Sondershausen Ger.
Søndre Strømfjord inlet Greenland see Kangerlussuaq
108 A1 Sondrio Italy
63 B2 Sông Câu Vietnam
62 B1 Sông Đa, Hồ resr Vietnam
119 D4 Songea Tanz.
65 B1 Sŏnggan N. Korea
65 B1 Songhua Hu resr China
65 B1 Songjianghe China
Sŏngjin N. Korea see Kimch'aek
63 B3 Songkhla Thai.
65 B2 Songnim N. Korea
120 A1 Songo Angola
121 C2 Songo Moz.
Songololo Dem. Rep. Congo see Mbanza-Ngungu
69 E1 Songyuan China
Sonid Youqi China see Saihan Tal
74 B2 Sonipat India
89 E2 Sonkovo Russia
62 B1 Sơn La Vietnam
74 A2 Sonmiani Pak.
74 A2 Sonmiani Bay Pak.
101 E2 Sonneberg Ger.
142 A2 Sonoita Mex.
144 A2 Sonora r. Mex.
135 B3 Sonora CA U.S.A.
143 C2 Sonora TX U.S.A.
146 B3 Sonsonate El Salvador
Soochow China see Suzhou
117 A4 Sopo watercourse S. Sudan
103 D2 Sopron Hungary
74 B1 Sopur India
108 B2 Sora Italy
130 C3 Sorel Can.
51 D4 Sorell Austr.
106 C1 Soria Spain
90 B2 Soroca Moldova
154 C2 Sorocaba Brazil
87 E3 Sorochinsk Russia
Soroki Moldova see Soroca
59 D2 Sorol atoll Micronesia
59 C3 Sorong Indon.
119 D2 Soroti Uganda
92 H1 Sørøya i. Norway
108 B2 Sorrento Italy
92 G2 Sorsele Sweden
64 B2 Sorsogon Phil.
86 C2 Sortavala Russia
92 G2 Sortland Norway
123 C2 Soshanguve S. Africa
89 E3 Sosna r. Russia
153 B3 Sosneado mt. Arg.
86 E2 Sosnogorsk Russia
86 D2 Sosnovka Russia
88 C2 Sosnovyy Bor Russia
103 D1 Sosnowiec Pol.
91 C1 Sosnytsya Ukr.
86 F3 Sos'va Russia
91 D2 Sosyka r. Russia
145 C2 Soto la Marina Mex.
118 B2 Souanké Congo
111 B3 Souda Greece
104 C3 Souillac France
Sŏul S. Korea see Seoul
104 B2 Soulac-sur-Mer France
104 B3 Soulom France
Soûr Lebanon see Tyre
107 D2 Sour el Ghozlane Alg.
129 D3 Souris Man. Can.
131 D3 Souris P.E.I. Can.
129 E3 Souris r. Can.
151 F3 Sousa Brazil
115 D1 Sousse Tunisia
104 B3 Soustons France
122 B3 South Africa country Africa
99 C4 Southampton U.K.
129 F1 Southampton, Cape Can.
129 F1 Southampton Island Can.
73 D3 South Andaman i. India
52 A1 South Australia state Austr.
140 B2 Southaven U.S.A.
142 B2 South Baldy mt. U.S.A.
130 B3 South Baymouth Can.
138 B2 South Bend U.S.A.
141 D2 South Carolina state U.S.A.
58 B2 South China Sea N. Pacific Ocean
South Coast Town Austr. see Gold Coast
136 C2 South Dakota state U.S.A.
99 C4 South Downs hills U.K.
159 E6 Southeast Indian Ridge Indian Ocean
55 O2 Southeast Pacific Basin S. Pacific Ocean
129 D2 Southend Can.
99 D4 Southend-on-Sea U.K.
54 B2 Southern Alps mts N.Z.
50 A3 Southern Cross Austr.
129 E2 Southern Indian Lake Can.
159 D7 Southern Ocean
141 E1 Southern Pines U.S.A.
Southern Rhodesia country Africa see Zimbabwe
96 B3 Southern Uplands hills U.K.
55 J2 South Geomagnetic Pole Antarctica
149 G8 South Georgia i. S. Atlantic Ocean
149 G8 South Georgia and the South Sandwich Islands terr. S. Atlantic Ocean
138 B2 South Haven U.S.A.
129 E1 South Henik Lake Can.
119 D2 South Horr Kenya

61 C2 Sumbawa i. Indon.
61 C2 Sumbawabesar Indon.
119 D3 Sumbawanga Tanz.
120 A2 Sumbe Angola
96 □ Sumburgh U.K.
96 □ Sumburgh Head hd U.K.
119 C2 Sumeih Sudan
67 D4 Sumisu-jima i. Japan
131 D3 Summerside Can.
138 C3 Summersville U.S.A.
141 D2 Summerville U.S.A.
137 D1 Summit U.S.A.
128 B2 Summit Lake Can.
103 D2 Šumperk Czech Rep.
81 C1 Sumqayıt Azer.
141 D2 Sumter U.S.A.
91 C1 Sumy Ukr.
75 D2 Sunamganj Bangl.
65 B2 Sunan N. Korea
79 C2 Şunaynah Oman
52 B3 Sunbury Austr.
139 D2 Sunbury U.S.A.
65 B2 Suncheon S. Korea
65 B2 Sunch'ŏn N. Korea
123 C2 Sun City S. Africa
93 H3 Sund Fin.
60 B2 Sunda, Selat str. Indon.
136 C2 Sundance U.S.A.
75 C2 Sundarbans coastal area Bangl./India
74 B1 Sundarnagar India
Sunda Strait Indon. see Sunda, Selat
Sunda Trench Indian Ocean see Java Trench
98 C2 Sunderland U.K.
128 C2 Sundre Can.
93 G3 Sundsvall Sweden
123 D2 Sundumbili S. Africa
60 B2 Sungailiat Indon.
60 B2 Sungaipenuh Indon.
60 B1 Sungai Petani Malaysia
80 B1 Sungurlu Turkey
75 C2 Sun Kosi r. Nepal
93 E3 Sunndalsøra Norway
134 C1 Sunnyside U.S.A.
135 B3 Sunnyvale U.S.A.
83 I2 Suntar Russia
74 A2 Suntsar Pak.
114 B4 Sunyani Ghana
92 I3 Suomussalmi Fin.
67 B4 Suō-nada b. Japan
86 C2 Suoyarvi Russia
142 A2 Superior AZ U.S.A.
137 D2 Superior NE U.S.A.
138 A1 Superior WI U.S.A.
138 B1 Superior, Lake Can./U.S.A.
63 B2 Suphan Buri Thai.
81 C2 Süphan Dağı mt. Turkey
89 D2 Suponevo Russia
81 C2 Süq ash Shuyūkh Iraq
70 B2 Suqian China
78 A2 Süq Suwayq Saudi Arabia
Suquţrā i. Yemen see Socotra
79 C2 Şūr Oman
74 A2 Surab Pak.
61 C2 Surabaya Indon.
61 C2 Surakarta Indon.
74 B2 Surat India
74 B2 Suratgarh India
63 A3 Surat Thani Thai.
89 D3 Surazh Russia
109 D2 Surdulica Serbia
100 C3 Sûre r. Lux.
74 B2 Surendranagar India
82 F2 Surgut Russia
64 B3 Surigao Phil.
64 B3 Surin Thai.
151 D2 Suriname country S. America
75 C2 Surkhet Nepal
Surt Libya see Sirte
Surt, Khalīj g. Libya see Sirte, Gulf of
60 B2 Surulangun Indon.
81 C2 Süsangerd Iran
89 F2 Susanino Russia
135 B2 Susanville U.S.A.
80 B1 Suşehri Turkey
139 D3 Susquehanna r. U.S.A.
131 D3 Sussex Can.
101 D1 Süstedt Ger.
100 C1 Sustrum Ger.
83 K2 Susuman Russia
111 C3 Susurluk Turkey
74 B1 Sutak India
53 D2 Sutherland Austr.
122 B3 Sutherland S. Africa
136 C2 Sutherland U.S.A.
134 B2 Sutherlin U.S.A.
Sutlej r. India/Pak. see Satluj
138 C3 Sutton U.S.A.
99 C3 Sutton Coldfield U.K.
98 C3 Sutton in Ashfield U.K.
66 D2 Suttsu Japan
49 I5 Suva Fiji
Suvalki Pol. see Suwałki
89 E3 Suvorov Russia
90 B2 Suvorove Ukr.
103 E1 Suwałki Pol.
141 D3 Suwanee Sound b. U.S.A.
63 B2 Suwannaphum Thai.
141 D3 Suwannee r. U.S.A.
Suways, Qanāt as canal Egypt see Suez Canal

Suweis, Qanâ el canal Egypt see Suez Canal
65 B2 Suwon S. Korea
79 C2 Sūzā Iran
89 D3 Suzemka Russia
70 B2 Suzhou Anhui China
70 C2 Suzhou Jiangsu China
67 C3 Suzu Japan
67 C3 Suzu-misaki pt Japan
82 B1 Svalbard terr. Arctic Ocean
90 A2 Svalyava Ukr.
92 H2 Svappavaara Sweden
91 D2 Svatove Ukr.
63 B2 Svay Riĕng Cambodia
93 F3 Sveg Sweden
88 C2 Švenčionys Lith.
93 F4 Svendborg Denmark
Sverdlovsk Russia see Yekaterinburg
111 B2 Sveti Nikole Macedonia
66 C1 Svetlaya Russia
88 B3 Svetlogorsk Russia
87 D4 Svetlograd Russia
88 B3 Svetlyy Russia
93 I3 Svetogorsk Russia
103 D2 Svidník Slovakia
111 C2 Svilengrad Bulg.
110 B2 Svinecea Mare, Vârful mt. Romania
110 C2 Svishtov Bulg.
88 B3 Svislach Belarus
103 D2 Svitavy Czech Rep.
91 C2 Svitlovods'k Ukr.
69 E1 Svobodnyy Russia
110 B2 Svoge Bulg.
92 F2 Svolvær Norway
88 C3 Svyetlahorsk Belarus
141 D2 Swainsboro U.S.A.
120 A3 Swakopmund Namibia
52 B3 Swan Hill Austr.
128 C2 Swan Hills Can.
129 D2 Swan Lake Can.
97 C1 Swanlinbar Ireland
129 D2 Swan River Can.
53 D2 Swansea Austr.
99 B4 Swansea U.K.
122 B3 Swartkolkvloer salt pan S. Africa
123 C2 Swartruggens S. Africa
Swatow China see Shantou
123 D2 Swaziland country Africa
93 G3 Sweden country Europe
143 C2 Sweetwater U.S.A.
136 B2 Sweetwater r. U.S.A.
122 B3 Swellendam S. Africa
103 D1 Świdnica Pol.
103 D1 Świdwin Pol.
103 D1 Świebodzin Pol.
103 D1 Świecie Pol.
129 D2 Swift Current Can.
97 C1 Swilly, Lough inlet Ireland
99 C4 Swindon U.K.
102 C1 Świnoujście Pol.
105 D2 Switzerland country Europe
97 C2 Swords Ireland
88 C3 Syanno Belarus
89 D1 Syas'stroy Russia
89 D2 Sychevka Russia
53 D2 Sydney Austr.
131 D3 Sydney Can.
131 D3 Sydney Mines Can.
91 D2 Syeverodonets'k Ukr.
111 B3 Sykia Greece
86 C2 Syktyvkar Russia
140 C2 Sylacauga U.S.A.
75 D2 Sylhet Bangl.
102 B1 Sylt i. Ger.
138 C2 Sylvania U.S.A.
51 C1 Sylvester, Lake imp. l. Austr.
111 C3 Symi i. Greece
91 D2 Synel'nykove Ukr.
90 C2 Synyukha r. Ukr.
109 C3 Syracuse Italy
136 C3 Syracuse KS U.S.A.
139 D2 Syracuse NY U.S.A.
77 C2 Syrdar'ya r. Asia
80 B2 Syria country Asia
80 B2 Syrian Desert Asia
111 B3 Syros i. Greece
91 C1 Syvash, Zatoka lag. Ukr.
91 C2 Syvas'ke Ukr.
87 D3 Syzran' Russia
102 C1 Szczecin Pol.
103 D1 Szczecinek Pol.
103 E1 Szczytno Pol.
Szechwan prov. China see Sichuan
103 E2 Szeged Hungary
103 D2 Székesfehérvár Hungary
103 D2 Szekszárd Hungary
103 E2 Szentes Hungary
103 D2 Szentgotthárd Hungary
103 D2 Szerencs Hungary
103 D2 Szigetvár Hungary
103 E2 Szolnok Hungary
103 D2 Szombathely Hungary
Sztálinváros Hungary see Dunaújváros

T

64 B2 Tabaco Phil.
78 B2 Tābah Saudi Arabia
108 A3 Tabarka Tunisia
76 B3 Ţabas Iran

79 C1 Tabāsīn Iran
81 D3 Tābask, Kūh-e mt. Iran
150 C3 Tabatinga Amazonas Brazil
154 C2 Tabatinga São Paulo Brazil
114 B2 Tabelbala Alg.
128 C3 Taber Can.
64 C3 Tablas i. Phil.
102 C2 Tábor Czech Rep.
119 D3 Tabora Tanz.
114 B4 Tabou Côte d'Ivoire
81 C2 Tabrīz Iran
48 L3 Tabuaeran atoll Kiribati
78 A2 Tabūk Saudi Arabia
93 G4 Täby Sweden
77 C2 Tacheng China
102 C2 Tachov Czech Rep.
64 B2 Tacloban Phil.
150 B4 Tacna Peru
134 B1 Tacoma U.S.A.
152 C3 Tacuarembó Uru.
142 B3 Tacupeto Mex.
114 C2 Tademaït, Plateau du Alg.
Tadjikistan country Asia see Tajikistan
117 C3 Tadjourah Djibouti
80 B2 Tadmur Syria
129 E2 Tadoule Lake Can.
Tadzhikskaya S.S.R. country Asia see Tajikistan
65 B2 Taebaek S. Korea
65 B2 Taebaek-sanmaek mts N. Korea/S. Korea
Taech'ŏn S. Korea see Boryeong
Taegu S. Korea see Daegu
Taejŏn S. Korea see Daejeon
107 C1 Tafalla Spain
152 B3 Tafí Viejo Arg.
79 D2 Taftān, Kūh-e mt. Iran
91 D2 Taganrog Russia
91 D2 Taganrog, Gulf of Russia/Ukr.
62 A1 Tagaung Myanmar
64 B2 Tagaytay City Phil.
64 B3 Tagbilaran Phil.
64 B2 Tagudin Phil.
51 E1 Tagula Island P.N.G.
64 B3 Tagum Phil.
106 B2 Tagus r. Port./Spain
60 B1 Tahan, Gunung mt. Malaysia
115 C2 Tahat, Mont mt. Alg.
69 E1 Tahe China
49 M5 Tahiti i. Fr. Polynesia
143 E1 Tahlequah U.S.A.
135 B3 Tahoe, Lake U.S.A.
135 B3 Tahoe, Lake U.S.A.
126 E2 Tahoe Lake Can.
115 C2 Tahoua Niger
79 C2 Tahrūd Iran
128 B3 Tahsis Can.
116 B2 Ţahţā Egypt
64 B3 Tahuna Indon.
70 B2 Tai'an China
70 A2 Taibai Shan mt. China
71 C3 Taibei Taiwan
Taibus Qi China see Baochang
70 B2 Taihang Shan mts China
54 C1 Taihape N.Z.
71 B3 Taihe China
70 A2 Tai Hu l. China
52 A3 Tailem Bend Austr.
71 C3 Tainan Taiwan
111 B3 Tainaro, Akra c. Greece
155 D1 Taiobeiras Brazil
Taiping China see Chongzuo
60 B1 Taiping Malaysia
Tairbeart U.K. see Tarbert
71 B3 Taishan China
70 B2 Tai Shan hills China
119 D3 Taita Hills Kenya
153 A4 Taitao, Península de pen. Chile
71 C3 T'aitung Taiwan
92 I2 Taivalkoski Fin.
92 H2 Taivaskero h. Fin.
71 C3 Taiwan country Asia
Taiwan Shan mts Taiwan see Zhongyang Shanmo
71 B3 Taiwan Strait China/Taiwan
77 C1 Taiynsha Kazakh.
70 B2 Taiyuan China
71 C3 Taizhong Taiwan
70 B2 Taizhou Jiangsu China
71 C3 Taizhou Zhejiang China
78 B3 Ta'izz Yemen
77 D3 Tajikistan country Asia
74 B2 Taj Mahal tourist site India
Tajo r. Spain see Tagus
145 C3 Tajumulco, Volcán de vol. Guat.
63 A2 Tak Thai.
54 B2 Takaka N.Z.
115 C2 Takalous, Oued watercourse Alg.
67 B4 Takamatsu Japan
67 C3 Takaoka Japan
54 B1 Takapuna N.Z.
67 C3 Takasaki Japan
122 B1 Takatokwane Botswana
122 B1 Takatshwaane Botswana
67 C3 Takayama Japan
Takefu Japan see Echizen
60 A1 Takengon Indon.
63 B2 Ta Khmau Cambodia
74 B1 Takht-i-Sulaiman mt. Pak.
66 D2 Takikawa Japan
128 B2 Takla Lake Can.
128 B2 Takla Landing Can.

Takla Makan des. China see Taklimakan Desert
77 E3 Taklimakan Desert des. China
Taklimakan Shamo China see Taklimakan Desert
128 A2 Taku r. Can./U.S.A.
63 A3 Takua Pa Thai.
115 C4 Takum Nigeria
88 C3 Talachyn Belarus
74 B1 Talagang Pak.
146 B4 Talamanca, Cordillera de mts Costa Rica
150 A3 Talara Peru
59 C2 Talaud, Kepulauan is Indon.
106 C2 Talavera de la Reina Spain
153 A3 Talca Chile
153 A3 Talcahuano Chile
89 E2 Taldom Russia
77 D2 Taldykorgan Kazakh.
Taldy-Kurgan Kazakh. see Taldykorgan
59 C3 Taliabu i. Indon.
64 B2 Talisay Phil.
61 C2 Taliwang Indon.
81 C2 Tall 'Afar Iraq
141 D2 Tallahassee U.S.A.
53 C3 Tallangatta Austr.
88 B2 Tallinn Estonia
140 B2 Tallulah U.S.A.
104 B2 Talmont-St-Hilaire France
90 C2 Tal'ne Ukr.
117 B3 Talodi Sudan
91 E1 Talovaya Russia
126 F2 Taloyoak Can.
88 B2 Talsi Latvia
152 A2 Taltal Chile
129 C1 Taltson r. Can.
60 A1 Talu Indon.
74 A1 Tālūqān Afgh.
53 C2 Talwood Austr.
114 B4 Tamale Ghana
115 C2 Tamanrasset Alg.
99 A4 Tamar r. U.K.
Tamatave Madag. see Toamasina
144 B2 Tamazula Mex.
145 C2 Tamazunchale Mex.
114 A3 Tambacounda Senegal
60 B1 Tambelan, Kepulauan is Indon.
86 G1 Tambey Russia
61 C1 Tambisan Malaysia
61 C2 Tambora, Gunung vol. Indon.
91 E1 Tambov Russia
119 C2 Tambura S. Sudan
62 A1 Tamenglong India
145 C2 Tamiahua, Laguna de lag. Mex.
Tammerfors Fin. see Tampere
141 D3 Tampa U.S.A.
141 D3 Tampa Bay U.S.A.
93 H3 Tampere Fin.
145 C2 Tampico Mex.
69 D1 Tamsagbulag Mongolia
102 C2 Tamsweg Austria
53 D2 Tamworth Austr.
99 C2 Tamworth U.K.
119 E3 Tana r. Kenya
Tana, Lake Eth. see Lake Tana
67 C4 Tanabe Japan
92 I1 Tana Bru Norway
60 A2 Tanahbala i. Indon.
61 C2 Tanahgrogot Indon.
61 D2 Tanahjampea i. Indon.
60 A2 Tanahmasa i. Indon.
50 C1 Tanami Desert Austr.
63 B2 Tân An Vietnam
126 B2 Tanana U.S.A.
Tananarive Madag. see Antananarivo
108 A1 Tanaro r. Italy
65 B1 Tanch'ŏn N. Korea
64 B2 Tandag Phil.
110 C2 Ţăndărei Romania
153 C3 Tandil Arg.
74 A2 Tando Adam Pak.
74 A2 Tando Muhammad Khan Pak.
52 B2 Tandou Lake imp. l. Austr.
67 B4 Tanega-shima i. Japan
114 B4 Tanezrouft reg. Alg./Mali
119 D3 Tanga Tanz.
75 C2 Tangail Bangl.
Tanganyika country Africa see Tanzania
119 C3 Tanganyika, Lake Africa
55 F3 Tange Promontory hd Antarctica
114 B1 Tanger Morocco
101 E1 Tangermünde Ger.
75 D1 Tanggulashan China
75 C1 Tanggula Shan mts China
Tangier Morocco see Tanger
70 B2 Tangra Yumco salt l. China
70 B2 Tangshan China
75 C1 Taniantaweng Shan mts China
59 C3 Tanimbar, Kepulauan is Indon.
64 B3 Tanjay Phil.
61 C2 Tanjung Indon.
60 A1 Tanjungbalai Indon.
Tanjungkarang-Telukbetung Indon. see Bandar Lampung
60 B2 Tanjungpandan Indon.
60 B1 Tanjungpinang Indon.
61 C1 Tanjungredeb Indon.
61 C1 Tanjungselor Indon.
74 B1 Tank Pak.
48 H5 Tanna i. Vanuatu
115 C3 Tanout Niger
75 C2 Tansen Nepal

Torrance

62 A2 **Thayetmyo** Myanmar
62 A1 **Thazi** Myanmar
146 C2 **The Bahamas** *country* West Indies
98 B2 **The Cheviot** *h.* U.K.
134 B1 **The Dalles** U.S.A.
136 C2 **Thedford** U.S.A.
53 D2 **The Entrance** Austr.
99 C3 **The Fens** *reg.* U.K.
114 A3 **The Gambia** *country* Africa
52 B3 **The Grampians** *mts* Austr.
The Great Oasis Egypt *see* **Wāḩāt al Khārijah**
79 C2 **The Gulf** Asia
79 C2 **The Gulf** Asia
100 B1 **The Hague** Neth.
129 E1 **Thelon** *r.* Can.
101 E2 **Themar** Ger.
123 C2 **Thembalihle** S. Africa
96 A1 **The Minch** *sea chan.* U.K.
97 A1 **The Mullet** *b.* Ireland
99 C4 **The Needles** *stack* U.K.
150 C3 **Theodore Roosevelt** *r.* Brazil
129 D2 **The Pas** Can.
111 B2 **Thermaïkos Kolpos** *g.* Greece
136 B2 **Thermopolis** U.S.A.
53 C3 **The Rock** Austr.
The Skaw *spit* Denmark *see* **Grenen**
99 C4 **The Solent** *str.* U.K.
130 B3 **Thessalon** Can.
111 B2 **Thessaloniki** Greece
99 D3 **Thetford** U.K.
131 C3 **Thetford Mines** Can.
62 A1 **The Triangle** *mts* Myanmar
147 D3 **The Valley** Anguilla
131 D2 **Thévenet, Lac** *l.* Can.
99 D3 **The Wash** *b.* U.K.
99 D4 **The Weald** *reg.* U.K.
143 D2 **The Woodlands** U.S.A.
140 B3 **Thibodaux** U.S.A.
129 E2 **Thicket Portage** Can.
137 D1 **Thief River Falls** U.S.A.
Thiel Neth. *see* **Tiel**
105 C2 **Thiers** France
114 A3 **Thiès** Senegal
119 D3 **Thika** Kenya
73 B4 **Thiladhunmathi** Maldives
75 C2 **Thimphu** Bhutan
105 D2 **Thionville** France
Thira *i.* Greece *see* **Santorini**
98 C2 **Thirsk** U.K.
73 B4 **Thiruvananthapuram** India
93 E4 **Thisted** Denmark
92 □B3 **Thjórsá** *r.* Iceland
129 E1 **Thlewiaza** *r.* Can.
63 B3 **Thô Chu, Đao** *i.* Vietnam
62 A2 **Thoen** Thai.
123 D1 **Thohoyandou** S. Africa
101 E1 **Thomasburg** Ger.
97 C2 **Thomastown** Ireland
140 C2 **Thomasville** *AL* U.S.A.
141 D2 **Thomasville** *GA* U.S.A.
100 C2 **Thommen** Belgium
129 E2 **Thompson** Can.
128 E3 **Thompson** *r.* U.S.A.
134 C1 **Thompson Falls** U.S.A.
128 B2 **Thompson Sound** Can.
51 D2 **Thomson** *watercourse* Austr.
141 D2 **Thomson** U.S.A.
63 B2 **Thon Buri** Thai.
63 A2 **Thongwa** Myanmar
Thoothukudi India *see* **Tuticorin**
142 B1 **Thoreau** U.S.A.
96 C3 **Thornhill** U.K.
136 C3 **Thornton** U.S.A.
55 F2 **Thorshavnheiane** *reg.* Antarctica
123 C2 **Thota-ea-Moli** Lesotho
104 B2 **Thouars** France
62 A1 **Thoubal** India
134 D1 **Three Forks** U.S.A.
128 C2 **Three Hills** Can.
63 A2 **Three Pagodas Pass** Myanmar/Thai.
114 B4 **Three Points, Cape** Ghana
138 B2 **Three Rivers** *MI* U.S.A.
143 D3 **Three Rivers** *TX* U.S.A.
73 B3 **Thrissur** India
63 B2 **Thu Dâu Môt** Vietnam
100 B2 **Thuin** Belgium
127 H1 **Thule Air Base** Greenland
121 B3 **Thuli** Zimbabwe
102 B2 **Thun** Switz.
130 B3 **Thunder Bay** Can.
130 B3 **Thunder Bay** *b.* Can.
63 A3 **Thung Song** Thai.
101 E2 **Thüringer Becken** *reg.* Ger.
101 E2 **Thüringer Wald** *mts* Ger.
Thuringian Forest *mts* Ger. *see* **Thüringer Wald**
97 C2 **Thurles** Ireland
59 D3 **Thursday Island** Austr.
96 C1 **Thurso** U.K.
96 C1 **Thurso** *r.* U.K.
55 R2 **Thurston Island** Antarctica
101 D1 **Thüster Berg** *h.* Ger.
121 C2 **Thyolo** Malawi
Thysville Dem. Rep. Congo *see* **Mbanza-Ngungu**
79 C2 **Tīāb** Iran
151 E3 **Tianguá** Brazil
70 B2 **Tianjin** China
70 B2 **Tianjin** *mun.* China
71 A3 **Tianlin** China
70 B2 **Tianmen** China
69 E2 **Tianshan** China
70 A2 **Tianshui** China
70 A2 **Tianzhu** China
107 D2 **Tiaret** Alg.
114 B4 **Tiassalé** Côte d'Ivoire
154 B2 **Tibagi** Brazil
154 B2 **Tibagi** *r.* Brazil
118 B2 **Tibati** Cameroon
108 B2 **Tiber** *r.* Italy
Tiberias, Lake Israel *see* **Galilee, Sea of**
115 D2 **Tibesti** *mts* Chad
75 C1 **Tibet** *aut. reg.* China
68 B2 **Tibet, Plateau of** China
115 D2 **Tībistī, Sarīr** *des.* Libya
52 B1 **Tibooburra** Austr.
144 A2 **Tiburón, Isla** *i.* Mex.
114 B3 **Tîchît** Maur.
114 B3 **Tîchît, Ḍahr** *hills* Maur.
114 A2 **Tichla** Western Sahara
105 D2 **Ticino** *r.* Italy/Switz.
139 E2 **Ticonderoga** U.S.A.
145 D2 **Ticul** Mex.
114 A3 **Tidjikja** Maur.
100 B2 **Tiel** Neth.
70 C1 **Tieling** China
75 B1 **Tielongtan** China
100 A2 **Tielt** Belgium
100 B2 **Tienen** Belgium
68 B2 **Tien Shan** *mts* China/Kyrg.
Tientsin China *see* **Tianjin**
Tientsin *mun.* China *see* **Tianjin**
93 G3 **Tierp** Sweden
145 C3 **Tierra Blanca** Mex.
145 C3 **Tierra Colorada** Mex.
153 B5 **Tierra del Fuego, Isla Grande de** *i.* Arg./Chile
106 B1 **Tiétar, Valle del** *val.* Spain
154 C2 **Tietê** Brazil
154 B2 **Tietê** *r.* Brazil
138 C2 **Tiffin** U.S.A.
Tiflis Georgia *see* **Tbilisi**
141 D2 **Tifton** U.S.A.
Tighina Moldova *see* **Bender**
75 C2 **Tigiria** India
118 B2 **Tignère** Cameroon
131 D3 **Tignish** Can.
150 B3 **Tigre** *r.* Ecuador/Peru
120 A2 **Tigres (abandoned)** Angola
81 C2 **Tigris** *r.* Asia
114 A3 **Tiguent** Maur.
114 B2 **Tiguesmat** *hills* Maur.
115 D3 **Tigui** Chad
145 C2 **Tihuatlán** Mex.
144 A1 **Tijuana** Mex.
91 E2 **Tikhoretsk** Russia
89 D2 **Tikhvin** Russia
89 D2 **Tikhvinskaya Gryada** *ridge* Russia
157 F7 **Tiki Basin** S. Pacific Ocean
54 C1 **Tikokino** N.Z.
81 C2 **Tikrīt** Iraq
83 J2 **Tiksi** Russia
100 B2 **Tilburg** Neth.
152 B2 **Tilcara** Arg.
52 B1 **Tilcha (abandoned)** Austr.
114 C3 **Tîlemsi, Vallée du** *watercourse* Mali
Tilimsen Alg. *see* **Tlemcen**
114 C3 **Tillabéri** Niger
134 B1 **Tillamook** U.S.A.
63 A3 **Tillanchong Island** India
111 C3 **Tilos** *i.* Greece
52 B2 **Tilpa** Austr.
86 F2 **Til'tim** Russia
89 E3 **Tim** Russia
86 D2 **Timanskiy Kryazh** *ridge* Russia
54 B2 **Timaru** N.Z.
91 D2 **Timashevsk** Russia
Timashevskaya Russia *see* **Timashevsk**
114 B3 **Timbedgha** Maur.
50 C1 **Timber Creek** Austr.
114 B3 **Timbuktu** Mali
115 C3 **Timia** Niger
114 C2 **Timimoun** Alg.
111 B2 **Timiou Prodromou, Akrotirio** *pt* Greece
110 B1 **Timiș** *r.* Romania
110 B1 **Timișoara** Romania
130 B3 **Timmins** Can.
89 E2 **Timokhino** Russia
151 E3 **Timon** Brazil
59 C3 **Timor** *i.* East Timor/Indonesia
Timor-Leste *country* Asia *see* **East Timor**
59 C3 **Timor Sea** Austr./Indon.
Timor Timur *country* Asia *see* **East Timor**
93 G3 **Timrå** Sweden
78 B2 **Tīn, Jabal at** *mt.* Saudi Arabia
114 B2 **Tindouf** Alg.
53 D1 **Tingha** Austr.
75 C2 **Tingri** China
93 F4 **Tingsryd** Sweden
92 E3 **Tingvoll** Norway
Tingzhou China *see* **Changting**
59 D2 **Tinian** *i.* N. Mariana Is
Tinnelvelly India *see* **Tirunelveli**
152 B2 **Tinogasta** Arg.
111 C3 **Tinos** Greece
111 C3 **Tinos** *i.* Greece
100 A3 **Tinqueux** France
115 C2 **Tinrhert, Hamada de** Alg.
62 A1 **Tinsukia** India
52 B3 **Tintinara** Austr.
136 C1 **Tioga** U.S.A.
107 D2 **Tipasa** Alg.
97 B2 **Tipperary** Ireland
151 E3 **Tiracambu, Serra do** *hills* Brazil
109 C2 **Tirana** Albania
Tiranë Albania *see* **Tirana**
108 B1 **Tirano** Italy
52 A1 **Tirari Desert** Austr.
90 B2 **Tiraspol** Moldova
122 A2 **Tiraz Mountains** Namibia
111 C3 **Tire** Turkey
96 A2 **Tiree** *i.* U.K.
Tîrgovişte Romania *see* **Târgovişte**
Tîrgu Frumos Romania *see* **Târgu Frumos**
Tîrgu Jiu Romania *see* **Târgu Jiu**
Tîrgu Lăpuş Romania *see* **Târgu Lăpuş**
Tîrgu Mureş Romania *see* **Târgu Mureş**
Tîrgu Neamţ Romania *see* **Târgu Neamţ**
Tîrgu Ocna Romania *see* **Târgu Ocna**
74 B1 **Tirich Mir** *mt.* Pak.
Tîrnăveni Romania *see* **Târnăveni**
155 C1 **Tiros** Brazil
118 C2 **Tiroungoulou** C.A.R.
73 B3 **Tiruchchirappalli** India
73 B4 **Tirunelveli** India
73 B3 **Tirupati** India
73 B3 **Tiruppattur** India
73 B3 **Tiruppur** India
Tisa *r.* Hungary *see* **Tisza**
109 D1 **Tisa** *r.* Serbia
129 D2 **Tisdale** Can.
107 D2 **Tissemsilt** Alg.
75 C2 **Tista** *r.* India
103 E2 **Tisza** *r.* Hungary
55 L1 **Titan Dome** Antarctica
Titicaca, Lago Bol./Peru *see* **Titicaca, Lake**
152 B1 **Titicaca, Lake** *l.* Bol./Peru
75 C2 **Titlagarh** India
Titograd Montenegro *see* **Podgorica**
Titova Mitrovica Kosovo *see* **Mitrovicë**
Titovo Užice Serbia *see* **Užice**
Titovo Velenje Slovenia *see* **Velenje**
Titov Veles Macedonia *see* **Veles**
Titov Vrbas Serbia *see* **Vrbas**
110 C2 **Titu** Romania
141 D3 **Titusville** U.S.A.
99 B4 **Tiverton** U.K.
108 B2 **Tivoli** Italy
79 C2 **Ţīwī** Oman
63 A2 **Ti-ywa** Myanmar
145 D2 **Tizimín** Mex.
107 D2 **Tizi Ouzou** Alg.
114 B2 **Tiznit** Morocco
92 G2 **Tjaktjajaure** *l.* Sweden
Tjirebon Indon. *see* **Cirebon**
145 C3 **Tlacotalpán** Mex.
144 B2 **Tlahualilo** Mex.
145 C3 **Tlalnepantla** Mex.
145 C3 **Tlapa** Mex.
145 C3 **Tlaxcala** Mex.
145 C3 **Tlaxiaco** Mex.
114 B1 **Tlemcen** Alg.
123 C1 **Tlokweng** Botswana
128 B2 **Toad River** Can.
121 □D2 **Toamasina** Madag.
60 A1 **Toba, Danau** *l.* Indon.
Toba, Lake Indon. *see* **Toba, Danau**
74 A1 **Toba and Kakar Ranges** *mts* Pak.
147 D3 **Tobago** *i.* Trin. and Tob.
59 C2 **Tobelo** Indon.
130 B3 **Tobermory** Can.
96 A2 **Tobermory** U.K.
129 D2 **Tobin Lake** Can.
60 B2 **Toboali** Indon.
86 F3 **Tobol'sk** Russia
Tobruk Libya *see* **Tubruq**
76 C1 **Tobyl** *r.* Kazakh./Russia
151 E3 **Tocantinópolis** Brazil
151 E3 **Tocantins** *r.* Brazil
141 D2 **Toccoa** U.S.A.
108 A1 **Toce** *r.* Italy
152 A2 **Tocopilla** Chile
53 C3 **Tocumwal** Austr.
59 C3 **Todeli** Indon.
108 B2 **Todi** Italy
144 A2 **Todos Santos** Mex.
128 B3 **Tofino** Can.
96 □ **Toft** U.K.
61 D2 **Togian** *i.* Indon.
61 D2 **Togian, Kepulauan** *is* Indon.
Togliatti Russia *see* **Tol'yatti**
114 C4 **Togo** *country* Africa
74 B2 **Tohana** India
141 D3 **Tohopekaliga, Lake** U.S.A.
126 C2 **Tok** U.S.A.
116 B3 **Tokar** Sudan
69 E3 **Tokara-rettō** *is* Japan
91 E1 **Tokarevka** Russia
80 B1 **Tokat** Turkey
49 J4 **Tokelau** *terr.* S. Pacific Ocean
91 D2 **Tokmak** Ukr.
77 D2 **Tokmok** Kyrg.
54 C1 **Tokoroa** N.Z.
68 B2 **Toksun** China
67 B4 **Tokushima** Japan
67 C3 **Tōkyō** Japan
121 □D3 **Tôlañaro** Madag.
Tolbukhin Bulg. *see* **Dobrich**
154 B2 **Toledo** Brazil
106 C2 **Toledo** Spain
138 C2 **Toledo** U.S.A.
106 C2 **Toledo, Montes de** *mts* Spain
140 B2 **Toledo Bend Reservoir** U.S.A.
121 □D3 **Toliara** Madag.
Toling China *see* **Zanda**
61 D1 **Tolitoli** Indon.
108 B1 **Tolmezzo** Italy
108 B1 **Tolmin** Slovenia
103 D2 **Tolna** Hungary
104 B3 **Tolosa** Spain
145 C3 **Toluca** Mex.
87 D3 **Tol'yatti** Russia
138 A2 **Tomah** U.S.A.
138 B1 **Tomahawk** U.S.A.
66 D2 **Tomakomai** Japan
61 D2 **Tomali** Indon.
61 C1 **Tomani** Malaysia
106 B2 **Tomar** Port.
103 E1 **Tomaszów Lubelski** Pol.
103 E1 **Tomaszów Mazowiecki** Pol.
144 B3 **Tomatlán** Mex.
154 C2 **Tomazina** Brazil
140 C2 **Tombigbee** *r.* U.S.A.
120 A1 **Tomboco** Angola
155 D2 **Tombos** Brazil
Tombouctou Mali *see* **Timbuktu**
142 A2 **Tombstone** U.S.A.
120 A2 **Tombua** Angola
123 C1 **Tom Burke** S. Africa
106 C2 **Tomelloso** Spain
53 C2 **Tomingley** Austr.
61 D2 **Tomini, Teluk** *g.* Indon.
109 C2 **Tomislavgrad** Bos. & Herz.
103 D2 **Tompa** Hungary
50 A2 **Tom Price** Austr.
82 G3 **Tomsk** Russia
93 F4 **Tomtabacken** *h.* Sweden
83 K2 **Tomtor** Russia
145 C3 **Tonalá** Mex.
150 C3 **Tonantins** Brazil
99 D4 **Tonbridge** U.K.
59 C2 **Tondano** Indon.
81 D2 **Tonekābon** Iran
49 J5 **Tonga** *country* S. Pacific Ocean
49 J6 **Tongatapu Group** *is* Tonga
71 B3 **Tongcheng** China
70 A2 **Tongchuan** China
71 A3 **Tongdao** China
100 B2 **Tongeren** Belgium
71 A3 **Tonghai** China
65 B1 **Tonghua** China
65 B2 **Tongjosŏn-man** *b.* N. Korea
62 B1 **Tongking, Gulf of** China/Vietnam
69 E2 **Tongliao** China
70 B2 **Tongling** China
52 B2 **Tongo** Austr.
Tongquan China *see* **Malong**
71 A3 **Tongren** China
Tongshan China *see* **Xuzhou**
Tongshi China *see* **Wuzhishan**
Tongtian He *r.* China *see* **Yangtze**
96 B1 **Tongue** U.K.
Tongxian China *see* **Tongzhou**
65 B3 **Tongyeong** S. Korea
69 E2 **Tongyu** China
65 A1 **Tongyuanpu** China
70 B2 **Tongzhou** China
71 A3 **Tongzi** China
119 C2 **Tonj** S. Sudan
74 B2 **Tonk** India
Tônlé Sab *l.* Cambodia *see* **Tonle Sap**
63 B2 **Tonle Sap** *l.* Cambodia
135 C3 **Tonopah** U.S.A.
93 F4 **Tønsberg** Norway
135 D2 **Tooele** U.S.A.
52 B3 **Tooleybuc** Austr.
53 D1 **Toowoomba** Austr.
137 D3 **Topeka** U.S.A.
144 B2 **Topia** Mex.
144 B2 **Topolobampo** Mex.
86 C2 **Topozero, Ozero** *l.* Russia
134 B1 **Toppenish** U.S.A.
111 C3 **Torbalı** Turkey
76 B3 **Torbat-e Ḩeydarīyeh** Iran
76 C3 **Torbat-e Jām** Iran
131 E3 **Torbay** Can.
106 C1 **Tordesillas** Spain
107 C1 **Tordesilos** Spain
92 H2 **Töre** Sweden
107 D1 **Torelló** Spain
100 B1 **Torenberg** *h.* Neth.
Toretam Kazakh. *see* **Baykonyr**
101 F2 **Torgau** Ger.
76 C2 **Torgay** Kazakh.
100 A2 **Torhout** Belgium
Torino Italy *see* **Turin**
67 D4 **Tori-shima** *i.* Japan
117 B4 **Torit** S. Sudan
154 B1 **Torixoréu** Brazil
89 D2 **Torkovichi** Russia
106 B1 **Tormes** *r.* Spain
92 H2 **Torneälven** *r.* Sweden
131 D2 **Torngat Mountains** Can.
92 H2 **Tornio** Fin.
106 B1 **Toro** Spain
130 C3 **Toronto** Can.
89 D2 **Toropets** Russia
119 D2 **Tororo** Uganda
Toros Dağları Turkey *see* **Taurus Mountains**
52 B3 **Torquay** Austr.
99 B4 **Torquay** U.K.
135 C4 **Torrance** U.S.A.

U

120 B2 Uamanda Angola
150 C3 Uarini Brazil
155 D2 Ubá Brazil
155 D1 Ubaí Brazil
151 F4 Ubaitaba Brazil
118 B3 Ubangi r. C.A.R./Dem. Rep. Congo
Ubangi-Shari country Africa see Central African Republic
67 B4 Ube Japan
106 C2 Úbeda Spain
154 C1 Uberaba Brazil
154 C1 Uberlândia Brazil
106 B1 Ubiña, Peña mt. Spain
63 B2 Ubombo S. Africa
119 C3 Ubundu Dem. Rep. Congo
150 B2 Ucayali r. Peru
100 B2 Uccle Belgium
74 B2 Uch Pak.
66 D2 Uchiura-wan b. Japan
76 C2 Uchquduq Uzbek.
83 J3 Uchur r. Russia
99 D4 Uckfield U.K.
128 B3 Ucluelet Can.
83 I2 Udachnyy Russia
74 B2 Udaipur India
91 C1 Uday r. Ukr.
93 F4 Uddevalla Sweden
92 G2 Uddjaure l. Sweden
100 B2 Uden Neth.
74 B1 Udhampur India
108 B1 Udine Italy
89 E2 Udomlya Russia
62 B2 Udon Thani Thai.
73 B3 Udupi India
83 K3 Udyl', Ozero l. Russia
67 C3 Ueda Japan
61 D2 Uekuli Indon.
118 C2 Uele r. Dem. Rep. Congo
83 N2 Uelen Russia
101 E1 Uelzen Ger.
119 C2 Uere r. Dem. Rep. Congo
87 E3 Ufa Russia
119 D3 Ugalla r. Tanz.
119 D2 Uganda country Africa
69 F1 Uglegorsk Russia
89 E2 Uglich Russia
89 D2 Uglovka Russia
89 D2 Uglovoye Russia
89 D3 Ugra Russia
103 D2 Uherské Hradiště Czech Rep.
Uibhist a' Deas i. U.K. see South Uist
Uibhist a' Tuath i. U.K. see North Uist
101 E2 Uichteritz Ger.
96 A2 Uig U.K.
120 A1 Uíge Angola
65 B2 Uijeongbu S. Korea
65 A1 Úiju N. Korea
135 D2 Uinta Mountains U.S.A.
65 B2 Uiseong S. Korea
120 A3 Uis Mine Namibia
123 C3 Uitenhage S. Africa
100 C1 Uithuizen Neth.
131 D2 Uivak, Cape Can.
Ujiyamada Japan see Ise
74 B2 Ujjain India
Ujung Pandang Indon. see Makassar
89 F3 Ukholovo Russia
Ukhta Respublika Kareliya Russia see Kalevala
86 E2 Ukhta Respublika Komi Russia
135 B3 Ukiah U.S.A.
127 I2 Ukkusissat Greenland
88 B2 Ukmergė Lith.
90 C2 Ukraine country Europe
Ukrainskaya S.S.R. country Europe see Ukraine
Ulaanbaatar Mongolia see Ulan Bator
68 C1 Ulaangom Mongolia
59 E3 Ulamona P.N.G.
70 A2 Ulan China
69 D1 Ulan Bator Mongolia
Ulanhad China see Chifeng
69 E1 Ulanhot China
87 D4 Ulan-Khol Russia
70 B1 Ulan Qab China
69 D1 Ulan-Ude Russia
75 C2 Ulan Ul Hu l. China
Uleåborg Fin. see Oulu
88 C2 Ülenurme Estonia
73 B3 Ulhasnagar India
69 D1 Uliastai China
68 C1 Uliastay Mongolia
Ulithi atoll Micronesia
65 B2 Uljin S. Korea
55 B2 Ulladulla Austr.
96 B2 Ullapool U.K.
65 C2 Ulleung-do i. S. Korea
98 B2 Ullswater l. U.K.
102 B2 Ulm Ger.
65 B2 Ulsan S. Korea
96 □ Ulsta U.K.
97 C1 Ulster reg. Ireland/U.K.
52 B3 Ultima Austr.
145 D3 Ulúa r. Hond.
111 C3 Ulubey Turkey
111 C3 Uluborlu Turkey
111 C2 Uludağ mt. Turkey
126 E2 Ulukhaktok Can.
123 D2 Ulundi S. Africa
77 C1 Ulungur Hu l. China
50 C2 Uluru h. Austr.
98 B2 Ulverston U.K.

90 C2 Ul'yanovka Ukr.
87 D3 Ul'yanovsk Russia
136 C3 Ulysses U.S.A.
90 C2 Uman' Ukr.
86 C2 Umba Russia
59 D3 Umboi i. P.N.G.
59 D3 Umbukul P.N.G.
92 H3 Umeå Sweden
92 H3 Umeälven r. Sweden
123 D2 uMhlanga S. Africa
127 J2 Umiiviip Kangertiva inlet Greenland
126 E2 Umingmaktok Can.
123 D2 Umlazi S. Africa
78 A2 Umm al Birak Saudi Arabia
79 C2 Umm as Samīm salt flat Oman
116 A3 Umm Keddada Sudan
78 A2 Umm Lajj Saudi Arabia
78 A2 Umm Maḥbār, Jabal mt. Saudi Arabia
116 B3 Umm Ruwaba Sudan
115 D1 Umm Sa'ad Libya
134 B2 Umpqua r. U.S.A.
120 A2 Umpulo Angola
Umtali Zimbabwe see Mutare
123 D3 Umtentweni S. Africa
154 B2 Umuarama Brazil
123 C3 Umzimkulu S. Africa
109 C1 Una r. Bos. & Herz./Croatia
155 E1 Una Brazil
154 C1 Unaí Brazil
126 B2 Unalakleet U.S.A.
78 B2 'Unayzah Saudi Arabia
136 B3 Uncompahgre Peak U.S.A.
52 B3 Underbool Austr.
136 C1 Underwood U.S.A.
89 D3 Unecha Russia
53 C2 Ungarie Austr.
52 A2 Ungarra Austr.
127 H2 Ungava, Péninsule d' pen. Can.
131 D2 Ungava Bay Can.
Ungeny Moldova see Ungheni
90 B2 Ungheni Moldova
Unguja i. Tanz. see Zanzibar Island
119 E3 Ungwana Bay Kenya
154 B3 União da Vitória Brazil
150 C3 Unini r. Brazil
134 C1 Union U.S.A.
140 C2 Union City U.S.A.
122 B3 Uniondale S. Africa
139 D2 Uniontown U.S.A.
79 C2 United Arab Emirates country Asia
United Arab Republic country Africa see Egypt
95 C3 United Kingdom country Europe
United Provinces state India see Uttar Pradesh
133 B3 United States of America country N. America
129 D2 Unity Can.
100 C2 Unna Ger.
96 □ Unst i. U.K.
101 E2 Unstrut r. Ger.
89 E3 Upa r. Russia
119 C3 Upemba, Lac l. Dem. Rep. Congo
122 B2 Upington S. Africa
74 B2 Upleta India
49 J5 'Upolu i. Samoa
134 B2 Upper Alkali Lake U.S.A.
128 C2 Upper Arrow Lake Can.
54 C2 Upper Hutt N.Z.
134 B2 Upper Klamath Lake U.S.A.
128 B1 Upper Liard Can.
97 C1 Upper Lough Erne l. U.K.
137 E1 Upper Red Lake U.S.A.
Upper Tunguska r. Russia see Angara
Upper Volta country Africa see Burkina Faso
93 G4 Uppsala Sweden
78 B2 'Uqlat aş Şuqūr Saudi Arabia
Urad Qianqi China see Xishanzui
76 B2 Ural r. Kazakh./Russia
53 D2 Uralla Austr.
87 E3 Ural Mountains Russia
76 B1 Ural'sk Kazakh.
Ural'skiy Khrebet mts Russia see Ural Mountains
119 D3 Urambo Tanz.
53 C3 Urana Austr.
129 D2 Uranium City Can.
86 F2 Uray Russia
98 C2 Ure r. U.K.
86 D3 Uren' Russia
82 G2 Urengoy Russia
144 A2 Ures Mex.
Urfa Turkey see Şanlıurfa
76 C2 Urganch Uzbek.
74 A1 Urgün-e Kalān Afgh.
100 B1 Urk Neth.
111 C3 Urla Turkey
110 C2 Urlaţi Romania
81 C2 Urmia Iran
81 C2 Urmia, Lake salt l. Iran
Uroševac Kosovo see Ferizaj
144 B2 Uruáchic Mex.
151 E4 Uruaçu Brazil
144 B3 Uruapan Mex.
150 B4 Urubamba r. Peru
151 D3 Urucará Brazil
151 E3 Uruçuí Brazil
151 D3 Uruçuí, Serra do hills Brazil
151 D3 Urucurituba Brazil
152 C2 Uruguaiana Brazil
153 C3 Uruguay country S. America
Urumchi China see Ürümqi

68 B2 Ürümqi China
Urundi country Africa see Burundi
53 D2 Urunga Austr.
119 D3 Uruwira Tanz.
110 C2 Urziceni Romania
67 B4 Usa Japan
86 E2 Usa r. Russia
111 C3 Uşak Turkey
120 A3 Usakos Namibia
88 C1 Ushachy Belarus
82 G1 Ushakova, Ostrov i. Russia
77 E2 Usharal Kazakh.
77 D2 Ushtobe Kazakh.
Ush-Tyube Kazakh. see Ushtobe
153 B5 Ushuaia Arg.
86 E2 Usinsk Russia
99 B4 Usk r. U.K.
88 C3 Uskhodni Belarus
89 E3 Usman' Russia
86 D2 Usogorsk Russia
104 C2 Ussel France
66 C1 Ussuri r. China/Russia
66 B2 Ussuriysk Russia
Ust'-Abakanskoye Russia see Abakan
Ust'-Balyk Russia see Nefteyugansk
91 E2 Ust'-Donetskiy Russia
108 B3 Ustica, Isola di i. Italy
83 H3 Ust'-Ilimsk Russia
86 E2 Ust'-Ilych Russia
102 C1 Ústí nad Labem Czech Rep.
Ustinov Russia see Izhevsk
103 D1 Ustka Pol.
83 L3 Ust'-Kamchatsk Russia
77 E2 Ust'-Kamenogorsk Kazakh.
86 F2 Ust'-Kara Russia
86 E2 Ust'-Kulom Russia
83 I3 Ust'-Kut Russia
91 D2 Ust'-Labinsk Russia
Ust'-Labinskaya Russia see Ust'-Labinsk
88 C2 Ust'-Luga Russia
83 K2 Ust'-Nem Russia
83 K2 Ust'-Nera Russia
83 I2 Ust'-Olenek Russia
83 K2 Ust'-Omchug Russia
83 H3 Ust'-Ordynskiy Russia
103 E2 Ustrzyki Dolne Pol.
86 E2 Ust'-Tsil'ma Russia
86 D2 Ust'-Ura Russia
76 B2 Ustyurt Plateau Kazakh./Uzbek.
89 E2 Ustyuzhna Russia
146 B3 Usulután El Salvador
Usumbura Burundi see Bujumbura
89 D2 Usvyaty Russia
135 D3 Utah state U.S.A.
135 D2 Utah Lake U.S.A.
88 C2 Utena Lith.
119 D3 Utete Tanz.
63 B2 Uthai Thani Thai.
74 A2 Uthal Pak.
123 D2 uThukela r. S. Africa
139 D2 Utica U.S.A.
107 C2 Utiel Spain
128 C2 Utikuma Lake Can.
93 G4 Utlängan i. Sweden
100 B1 Utrecht Neth.
123 D2 Utrecht S. Africa
106 B2 Utrera Spain
92 I2 Utsjoki Fin.
67 C3 Utsunomiya Japan
87 D4 Utta Russia
62 B2 Uttaradit Thai.
75 B2 Uttarakhand state India
75 B2 Uttar Pradesh state India
127 I2 Uummannaq Greenland
127 I2 Uummannaq Fjord inlet Greenland
93 H3 Uusikaupunki Fin.
143 D3 Uvalde U.S.A.
119 D3 Uvinza Tanz.
123 D3 Uvongo S. Africa
68 C1 Uvs Nuur salt l. Mongolia
67 B4 Uwajima Japan
78 A2 'Uwayriḍ, Ḥarrat al lava field Saudi Arabia
116 A3 Uweinat, Jebel mt. Sudan
83 H3 Uyar Russia
115 C4 Uyo Nigeria
79 B2 Uyun Saudi Arabia
152 B2 Uyuni Bol.
152 B2 Uyuni, Salar de salt flat Bol.
76 C2 Uzbekistan country Asia
Uzbekskaya S.S.R. country Asia see Uzbekistan
Uzbek S.S.R. country Asia see Uzbekistan
104 C2 Uzerche France
105 C3 Uzès France
90 C1 Uzh r. Ukr.
90 A2 Uzhhorod Ukr.
Uzhorod Ukr. see Uzhhorod
109 C2 Užice Serbia
89 E3 Uzlovaya Russia
111 C3 Üzümlü Turkey
111 C2 Uzunköprü Turkey

V

123 B2 Vaal r. S. Africa
92 I3 Vaala Fin.
123 C2 Vaal Dam S. Africa
123 C1 Vaalwater S. Africa
92 H3 Vaasa Fin.

103 D2 Vác Hungary
152 C2 Vacaria Brazil
154 B2 Vacaria, Serra hills Brazil
135 B3 Vacaville U.S.A.
74 B2 Vadodara India
92 I1 Vadsø Norway
105 D2 Vaduz Liechtenstein
94 B1 Vágar i. Faroe Is
94 B1 Vágur Faroe Is
103 D2 Váh r. Slovakia
49 I4 Vaiaku Tuvalu
88 B2 Vaida Estonia
136 B3 Vail U.S.A.
77 C3 Vakhsh Tajik.
Vakhstroy Tajik. see Vakhsh
79 C2 Vakīlābād Iran
108 B1 Valdagno Italy
Valdai Hills Russia see Valdayskaya Vozvyshennost'
89 D2 Valday Russia
89 D2 Valdayskaya Vozvyshennost' hills Russia
106 B2 Valdecañas, Embalse de resr Spain
93 G4 Valdemarsvik Sweden
106 C2 Valdepeñas Spain
153 B4 Valdés, Península pen. Arg.
126 C2 Valdez U.S.A.
153 A3 Valdivia Chile
130 C3 Val-d'Or Can.
141 D3 Valdosta U.S.A.
128 C2 Valemount Can.
152 E1 Valença Bahia Brazil
155 D2 Valença Rio de Janeiro Brazil
105 C2 Valence France
107 C2 Valencia Spain
107 C2 Valencia reg. Spain
150 C1 Valencia Venez.
107 C2 Valencia, Golfo de g. Spain
106 B1 Valencia de Don Juan Spain
97 A3 Valencia Island Ireland
105 C1 Valenciennes France
136 C2 Valentine U.S.A.
64 B2 Valenzuela Phil.
150 B2 Valera Venez.
88 C2 Valga Estonia
109 C2 Valjevo Serbia
88 C2 Valka Latvia
93 H3 Valkeakoski Fin.
100 B2 Valkenswaard Neth.
91 D2 Valky Ukr.
55 G2 Valkyrie Dome Antarctica
145 C3 Valladolid Mex.
106 C1 Valladolid Spain
93 E4 Valle Norway
145 C2 Vallecillos Mex.
150 C2 Valle de la Pascua Venez.
150 B1 Valledupar Col.
145 C2 Valle Hermoso Mex.
135 B3 Vallejo U.S.A.
152 A2 Vallenar Chile
84 F5 Valletta Malta
137 D1 Valley City U.S.A.
134 B2 Valley Falls U.S.A.
128 C2 Valleyview Can.
107 D1 Valls Spain
129 D3 Val Marie Can.
88 C2 Valmiera Latvia
104 B2 Valognes France
88 C3 Valozhyn Belarus
154 B2 Valparaíso Brazil
153 A3 Valparaíso Chile
105 C2 Valréas France
59 D3 Vals, Tanjung c. Indon.
74 B2 Valsad India
122 B2 Valspan S. Africa
91 D1 Valuyki Russia
106 B2 Valverde del Camino Spain
81 C2 Van Turkey
81 C2 Van, Lake salt l. Turkey
81 C2 Vanadzor Armenia
83 H2 Vanavara Russia
140 B1 Van Buren AR U.S.A.
139 F1 Van Buren ME U.S.A.
Van Buren OH U.S.A. see Kettering
128 B3 Vancouver Can.
134 B1 Vancouver U.S.A.
128 B3 Vancouver Island Can.
138 B3 Vandalia IL U.S.A.
138 C3 Vandalia OH U.S.A.
123 C2 Vanderbijlpark S. Africa
Vanderhoof Can.
122 B3 Vanderkloof Dam dam S. Africa
50 C1 Van Diemen Gulf Austr.
88 C2 Vändra Estonia
Väner, Lake Sweden see Vänern
93 F4 Vänern l. Sweden
93 F4 Vänersborg Sweden
121 □D3 Vangaindrano Madag.
Van Gölü salt l. Turkey see Van, Lake
142 C2 Van Horn U.S.A.
59 D3 Vanimo P.N.G.
83 K3 Vanino Russia
104 B2 Vannes France
Vannovka Kazakh. see Turar Ryskulov
59 D3 Van Rees, Pegunungan mts Indon.
177 A3 Vanrhynsdorp S. Africa
93 H3 Vantaa Fin.
49 I5 Vanua Levu i. Fiji
48 H5 Vanuatu country S. Pacific Ocean
138 C2 Van Wert U.S.A.
122 B3 Van Wyksvlei S. Africa
122 B3 Van Zylsrus S. Africa
75 C2 Varanasi India
92 I1 Varangerfjorden sea chan. Norway

X

Acknowledgements

pages 34-35
Climatic map data:
Kottek, M., Grieser, J., Beck, C., Rudolf, B., and Rubel, F., 2006: World Map of the Köppen-Geiger climate classification updated.
Meteorol. Z., 15, 259–263.
http://koeppen-geiger.vu-wien.ac.at

pages 36-37
World land cover map data:
© ESA 2010 and UCLouvain
Arino, O., Ramos, J., Kalogirou, V., Defourny, P., Achard, F., 2010.
GlobCover 2009. ESA Living Planet Symposium 2010, 28th June - 2nd July, Bergen, Norway, SP-686, ESA.
www.esa.int/due/globcover
http://due.esrin.esa.int/prjs/Results/20110202183257.pdf

pages 38-39
Population map data:
Center for International Earth Science Information Network (CIESIN), Columbia University; and Centro Internacional de Agricultura Tropical (CIAT). 2005. Gridded Population of the World Version 3 (GPWv3). Palisades, NY: Socioeconomic Data and Applications Center (SEDAC), Columbia University.
Available at: http://sedac.ciesin.columbia.edu/gpw
http://www.ciesin.columbia.edu

Cover
Tana Glacier, Alaska, USA © Frans Lanting Studio/Alamy

KEY TO MAP PAGES AFRICA, NORTH AMERICA, SOUTH AMERICA

(see front endpapers for Oceania, Asia and Europe)

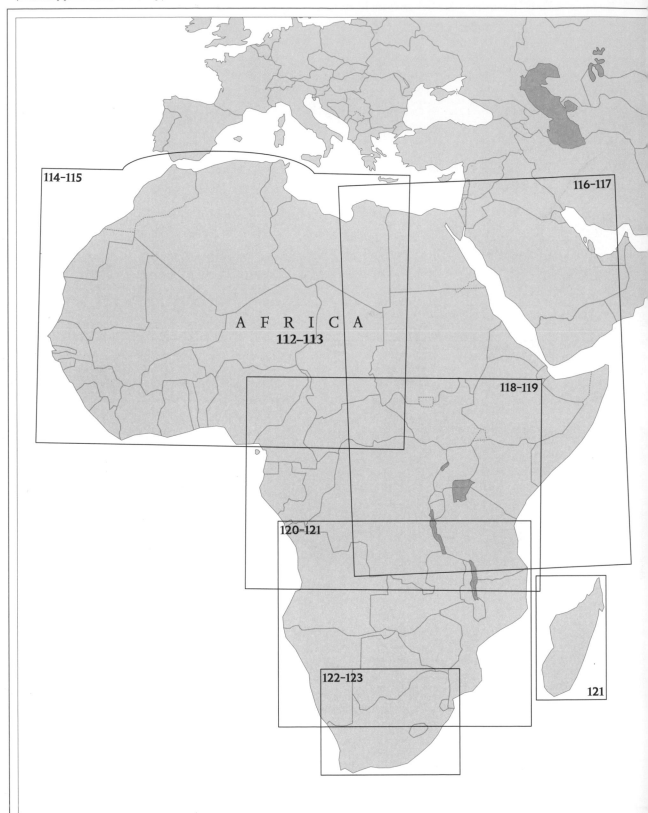

114–115

116–117

A F R I C A
112–113

118–119

120–121

122–123

121